TECHNIQUES
IN MICROBIAL
ECOLOGY

# TECHNIQUES IN MICROBIAL ECOLOGY

*Edited by*

Robert S. Burlage
Environmental Sciences Division
Oak Ridge National Laboratory

Ronald Atlas
Department of Biology
University of Louisville

David Stahl
Department of Civil Engineering
Northwestern University

Gill Geesey
Center for Biofilm Engineering
Montana State University

Gary Sayler
Center for Environmental Biotechnology
University of Tennessee

New York Oxford
Oxford University Press
1998

Oxford University Press

Oxford New York
Athens Auckland Bangkok Bogota Bombay Buenos Aires
Calcutta Cape Town Dar es Salaam Delhi Florence Hong Kong
Istanbul Karachi Kuala Lumpur Madras Madrid Melbourne
Mexico City Nairobi Paris Singapore Taipei Tokyo Toronto Warsaw

and associated companies in
Berlin Ibadan



Published by Oxford University Press, Inc.
198 Madison Avenue, New York, New York 10016



Library of Congress Cataloging-in-Publication Data
Techniques in microbial ecology / edited by Robert S. Burlage . . . [et al.].
p. cm. Includes bibliographical references and index.
ISBN 0-19-509223-6
1. Microbial ecology—Laboratory manuals.
2. Molecular microbiology—Laboratory manuals.
I. Burlage, Robert S.
QR100.T44 1997 579'.17—dc21
DNLM/DLC for Library of Congress 96-49666

9 8 7 6 5 4

Printed in the United States of America
on acid-free paper

# Preface

Interest in microbial ecology and environmental microbiology has increased steadily with the realization that microbial interactions offer some of the most challenging and rewarding research opportunities available. Although much is known about microbial genetics and physiology in pure cultures, relatively little is known about these processes in microbial communities. Even less is known about interactions between microorganisms under normal environmental conditions (e.g., commensalism, predation, parasitism, and communication). This lack of knowledge can be remedied by well-designed experiments, which can yield dramatic developments in our understanding of the microbial world.

This book is designed to provide a microbiologist with the background and protocols necessary for examining key environmental arenas. It does not aspire to be comprehensive; such a work would be truly voluminous. However, by providing the most useful techniques for examining microorganisms of great experimental interest, it provides a good starting place for those who are planning to investigate some aspect of a microbial community. The most useful techniques for obtaining information on the most useful parameters are presented here. It also serves as a reference for workers at all levels of interest in microbial ecology.

The contributors of the individual chapters were given great latitude in their presentation of techniques. This was done because imposing standard formats for such a varied science is most likely counterproductive. I recommend that you first read each chapter completely to gain an overall understanding of the particular field before concentrating on the particular technique of interest.

The chapters in the first part of the book describe microorganisms that are associated with nutrient cycles in nature (e.g., the nitrogen cycle) or that are otherwise related (e.g., viruses). Part II describes techniques that are generally applicable to environmental research (e.g., sampling techniques and methods for isolating and analyzing nucleic acids). This part includes a chapter on modeling because of its importance in describing complex ecosystems, as well as a chapter on bioremediation because of the substantial interest in this field and because the many techniques described in this book all find common purpose in bioremediation work.

References are provided to enable the investigator to become better acquainted with the topic. There are any number of ways in which an experiment might be constructed, and techniques that have been described in the literature may or may not be helpful. You may find that some tinkering is required. I and all the contributors hope that this book stimulates your creativity and wish you success in your experiments.

R. S. B.

Oak Ridge, TN
August, 1997

# Contents

## PART III SPECIAL TOPICS

# Contributors

Michel Aragno
Laboratoire de Microbiologie
Institut de Botanique de
l'Université
P.O. Box 2
CH 2007 Neuchâtel, Switzerland

D. L. Balkwill
Florida State University
Tallahassee, Florida 32306

Lynne Boddy
School of Pure and Applied
Biology
University of Wales College of
Cardiff
P.O. Box 915
Cardiff CF1 1TL, United Kingdom

Robert S. Burlage
Environmental Sciences
Division
Oak Ridge National Laboratory
Oak Ridge, Tennessee 37831

K. J. Clarke
Institute of Freshwater Ecology
The Ferry House, Far Sawrey
Ambleside, Cumbria LA22 0LP,
United Kingdom

Rita R. Colwell
Department of Microbiology
University of Maryland
College Park, MD 20742
*and* University of Maryland
Biotechnology Institute
4321 Hartwick Road
College Park, Maryland 20740

G. F. Esteban
Institute of Freshwater Ecology
The Ferry House, Far Sawrey
Ambleside, Cumbria LA22 0LP,
United Kingdom

B. J. Finlay
Institute of Freshwater Ecology
The Ferry House, Far Sawrey
Ambleside, Cumbria LA22 0LP,
United Kingdom

J. K. Fredrickson
Pacific Northwest Laboratory
Richland, Washington 99352

Richard S. Hanson
Department of Microbiology
University of Minnesota Medical
School
Box 196
1460 Mayo Memorial Building
420 Delaware St., S.E.
Minneapolis, Minnesota 55455

Don P. Kelly
Institute of Education
The University of Warwick
Conventry CV4 7AL, United Kingdom

Joel Kostka
Center for Great Lakes Studies
University of Wisconsin-Milwaukee
600 E. Greenfield Ave.
Milwaukee, Wisconsin 53204

Eugene L. Madsen
Department of Biological Sciences
Section of Microbiology
Wing Hall
Ithaca, New York 14853-8101

Robert V. Miller
Department of Microbiology and Molecular Genetics
Oklahoma State University
Stillwater, Oklahoma 74078

M. W. Mittelman
Centre for Infection and Biomaterials Research
University of Toronto
Toronto, Ontario, Canada M5G 2C4

Kenneth H. Nealson
Center for Great Lakes Studies
University of Wisconsin-Milwaukee
600 E. Greenfield Ave.
Milwaukee, Wisconsin 53204

Andrew Ogram
Department of Crop and Soil Sciences
University of Florida
Gainesville, Florida 32611

Hans W. Paerl
Institute of Marine Sciences
University of North Carolina at Chapel Hill
Morehead City, North Carolina 28557

J. I. Prosser
Department of Molecular and Cell Biology
Marischal College
University of Aberdeen
Aberdeen AB9 1AS
Scotland

Alan D. M. Rayner
School of Biology and Biochemistry
University of Bath
Claverton Down
Bath BA2 7AY, United Kingdom

David B. Ringelberg
Center for Environmental Biotechnology
University of Tennessee
10515 Research Drive
Suite 300
Knoxville, Tennessee 37932-2575

Estelle Russek-Cohen
Department of Microbiology
University of Maryland
College Park, Maryland 20742

David C. White
Center for Environmental Biotechnology & Department of Microbiology
University of Tennessee
10515 Research Drive
Suite 300
Knoxville, Tennessee 37932-2575
*and* Environmental Sciences Division
Oak Ridge National Laboratory
Oak Ridge, Tennessee 37831

Ann P. Wood
Division of Life Sciences
King's College London
Campden Hill Road
London W8 7AH, United Kingdom

Stephen H. Zinder
Section of Microbiology
Cornell University
Ithaca, New York 14850

# PART I

# ENVIRONMENTAL MICROBIOLOGY

# 1

HANS W. PAERL

# Microbially Mediated Nitrogen Cycling

Nitrogen (N) is a fundamentally important element in biologically mediated production and nutrient cycling processes [20]. Nitrogen-containing constituents of organic molecules often confer bioactivity to these molecules. Major cellular, structural, and functional constituents have essential and often highly specific requirements for N. Most important are proteins, the cell's building blocks, and enzymes, biochemical "catalysts." In addition, deoxyribonucleic acid (DNA) and ribonucleic acid (RNA), the macromolecules that encode and transmit genetic information, exhibit a strong reliance on N. N-sufficiency is essential for synthesis, activity, reproduction, and growth of living matter. In this regard, N can be considered the "meat" in the "meat and potatoes" when it comes to providing cell structure, function, and growth.

Because of appreciable and broad requirements in cellular metabolism, growth, and elemental cycling, N supply rates can be outstripped by biologic demand, rendering it a limiting nutrient in natural ecosystems. This is particularly true in the marine environment, where the need for biologically usable N frequently exceeds its availability [22, 69]. It follows that inputs of newly supplied or "new" N often control primary and secondary production in vast segments of the world's estuarine, coastal, and open ocean waters [25, 53].

The N cycle is complex because this element exists as diverse gaseous, dissolved ionic, and particulate forms. Exchange among these forms is mediated by a wide variety of microbiologic transformations (Fig. 1-1) [20, 21]. In natural ecosystems, new N can originate from external (allochthonous) sources of either natural or anthropogenic origins. These external sources include both point- and non-point sources, such as wastewater discharge, agricultural (fertilizer) runoff, atmospheric deposition, and groundwater. In the marine environment, new N can be advected and upwelled from deep ocean water, where it may have resided for decades to centuries.

Nitrogen availability is additionally controlled by within-system or internal cycling processes. Among these processes, $N_2$ fixation, the prokaryotic (bacterial and cyanobacterial) enzyme-(nitrogenase) mediated reduction of atmospheric di-

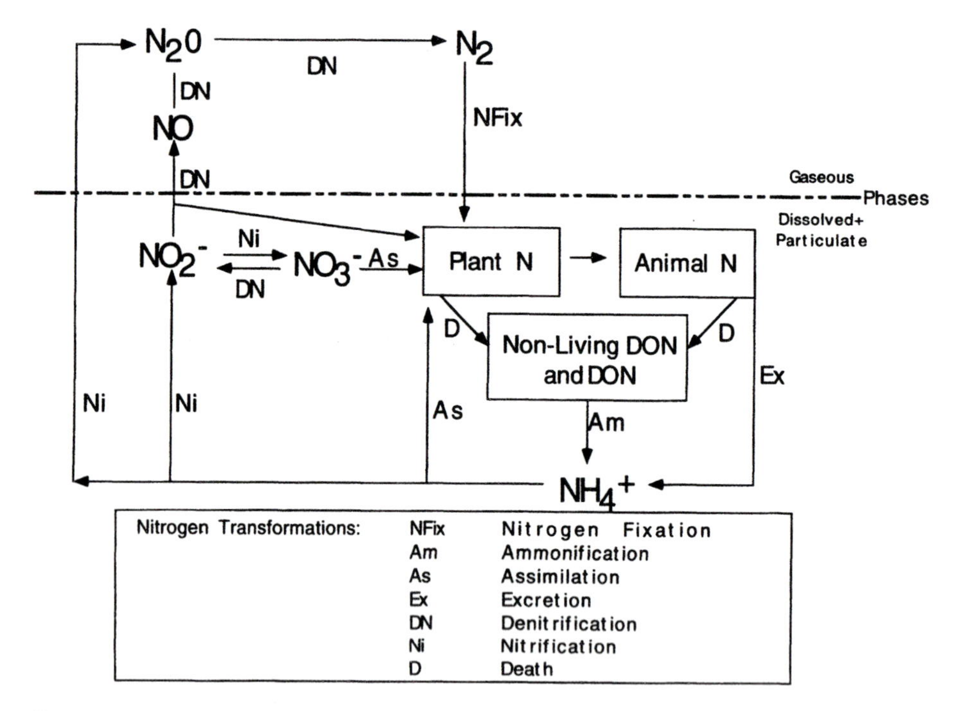


Figure 1-1 The nitrogen (N) cycle, emphasizing key microbially mediated N transformations.

nitrogen ($N_2$) to ammonia ($NH_3$), represents a within-system or autochthonous source of new N [15, 22, 30, 61, 85]. Conversely, bacterial denitrification, in which the oxidized form of inorganic N, nitrate ($NO_3^{-1}$), is reduced to $N_2$ gas by various heterotrophic bacterial genera, represents an autochthonous N loss mechanism [38, 51, 79]. The balance between these biologically mediated "bottleneck" processes can determine ecosystem-level N availability and (in N-limited systems) fertility [10, 20, 23].

Intermediate N transformations also determine N flux and availability. Denitrification is dependent on $NO_3^{-1}$ supply. $NO_3^{-1}$ availability, in turn, is a product of ammonium ($NH_4^{+1}$) oxidation. In unfertilized ecosystems, bacterial $NH_4^{+1}$ oxidation, initially to $NO_2^{-1}$ (*Nitrosomonas, Nitrosococcus*) and then to $NO_3^{-1}$ (*Nitrobacter*), is often the rate-limiting step for denitrification because it controls the formation of $NO_3^{-1}$ [16, 51, 81, 87]. Furthermore $NH_4^{+1}$ availability is largely a product of bacterial decomposition, mineralization, and ammonification of organic N [10, 20]. The spatial and temporal interactions of new N inputs, N losses via denitrification, and intermediate N transformations closely control ecosystem trophodynamics.

Because of the strong interdependence of N transformations, it is essential that individual yet linked (in time and space) measurements and assessments be made contemporaneously and contiguously in order to understand the physiology and ecology of microorganisms that mediate N cycling. Fortunately, a reasonably large knowledge base exists concerning the microbiology and biogeochemistry of N cycling. A set of suitable (for laboratory and field studies), highly specific, sensitive, and complementary molecular, immunochemical, low-level analytical, and microbiologic techniques are evolving to help us gain conceptual and functional insights into the roles and significance of microbes involved in N cycling. To achieve these goals, we must be able to identify, characterize, and, in many instances, quantitate (1) individual N transformations, (2) the organism(s) responsible, (3) environmental controls of the processes and organisms involved, (4) interactions and interdependence of relevant processes and biota, and (5) the contributions of these processes and relevant biota to large-scale regional and global production and cycling processes.

In this chapter, the utility and applicability of biogeochemical, molecular, immunochemical, and analytical techniques for assessing microbial N transformations are discussed and evaluated. Specific emphasis is placed on state-of-the-art techniques currently available to a broad range of microbiologists, aquatic and terrestrial ecologists, and biogeochemists addressing production, nutrient cycling, biotic resource (i.e., biodiversity), evolution, and environmental quality issues. Because the N cycle is complex, it is impossible to detail the entire suite of methodologies currently available for assessing all N transformations. Rather, specific examples and protocols of some of the more widely used techniques are presented. For more detailed protocols, the reader is referred to individual studies.

Over the past decade there has been significant progress in the development and application of techniques suitable for examining rates of and microorganisms responsible for N transformations. Despite these advances, N transformations that mediate the flux and availability of N are elusive from analytic, physical-chemical, and biologic perspectives. In particular, denitrification and organic N cycling remain enigmatic, in large part because of analytic constraints and interpretational pitfalls. Molecular approaches to these problems have yielded substantial improvement in detecting, quantitating, and more precisely characterizing N transformations and the microorganisms responsible for them. With the maturation of nucleic acid technologies, such as the detection and characterization of mRNA synthesis and turnover, our knowledge of rates and environmental constraints on these microbial processes will increase dramatically. This understanding, in turn, will enhance our conceptual and functional knowledge of microbial communities mediating N-cycling dynamics in aquatic and terrestrial ecosystems.

## BIOLOGICALLY AVAILABLE FORMS OF NITROGEN: ANALYSES

Biologically available N occurs in inorganic and organic forms. The dominant inorganic forms—ammonia/ammonium ($NH_3/NH_4^{+1}$), nitrate ($NO_3^{-1}$), and nitrite ($NO_2^{-1}$)—are soluble. Accordingly, colorimetric methods outlined in this section determine aqueous concentrations. Biologically available organic N can occur in either dissolved or particulate forms [2, 65]. Organic N is determined by first oxidizing the samples of interest. Common oxidation methods include "wet" oxidation by a strong oxidizer such as potassium persulfate, ultraviolet oxidation, and high-temperature catalytic oxidation. After oxidation, the released forms of N (either $NO_3^{-1}$ or $NH_4^{+1}$) are measured by colorimetric techniques. These oxidation methods are discussed in detail by Parsons et al. [64] and Williams [84]. Virtually all the nitrogen transformations discussed in this chapter use these techniques to quantify soluble and particulate N reactants and products.

### Colorimetric Determinations of Ammonia/Ammonium ($NH_3/NH_4^{+1}$)

#### Methodology

Ammonia/ammonium in natural waters is quantified by conducting an oxidation reaction with hypochlorite in alkaline citrate medium in the presence of sodium nitroprusside as a catalyst. The resulting blue indophenol color formed with ammonia is measured spectrophotometrically.

Equipment/Specialized Apparatus

A spectrophotometer, having either 5 or 10 cm absorbance cells, capable of reading extinction at 640 nm
125-ml Erlenmeyer flasks
Several graduated cylinders (50–500 ml) and volumetric flasks (100–1000 ml)
Automatic pipettes for dispensing reagents and samples

Reagents

1. Deionized water.
2. Phenol solution: Dissolve 20 g of analytic-grade phenol in 200 ml 95% ethyl alcohol.
3. Sodium nitroprusside solution: Dissolve 1.0 g sodium nitroprusside, $Na_2[Fe(CN)_5NO]\cdot 2H_2O$, in 200 ml deionized water. Store in a dark glass bottle (1 month shelf-life at room temperature).
4. Alkaline reagent: Dissolve 100 g sodium citrate and 5 g sodium hydroxide in 500 ml deionized water.
5. Sodium hypochlorite solution: Use commercially available hypochlorite (e.g., "Chlorox"). Discard after 1 month.
6. Oxidizing solution: Mix 100 ml of reagent 4 and 25 ml of Reagent 5. Keep stoppered and make fresh on the day of analysis.

Procedure

1. Add 50 ml of aqueous sample to Erlenmeyer flask, using a 50-ml measuring cylinder. Add 2 ml of phenol solution, mix by swirling, and then add sequentially (followed by swirling) 2 ml nitroprusside and 5 ml oxidizing solution.
2. Keep flasks covered and in a dark place at room temperature (~25° C) for at least 1 h (color is stable for up to 24 h).
3. Read extinction at 640 nm, using either a 5 or 10 cm absorbance cell (depending on concentrations, color development, and spectrophotometer sensitivity).
4. Correct the measured extinction for the reagent blank (deionized water), and calculate ammonia-N using this formula:

$$\mu\text{g-at N/l } (\mu\text{M N}) = F \times E$$

where E is the corrected extinction and F is the factor as determined below.

Calibration

1. Prepare 50 ml ammonium (chloride) standards (choose a range of concentrations approximating those found in samples).
2. Correct the extinction for the reagent blank and calculate F as: $F = 3.0/E_S$.

## Colorimetric Determinations of Nitrate ($NO_3^{-1}$) and Nitrite ($NO_2^{-1}$)

Methodology

Dissolved nitrite can be analyzed directly by diazotizing with sulfanilamide and coupling N-(1-naphthyl)-ethylenediamine to form a red-colored azo dye that is measured spectrophotometrically. Nitrate must first be reduced by passing the sample through a cadmium reduction column. The nitrite produced is then analyzed as outlined in this section. Nitrite initially present in the reduced sample must be corrected for when using the cadmium reduction step.

### Equipment and Specialized Apparatus

A spectrophotometer, with a 1-cm cuvette and capable of reading samples at 543 nm
50-ml graduated cylinders
125-ml Erlenmeyer flasks
A reduction column (see below)

### Reagents

1. Concentrated ammonium chloride solution: Dissolve 125 ml analytic reagent-grade ammonium chloride in 500 ml deionized water. Store in glass bottle.
2. Dilute ammonium chloride solution: Dilute 50 ml concentrated ammonium chloride solution to 200 ml with deionized water. Store in glass bottle.
3. Cadmium-copper filings (for reduction column): Generate cadmium filings, usually by filing cadmium metal with a coarse wood file, and pass through a 2-mm sieve. Stir about 100 g of filings (enough for two columns) with 500 ml of 2% w/v copper sulfate ($CuSO_4 \cdot 5H_2O$) until the blue color disappears from the solution. Place a plug of copper wool in the bottom of the reduction tube (10-mm I.D., 30-cm length) and fill the column to the top with dilute ammonium chloride and a slurry of $CuS0_4$-activated cadmium filings. Wash the column with ammonium chloride several times. Attach a collection cup (approximately 100 ml volume) to the top of the column. Using a flow restrictor at the end of the column (if necessary), a flow rate of approximately 10 ml/min is established. If the flow rate falls below this, the columns require recharging and repacking (see [64] for details).
4. Sulfanilamide solution: Dissolve 5 g of sulfanilamide in a mixture of 50 ml concentrated hydrochloric acid and 300 ml deionized water. This solution is stable for several months.
5. N-(1-naphthyl)-ethylenediamine dihydrochloride solution: Dissolve 0.5 g of the dihydrochloride in 500 ml deionized water. Store in a dark glass bottle. This solution is stable for 1 month at room temperature.

### Procedure

1. Dispense 100 ml filtered (to remove any turbidity) water samples into 125-ml Erlenmeyer flasks.
2. Add (with mixing) 2.0 ml of concentrated ammonium chloride. Pour the sample onto the reduction column and pass it through.
3. Collect about 40 ml of the eluant from the column and discard. Collect the next 50 ml in a graduated cylinder and dispense in the original 125-ml Erlenmeyer sample flask. Repeat the procedure for the next sample.
4. Add 1.0 ml sulfanilamide to the 50 ml sample; mix, and let stand for 2 to 8 minutes. Add 1 ml of naphthyl-ethylenediamine solution and mix immediately. Allow for color development for at least 20 min, and read absorbance at 543 nm in a 1-cm cuvette.

5. Correct extinction from samples with a reagent blank (dilute ammonium chloride solution plus reagents). The following equation should be used to correct for the reagent blank:

$$\mu\text{g-at N/L } (\mu\text{M N}) = (\text{corrected extinction} \times \text{F}) - 0.95\text{C}$$

where C is the original concentration of nitrite in the sample.

Calibration

1. Make up synthetic seawater: Dissolve 310 g analytic reagent-quality sodium chloride (NaCl), 100 g reagent-grade magnesium sulfate ($MgSO_4 \cdot 7H_2O$), and 0.5 g sodium bicarbonate ($NaHCO_3 \cdot H_2O$) in 10 liters of deionized water.
2. Standard nitrate solution: Dissolve 1.02 g analytic reagent-quality potassium nitrate ($KNO_3$) in 1000 ml deionized water. Dilute 4.0 ml of this solution to 2000 ml with synthetic seawater, which should yield a $NO_3^{-1}$ concentration of 20 μg-at N/L (μM N).
3. Add 100 ml of the dilute standard $NO_3^{-1}$ solution to a 125-ml Erlenmeyer flask, and analyze as outlined above. Read the extinction for each reduction column. The factor F is $F = 20/E_S$, where $E_S$ is the extinction of the standard, corrected for the blank.

## NITROGEN TRANSFORMATIONS: CHARACTERIZING RELEVANT MICROBIOTA AND THEIR ACTIVITIES

### Nitrogen Fixation

#### *Cultivation*

Nitrogen fixation is the enzymatic reduction of $N_2$ to $NH_3$, which is mediated by the multimeric enzyme complex nitrogenase [66] (Fig. 1-2). It is confined to prokaryotes. Representative $N_2$-fixing groups include (1) anoxygenic phototrophs (photosynthetic bacteria; *Chlorobium, Chromatium, Rhodospirillum*); (2) oxygenic phototrophs (cyanobacteria), including all heterocystous filamentous (*Anabaena,*

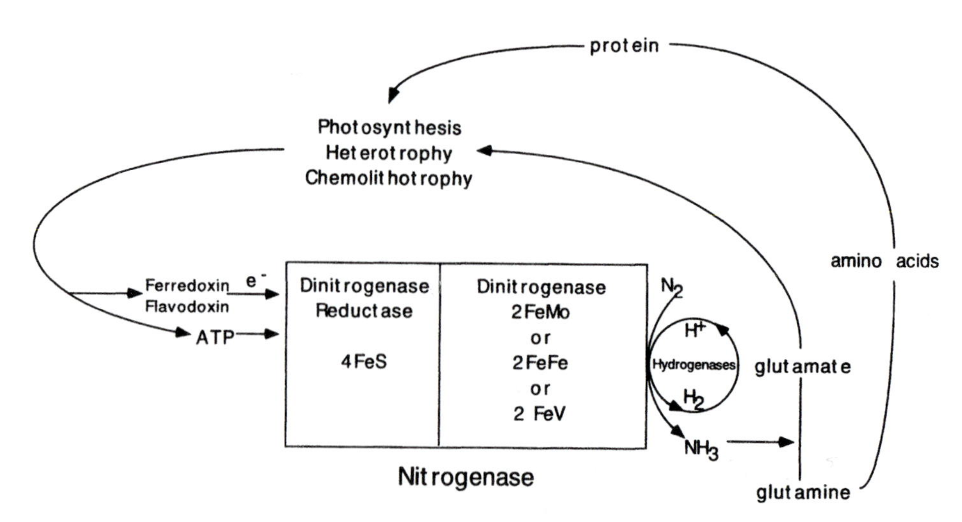


Figure 1-2 The energetics, enzymology, and assimilatory processes associated with $N_2$ fixation.

*Aphanizomenon*, *Nostoc*), some non-heterocystous filamentous (*Oscillatoria*, *Lyngbya*, *Trichodesmium*), and non-filamentous (*Gloeocapsa*, *Synechococcus*) genera; (3) some anaerobic heterotrophic bacterial genera (*Clostridium*, *Desulfovibrio*); (4) numerous microaerophilic heterotrophic bacterial genera (*Klebsiella*, *Vibrio*); (5) a few aerobic genera, most notably *Azotobacter*; and (6) chemolithoautotrophic bacterial genera (*Thiobacillus*).

Because $N_2$ fixation is widespread among physiologically distinct microbial groups, no general culturing technique is effective, either qualitatively or quantitatively, in isolating and maintaining a broad range of diazotrophs from nature. Instead, specific isolation and culturing techniques have been applied in efforts to characterize and enumerate $N_2$ fixers from diverse environments. Most techniques involve organic matter enrichment under microaerophilic or anaerobic N-free conditions. $N_2$ fixers are highly fastidious and are difficult to grow and maintain in culture [35, 36, 48, 55, 78]. Therefore, no single universal culturing technique can be prescribed.

These media are commonly used to isolate and culture $N_2$-fixing microorganisms (adapted from [1]).

*Azotobacter* Medium

| | |
|---|---|
| Mannitol | 2.0 g |
| $K_2HPO_4$ | 0.5 g |
| $MgSO_4 \cdot 7H_2O$ | 0.2 g |
| $FeSO_4 \cdot 7H_2O$ | 0.1 g |
| Distilled water | 1000 ml |

Adjust the pH to 7.3 to 7.6. Sterilize at 121° C for 15 min.

ASN-III (-N) Medium

| | |
|---|---|
| NaCl | 25.0 g |
| $MgCl_2 \cdot 6H_2O$ | 2.0 g |
| KCl | 0.5 g |
| $K_2HPO_4 \cdot 3H_2O$ | 0.02 g |
| $MgSO_4 \cdot 7H_2O$ | 3.5 g |
| $CaCl \cdot 2H_2O$ | 0.5g |
| Citric acid | 0.003 g |
| Ferric ammonium citrate | 0.003 g |
| Ethylenediaminetetraacetic acid (EDTA), disodium magnesium salt | 0.0005 g |
| $Na_2CO_3$ | 0.02 g |
| Trace metal mix A5* (see BG-11 Medium) | 1.0 ml |
| Distilled water | 1000 ml |

After autoclaving and cooling, the pH should be 7.5.
*Trace metal mix should be added aseptically after cooling and pH adjustment.

BG-11 (-N) Medium

| | |
|---|---|
| $K_2HPO_4 \cdot 3H_2O$ | 0.04 g |
| $MgSO_4 \cdot 7H_2O$ | 0.075 g |
| $CaCl_2 \cdot 2H_2O$ | 0.036 g |
| Citric acid | 0.006 g |
| Ferric ammonium citrate | 0.006 g |
| EDTA, disodium magnesium salt | 0.001 g |

| | |
|---|---|
| $Na_2CO_3$ | 0.02 g |
| Trace metal mix A5* | 1.0 ml |
| Deionized water | 1000 ml |

After autoclaving and cooling, the pH should be 7.4.
*Trace metal mix should be added aseptically after cooling and pH adjustment.

Trace Metal Mix A5*

| | |
|---|---|
| $H_3BO_3$ | 2.86 mg/ml |
| $MnC1_2 \cdot 4H_2O$ | 1.81 mg/ml |
| $ZnSO_4 \cdot 7H_2O$ | 0.222 mg/ml |
| $Na_2MoO_4 \cdot 2H_2O$ | 0.39 mg/ml |
| $CuSO_4 \cdot 5H_2O$ | 0.079 mg/ml |
| $CoCl_2 \cdot 6H_2O$ | 0.0494 mg/ml |

*Trace metal mix should be added aseptically after cooling and pH adjustment.

*Clostridium pasteurianum* N-Free Medium

| | |
|---|---|
| $MgSO_4 \cdot 7H_2O$ | 0.0493 g |
| $FeCl_3 \cdot 6H_2O$ | 0.0541 g |
| $MnSO_4 \cdot H_2O$ | 0.0034 g |
| $Na_2MoO_4 \cdot 2H_2O$ | 0.0048 g |
| $ZnSO_4 \cdot 7H_2O$ | 0.00058 g |
| $CuSO_4 \cdot 5H_2O$ | 0.00050 g |
| $CoCl_2 \cdot 6H_2O$ | 0.00048 g |
| $K_2HPO_4$ | 1.132 g |
| Biotin | Trace |
| Sucrose | 20.0 g |
| $CaCO_3$ | 3.0 g (can be increased to 30 g for increased buffering) |
| Distilled water | 1000 ml |

Combine ingredients and autoclave. The medium must be made anaerobic before inoculating by bubbling oxygen-free nitrogen through it.

Yoch and Pengra N-Free Medium for *Klebsiella*

*Solution A*

| | |
|---|---|
| $Na_2HPO_4$ | 6.25 g |
| $KH_2PO_4$ | 0.75 g |
| Distilled water | 500 ml |

*Solution B*

| | |
|---|---|
| $MgSO_4 \cdot 7H_2O$ | 6.25 g |
| $FeSO_4 \cdot 7H_2O$ | 0.04 g |
| $Na_2MoO_4$ (anhydrous) | 0.005 g |
| Sucrose | 20.0 g |
| NaCl | 8.5 g |
| Distilled water | 500 ml |

Autoclave Solutions A and B separately. When cool, combine equal volumes aseptically.

Hino and Wilson N-Free Medium for *Bacillus*

*Solution A*

| | |
|---|---|
| Sucrose | 20.0 g |
| $MgSO_4 \cdot 7H_2O$ | 0.5 g |
| NaCl | 0.01 g |
| $FeSO_4 \cdot 7H_2O$ | 0.015 g |
| $Na_2MoO_4 \cdot 2H_2O$ | 0.005 g |
| $CaCO_3$ | 10.0 g |
| Distilled water | 500 ml |

*Solution B*

| | |
|---|---|
| p-Aminobenzoic acid | 10 μg |
| Biotin | 5 μg |
| $K_2HPO_4$-$KH_2PO_4$ buffer (pH 7.7) | 0.1 M |
| Distilled water | 500 ml |

Autoclave solutions A and B separately. When cool, combine equal volumes aseptically.

### *Molecular Detection*

Nitrogen fixation is an ancient process that in all likelihood has been present since the evolution of eubacteria [67]. It is an anaerobic process that is readily inactivated by $O_2$. Inactivation may be reversible in intact microorganisms, but is irreversible in cell-free extracts [66]. Because of its ancient roots, the nitrogenase complex is highly conserved, both on structural (protein) and genetic (*nif* genes encoding for specific protein subunits) levels [67].

The Fe-protein (dinitrogenase reductase) subunit of nitrogenase is identical in terms of peptide sequences among all diazotrophic groups [11]. This quality makes it an ideal target for immunochemical probing for characterization and potentially for enumeration of diazotrophs; as such it is a much-needed and welcome alternative to culturing. Because nitrogenase is an intracellular enzyme, immunoassays must employ either cell disruption (lysis), followed by protein isolation and blotting (i.e., immunoblotting) [40, 52], or permeation of cell walls and membranes prior to whole cell immunoassays [18]. In addition, immunoelectron (immunogold) microscopy has been employed on sectioned cells to localize nitrogenase in a variety of microorganisms, most notably cyanobacteria [4, 62, 63].

Substantial conservation exists at the level of genes that encode nitrogenase [67], although the degrees of conservation are lower for the proteins expressed because of degeneration among bases that encode for individual amino acids. The *nif*H gene, which encodes for the nitrogenase reductase subunit, has been used as a target for nucleic acid probe hybridization (i.e., Southern blots) [41]. The polymerase chain reaction (PCR), using degenerate primers, is effective in detecting and characterizing $N_2$-fixing microorganisms in culture and in nature [3, 49, 50, 87]. In particular, PCR has proven useful for phylogenetic characterization of $N_2$ fixers through the examination of DNA sequences of the approximately 360-bp amplification product. These sequences have proven to be diagnostic of specific $N_2$-fixing taxa [3, 89].

Dinitrogenase, the other subunit of nitrogenase, is not as highly conserved as dinitrogenase reductase [11]. Dinitrogenase has alternative forms, each having a unique suite of metal co-factors, including the most common FeMo, but also FeV and FeFe forms [9]. The ecologic and biogeochemical significance and ramifications of these alternative nitrogenases are poorly understood. The genes encoding for alternative nitrogenases are known, and gene probing (i.e., Southern hybridization) and PCR techniques are available for potentially detecting and characterizing these nitrogenases (See Chapter 13 for more information on molecular techniques).

There are at least two *primers* for conducting PCR detection and characterization of $N_2$-fixing microorganisms in culture and nature (adapted from [3, 49, 88]):

Primer 1: 5′-**GGAATTC**CTGYGAYCCNAARGCNGA-3′

Primer 2: 5′-C**GGATCC**GDNGCCATCATYTCNCC-3′

Note: Primer 1 is located at the upstream end and has a *Eco*RI site (**bold**). Primer 2 is located at the downstream end of the opposite strand and has a *Bam*HI site (**bold**). The 324-bp fragment between these primers corresponds to positions 336–660 in *nif*H from *Anabaena* sp. strain PCC7120. Y = T or C; R = A or G; D = A, G, or T; N = A, C, G, or T.

Conditions

1. The 359-bp fragment of the *nif*H gene is amplified using Taq polymerase, with 35 cycles of denaturation (93° C, 1.2 min), annealing (50° C, 1.0 min), and extension (70° C, 1.5 min).
2. A single band is visualized on a 4% NuSieve agarose gel, stained with ethidium bromide. The band is then isolated with diethylaminoethanol paper.
3. Amplified DNA fragments from individual strains/species are cloned into phage M13 mpl8 and mpl9. Transformation is into *E. coli* JM 101.
4. Single-stranded DNA is extracted from recombinant clones and sequenced by the dideoxynucleotide chain termination method, using Sequenase version 2.0 and [ $^{35}S$] deoxyadenosine triphosphate.
5. The region where nucleotide sequence differences are found between strains/species is confirmed by sequencing again with inosine to eliminate sequencing artifacts. The DNA sequence data are analyzed using Genetic Computer Group computer programs.

### *$N_2$-Fixation Assays*

Nitrogenase activity, the common indicator of $N_2$ fixation, can be assessed using a variety of techniques, including the acetylene reduction assay, $^{15}N_2$ assimilation, and the accumulation of N in microbial biomass and associated particulate matter (i.e., Kjeldahl or C,H,N analyses).

**Acetylene Reduction** The most commonly used technique is the acetylene reduction assay (ARA) [14, 76], which is based on the well-documented ability of the nitrogenase enzyme complex to reduce a variety of triple-bonded substrates, including acetylene, as analogs to $N_2$ gas [77]. Nitrogenase reduces acetylene to ethylene versus $N_2$ to $NH_3$ in an approximate ratio of 3:1. This ratio varies, depending on the extent to which nitrogenase reduces $H_2$ from $H^+$ [27, 37], which occurs in parallel with $N_2$ reduction to $NH_3$ [11, 66]. The ARA technique, although indirect, is often preferred, because it is easily and rapidly deployed in the field and laboratory; is highly sensitive and readily quantified by gas chromatography (flame ionization detector; see Capone [14] or Paerl and Kellar [60]); and is amenable to experimental manipulations (e.g., variations in $pO_2$ and other gases, light and temperature effects, and nutrient additions). ARA procedures are outlined in this section.

Materials and Equipment

Purified acetylene (either supplied as compressed gas or generated from $CaC_2$)
Purified ethylene ($C_2H_4$) for standards
Gas-tight syringes (100 μl, 300 μl, and 5–10 ml)
Serum bottles (5–50 ml) with natural rubber serum stoppers
For aqueous samples, distilled or 0.20-filtered "blank" seawater/freshwater
Gas chromatograph (GC) with flame ionization detector (FID) and gases (GC grade purity $N_2$, high purity $H_2$, dry grade high purity air or $O_2$)
2-m Poropak types T, R, or N (80/100) GC column

Procedure

1. Place samples (either suspended in water, sediment, soil cores, or solids) in serum bottles. For aqueous (including sediment) samples, the aqueous to vapor phase is usually approximately 60:40 (v:v). For solid samples, no aqueous phase is required. Seal the serum bottles with stoppers. Either leave the samples exposed to ambient atmospheric conditions or flush with inert gases (Ar or He, most often with a trace of $CO_2$ to exclude $O_2$ if necessary). Inject acetylene at approximately 15% of headspace volume; the exact percentage should be determined after examining acetylene "saturation" of nitrogenase over a range of concentrations.
2. Incubate samples (preferably under in situ conditions, unless laboratory populations are being examined) over a range of time intervals (usually 1–6 h). Determine the preferred incubation time by withdrawing small volumes (100–300 μl) of headspace periodically and injecting them into the GC to check for amounts and linearity of $C_2H_4$ production. Ideally, at least 10 times background (distilled or filtered water "blanks") $C_2H_4$ production should be detected, and the incubation period should fall within the linear range of $C_2H_4$ production. If inconvenient (i.e., in the field), headspace samples need not be injected immediately into the GC, but can be stored in small (1–5 ml), well-sealed evacuated tubes (e.g., Vacutainers) or serum bottles.
3. Make up a parallel set of $C_2H_4$ standards and inject into the CG. Standards are used to calibrate GC sensitivity, linearity, and concentrations of $C_2H_4$ produced in samples. After calibration and standardization, $C_2H_2$ reduction is most commonly expressed as nmol $C_2H_4$ produced per volume of suspension, dry weight, or surface area of sample. For plant materials, it can be expressed per amount of protein or chlorophyll a.

**$^{15}N$ Assimilation** $N_2$ fixation can also be assessed by measuring $^{15}N_2$ assimilation. In addition to being a direct assay, this approach is sometimes favored over the ARA when certain environmental conditions prohibit the use of the latter. The ARA is suspect under anaerobic, sulfidic (high ambient $H_2S$) conditions, when acetylene and ethylene may either be consumed or produced by processes unrelated to nitrogen fixation. Although the details of the $^{15}N$ technique are discussed more thoroughly elsewhere [5, 13], this approach warrants a brief overview here. As with the ARA, natural and cultured samples are placed in stoppered vessels. However, for aquatic samples, whereas an aqueous to vapor phase of 3 to 1 is usually preferred for the ARA for $^{15}N_2$-fixation assays, a complete aqueous phase is preferred. $^{15}N_2$ gas (over 90% atom excess $^{15}N$) is injected into an inverted vessel (~1 ml per 20 ml of water). Displaced water is allowed to escape the vessel by placing a syringe needle in the serum stopper while injecting $^{15}N_2$. For terrestrial samples, $N_2$ gas in the headspace is first removed by flushing the sealed vessel (using two syringe needles placed in the serum stopper) with an inert gas-

(Ar or He) $CO_2$ mixture (1–5% $CO_2$). $^{15}N_2$ is then injected into the headspace to obtain at least 90% atom excess $^{15}N$.

The incorporation of $^{15}N_2$ can be measured in particulate matter by either mass spectrometry (when high precision and sensitivity are desired, as is the case when rates of $N_2$ fixation and hence $^{15}N$ enrichment are low relative to diazotropic biomass) or by emission spectrophotometry, when sensitivity is not an issue and ease of sample preparation and analysis is preferred. Because measuring $^{15}N$ assimilation involves considerable expense—purchasing $^{15}N_2$ gas, requiring access to either a mass spectrometer or emission spectrophotometer—and preparation for analysis is time-consuming and requires specialized equipment, microbial ecologists tend to favor the ARA with periodic backup $^{15}N$ calibration checks. The ARA is calibrated in terms of the ratio of acetylene reduced (to ethylene) by conducting parallel $^{15}N_2$-fixation assays. This process allows for quantitative estimates of $N_2$ fixation while minimizing sample and analytic costs.

**N Accumulation into Particulate Matter** Accumulation of N in the organisms and particulate matter under consideration is a direct measure of $N_2$ fixation, which can be assessed by conducting either Kjeldahl digestion/oxidation or C,H,N elemental analyses on the material in question. However, this method has several limitations. It relies on high rates of $N_2$ fixation relative to the high "background" N in particulate matter. Furthermore, corrections need to be made for contemporaneous losses of N from particulates by mineralization, ammonification, and denitrification. To make these corrections, one often needs to know the rates of these processes, and this information may not always be available, especially in field studies.

## Denitrification: Assimilatory and Dissimilatory Nitrate Reduction

The dominant oxidized inorganic N form is nitrate ($NO_3^{-1}$), which is used by diverse phototrophs, chemoautotrophs, and heterotrophs as a N source to support growth or cellular redox reactions (i.e., electron acceptor). After uptake, $NO_3^{-1}$ is reduced via several pathways, the most common of which are assimilatory and dissimilatory nitrate reduction (see Fig. 1-1). In assimilatory nitrate reduction, $NO_3^{-1}$ is reduced to ammonium ($NH_4^{+1}$), followed by incorporation into a variety of organic molecules supporting cell growth and metabolism. The ubiquitous enzymes that mediate $NO_3^{-1}$ reduction, initially to nitrite ($NO_2^{-1}$) and then to $NH_4^{+1}$, are nitrate reductase and nitrite reductase, respectively. The in vivo reduction of $NO_3^{-1}$ to $NO_2^{-1}$ is coupled closely to the oxidation of NADPH or NADH, whereas the reduction of $NO_2^{-1}$ is coupled to ferrodoxin (oxidation). Both sets of reactions are energy-dependent, with the energy derived from photosynthesis in algae, cyanobacteria, and photosynthetic bacteria, oxidation of inorganic compounds in chemoautotrophs, or from the oxidation of organic matter in heterotrophs [26].

Nitrate reductase is present in all microorganisms that are able to reduce $NO_3^{-1}$. This enzyme is a large (~300 kd) soluble protein that appears to be very similar, but not identical, in structure among diverse microbial groups. The general scheme for $NO_3^{-1}$ reduction is depicted in Figure 1-3. Several assays have been developed and broadly applied to assess nitrate reductase activity in aquatic and terrestrial microorganisms [26]. One assay assesses the coupled oxidation of NADH (via NADH oxidase). It does not, however, ensure coupling to $NO_3^{-1}$ reduction since it only measures diaphorase activity, and thus its use in microbial ecologic and biogeochemical studies is limited. An alternative and widely used assay is based on methyl viologen or FADH-dependent $NO_3^{-1}$ reduction. A third measurement is based on NADH-dependent $NO_3^{-1}$ reduction.

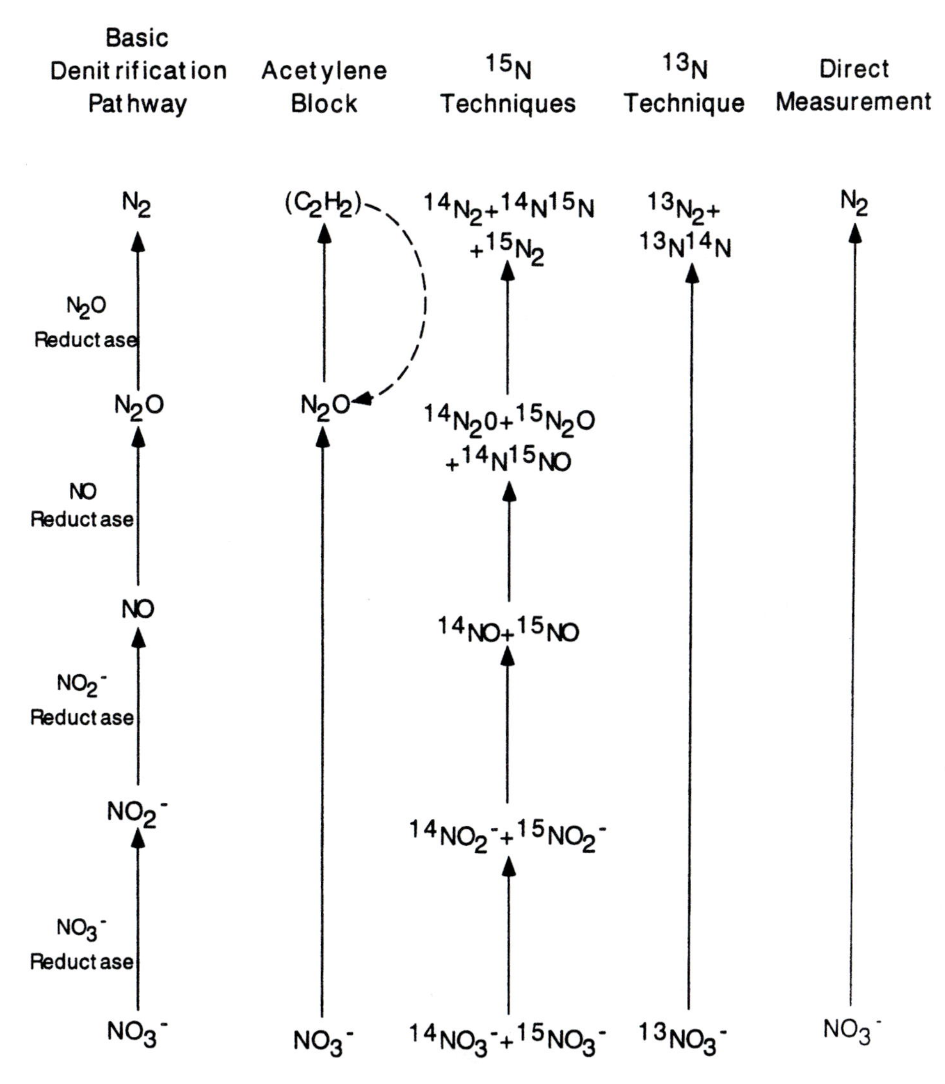


Figure 1-3 Denitrification dynamics, showing the application of various techniques for assessing rates of denitrification.

All assays are colorimetric and hence require a spectrophotometer. The coupling of NADH oxidation to diaphorase activity results in a reduction of FAD and cytochrome in cell extracts [31]. When $NO_3^{-1}$ is added, cytochrome reoxidizes and $NO_2^{-1}$ is produced. Nitrite can be detected colorimetrically by reacting it with sulfanilamide and *N*-(1-naphthyl)-ethylenediamine dihydrochloride [24, 64]. Following $NO_2^{-1}$ production over time is a common way to assess nitrate reduction activity in the laboratory and field. For kinetic assays (over close time intervals), $NO_3^{-1}$-dependent oxidation of NADH can be followed spectrophotometrically. Ion-specific electrodes have also been used for this purpose; however, unless fairly high substrate and reactant concentrations are encountered, the use of ion-specific electrodes may be restricted by relatively low sensitivity.

In contrast to highly conserved components of the nitrogenase enzyme complex, nitrate reductase structural subunits and the genes encoding for them do differ among microbial groups. Thus, the scope and application of immunochemical and molecular (i.e., Southern hybridization, PCR amplification) detection and characterization of this enzyme in complex communities are limited. The genes for nitrite reductase seem to be conserved more highly among bacterial genera than those for nitrate reductase. Smith and Tiedje [74] recently isolated and characterized a nitrite reductase gene (*nir*) from the terrestrial denitrifying bacterium, *Pseudomonas stutzeri* JM300. Gene probes made from restriction fragments within the *nir* operon were shown to hybridize to bacterial strains from at least five different denitrifying genera, whereas no hybridization to non-denitrifyers (strains representing at least six genera) took place. Clearly, more testing across numerically diverse and functionally dominant bacterial genera representing natural communities is needed before such probes can be deemed to be ecologically and biogeochemically useful and relevant. Results obtained thus far seem encouraging, however. A DNA probe (Southern hybridization) for the gene encoding for *nir*, and a polyclonal antibody to this enzyme have also been described for *P. stutzeri* (ATCC 14405) [83]. The DNA probe showed a higher degree of hybridization to the *nir* gene among *P. stutzeri* and other denitrifying strains than did the antibody to the NiR protein. This finding indicates that the DNA probe may be more useful for detecting *nir* among genetically diverse strains, whereas immunodetection and quantification may be more effective on the single strain level [83].

In dissimilatory $NO_3^{-1}$ reduction, products of nitrate reductase activity are not used as a cellular N source, but rather are excreted, either as dissolved solids ($NO_2^{-1}$ as in $NO_3^{-1}$ respiration) or gases (NO, $N_2O$, and $N_2$ as in denitrification). Microbial groups known to conduct $NO_3^{-1}$ respiration (as a source of oxidant) include a variety of heterotrophic bacteria (*Aeromonas*, *Pseudomonas*) and facultative anaerobes (*Bacillus*, *Klebsiella*, *Vibrio*). The excreted $NO_2^{-1}$ can be reassimilated by a variety of heterotrophs, chemoautotrophs, and photoautotrophs. Denitrifiers include taxonomically and ecologically diverse facultative anaerobic and obligate anaerobic heterotrophic genera. Among facultative anaerobes, *Pseudomonas* and *Thiobacillus* are often cited, whereas obligate anaerobes include *Bacillus* and *Spirillum*. Whether or not facultative anaerobes conduct denitrification depends largely on the presence of ambient $pO_2$ [51, 79, 80]. When $O_2$ is present, these organisms preferentially reduce $O_2$ over $NO_3^{-1}$. Molecular $O_2$ can also be inhibitory to denitrification. Accordingly, the preferred habitats for denitrifiers include anoxic zones near $NO_3^{-1}$ sources. In aquatic ecosystems, organically enriched surface sediments harboring $O_2$-devoid microenvironments (microzones) that are close to $NO_3^{-1}$ influx from overlying oxygenated waters are ideal niches for denitrifiers. Analogously, organic-matter-rich soils supporting anoxic microzones that are exposed to $NO_3^{-1}$ supplied by surface or groundwater are ideal sites for denitrification. Denitrification proceeds in a series of enzymatically mediated steps (Fig. 1-4):

1. $NO_3^{-1} \rightarrow NO_2^{-1}$ (nitrate reductase)
2. $NO_2^{-1} \rightarrow NO$ (nitrite reductase)
3. $NO \rightarrow N_2O$ (nitric oxide reductase)
4. $N_2O \rightarrow N_2$ (nitrous oxide reductase)

Depending on environmental conditions ($pO_2$, organic matter supply, pH, $E_h$), the reduction of $NO_3^{-1}$ can be terminated at any of these steps. $N_2O$ and $N_2$ are the most common end products of denitrification (see Fig. 1-1) [38, 51]. Since both are gases, they can be assayed by a variety of gas chromatographic, mass spectrometric, and volumetric techniques. Commonly used methods for determining denitrification are outlined in this section.

### The Acetylene Block

Among its many interactions with biochemical transformations (e.g., acetylene reduction by nitrogenase), acetylene ($C_2H_2$) also blocks the reduction of $N_2O$ to

$N_2$ (see Fig. 1-4). It does so by noncompetitive inhibition of $N_2O$ reductase, at a $K_i$ value of approximately 28 μM. As a result, $N_2O$ can accumulate in the presence of reasonably small amounts of $C_2H_2$, and this characteristic has been used as the basis of an assay for denitrification [75]. The acetylene block technique has several advantages over more tedious, impractical, and costly assays described later in this section. Like the ARA for nitrogenase activity, the acetylene block can be deployed easily in field and laboratory studies. The $N_2O$ evolved can be measured readily and sensitively by gas chromatography (electron capture is the preferred detection system). The assay is commonly executed over 1 to 4 hours, making it amenable to laboratory kinetic, short-term, and longer term (e.g., 24h diel) field studies [75]. $N_2O$ production is also measured directly by electron capture gas chromatography, serving as "blank" and "background" $N_2O$ production/consumption measurements. There is considerable interest in such measurements, because $N_2O$ is an infrared light-absorbing greenhouse gas, the concentrations of which may be increasing in response to naturally and anthropogenically induced alterations of aquatic and terrestrial ecosystems [86].

Although the acetylene block technique is attractive from analytic and logistic perspectives, it suffers from several, occasionally serious methodologic and interpretive pitfalls. Most problematic is the fact that acetylene also inhibits nitrification [42, 44], and it is well known that in many instances denitrification can be closely coupled in time and space to $NO_3^{-1}$ produced by nitrification [44, 45]. Under such conditions, reasonable estimates of denitrification may be obtained by simply measuring nitrification (see the section, Nitrification) and assuming that the $NO_3^{-1}$ produced is denitrified. Close coupling of nitrification and denitrification occurs in surface sediments, biofilms, and soils where $NO_3^{-1}$ concentrations are low due to either low supply rates or competing $NO_3^{-1}$ uptake processes (assimi-

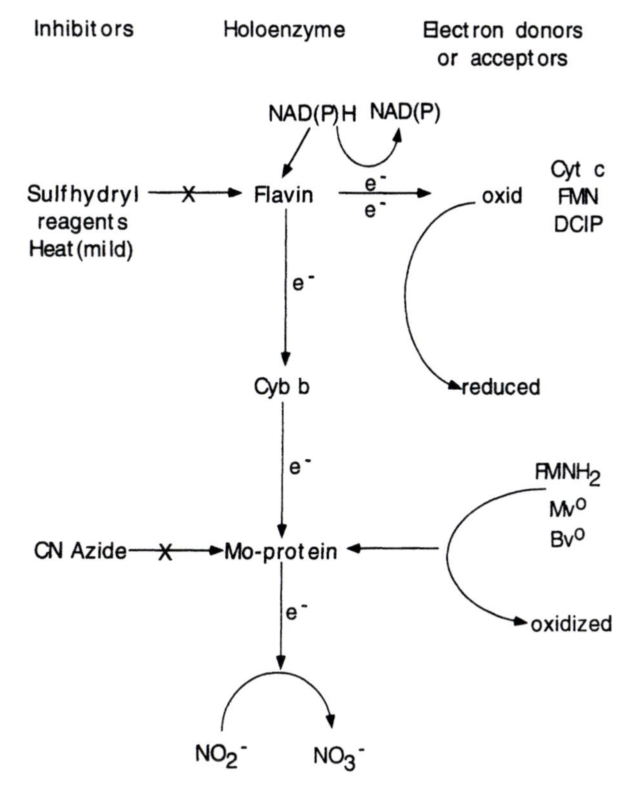


Figure 1-4 The enzymology and electron flow involved in $NO_3^{-1}$ reduction ($NO_3^{-1}$ reductase; adapted from [26]).

latory $NO_3^{-1}$ reduction) that occur contemporaneously and contiguously with denitrification [45]. Under these conditions, the acetylene block can be expected to underestimate denitrification rates. Conversely, when a majority of the $NO_3^{-1}$ used in denitrification is supplied from external sources, such as the overlying water column or groundwater, the acetylene block can be an effective measure of true denitrificaton rates [46].

Another problem encountered with the acetylene block technique is that at times it may incompletely block $N_2O$ reductase. Acetylene inhibition of $N_2O$ reduction is ineffective below $NO_3^{-1}$ concentrations of 10 μM [58]. Also, in sediments and soils rich in $S^{-2}$, $N_2O$ reduction is blocked either partially or totally [58, 75]. In such cases, there are serious kinetic and quantitative constraints on using the technique. In addition, $N_2O$ consumption rates and indirect effects of $N_2O$ production on microbial metabolism during the course of assays may vary during the incubation period, which is difficult to evaluate and prevent.

The acetylene inhibition technique for estimating denitrification in marine or freshwater sediments (adapted from [45]) is described below.

### Materials and Equipment

Plastic core tubes with silicon filled side ports at 1 cm depth intervals and rubber stoppers
Purified acetylene (either supplied as compressed gas or generated from $CaC_2$)
Serum bottles (30 ml) with natural rubber serum stoppers
1 N KCl solution
Gas-tight syringes (1, 10, 30, and 60 ml)
Evacuated serum bottles (15 ml) with plug stoppers and aluminum seals
Purified nitrous oxide ($N_2O$) for standards
Gas chromatograph (GC) with electron capture detector (ECD) and carrier gas (ultra high purity 5% methane/95% argon mix)
2-m Hayes Sep D GC column

### Procedure

1. Collect sediment cores to a desired depth and transport them to the laboratory within 2 hours. Prepare acetylene incubation solution by adding 20 ml seawater or freshwater (amended with $NO_3^{-1}$ if necessary) to a 50-ml serum vial, evacuating the headspace air with a 30-ml syringe and injecting 40 ml acetylene to saturate the solution. Inject sediment cores with the acetylene-saturated solution at a volume equal to 10% of the porewater in a 1-cm interval of the core and 10% of the volume of overlying water (collected in situ). Measure background production of $N_2O$ in replicate cores injected with $N_2$-saturated incubation solution. Seal cores with stoppers, and incubate at ambient field temperature for a suitable time interval (usually 0.75–4 h). The incubation period should fall within a linear range of $N_2O$ production, experimentally determined from time series analyses.
2. Place large cores that are extruded and subcored through a side port and subcores (or small sample cores) into previously weighed 30-ml serum vials containing 10 ml 1 N KCl. Seal these vials immediately, and weigh them again.
3. Place the serum vials on a rotary shaker table at 1000 rpm for 5 minutes to equilibrate $N_2O$ between gas and water phases. Remove 15 ml gas samples with a gas-tight syringe by displacement with distilled water, and place in sealed evacuated 15 ml serum vials for storage.

4. Analyze $N_2O$ by gas chromatography. Determine $N_2O$ concentration by comparison to a standard curve generated with purified $N_2O$. Calculate denitrification rates on a gram wet sediment (gws) basis; calculate integrated areal rates by using sediment wet density (gws/cc) and the depth interval.

### $^{15}N$ Isotope Techniques

Another way to quantitate denitrification is to measure the production of $^{15}N_2$ from $^{15}NO_3^{-1}$ added in trace quantities [34, 38] (see Fig. 1-4). The $^{15}N_2$ produced is collected by a variety of means (usually by being cryogenically "trapped" and concentrated) and measured by either mass spectrometry or emission spectrophotometry. Although this is a direct method, it suffers from several problems, including low sensitivity of $^{15}N$ detection (however, sensitivity may be increased through the development of more sensitive detectors and the use of $^{15}N$:$^{14}N$ mass spectrometry). The current low level of sensitivity requires the addition of relatively large amounts of $^{15}NO_3^{-1}$, which alters the natural substrate concentrations, hence no longer allowing this to be a tracer technique. Therefore, the $^{15}N$ technique is often the method of choice when relatively large amounts of $NO_3^{-1}$ are available for denitrification.

The radioisotope, $^{13}N$, has been used to develop a highly sensitive tracer measure of denitrification [32] (Fig. 1-4). This technique, however, is not amenable to field studies, since the relatively short half-life (~7 min) of $^{13}N$ dictates that determinations be made near a cyclotron facility.

### Direct Measurement of $N_2$ Accumulation

Another approach is to simply measure the accumulation of $N_2$ gas in soil or water samples (Fig. 1-4). $N_2$ evolved by denitrification is then measured either (in order of increased sensitivity) volumetrically (i.e., nitrometer), by gas chromatography (most often by thermal conductivity detection), or by mass spectrometry. The last approach has been used extensively by Seitzinger and colleagues [72, 73]. However, this method, although also direct, has specific requirements that may rule it out for some studies. It requires that all $N_2$ gas be flushed from samples before incubation, which may be very difficult in soils and sediments where $N_2$ can diffuse at very slow rates. The method is susceptible to external $N_2$ contamination, so one has to ensure isolation of the sample from the ambient environment during incubations. Relatively long incubation periods (i.e., days to weeks) may be required to obtain measurable amounts of $N_2$. It is therefore difficult to ascertain short-term, instantaneous in situ denitrification rates.

### $^{15}N$ Dilution and Isotope Pairing Technique

A $^{15}N$:$^{14}N$ isotope pairing technique has recently been developed that is capable of measuring denitrification in association with (i.e., coupled to) nitrification [57, 68]. Here, $^{15}NO_3^{-1}$ is added to water overlying sediments or soils. If denitrification of this $^{15}NO_3^{-1}$ source occurs, $^{15}N$ atoms will appear in the denitrification product, $N_2$. If all the denitrification taking place is supported by $^{15}NO_3^{-1}$ in overlying water, then the $N_2$ evolved should be $^{15}N$:$^{15}N$. However, if a fraction of the denitrification is supported by internal nitrification processes, in which the naturally abundant isotope $^{14}N$ should dominate, then $N_2$ evolved by denitrification should have a "split" isotopic composition (i.e., $^{15}N$:$^{14}N$). The proportion of $^{15}N$ to $^{14}N$ in the $N_2$ evolved is thus indicative of the $NO_3^{-1}$ source(s) supporting denitrification. Isotopic composition of the various sources and products is determined by mass spectrometry. The isotope pairing technique can be used to determine rates of gross $NO_3^{-1}$ uptake, denitrification (based on $NO_3^{-1}$ supplied in overlying water), the proportion of denitrification attributable to coupling to nitrification versus externally supplied $NO_3^{-1}$, nitrification, and, to a limited extent, N mineralization in sediments and water-saturated soils. The methodological details are provided in Nielsen [57] and Rysgaard et al. [68].

### *$N_2$: Ar Ratio Determination of Denitrification*

A relatively new technique for measuring rates of denitrification is the detection of short term (<24 h) changes in $N_2$:Ar ratio in the water of in situ sediment chambers with a membrane inlet high precision mass spectrometer [46]. This method, which is under development, is based on the premise that the inert trace gas argon is not biologically altered and hence remains conserved relative to $N_2$ evolved by denitrification. Potential advantages of this approach over previously discussed methods include: simplicity in concept and execution, high sensitivity, enabling relatively short-term measurements, and a direct measure of changes in $N_2$ flux relative to Ar. The method should also be adaptable to soils and water samples.

## Nitrification

Nitrification is a critically important "intermediate" N transformation (see Fig. 1-1). Under aerobic conditions, $NH_4^{+1}$ liberated from microbial decomposition is oxidized, initially to $NO_2^{-1}$ and then to $NO_3^{-1}$. Both combined N forms are readily assimilated by autotrophs and heterotrophs. In addition, $NO_3^{-1}$ serves as the substrate for denitrification. Nitrification is mainly mediated by aerobic chemolithoautotrophs of the genera *Nitrosomonas*, *Nitrosococcus*, and *Nitrobacter* and as such takes place in well-oxygenated soils and waters. Nitrification can be particularly intense near sources of $NH_4^{+1}$, including the oxic/anoxic interphases of sediments, soils, and biofilms, and near the surface microenvironments of suspended detrital particles. There is evidence that heterotrophic nitrification may also exist; however, its ecologic significance is uncertain. This section focuses on methods of assessing chemoautolithotrophic nitrification.

It is generally agreed that nitrifiers grow and oxidize substrates slowly, and therefore culturing techniques and activity measurements can take substantial periods of time (hours to days for activity; days to weeks for growth measurements) [47]. In contrast, ambient $NO_3^{-1}$ concentrations in soils and natural waters frequently exceed concentrations of $NH_4^{+1}$ by orders of magnitude, a strong indication that this process is highly efficient (i.e., high affinities for low concentrations of $NH_4^{+1}$) and that high standing stocks of nitrifiers must be present.

### *Cultivation*

Culturing efforts indicate a widespread presence of $NH_4^{+1}$-oxidizing bacteria, with *Nitrosomonas* and *Nitrobacter* being predominant in both soil and aquatic ecosystems. These genera mediate the following reactions:

$$(1)\ NH_4^{+1} + 1.5\ O_2 \rightarrow NO_2^{-1} + H_2O + 2H^+\ (\textit{Nitrosomonas})$$

$$(2)\ NO_2^{-1} + 0.5\ O_2 \rightarrow NO_3^{-1}\ (\textit{Nitrobacter})$$

Molecular oxygen serves as the terminal electron acceptor. Hydroxylamine is known to be an intermediate of the oxidation; however, it decomposes rapidly to either $N_2$, $NH_4^{+1}$, or $N_2O$. The evolution of the last intermediate—$N_2O$—is of considerable interest, since it is a greenhouse gas that is emitted in nature [16,

86]. Functionally, there are similarities between nitrifiers and methane ($CH_4$) oxidizers. Both contain well-defined and complex membranous structures that harbor the Cu-containing $CH_4$- and $NH_4^{+1}$-oxidizing enzymes, respectively. Acetylene inhibits both processes. Ammonium inhibits $CH_4$ oxidation in a variety of methane oxidizers, and conversely, $CH_4$ inhibits $NH_4^{+1}$ oxidation in *Nitrosomonas*. A variety of other inhibitors selectively arrest both processes [58].

Medium Commonly Used for Isolating and Culturing Nitrifying Bacteria (after [1])

| | |
|---|---|
| $Na_2HPO_4$ | 13.5 g |
| $KH_2PO_4$ | 0.7 g |
| $MgSO_4 \cdot 7H_2O$ | 0.1 g |
| $NaHCO_3$ | 0.5 g |
| $FeCl_3 \cdot 6H_2O$ | 0.014 g |
| $CaCl_2 \cdot 2H_2O$ | 0.18 g |
| Distilled water | 1000 ml |

Procedure

1. Place 75 ml amounts of medium into 250-ml Erlenmeyer flasks, and sterilize at 121° C for 15 min.
2. For nitrite oxidizers add 0.5 g of $NaNO_2$ per liter before sterilization. For ammonium oxidizers, sterilize a stock solution of $(NH_4)_2SO_4$ separately from the basal medium, and add aseptically to give a final concentration of 0.5 g per liter.

### *Nitrification Measurements*

**Dark $CO_2$ Fixation for Nitrification Measurement in Sediments** During nitrification $CO_2$ is reduced by the pentose phosphate (Calvin) cycle [7]. From a methodologic perspective, this is a useful characteristic since dark-mediated $^{14}CO_2$ fixation can be used as a specific measure of nitrification. Chlorate specifically inhibits $NO_2^{-1}$ oxidation by autotrophic nitrifiers, and nitrapyrin or N-serve [2-chloro-6-(trichloromethylpyridine)] also blocks nitrification [58]. N-serve also inhibits sulfate reduction, methanogenesis, and algal photosynthesis. Dark $CO_2$ fixation has found general use in ecologic studies. Measurements of dark $^{14}CO_2$ fixation with and without N-serve are used to assess $CO_2$ fixation by nitrifiers. N-serve (at 20 μg per gws) has also been used to examine the coupling of nitrification to denitrification (i.e., nitrification-dependent denitrification) [42].

Materials and Equipment

Sediment cores with silicon-filled side ports at 0.5-cm intervals
N-serve dissolved in acetone
Core aeration apparatus (needles connected with tubing to an aquarium pump)
2 N KCl

Procedure

1. Collect replicate sediment cores to desired depth, and transport them to the laboratory. Inject one-half of the cores with N-serve at a concentration of 20 ppm ($\mu g \cdot g^{-1}$ wet sediment) at 0.5-cm intervals and 5 ppm in overlying water. Inject the remaining cores with a similar volume of acetone without

N-serve. Incubate cores under continuous aeration in the dark for 48–72 h at ambient, field temperature. Change the overlying water daily; at that time add fresh N-serve to the water to ensure full inhibition during the incubation period. Filter and freeze the water for analysis of $NH_4^{+1}$ and $NO_3^{-1}$.

2. After the incubation, measure porewater KCl-extractable $NH_4^{+1}$. Section cores at desired depth intervals, and place the sections into centrifuge tubes. Add 2 N KCl to each section, and shake the tubes on a rotary shaker at 1000 rpm for 10 min. Filter porewater from each section through a glass fiber filter (Whatman GF/C or equivalent), and analyze for $NH_4^{+1}$. Calculate the nitrification rate by noting the difference between the means of $NH_4^{+1}$ content for N-serve treatments and acetone controls. Ample replication of sediment cores is necessary due to considerable variation among cores, particularly in vegetated or bioturbated sediments.

**$N_2O$ Evolution** The evolution of $N_2O$ has also been used as an index of nitrification [16]. Evolved $N_2O$ is most often measured by electron capture GC. During the oxidation of $NH_4^{+1}$ to $NO_2^{-1}$, a relatively constant percentage (0.1% to 0.3%) of $NO_2^{-1}$ produced is believed to be released as $N_2O$. This proportion is strongly dependent on $pO_2$, so in waters and soils supporting strong $O_2$ gradients, hypoxia, and periodic anoxia, there may be substantial variations in the amounts of $N_2O$ produced.

**$^{15}NH_4^{+1}$ Oxidation** A direct assay for nitrification activity involves adding $^{15}N$-labeled $NH_4^{+1}$ and following its oxidation, initially to $NO_2^{-1}$, then to $NO_3^{-1}$ [39]. The $^{15}N$-labeled products are detected and quantified by mass spectrometry. This method, although direct and very useful in waters containing relatively high amounts of $NH_4^{+1}$, suffers from similar pitfalls as the $^{15}N$ denitrification assay; namely, relatively large concentrations of $^{15}NH_4^{+1}$ are required to meet analytic sensitivity limitations. This requirement may call for $^{15}NH_4^{+1}$ additions far in excess of naturally occurring $NH_4^{+1}$ concentrations, which potentially alters the kinetics of nitrification [47]. Incubation periods required to detect significant $^{15}NO_2^{-1}$ or $^{15}NO_3^{-1}$ formation may be lengthy, leading to a "bottle effect" problem in which the enclosed microflora (and grazers) may undergo substantial changes in community structure and function.

Marine and soil nitrifiers have been identified and enumerated through immunochemical techniques [71, 82]. Immunofluorescence assays (against specific cell surface antigens) have been used to investigate community structure and distributions of marine nitrifiers, both in culture and natural environments [81, 82]. Recently, an enzyme immunoassay for the detection of the ubiquitous nitrifying bacterium, *Nitrosomonas europaea*, has been described and tested [70]. Molecular approaches are currently under development, including a PCR technique making use of degenerative primers for detecting key nitrifiers from the marine environment.

## Organic Nitrogen Utilization and Metabolism

By far, the largest pool of combined nitrogen in aquatic and terrestrial ecosystems is organic nitrogen (ON), present either as dissolved organic nitrogen (DON) or particulate organic nitrogen (PON). The distinction between DON and PON is based on size, with DON constituting a less than 0.2 μm fraction (based on membrane filtration). DON therefore includes truly dissolved substances, as well as macromolecules (e.g., proteins, nucleic acids, complex carbohydrates) and colloidal matter less than 0.2 μm in size. In aquatic ecosystems, the water column usually contains far more DON than PON. In aquatic sediments, soils, and above ground biomass, most of the ON can reside in the PON fraction. Virtually all the ON results from biologic assimilation and incorporation of inorganic nitrogen. Some fraction of DON results from incomplete decomposition (including lysis of dying or decaying biota) of PON or excretion of DON by actively metabolizing plants, animals, and microorganisms.

Evidence is mounting that DON and PON, for many years considered "the ashes of the fire," may not be the collection of recalcitrant molecules they have long been believed to be. DON utilization by heterotrophic and photosynthetic organisms can provide an important source of N [2, 8, 12, 59]. It is intuitively obvious that the relatively large amounts of DON and PON being discharged (either from natural or anthropogenic sources) in terrestrial and aquatic ecosystems must be used by resident microflora or we would experience vast excesses of these N pools in the biosphere. Elucidating and quantifying the routes of DON utilization are of fundamental importance in fully understanding the dynamics of microbially mediated N-cycling.

In soils, lakes, rivers, estuaries, and coastal waters known to contain relatively high concentrations of DON, bacteria, fungi, and microalgae are capable of using a wide variety of DON compounds. These compounds can serve as either energy or growth sources. Uptake and assimilation processes are, not unexpectedly, quite diverse, as are the proportions of DON used as energy, cell maintenance, and growth sources [2, 28, 65]. The proportion of DON utilization attributable to bacterial heterotrophs, fungi, microalgae, and higher plants is dependent on a suite of environmental conditions, including temperature, pH, irradiance, and, most importantly, the origin and composition of the DON. It is generally agreed that, because of their small size, efficient uptake systems, and exclusive requirements for organic matter for growth, heterotrophic microorganisms dominate DON utilization in natural waters [8, 17].

Under dark conditions, heterotrophic bacteria and fungi are responsible for a large fraction of DON uptake; some dark-mediated DON uptake by microalgae has been documented in aquatic ecosytems [2]. Under illuminated conditions, microalgae possess the ability to use a range of dissolved organic substrates, including such DON compounds as amino and nucleic acids [2, 56]. Algal utilization of DON seems to be controlled by several mechanisms, including dark-mediated uptake assimilation similar to that described for bacterial heterotrophs and light-mediated photoheterotrophy, a poorly understood process that is distinguished by being insensitive to the photosynthetic electron transport (photosystem II) inhibitor 3 (3,4-dichloro-phenyl)-1,1-dimethylurea (DCMU) [56, 59]. Light-mediated DON utilization by phytoplankton may be of considerable ecologic and biogeochemical importance in N-deplete (and hence N-limited) oligotrophic systems, where the DON pool represents a relatively (to inorganic N) large source of nitrogen.

Berman [6] postulated that, among microalgae, differential abilities to utilize DON could be an important factor for determining natural species composition. Since sediments often contain DON at higher concentrations than in overlying waters [19], it is perhaps not surprising that sediment-associated microalgae grew more rapidly than planktonic algae when using amino acids. Benthic microalgal communities have recently been implicated in the control of N fluxes between the sediment and overlying water [19]. However, studies on biotic regulation of sediment flux have largely concentrated on dissolved inorganic nitrogen (DIN) rather than DON compounds, despite the fact that benthic microalgal communities are in an environment supplied with DON both from the underlying sediment [19] and overlying water column [54]. The role of DON in algal nutrition may have been overlooked by experimentation that has been focused exclusively on DIN assimilation.

Although sediment DON concentrations usually exceed those in overlying waters, sediment-associated DON is often dominated by high molecular weight (HMW) molecules [19]. The tacit assumption has been that, since HMW species dominate the DON pool, it is relatively unreactive and hence of less consequence to biologically mediated N cycling than DIN. The ecologic and biogeochemical importance of HMW DON compounds as a N-source for bacteria and microalgae remains unclear, however, particularly in light of suggestions that previous assumptions of recalcitrance and biotic utilization of HMW DON are in need of reevaluation [2].

Even less is known about PON utilization and cycling dynamics, in part because it is more difficult to experimentally determine the uptake, assimilation, and fate of PON in diverse microbial communities. Microorganisms able to excrete enzymes capable of hydrolytically cleaving PON into assimilable DON compounds may be of particular importance in the mediation of PON utilization and biodegradation. It is known that fungi and certain bacterial groups produce such enzymes. Fungi are known to be effective decomposers of such "refractile" substances as cellulose, tannins, and lignins. They may function in a consortial fashion, with bacteria, microalgae, and protozoans playing integral and complementary roles in the degradation process.

### *Methods for Determining DON and PON Utilization and Cycling*

**Assessing Growth on Specific Substrates** If specific DON/PON substrates of interest are identifiable and quantitated, they can be administered over a range of concentrations, and their utilization and breakdown in the natural environment can then be monitored. In addition, the growth responses of specific microbial taxa or major functional groups to DON and PON additions can be followed with microscopic enumeration and biomass determination techniques. These techniques include acridine orange direct counting (AODC), 4′,6-diamidino-2-phenylindole (DAPI) nucleic acid staining, immunofluorescence counting, and chlorophyll assays for algal biomass. Growth measurements can be coupled to metabolic activity determinations, such as $O_2$ consumption and production, $CO_2$ consumption and production, electron transport system (ETS) activity, and ATP/ADP concentrations (Table 1-1).

**Uptake, Assimilation and Degradation of Radiolabeled Substrates** A widely used approach for assessing DON/PON uptake, assimilation, and degradation is to examine the fate of radioisotopically labeled compounds in natural or cultured microbial assemblages. In particular, either $^{14}C$- or $^{3}H$-labeled DON/PON substrates are broadly used for these purposes [17, 43, 65] (Table 1-1). There are several advantages to using radioisotope assays. They are extremely sensitive and hence can be administered at true tracer (no significant alteration of natural substrate concentrations) levels or over a broad range of concentrations (for kinetic assays). In addition, high sensitivity facilitates short-term uptake and assimilation assays. Radioassays can be coupled to localization detection methods, such as autoradiography, which enables the investigator to identify the microorganisms involved in DON/PON processing. Products of DON/PON metabolism, such as $CO_2$ and $H_2O$, can be collected and quantified readily. If the natural concentration of a substrate is known, then turnover and flux of that substrate can be determined in the natural environment. Finally, the metabolic route or pathway of DON/PON breakdown can often be elucidated by conducting a series of short-term (minutes to hours) incubations, followed by recovery of the radioisotope in purified molecules.

One can examine parallel uptake and assimilation of a variety of DON/PON substrates by conducting double labeling experiments (see Table 1-1). In this case, one substrate is exclusively labeled with $^{14}C$, whereas the other is $^{3}H$ labeled. It is also possible to label different moieties (active groups) of single molecules with either of these radioisotopes to examine various aspects of the degradation of these molecules.

Radioassays have their limitations, however, and these limitations can at times lead to serious methodologic and interpretational pitfalls and shortcomings. If ambient concentrations of a substrate of interest are not known, it is difficult to conduct kinetic assays. One can perform "saturation" types of experiments in which very high concentrations of the substrate (a portion of it radioisotopically labeled) are administered. These types of experiments are useful in describing the potential heterotrophic uptake and assimilation of a particular substrate or group of substrates under enzyme-saturated conditions. Another problem is related to the fact that, when using either $^{14}C$- or $^{3}H$-labeled substrates, one is only able to monitor the fate of the carbon or hydrogen or both, as opposed to nitrogen, constituents of DON/PON molecules (see Table 1-1). One must therefore make assumptions as to whether N behaves similarly to the C and H constituents. Some microorganisms preferentially cleave N off DON/PON before uptake, and the C and H constituents are not utilized during the N uptake process. It is therefore conceivable that DON/PON uptake and assimilation determinations based on C

Table 1-1 Methods for Determining Organic Nitrogen (ON) Utilization, Metabolism, and Decomposition in Microbial Assemblages

**Uptake and fate of $^{14}$C-labeled ON**

A. $^{14}C\text{-ON} \xrightarrow{\text{uptake}} {}^{14}C\text{-labeled cells}$

(measured by liquid scintillation counting or microautoradiography)

B. $^{14}C\text{-labeled cells} \xrightarrow[\text{decomposition}]{\text{excretion, respiration}} {}^{14}C\text{-organic metabolites} + {}^{14}CO_2$

**Uptake and fate of $^{3}$H-labeled ON**

$^{3}H\text{-ON} \xrightarrow{\text{uptake}} {}^{3}H\text{-labeled cells}$

(measured by liquid scintillation counting or microautoradiography)

$^{3}H\text{-labeled cells} \xrightarrow[\text{decomposition}]{\text{excretion, respiration}} {}^{3}H\text{-organic metabolites} + {}^{3}H_2O$

**Uptake and fate of $^{15}$N-labeled ON**

$^{15}N\text{-ON} \xrightarrow{\text{uptake}} {}^{15}N\text{-labeled cells}$

(determined by mass spectrometry or emission spectrophotometry)

$^{15}N\text{-labeled cells} \xrightarrow[\text{decomposition}]{\text{excretion}} \text{soluble } {}^{15}N\text{-labeled metabolites}$

**Double- or triple-labeled combinations of 1, 2, and 3**

**$CO_2$ evolution coupled to DON/PON metabolism/decomposition**

**$O_2$ uptake coupled to DON/PON metabolism/decomposition**

**Electron transport system (ETS) activity associated with DON/PON metabolism**

**Cellular ATP/ADP ratios (energy charge) associated with DON/PON utilization and metabolism**

or H radioisotopes may not yield qualitatively and quantitatively meaningful results. One way to circumvent this problem is to conduct $^{14}C$ (or $^{3}H$) uptake and assimilation determinations in parallel with $^{15}N$ stable isotope-labeled DON/PON utilization experiments (see Table 1-1).

**Use of Stable Isotopes** Like radioisotopes, the stable isotope of nitrogen, $^{15}N$, has been used to examine DON/PON uptake and assimilation. An obvious advantage of using $^{15}N$ for this purpose is that the fate of the nitrogen component of DON/PON can be followed *directly*, unlike when $^{14}C$ or $^{3}H$ are used. In many instances the incubation and some of the sample preparation procedures, as well as applications of results, are similar to those discussed for radioisotopically labeled substrates. Because of technical limitations in the synthesis of diverse $^{15}N$-labeled DON/PON substrates, the choice of substrate is far more limited than for those labeled with radioisotopes. Consequently, our knowledge of DON cycling based on $^{15}N$ substrates is limited to a few amino acids (including glycine, serine, and glutamine), urea, and methylamine [65]. In addition, mass spectrometric or emission spectrophotometric measurements of $^{15}N$ are far less sensitive than those for radioisotopes. This low sensitivity often necessitates the addition of relatively large quantities of $^{15}N$-labeled substrates for uptake/assimilation determinations, possibly exceeding the naturally occurring concentrations of the substrate in question. Therefore, there are limitations to using $^{15}N$ in uptake and turnover experiments.

Recent advances in mass spectrometric detection of low levels and ratios of the stable isotopes of nitrogen, $^{14}N$ and $^{15}N$, have led to increased use and popularity of natural abundance stable isotope measurements in microbial ecology. The ratio of $^{14}N$ and $^{15}N$ ($\delta\ ^{15}N$) in various dissolved and particulate N fractions can be indicative of specific N sources, as well as the manner in which these sources are utilized, cycled, and transported in aquatic and terrestrial ecosystems. Stable N isotope techniques and their applications have clearly found a place in microbial ecology and biogeochemistry [29].

## Ammonification and N Regeneration

Stable isotope enrichment and dilution techniques are used to examine the release of ammonium from organic matter, commonly referred to in microbiologic terminology as ammonification and ecologically as N regeneration. Techniques employed include (1) measuring the rate of $^{15}N$ release, as $NH_4^{+1}$, from $^{15}N$-labeled organic matter and (2) determining new $NH_4^{+1}$ release in relation to the existing background $NH_4^{+1}$ pool by N isotope dilution [33]. In the former procedure, $^{15}N$-labeled organic matter is usually produced by supplying $^{15}N$-labeled $NH_4^{+1}$ or $NO_3^{-1}$ as an N source to plants or microbes. After assimilating the $^{15}N$-labeled N sources, the organic matter produced is either artificially converted to detritus (sterilized, irradiated, etc.) or naturally allowed to decompose. $^{15}N$ losses, as $^{15}NH_4^{+1}$, are then determined. In the isotope dilution approach, changes (decreases) in the atom percent excess of $^{15}N$ in a previously enriched $NH_4^{+1}$ pool are monitored over time in order to examine the rate of new $^{14}NH_4^{+1}$ production originating from $^{14}N$-labeled organic N sources. The amount of new $^{14}NH_4^{+1}$ that is produced by ammonification and regeneration should be proportional to a decrease in the atom percent excess $^{15}N$ of the total $NH_4^{+1}$ pool over time.

Complications in either of these two methods for measuring $NH_4^{+1}$ production can arise from the simultaneous uptake and assimilation of the $NH_4^{+1}$ being released. These confounding effects can be minimized by blocking the uptake of $NH_4^{+1}$ or spatially and temporally separating $NH_4^{+1}$ uptake from release.

*Acknowledgments* The author acknowledges research and logistic support by the National Science Foundation (OCE 9115706, DEB 9210495, and DEB 9408471), the North Carolina Sea Grant College Program/NOAA (R/MER-23), and the University of North Carolina Water Resources Research Institute/U.S. Geological Survey (790–2). Technical assistance, reviews, and discussions provided by M. Go, J. Olson, B. Peierls, M. Piehler, J. Pinckney, and S. Thompson are much appreciated.

## REFERENCES

1. American Society for Microbiology. 1981. Manual of Methods for General Bacteriology, American Society for Microbiology, Washington, DC.
2. Antia, N. J., Harrison, P. J. and Oliveira, L. 1991. The role of dissolved organic nitrogen in phytoplankton nutrition, cell biology, and ecology. Phycologia 30:1–89.
3. Ben-Porath, J. and Zehr, J. P. 1994. Detection and characterization of cyanobacterial *nif*H genes. Appl. Environ. Microbiol. 60:880–887.
4. Bergman, B., Lindblad, P. and Rai, A. N. 1986. Nitrogenase in free-living and symbiotic cyanobacteria: immunoelectron microscopic localization. FEMS Microbiol. Lett. 35:75–78.
5. Bergerson, F. J. 1980. Measurement of nitrogen fixation by direct means, In Methods for Evaluating Biological Nitrogen Fixation (Bergerson, F. J., Ed.), pp. 65–75. Wiley and Sons, Chichester.
6. Berman, T. 1991. Dissolved organic substrates as phosphorus and nitrogen sources for axenic batch cultures of freshwater green algae. Phycologia 30:339–345.
7. Billen, G. 1976. Evaluation of nitrifying activity in the sediments by dark $^{14}C$ bicarbonate incorporation. Water Res. 10:51–57.
8. Billen, G. 1981. Heterotrophic utilization and regeneration of nitrogen, In Heterotrophic Activity in the Sea (Hobbie, J. E. and Le. B. Williams, P. J., Eds.), pp. 315–335. Plenum Publishing, New York.
9. Bishop, P. E., Premakumar, R., Joerger, R. D., Jacobson, M. R., Dalton, D. A., Chisnell, J. R., and Wolfinger, E. D. 1988. Alternative nitrogen fixation systems in *Azotobacter vinelandii*. In Nitrogen Fixation: Hundred Years After (Bothe, H., de Bruijn, F. J., and Newton, W. E., Eds.), pp. 71–79. Gustav Fischer, Stuttgart.
10. Blackburn, T. H. and Sorensen, J. 1988. Nitrogen Cycling in Coastal Marine Environments. John Wiley and Sons, New York.
11. Bothe, H. 1982. Nitrogen fixation. In The Biology of Nitrogen Fixation (Carr, N. G. and Whitton, B. A., Eds.), pp. 87–104. Blackwell Scientific Publications, Oxford.
12. Bronk, D. A. and Glibert, P. M. 1991. A $^{15}N$ tracer method for the measurement of dissolved organic nitrogen release by phytoplankton. Mar. Ecol. Progr. Ser. 77:171–182.
13. Burris, R. H. 1974. Methodology. In The Biology of $N_2$ Fixation (Quispel, A., Ed.), pp. 9–21. North-Holland Publishers, Amsterdam.
14. Capone, D. G. 1993. Determination of nitrogenase activity in aquatic samples using the acetylene reduction procedure. In Current Methods in Aquatic Microbial Ecology (Kemp, P. F, Sherr, B. F., Sherr, E. B., and Cole, J. J., Eds.), pp. 621–632. Lewis Publishers, New York.
15. Capone, D. G. and Carpenter, E. J. 1982. Nitrogen fixation in the marine environment. Science 217:1140–1142.
16. Codispoti, L. A. and Christensen, J. P. 1985. Nitrification, denitrification and nitrous oxide cycling in the eastern tropical Pacific. Mar. Chem. 16:277–300.
17. Crawford, C. C., Hobbie, J. E., and Webb, K. L. 1974. The utilization of dissolved free amino acids by estuarine microorganisms. Ecology 55:551–563.
18. Currin, C. A., Paerl, H. W., Suba, G. K., and Alberte, R. S. 1990. Immunofluorescence detection and characterization of $N_2$-fixing microorganisms from aquatic environments. Limnol. Oceanogr. 35:59–71.
19. Daumas, R. 1990. Contribution to the sediment-water interface to the transformations of biogenic substances: application to nitrogen compounds. Hydrobiologia 207:15–29.

20. Delwiche, C. C. 1970. The nitrogen cycle. Sci. Am. 223:136–146.
21. Delwiche, C. C. 1981. The nitrogen cycle and nitrous oxide, In Dentrification, Nitrification and Atmospheric Nitrous Oxide (Delwiche, C. C., Ed.), pp. 1–15. Wiley Interscience, New York.
22. Dugdale, R. C. 1967. Nutrient limitation in the seas: dynamics, identification, and significance. Limnol. Oceanogr. 12:685–695.
23. Dugdale, R. C., Menzel, D. W., and Ryther, J. H. 1961. Nitrogen fixation in the Sargasso Sea. Deep Sea Res. 7:298–300.
24. Eppley, R. W. 1978. Nitrate reductase in marine phytoplankton. In Handbook of Phycological Methods (Hellbust, J. A. and Cragie, J. S., Eds.), pp. 217–223. Cambridge University Press, Cambridge.
25. Eppley, R. W. and Peterson, B. J. 1979. Particulate organic matter flux and planktonic new production in the deep ocean. Nature 282:677–680.
26. Falkowski, P. G. 1983. Enzymology of nitrogen assimilation, In Nitrogen in the Marine Environment (Carpenter, E. J. and Capone, D. G., Eds.), pp. 839–868. Academic Press, New York.
27. Flett, R., Hamilton, R., and Campbell, N. 1976. Aquatic acetylene reduction techniques: solutions to several problems. Can. J. Microbiol. 22:43–51.
28. Flynn, K. J. and Butler, I. 1986. Nitrogen sources for the growth of marine microalgae: role of dissolved free amino acids. Mar. Ecol. Progr. Ser. 34:281–304.
29. Fogel, M. L. and Cifuentes, L. A. 1993. Isotopic fractionation during primary production. In Organic Geochemistry (Engel, M. H., Macko, S. A., Eds.), pp. 73–98. Plenum Press, New York.
30. Fogg, G. E. 1942. Studies on nitrogen fixation by blue-green algae. I. Nitrogen fixation by *Anabaena cylindrica* lemn. J. Exp. Biol. 19:78–87.
31. Garrett, R. H. and Nason, A. 1969. Further purification and properties of *Neurospora* nitrate reductase. J. Biol. Chem. 244:2870–2882.
32. Gersberg, R., Krohn, K., Peek, N., and Goldman, C. R. 1976. Denitrification studies with $^{13}$N-labeled nitrate. Science 192:1229–1231.
33. Glibert, P. M., Lipschultz, F., McCarthy, J. J., and Altabet, M. A. 1982. Isotope dilution models of uptake and remineralization of ammonium by marine plankton. Limnol. Oceanogr. 27:639–650.
34. Goering, J. J. and Dugdale, R. C. 1966. Denitrification rates in an island bay of the equatorial Pacific Ocean. Science 154:404–206.
35. Guerinot, M. L. and Colwell, R. R. 1985. Enumeration, isolation, and characterization of $N_2$-fixing bacteria from seawater. Appl. Environ. Microbiol. 50:350–355.
36. Guerinot, M. L. and Patriquin, D. G. 1981. $N_2$-fixing vibrios isolated from the gastrointestinal tract of sea urchins. Can. J. Microbiol. 27: 311–347.
37. Hardy, R. W. F., Holsten, R. D., Jackson, E. K., and Burns, R. C. 1968. The acetylene-ethylene assay for $N_2$ fixation: laboratory and field evaluation. Plant Physiol. 43: 1185–1207.
38. Hattori, A. 1983. Denitrification and dissimilatory nitrate reduction. In Nitrogen in the Marine Environment (Carpenter, E. J. and Capone, D. G., Eds.), pp. 191–232. Academic Press, New York.
39. Hattori, A., Goering, J. J., and Boisseau, D. B. 1978. Ammonium oxidation and its significance in the summer cycling of nitrogen in oxygen depleted Skan Bay, Unalaska Island, Alaska. Mar. Sci. Commun. 4:139–151.
40. Hawkes, R., Niday, E., and Gordon, J. 1982. A dot immunobinding assay for monoclonal and other antibodies. J. Anal. Biochem. 119:142–147.
41. Hazelkorn, R. 1986. Organization of the genes for nitrogen fixation in photosynthetic bacteria and cyanobacteria. Annu. Rev. Microbiol. 40:525–547.
42. Henriksen, K. and Kemp, W. M. 1988. Nitrification in estuarine and coastal marine sediments. In Nitrogen Cycling in Coastal Marine Environments, SCOPE 33 (Blackburn, T. H. and Sorensen, J., Eds.), pp. 207–249. John Wiley and Sons, New York.
43. Hobbie, J. E., Crawford, C. C., and Webb, K. L. 1968. Amino acid flux in an estuary. Science 159:1463–1464.
44. Jenkins, M. C. and Kemp, W. M. 1984. The coupling of nitrification and denitrification in two estuarine sediments. Limnol. Oceanogr. 29:609–619.

45. Joye, S. B. and Paerl, H. W. 1993. Contemporaneous nitrogen fixation and denitrification in intertidal microbial mats: rapid response to runoff events. Mar. Ecol. Progr. Ser. 94:267–274.
46. Kana, T. M., Darkangelo, C., Hunt, M. D., Oldham, J. B., Bennett, G. E., and Cornwell, J. C. 1994. Membrane inlet mass spectrometer for rapid high-precision determination of $N_2$, $O_2$, and Ar in environmental water samples. Anal. Chem. 66:4166–4170.
47. Kaplan, W. A. 1983. Nitrification. In Nitrogen in the Marine Environment (Carpenter, E. J. and Capone, D. G., Eds.), pp. 139–190. Academic Press, New York.
48. Kawai, A. and Sugahara, I. 1971. Microbiological studies on nitrogen fixation in aquatic environments. 2. On the nitrogen fixing bacteria in offshore regions. Bull. Jpn. Soc. Sci. Fish. 37:981–985.
49. Kirshtein, J. D., Paerl, H. W., and Zehr, J. 1991. Amplification, cloning and sequencing of a *nif*H segment from aquatic microorganisms and natural communities. Appl. Environ. Microbiol. 57:2645–2650.
50. Kirshtein, J. D., Zehr, J. P., and Paerl, H. W. 1993. Determination of $N_2$ fixation potential in the marine environment: application of the polymerase chain reaction. Mar. Ecol. Prog. Ser. 95:305–309.
51. Knowles, R. 1982. Denitrification. Microbiol. Rev. 46:43–70.
52. Laemmli, U. K. 1970. Cleavage of structural proteins during the assembly of the head of bacteriophage T4. Nature 227:680–685.
53. Legendre, L. and Gosselin, M. 1989. New production and export of organic matter to the deep ocean: consequences of some recent discoveries. Limnol. Oceanogr. 34: 1374–1380.
54. Mantoura, R. C. F., Owens, N. J. P., and Burkill, P. H. 1988. Nitrogen biogeochemistry and modeling of Carmarthen Bay. In Nitrogen Cycling in the Marine Environment (Blackburn, T. H. and Sorensen, J., Eds.), pp. 415–444. John Wiley and Sons, New York.
55. Maruyama, Y., Toga, N., and Matsuda, O. 1970. Distribution of nitrogen fixing bacteria in the Central Pacific Ocean. J. Oceanogr. Soc. Jpn. 26:360–366.
56. Neilson, A. H. and Lewin, R. A. 1974. The uptake and utilization of organic carbon by algae: an essay in comparative biochemistry. Phycologia 13:227–264.
57. Nielsen, L. P. 1992. Denitrification in sediment determined from nitrogen isotope pairing. FEMS Microbiol. Ecol. 86:357–362.
58. Oremland, R. S. and Capone, D. G. 1988. Use of "specific" inhibitors in biogeochemistry and microbial ecology. In Advances in Microbial Ecology (Marshall, K. C., Ed.), pp. 285–383. Plenum Press, New York.
59. Paerl, H. W. 1991. Ecophysiological and trophic implications of light-stimulated amino acid utilization in marine picoplankton. Appl. Environ. Microbiol. 57:473–479.
60. Paerl, H. W. and Kellar, P. E. 1978. Significance of bacterial (Cyanophyceae) *Anabaeana* associations with respect to $N_2$ fixation in freshwater. J. Phycol. 14:254–260.
61. Paerl, H. W., Webb, K. L., Baker, J., and Wiebe, W. J. 1981. Nitrogen fixation in waters. In Nitrogen Fixation, Vol. 1 (Broughton, W. J., Ed.), pp. 193–241. Ecology, Clarendon, New York.
62. Paerl, H. W., Priscu, J. C., and Brawner, D. L. 1989. Immunochemical localization of nitrogenase in marine *Trichodesmium* aggregates: relationship to $N_2$ fixation potential. Appl. Environ. Microbiol. 55:2965–2975.
63. Paerl, H. W., Prufert, L. E., and Ambrose, W. W. 1991. Contemporaneous $N_2$ fixation and oxygenic photosynthesis in the non-heterocystous mat-forming cyanobacterium *Lyngbya aestuarii*. Appl. Environ. Microbiol. 57:3086–3092.
64. Parsons, T. R., Maita, Y., and Lalli, C. M. 1984. A Manual of Chemical and Biological Methods for Seawater Analysis. Pergamon Press, New York.
65. Paul, J. H. 1983. Uptake of organic nitrogen. In Nitrogen in the Marine Environment (Carpenter, E. J. and Capone, D. G., Eds.), pp. 275–398. Academic Press, New York.
66. Postgate, J. R. 1978. Nitrogen fixation. In Studies in Biology, No. 92, p. 239. E. Arnold, London.
67. Postgate, J. R. and Eady, R. R. 1988. The evolution of biological nitrogen fixation. In Nitrogen Fixation: Hundred Years After (Bothe, H., de Bruijn, F. J., and Newton, W. E., Eds.), pp. 31–40. Gustav Fischer, Stuttgart.

68. Rysgaard, S., Risgaard-Petersen, N., Nielsen, L. P., and Revsbech, N. P. 1993. Nitrification and denitrification in lake and estuarine sediments measured by the $^{15}N$ dilution technique and isotope pairing. Appl. Environ. Microbiol. 59:2093–2098.
69. Ryther, J. H. and Dustan, W. M. 1971. Nitrogen, phosphorus and eutrophication in the coastal marine environment. Science 171:1008–1012.
70. Saraswat, N., Alleman, J. E., and Smith, T. J. 1994. Enzyme immunoassay detection of *Nitrosomonas europaea*. Appl. Environ. Microbiol. 60:1969–1973.
71. Schmidt, E. L. 1974. Quantitative autecological study of microorganisms in soil by immunofluorescence. Soil Sci. 118:141–149.
72. Seitzinger, S. P. 1987. Nitrogen biochemistry in an unpolluted estuary: the importance of benthic denitrification. Mar. Ecol. Progr. Ser. 41:177–186.
73. Seitzinger, S. P., Nielsen, L. P., Caffrey, J., and Christensen, P. B. 1993. Denitrification measurements in aquatic sediments: a comparison of three methods. Biogeochemistry 23:147–167.
74. Smith, G. B. and Tiedje, J. M. 1992. Isolation and characterization of a nitrite reductase gene and its use as a probe for denitrifying bacteria. Appl. Environ. Microbiol. 58: 376–384.
75. Sorensen, J. 1987. Denitrification rates in a marine sediment as measured by the acetylene inhibition technique. Appl. Environ. Microbiol. 36:139–143.
76. Stewart, W. D. P., Fitzgerald, G. P., and Burris, R. H. 1967. In situ studies on $N_2$ fixation, using the acetylene reduction technique. Proc. Natl. Acad. Sci. USA 58: 2071–2078.
77. Stiefel, E. I. 1977. The mechanism of nitrogen fixation, In Recent developments in Nitrogen Fixation (Newton, W., Postgate, J. R., and Rodriguez-Barrueco, C., Eds.), pp. 69–108. Academic Press, London.
78. Tibbles, B. J. and Rawlings, D. E. 1994. Characterization of nitrogen-fixing bacteria from a temperate saltmarsh lagoon, including isolates that produce ethane from acetylene. Microb. Ecol. 27:65–80.
79. Tiedje, J. M. 1988. Ecology of denitrification and dissimilatory nitrate reduction to ammonium. In Biology of Anaerobic Microorganisms (Zehnder, A. J. B., Ed.), pp. 179–243. John Wiley and Sons, New York.
80. Tiedje, J. M. 1994. Denitrifiers. In Methods in Soil Analysis, Part 2. Microbiological and Biochemical Properties (Weaver, R. W., Ed.), p. 245–267. Soil Science Society of America Book Series, No. 5, Madison, WI.
81. Ward, B. B. 1986. Nitrification in marine environments. In Nitrification (Prosser, J. L., Ed.), pp. 157–184. IRL Press, Oxford.
82. Ward, B. B. and Carlucci, A. F. 1985. Marine ammonia- and nitrate-oxidizing bacteria: serological diversity determined by immunofluorescence in culture and in the environment. Appl. Environ. Microbiol. 50: 194–201.
83. Ward, B. B., Cockroft, A. R., and Kilpatrick, K. A. 1993. Antibody and DNA probes for detection of nitrite reductase in seawater. J. Gen. Microbiol. 139:2285–2293.
84. Williams, P. M. 1992. Measurement of dissolved organic carbon and nitrogen in natural waters. Oceanography 5:107–116.
85. Wilson, P. W. and Burris, R. H. 1947. The mechanism of biological nitrogen fixation. Bacteriol. Rev. 11:41–73.
86. Ye, R. W., Averill, B. A., and Tiedje, J. M. 1994. Denitrification: production and consumption of nitric oxide. Appl. Environ. Microbiol. 60:1053–1058.
87. Yoshida, N. H., Morimoto, M., Hirano, M., Koike, I., Matsuo, S., Wada, E., Saino, T., and Hattori, A. 1989. Nitrification rates and abundances of $N_2O$ and $NO_3^{-1}$ in the western Pacific. Nature 342:895–897.
88. Zehr, J. P. and McReynolds, L. 1989. Use of degenerate oligonucleotides for amplification of the *nif*H gene from the marine cyanobacterium *Trichodesmium thiebautii*. Appl. Environ. Microbiol. 55:2522–2526.
89. Zehr, J. P., Mellon, M., Braun, S., Litaker, W., Steppe, T., and Paerl, H. W. 1995. Diversity of heterotrophic nitrogen fixation genes in a marine cyanobacterial mat. Appl. Environ. Microbiol. 61: 2527–2532.

2

DON P. KELLY
ANN P. WOOD

# Microbes of the Sulfur Cycle

This chapter outlines the recovery from the natural environment of those aerobic and facultatively anaerobic chemolithotrophic bacteria the energy metabolism of which is dependent on the oxidation or reduction of sulfur in its inorganic or organic combinations. It also describes the methods for growth and maintenance of these organisms in pure culture, the chemical analysis and enzyme assay procedures used in assaying their activities, and some procedures for identifying biochemical characteristics typical of the groups.

The turnover of sulfur in nature is driven largely by microbial processes: Principal compounds involved as substrates are hydrogen sulfide, dimethyl sulfide, and sulfate, with the transient formation or accumulation of inorganic (polythionates) and organic compounds (e.g., sulfonates) exhibiting intermediate states of chemical reduction of sulfur. Organisms driving the reactions leading to the cyclic interconversion of sulfate/sulfide/sulfate (via polythionates, sulfonate, sulfur, sulfite, and thiols) seem to be ubiquitous in nature, occupying all habitats so far studied: soil, marine and freshwater, hot springs, solfataras, acidic metal-rich locations, haline environments, and animal and plant symbioses or associations. The sulfate-reducing genera are normally restricted to oxygen-free environments, whereas the sulfur-oxidizing types require oxygen or oxidized nitrogen compounds as respiratory oxidants. The basic media recipes described in this chapter are suitable for organisms from all such habitats, but need to be modified to mimic the source environment (e.g., by addition of salt or other ions), and incubation temperatures, aeration, and other physicochemical conditions would also need to take into account the normal habitat.

The aim of this chapter is to provide an introduction to the basic media and procedures for isolating and culturing a huge diversity of sulfur bacteria. It is not exhaustive of all sulfur bacteria (e.g., phototrophs are omitted), media, or techniques (in particular, molecular biology procedures), but is a working basis from which the more specialized literature and methodology may be accessed and developed.

## TECHNIQUES

### Enrichment Techniques

Depending on the habitat sampled, enrichment techniques can include shaken and static flask cultures, continuous flow chemostat systems, semisolid and sulfide-gradient cultures, and filled bottles or tubes (for anaerobes or microaerophiles). Chemostat procedures are useful for isolating strains that would otherwise be overgrown by fast growing species in batch culture enrichment in flasks or bottles. Standard microbiologic enrichment methods are all applicable to these groups of organisms, including extinction dilution and most probable number (MPN) procedures for estimating numbers in a sample and the use of a standard spread plate (aerobic and anaerobic) for single colony isolation.

### Sulfur-Oxidizers

Substrates typically used for enrichment, isolation, and culture of sulfur-oxidizers are thiosulfate (at 5–40 mM), tetrathionate (at 2–20 mM), elemental sulfur (1–10 g/100 ml medium), and sulfide (0.5–2.0 mM), all of which are available commercially at low cost. Thiosulfate is stable around neutral pH and in alkaline solutions and for several days at pH 4.0. It is stable to autoclaving at pH 7.0–9.0, nontoxic, very soluble, and easy to handle. Tetrathionate is stable in dilute acid solution (pH 3.0), can be autoclaved in acid solution, is readily soluble, and is easy to handle. Sulfide (either as the gas or more usually as the sodium salt) is auto-oxidized in solution, and solutions of the sodium salt need to be prepared from washed and dried crystals and sterilized by filtration. It is a very toxic compound, and $H_2S$ gas should not be allowed to escape from culture vessels. It is also toxic to the bacteria able to use it and is normally used at 2–3 mM at most. Elemental sulfur is virtually insoluble in water, is usually provided as "flowers of sulfur" (10 g/l), and is sterilized by Tyndallization after addition to the medium. Sulfur is used mainly for enrichment of *Thiobacillus thiooxidans*. Other polythionates have been used in cultures, but their instability in solution at around neutral pH and their commercial unavailability limit their routine use.

Many of the recipes given are strongly buffered with phosphates, because of the substantial amounts of acid produced from substrate oxidation by some sulfur-oxidizers. Some strains are sensitive to phosphate, so enrichments with low substrate concentration, or by chemostat procedures, or in media with alternative buffers would then need to be considered.

Successful enrichment cultures on thiosulfate or tetrathionate exhibit decreases in pH and substrate concentration, both of which should be monitored regularly. Changes in pH can be monitored visually by including an indicator in liquid (or agar) media. An increase in visible turbidity, which may be due in part to precipitated sulfur, may be observed if the enrichment is not already turbid or too highly colored. Static enrichments for aerobic thiobacilli (e.g., *T. thioparus*) frequently develop a surface film or pellicle of bacteria often mixed with a fine precipitate of elemental sulfur. Enrichments with elemental sulfur as substrate are best monitored by the measurement of pH, which can fall to 1.0 or below if *T. thiooxidans* is present. Typically 3 to 10 days should be sufficient for activity to be observed on the initial enrichment culture, depending on the source and initial population of the target organism. Sequential subculture in liquid media will normally result in progressively faster substrate consumption and enrichment of the target organism, which can then be purified by routine single colony isolation procedures. Occasionally sensitivity to agar has been found among thiobacilli (particularly *T.*

*ferrooxidans* on ferrous iron media), and purification has then required other gelling agents or the use of extinction dilution culture.

A small number of species of *Thiobacillus* can also use the oxidation of some organosulfur compounds as a source of energy for growth (used at 2–3 mM in media). These substrates include methanethiol (MT), dimethyl sulfide (DMS), dimethyl disulfide (DMDS), carbonyl sulfide (COS), carbon disulfide (CDS), and simple substituted thiophenes (thiophene-2- and -3-carboxylates and acetates). The thiophenes primarily support heterotrophic growth, but sulfate may also be formed. Some of these organosulfur substrates are toxic and volatile and need special handling. Thus, DMS, DMDS, COS (a gas), and CDS need to be dispensed by glass syringe; MT (available as a compressed gas) can be dispensed by syringe as an aqueous solution. Culture flasks need to be sealed with red or black rubber Suba seals: some of these compounds are strongly absorbed by red rubber, and the use of black butyl rubber (as stoppers and in tubing to chemostats if used) is advised to minimize losses from the cultures. These compounds are evil-smelling as well as toxic, and handling should be in extractor hoods or sealed vessels, taking note of specific safety guidelines and regulations.

## Sulfate-Reducers

These organisms may be heterotrophs, reducing sulfate to $H_2S$ with such substrates as lactate, or chemolithotrophs using hydrogen as the reductant. Some species can use elemental sulfur or reduced inorganic sulfur compounds as oxidants, and different types can use diverse organic reductants, from acetate to complex molecules. Effective enrichments can be achieved using bottles filled with medium and inoculated with a suitable environmental sample, such as black mud from a sediment, salt marsh, or anaerobic bog below the water table. Active enrichments are easily identified by the emission of $H_2S$ (*evil-smelling and toxic!*) and precipitation of iron sulfides in media containing sufficient ferrous iron.

## Procedural Modifications for Specific Types

For the isolation of marine strains it may be necessary to include a vitamin B mixture in the standard medium recipes. For the isolation of halophilic or halotolerant sulfur bacteria from saline sources, sodium chloride is added to the standard medium recipes at a concentration determined by either the natural environment or by the investigator (e.g., above and below the environmental level). To isolate bacteria from environments rich in solutes other than NaCl (e.g., sulfate or carbonate), appropriate additions need to be made to the enrichment media to ensure that any specific requirements for those solutes are not overlooked. Neutrophiles and acidophiles are isolated by the choice and use of media of appropriate pH (see recipes). To date no extremely alkaliphilic sulfur-oxidizing bacteria have been isolated, but standard recipes may be modified by the addition of sodium hydroxide or carbonate and the alteration of phosphate balance. These changes raise the initial pH of the medium, but the metabolic activity of the organisms will reduce the pH as acid is produced as a consequence of sulfur oxidation.

Anaerobic isolations use filled bottles and medium supplemented with nitrate as the alternative final electron acceptor for sulfur-oxidizers and the appropriate media for sulfate-reducers.

The choice of temperature for enrichment culture should reflect that typical of the habitat sampled, but initially several incubation temperatures that range at least 10°C around that of the source should be used. Frequently the optimum temperature for organisms isolated from a particular habitat may be significantly different from the environmental temperature. For example, *T. tepidarius* has an optimum temperature of about 43°C, which is the constant temperature of the source of the type strain, but isolates of sulfur bacteria from cool marine waters (5°C) may have optimal temperatures of 20–35°C, and bacteria from extremely hot environments (e.g., hot springs and marine vents at 80 to over 100°C) may exhibit optima 10 or 20°C lower than the ambient temperature of those sources.

In preparing enrichment cultures for sulfate-reducing bacteria, temperature can be very selective as the gram-negative genera are non-sporeforming, whereas *Desulfotomaculum* is sporeforming and contains species with temperature optima of 54–65°C [23]. Thus, heating a sample (10 ml) from an anoxic environment at 80°C for 10 to 20 minutes eliminates non-sporeformers. Similarly incubation at 50–60°C is selective for species with high optima and would discriminate against the gram-negative genera and those species of *Desulfotomaculum* with low (20–38°C) temperature optima.

## TYPES OF BACTERIA

### Aerobic, Non-Filamentous Sulfur Bacteria

These microorganisms are primarily the gram-negative bacteria currently classified as species of the genera *Thiobacillus*, *Thiomicrospira*, and *Thiosphaera*, but heterotrophs, such as some species of *Paracoccus*, *Xanthobacter*, *Alcaligenes*, and *Pseudomonas*, can also exhibit chemolithotrophic growth on inorganic sulfur compounds.

Two clear metabolic types exist in this group: the obligate chemolithotrophs, which can only grow when supplied with oxidizable sulfur compounds (and carbon dioxide as the source of metabolic carbon) and heterotrophs that can also use the chemolithoautotrophic mode of growth. The obligate chemolithotrophs include *Thiobacillus thioparus*, *T. neapolitanus*, *T. denitrificans* (facultative denitrifier), *T. thiooxidans* (extreme acidophile), *T. ferrooxidans* (acidophilic ferrous iron-oxidizer), *T. tepidarius* (moderate thermophile), *T. halophilus* (halophile), and some species of *Thiomicrospira*. The heterotrophs include *Thiobacillus novellus*, *T. acidophilus* (acidophile), *T. aquaesulis* (moderate thermophile), *T. intermedius*, *Paracoccus denitrificans*, *P. versutus*, *Xanthobacter tagetidis*, *Thiosphaera pantotropha*, and *Thiomicrospira thyasirae*.

A variety of archaeal genera also derive energy from sulfur (and sulfur compound oxidation). Best known of these are the *Sulfolobales* [18], of which the best-known genus is *Sulfolobus*. The sulfur-oxidizing genera of *Sulfolobales* are all extreme thermophiles with temperature optima ranging between 70–90°C.

### Aerobic Filamentous Sulfur Bacteria

There are diverse examples of these bacteria from freshwater and marine sources, including genera that have so far not been isolated into pure culture for laboratory cultivation. This chapter describes media for the two genera that have been most studied—*Beggiatoa* and *Thiothrix*—both of which occur in a wide range of aqueous habitats, including marine hydrothermal vents. The occurrence of these organisms in a stream or vent environment is easily observed because of the macroscopic scale of the mats they form, which in some habitats may be several centimeters thick. Both genera are typified by strains growing as heterotrophs or mixotrophs, which also carry out oxidation of sulfide or thiosulfate to sulfate or sulfur. They also contain species capable of autotrophic growth using energy from sulfide or thiosulfate oxidation as the sole source of energy. Purification from natural mats or enrichments is achieved by micromanipulation and selection of filament tips. The strains cultured to date are all mesophilic.

### Anaerobic Sulfur- and Sulfate-Reducing Bacteria

These bacteria are phylogenetically very diverse and include archaea as well as gram-negative and gram-positive Bacteria. The best-known Bacterial genera are *Desulfovibrio*, *Desulfotomaculum*, and *Desulfuromonas*, but the first of these is probably a conglomerate of several genera [23, 25]. Mesophilic (20–30°C), thermophilic (45–70°C), and extremely thermophilic (80–110°C) sulfate-reducers exist. Among Bacterial sulfur-reducers, *Thermodesulfurobacterium* is quite distinct from other Bacteria [24] and has an optimum temperature of 70°C (and maximum at 82°C). Sulfur and sulfate reduction that occurs above these temperatures is due to archaea; for example, by species of *Archaeoglobus* with sulfate-reducing optima above 80°C [3, 21].

Although sulfate-reducers have been known for many years, only relatively recently have species been found that grow chemolithotrophically by means of the anaerobic "disproportionation" of inorganic sulfur compounds (such as thiosulfate or sulfite) to produce mixtures of sulfide and sulfate [1]. These microorganisms seem to derive energy from coupling the oxidation of partially reduced sulfur to sulfate to the production of sulfide from the same substrate. Similarly there is now a substantial literature on archaeal sulfur- and sulfate-reducers, some of which are capable of either aerobic sulfur-oxidizing growth or anaerobic sulfur-reducing metabolism and some of which are obligatorily dependent on sulfur reduction with molecular hydrogen as their means of generating energy [3, 6, 19].

## STOCK SOLUTIONS AND CULTURE MEDIA

This section presents the basic recipes that have been used for the culture of various types of filamentous and non-filamentous sulfur-oxidizers and for sulfate-reducing bacteria. Additional supplements may be needed if the organisms to be isolated and studied are halophilic or have known growth requirements. Other modifications include the substitution of alternative energy or carbon substrates and modification of pH by varying the phosphate ratios.

### Trace Metal Solution T (For thiobacilli, *Thiothrix*)

Procedure

1. Dissolve 50 g ethylenediaminetetraacetic acid (EDTA, disodium salt) in about 400 ml water. Dissolve 9 g NaOH in the EDTA solution. It is best to do this in a 1- to 2-liter beaker on a magnetic stirrer.
2. Dissolve the following salts individually in 30–40 ml lots of water, and add to the EDTA solution (plus 5–10 ml washings).

| | |
|---|---|
| Zinc sulfate ($ZnSO_4 \cdot 7H_2O$) | 5 g |
| Calcium chloride ($CaCl_2$) (or $CaCl_2 \cdot 2H_2O$, 7.34 g) | 5 g |
| Manganese chloride ($MnCl_2 \cdot 6H_2O$) | 2.5 g |
| Cobalt chloride ($CoCl_2 \cdot 6H_2O$) | 0.5 g |
| Ammonium molybdate | 0.5 g |
| Ferrous sulfate ($FeSO_4 \cdot 7H_2O$) | 5.0 g |
| Copper sulfate ($CuSO_4 \cdot 5H_2O$) | 0.2 g |

3. Adjust pH to 6.0 with 1 N NaOH (≈24 ml), adding gradually with stirring.
4. Make up to 1 liter with distilled water. Store in a dark bottle at room temperature. This solution is pale green, but turns brown on storage; which does not affect its stability. Do not autoclave the undiluted stock solution: it is normally added to culture media (see below) and sterilized at that stage. If necessary, filter-sterilized solutions can be used for addition to sterile media.

## Vitamin B Mixture (General Applicability)

Procedure

1. To make a final volume of 1 liter, mix these substances. Dissolve the last three components separately before adding to the bulk solution.

| | |
|---|---|
| Thiamine-HCl | 10 mg |
| Nicotinic acid | 20 mg |
| Pyridoxine-HCl | 20 mg |
| p-Aminobenzoic acid | 10 mg |
| Riboflavin | 20 mg |
| Calcium pantothenate | 20 mg |
| Biotin | 1 mg |
| Cyanocobalamin | 0.5–1.0 mg |

2. Adjust the final pH to 4.0 by adding 0.1 M NaOH before making up to 1 liter. Do this with care as it is very easy to overshoot.
3. Store the mixture frozen (unsterile) or refrigerate it after filter-sterilization. This solution is used in the *Thiothrix* medium (at 1 ml per liter) and may be required by marine sulfur bacteria.

## *Paracoccus versutus* Types (High Phosphate Concentration; pH 7.3–8.4)

### *Stock Solutions*

There are three stock solutions which can be stored indefinitely without sterilization.

1. TP5 solution:

| | |
|---|---|
| $Na_2HPO_4 \cdot 2H_2O$ | 39.5 g |
| $KH_2PO_4$ | 7.5 g |

Dissolve with vigorous stirring in distilled water to make up to 1 liter. The dodecahydrate (79.5 g) and anhydrous (31.5 g) salts of disodium hydrogen phosphate may also be used.

2. 4% w/v $NH_4Cl$
3. 4% w/v $MgSO_4 \cdot 7H_2O$

### *Thiosulfate-Agar Medium for Maintenance Slopes or Plates* (pH 8.4)

Procedure

1. Use distilled water to make up to 1 liter.

| | |
|---|---|
| TP5 solution | 200 ml |
| 4% $NH_4Cl$ | 10 ml |

| | |
|---|---|
| 4% $MgSO_4 \cdot 7H_2O$ | 2.5 ml |
| Solution T | 10 ml |
| 1N NaOH | 11.5 ml |
| $Na_2S_2O_3 \cdot 5H_2O$ | 5.0 g |
| Agar | 15.0 g |
| Phenol red (saturated solution) | 10 ml |

2. Sterilize at 10 psi/10 min, and mix well before pouring plates or slopes.

### *Liquid Culture Media*

These media are best prepared in three parts to avoid precipitation of phosphates and any decomposition of growth substrates. The recipe given is for a final volume of 1 liter.

Part I (in bottles or flasks)

| | |
|---|---|
| 4% $NH_4Cl$ | 20 ml |
| 4% $MgSO_4 \cdot 7H_2O$ | 2.5 ml |
| Solution T | 10 ml |
| Distilled water | 550 ml |

Part II

| | |
|---|---|
| TP5 solution | 200 ml |
| 1 N NaOH | 10 ml |

Sterilize separately at 10 psi/10 min and mix when cool.

Part III

1. This consists of the growth substrate dissolved in 200 ml water. The solution can thus be made at five times the concentration required in the final volume of 1-liter of medium. This recipe allows for addition of mixtures of substrates to the medium: 100 ml each of 10 times strength solutions of two substrates could be used, for example.
2. The pH of this medium may be modified by omitting the NaOH which lowers the pH to 7.3. Intermediate pH values are obtained using intermediate volumes of NaOH. Bulk quantities of medium prepared for use with chemostat cultures should omit the NaOH in order to keep the pH as low as possible since at higher pH, salts may deposit in the tubing before reaching the culture vessel. The culture pH is maintained in the growth vessel by the addition of alkali (usually NaOH or $Na_2CO_3$).

## *Thiobacillus aquaesulis* Types (Low Phosphate Concentration; pH 8.0 and Above)

Part I

| | |
|---|---|
| 1.1% (w/v) $K_2HPO_4$ | 45 ml |
| $H_2O$ | 200 ml |

Part II

| | |
|---|---|
| $Na_2S_2O_3$ | 2.5 g |
| 0.11% (w/v) $MgSO_4$ | 45 ml |
| 0.44% (w/v) $NH_4Cl$ | 45 ml |
| Solution T | 1.1 ml |
| Agar | 7.5 g |
| Phenol red (saturated) | 5.0 ml |
| Water | 160 ml |

Procedure

1. Sterilize the two solutions separately at 10 psi/10 min and mix when cool.
2. To modify the initial pH of 8.0, add sodium hydroxide as follows:

   pH 9.0: 12 ml 0.1 N NaOH/500 ml
   pH 9.5: 22 ml 0.1 N NaOH/500 ml
   pH 10.0: 32 ml 0.1 N NaOH/500 ml

3. To avoid drying out, place at least 30 ml per Petri dish of moderate thermophile media incubated at elevated temperatures.

## *Thiobacillus neapolitanus* Types (pH 6.6–7.2)

There are three stock solutions: Solution A, Solution B, and trace metal Solution T (described earlier). Each should be made up to 1 liter with distilled water.

1. Solution A (final pH 6.9–7.0)

| | |
|---|---|
| $KH_2PO_4$ | 16 g |
| $K_2HPO_4$ | 16 g |

2. Solution B

| | |
|---|---|
| $NH_4Cl$ | 10 g |
| $MgSO_4 \cdot 7H_2O$ | 20 g |

3. Solution T

To prepare a medium for agar slants or plates, mix and autoclave the following components at 10 psi/10 min. The thiosulfate may be replaced by potassium tetrathionate (0.6 g) for *T. tepidarius* types.

| | |
|---|---|
| $Na_2S_2O_3 \cdot 5H_2O$ | 1 g |
| Solution A | 25 ml |
| Solution B | 4 ml |
| Solution T | 1 ml |
| Agar | 1.5 g |
| Bromo-cresol purple (saturated) | 0.2 ml |
| Distilled water | 70 ml |

To prepare a medium for liquid culture, dissolve these components in distilled water:

| | |
|---|---|
| $Na_2S_2O_3 \cdot 5H_2O$ | 5 g |
| Solution B | 40 ml |
| Solution T | 10 ml |
| Distilled water | 700 ml |

Sterilize this solution and 250 ml of Solution A separately at 10 psi/10 min and combine when cool.

Alternatively, omit the thiosulfate, reduce the volume of water by 10% of the final volume, and make up the volume at the time of inoculation by adding the appropriate concentration of thiosulfate (or tetrathionate). This medium may be modified to provide different pH values by varying the balance of acid and basic phosphates, as follows:

Stock solutions of 0.2 M $K_2HPO_4$ and 0.2 M $KH_2PO_4$, which may be combined in 1-liter medium recipe as Solution A

| pH | $K_2HPO_4$ (ml) | $KH_2PO_4$ (ml) |
|---|---|---|
| 5.0 | 0 | 250 |
| 5.8 | 20 | 230 |
| 6.4 | 66 | 184 |
| 7.0 | 152 | 98 |
| 7.7 | 217 | 33 |
| 8.0 | 237 | 13 |
| 8.5 | 250 | 0 |

## *Thiobacillus ferrooxidans* Types (pH 4.0–4.5)

This organism is an extreme acidophile, best known for its ability to grow at an acid pH (pH 1.0–3.5) on ferrous iron or pyrite and other metal sulfides. However, it is also capable of good growth on soluble sulfur compounds, such as tetrathionate and thiosulfate at pH values above about pH 3.5.

Procedure

1. To make Salts 1, dissolve these components in distilled water to make up to 1 liter.

| | |
|---|---|
| $(NH_4)_2SO_4$ | 30 g |
| $KH_2PO_4$ | 30 g |
| $MgSO_4 \cdot 7H_2O$ | 5 g |

2. To make Salts 2, dissolve this compound in distilled water to make up to 1 liter.

$CaCl_2 \cdot 2H_2O$ 2.5 g

3. For basal salts medium, mix 100 ml Salts 1 + 100 ml Salts 2 + 800 ml water.
4. For medium S2, add 5.0 g $Na_2S_2O_3 \cdot 5H_2O$/liter.
5. For medium S4, add 3.02 g $K_2S_4O_6$/liter.
6. Autoclave at 10 psi/10 min.
7. Dispense 100 ml into 250-ml flasks and add 0.5 ml $FeSO_4 \cdot 7H_2O$ (1 g in 1 liter 0.1N HCl).
8. To make the agar medium, make the following two solutions:

| Solution 1 | |
|---|---|
| Salts 1 | 50 ml |
| Salts 2 | 50 ml |
| $K_2S_4O_6$ | 1.51 g |
| Distilled water | 300 ml |

| Solution 2 | |
|---|---|
| Agar | 5 g |
| Distilled water | 100 ml |

9. Sterilize separately at 10 psi/10 min. Add 1.25 ml 0.2% $FeSO_4$ in 0.1 N HCl. Then add the agar solution, mix, and dispense in 30-ml aliquots to Petri plates.

## *Thiobacillus denitrificans* Types (Anaerobic, pH 7.0)

This species is facultatively anaerobic, growing either with nitrate (or nitrite or nitrous oxide) or oxygen as a respiratory oxidant. For anaerobic enrichment or batch culture of pure cultures, bottles completely filled with medium are used, but with loosely tightened caps to allow escape of the large amounts of nitrogen generated by denitrification.

Procedure

1. Dissolve in distilled water to make up to 900 ml.

| | |
|---|---|
| $Na_2S_2O_3 \cdot 5H_2O$ | 5.0 g |
| $KNO_3$ | 2.0 g |
| $KH_2PO_4$ | 2.0 g |
| $NH_4Cl$ | 1.0 g |
| $MgSO_4 \cdot 7H_2O$ | 0.8 g |
| $FeSO_4 \cdot 7H_2O$ (2% w/v in 1M HCl) | 1 ml |
| Solution T | 1 ml |

2. Autoclave at 10 psi/10 min.
3. When cool, add 100 ml of 2% $NaHCO_3$ separately sterilized by membrane filtration.
4. Alternatively filter-sterilize the complete medium without autoclaving any components. For agar plates, add 1.5% (w/v) Difco agar, and incubate under oxygen-free atmospheres.

## *Sulfolobus* and *Acidianus*

Several media have been used for these organisms, but in general they contain relatively low concentrations of dissolved solutes (e.g, ammonium sulfate 0.4–3.0 g/liter and low phosphate concentrations, in different recipes). Stock cultures can be maintained in gently agitated tubes of liquid medium (some strains are quite microaerophilic) or in flasks shaken at relatively low speeds. The following medium is very effective for the autotrophic culture of *Sulfolobus* spp. and *Acidianus brierleyi* on tetrathionate [27]:

Procedure

1. Dissolve in distilled water:

| | |
|---|---|
| $(NH_4)_2SO_4$ | 0.4 g |
| $MgSO_4 \cdot 7H_2O$ | 0.4 g |
| KCl | 0.2 g |
| $K_2HPO_4$ | 0.2 g |
| $FeSO_4 \cdot 7H_2O$ | 0.01 g |

2. Adjust to pH 3.0 with $H_2SO_4$.
3. Then make the mixture up to 900 ml and autoclave at 15 psi for 15 min.
4. After cooling add 3.0 g $K_2S_4O_6$ dissolved in 100 ml of dilute $H_2SO_4$ at pH 3.0, and sterilize by filtration through a 0.2-μm membrane filter. The organisms may also be grown on elemental sulfur (5–10 g/liter), and some strains give much better growth if supplemented with yeast extract (0.05–0.2 g/liter).

### *Beggiatoa*

Within the family *Beggiatoaceae* are several genera showing filamentous morphology and gliding motility. The recipes here have been used for *Beggiatoa* and *Thiothrix*. They are both dilute basal salts media, having low phosphate levels (and therefore low buffering capacity), high calcium levels, and low substrate concentrations, because the organisms grow best under oligotrophic conditions. The distinctive feature of these organisms arising from their filamentous and gliding abilities is the presence of mat or tuft formations in their natural environments, which normally include flowing water. Thus their enrichment and isolation procedures require micromanipulation methods to visualize and separate the organisms from each other and from other "normal" sulfur bacteria. This procedure is a typical general approach.

Procedure

1. Serially wash filaments with sterile basal salts medium supplemented with neutralized sulfide (5 mM), transferring tufts/filaments from one wash to the next using microforceps and removing contaminating matter in the process.
2. Place washed tufts onto a dry 1.6% (w/v) agar plate for a minute to absorb excess fluid, before transferring to prescored 1.4% (w/v) agar plates containing 2.5 mM sulfide, yeast extract, and sodium acetate (0.1 g/l each) and basal salts.
3. Incubate for 24–48 h, use a dissecting microscope to observe pure filaments that can be removed with a small segment of agar beneath them.

*Beggiatoa alba* has been grown successfully in batch and chemostat culture [5] in the following medium (g/liter):

| | |
|---|---|
| $NH_4Cl$ | 0.2 |
| $K_2HPO_4$ | 0.01 |
| $MgSO_4 \cdot 7H_2O$ | 0.01 |
| Saturated $CaSO_4$ solution | 20 ml |
| Microelement solution (see below) | 5 ml |

After adjusting the medium to pH 7.5 and autoclaving, acetate (0.5 g/l), thiosulfate (0.5 g/l), or sulfide (1 mM) can be added. The microelement solution contains (g/l of distilled water acidified with 0.5 ml concentrated $H_2SO_4$) $MnSO_4 \cdot H_2O$, 2.28; $ZnSO_4 \cdot 7H_2O$, 0.05; $H_3BO_3$, 0.05; $CuSO_4 \cdot 5H_2O$, 0.025; $Na_2MoO_4 \cdot 2H_2O$, 0.025; $CoCl_2 \cdot 6H_2O$, 0.045.

### *Thiothrix*

We have used the following medium successfully to support heterotrophic, mixotrophic, and autotrophic growth of *Thiothrix ramosa*, when supplemented with the appropriate energy and carbon substrates.

Procedure

1. Dissolve in distilled water to make up to 1 liter:

| | |
|---|---|
| $(NH_4)_2SO_4$ | 0.5 g |
| $MgSO_4 \cdot 7H_2O$ | 0.1 g |
| $CaCl_2 \cdot 2H_2O$ | 0.05 g |
| $K_2HPO_4$ | 0.11 g |
| $KH_2PO_4$ | 0.085 g |

2. Sterilize 1 ml vitamin B mixture and 1 ml trace metal solution T separately by filtration and add to the autoclaved salts (10 psi/10 min) when cooled.
3. Prepare solid media by adding 1.5% (w/v) agar (Oxoid Bacteriological No. 1).
4. Sterilize carbon and energy substrates separately and add when cool. Final concentrations that we found successful were sodium lactate 0.4–1.0 mM and sodium thiosulfate, 2.0–6.0 mM.

## Sulfide Gradients

These media may be used for organisms that exist in the natural environment in sulfide gradients, such as *Beggiatoa* and *Thiothrix* [14, 15]. These bacteria may be able to grow at only one particular level in the sulfide/oxygen gradient or may be gliding bacteria that are able to move through the gradient. The medium is semisolid to facilitate the gliding potential. *Beggiatoa* may be grown in this way as well as on solid surfaces as mats or tufts.

Procedure

1. Load test tubes (15 × 120 mm) with 4 or 5 ml 1.5% (w/v) molten agar, containing 1–8 mM neutralized $Na_2S$ (pH 8.0), and allow this solution to set as a plug in the bottom of the tube.
2. Overlay the agar plug plus sulfide with 10 ml appropriate "slush" medium containing 0.2 or 0.25% (w/v) agar. For marine strains a medium supplemented with NaCl is necessary, and vitamin or organic substrate supplements are needed in some cases, particularly for mixotrophic growth of *Beggiatoa*.
3. Seal the tube with a rubber stopper, and allow a gradient to establish over time. Alternatively screw-topped culture tubes may be used.
4. Inoculate within 10–24 h by insertion midway into the tube, using a fine Pasteur pipette or long syringe needle.

## Sulfate-Reducing Bacteria

Numerous media have been described for sulfate-reducers [16, 17, 25, 26]. The most widespread types of sulfate-reducers are the mesophilic non-sporeforming species, of which the greatest variety occurs in marine sediments. All genera are anaerobes, requiring anoxia for development; some are more sensitive than others to oxygen. Special techniques and devices for the preparation and dispensing of anoxic media have been described, and the reader is referred to the specialized literature [25, 26]. The media described in this section can be prepared and dispensed by conventional means for use in enrichment culture and for the cultivation of many isolates. Oxygen should be excluded as far as possible from the vessels used. Sealed bottles can be made anoxic by expelling air with nitrogen or argon, and the residual oxygen in the media can be reduced by sulfide in inoculum cultures and by the addition of reductants. The reductants are usually dithionite or ascorbate plus thioglycollate.

### *Lactate Media for* Desulfovibrio *and* Desulfotomaculum [25]

Procedure

1. Dissolve the components sequentually in 1 liter distilled or tapwater. Adjust the solution to approximately pH 7.2 before autoclaving.

| | Medium B | Medium C |
|---|---|---|
| Sodium lactate (50% solution) | 5.5 ml (7.0 g) | 9.5 ml (12.0 g) |
| $Na_2SO_4$ | — | 4.5 g |
| $CaSO_4$ | 1.0 g | — |
| $CaCl_2 \cdot 2H_2O$ | — | 0.06 g |
| Sodium citrate | — | 0.3 g |
| $NH_4Cl$ | 1.0 g | 1.0 g |
| $KH_2PO_4$ | 0.5 g | 0.5 g |
| $MgSO_4 \cdot 7H_2O$ | 2.0 g | 2.0 g |
| NaCl (if needed: for marine strains) | (2.5 g) | (2.5 g) |
| Yeast extract | 1.0 g | 1.0 g |

2. Add the following components aseptically to the autoclaved media:

   10 ml (medium B) or 0.08 ml (medium C) of a 5% (w/v) solution of $FeSO_4 \cdot 7H_2O$ (stored under nitrogen or dissolved in 0.1 M $H_2SO_4$ to prevent auto-oxidation)

   10 ml (both media) of reductant solution, prepared by dissolving 0.1 g each of ascorbic acid and sodium thioglycollate in 10 ml water and autoclaving under nitrogen

Both media can be solidified with 1% (w/v) agar for preparation of plates, deeps, or slants for anaerobic incubation. Medium B is a good diagnostic medium for enrichment culture because the excess iron precipitates as black FeS when $H_2S$ is generated by a successful enrichment of sulfate-reducers. It is also recommended for the maintenance of stock cultures, as survival is improved by the presence of the precipitated material. Medium C is virtually clear and does not produce a precipitate: It is thus useful for mass culture to produce large amounts of "clean" organisms for experimentation. For growth with hydrogen as a reductant, medium B or C is supplemented with 0.2–0.4% (w/v) sodium acetate, and $CO_2$ (10 to 20%) should be provided in the $H_2$ gas phase.

Numerous variations on these media have been used, and a great range of organic substrates are used for growth by different sulfate-reducers; the reader is referred to Widdel and Bak [25] and Widdel and Pfennig [26]. For some genera, special conditions must be provided: For example, the filamentous gliding *Desulfonema* needs an insoluble substratum for optimum development, which can be provided as a sloppy agar or as precipitated aluminum phosphate. *Desulfotomaculum* has been enriched and grown in the standard media, but can also be grown (sometimes more effectively) in media reduced with dithionite just sufficient to decolorize resazurine (1 ml of 0.1% w/v sodium salt per liter).

## ASSAYS OF SUBSTRATES AND INTERMEDIATES

These assays enable measurement of sulfur substrates (and products) in growing cultures and the use of nitrate or nitrite by denitrifying sulfur-oxidizers.

### Nitrite Estimation: Griess-Ilosvay Reagent

The Griess-Ilosvay reagent (also available commercially) is a mixture of equal volumes of (1) 0.5 g sulfanilic acid dissolved in 70 ml 30% (v/v) acetic acid and (2) 0.1 g α-naphthylamine boiled *in a fume hood* with 20 ml water to give a clear liquid. Add 150 ml 30% (v/v) acetic acid. Note: α-naphthylamine is a carcinogenic compound and can be purchased (Sigma Chemical Company) as a preweighed solid in a serum bottle with a butyl rubber stopper, enabling solution preparation and handling without contact with the solid.

Procedure

1. Add the sample (1 ml) containing nitrite to 1 ml of mixed Griess-Ilosvay reagent in a 10-ml-graduated tube. This tube should be acid-washed glass (or equivalent) or previously unused plastic ware.
2. Make up to 10 ml with distilled water and mix.
3. Read the absorbance at 520 nm after 20 min using 1-cm cuvettes.
4. Prepare a calibration curve using sodium nitrite, using dilutions between 0–0.7 μmoles $NaNO_2$ per 10-ml assay volume.
5. The presence of thiosulfate will interfere with this assay. To eliminate thiosulfate, add 1 ml 0.3 M cadmium sulfate to the sample before adding Griess-Ilosvay reagent. For very dilute samples, mix 9 ml with 1 ml of reagent, allowing detection of concentrations around 50–100 nM.

## Nitrate Estimation

Procedure

1. Place sample (50 or 100 μl) in a tube.
2. Add 1 ml 2% (w/v) sulfamic acid. This substance destroys nitrite, which interferes with the assay, as follows:

$$NH_2SO_3H + HNO_2 = N_2 + H_2SO_4 + H_2O$$

3. Mix and then add 9 ml 5% (v/v) perchloric acid, which reduces interference by iron or nitrite.
4. Measure absorbance at 210 nm.
5. Prepare calibration curve using $KNO_3$. Range is up to ≈0.2 mM nitrate (i.e., 2 μmoles per 10-ml assay volume and 2 μmoles per 50- or 100-μl sample).

## Sulfate Estimation

### *Barium Sulfate Turbidity Method* [4]

Procedure

1. Add 5 ml of 20% (v/v) glycerol in water to 2-ml liquid sample containing 1–10 μmoles sulfate, and shake the mixture vigorously.
2. Add 1 ml 10% (w/v) barium chloride in 5% (v/v) HCl, and shake the mixture vigorously to ensure that the barium sulfate precipitate that is formed is homogeneously suspended throughout the glycerol solution.
3. Make up the final volume to 10 ml with distilled water.
4. After shaking the mixture, measure turbidity at 460 nm or in a colorimeter with a dark blue filter.

### *Measurement of Residual Barium after Barium Sulfate Precipitation*

Principle

Barium sulfate has very limited solubility in dilute acid solution. Small amounts of sulfate in a sample can be measured by precipitation of sulfate by excess barium ions and measurement of the residual barium by atomic absorption spectrometry.

Reagents

1. Barium chloride in 1% HCl: 200 mg Ba/liter
2. 1% (v/v) hydrochloric acid
3. Sulfate standards: 1.0 and 3.0 mM potassium hydrogen sulfate in 1% HCl, which provide sulfate at 1.0 and 3.0 μmoles/ml, respectively

Procedure

1. Dilute samples if necessary so that the amount of sulfate in the final 10-ml assay volume is less than 8 μmoles. Centrifuge samples of bacterial cultures to remove cells, and use the supernates for analysis.
2. Mix 2-ml samples (containing 2.0–8.0 μmoles sulfate) in test tubes with 3 ml 1% HCl (to give a volume of 5 ml), and add 5 ml of barium chloride solution. Shake the mixtures vigorously or vortex them for a few minutes. Then leave the tubes at room temperature overnight to allow the barium precipitate to deposit.
3. Measure the residual barium in the clear solution directly (without further treatment of the tubes, but taking care not to agitate them), using atomic absorption spectrometry at 553.6 nm. We have used a Varian AA-1275 spectrophotometer for this purpose with a barium lamp to measure barium.
4. Prepare the standard curve with potassium hydrogen sulfate solutions to give 0–8.0 μmoles sulfate/assay tube, and deduce the amount of sulfate in a sample from a standard curve produced by plotting absorbance against sulfate content. The highest barium value is obviously that when no sulfate is present. Typical readings are 2.60 in the absence of sulfate, 2.20 (with 2 μmoles sulfate), 1.80 (3), 1.50 (4), 1.30 (5), 1.05 (6), 0.85 (7), and 0.65 (8). Other ions in samples (e.g., high phosphate content) might affect the readings, and standards should be checked both by preparing standards in the same solutions as culture samples and by including standard amounts of sulfate in experimental or field samples as internal checks on the calibration.

## Sulfide Estimation [13]

Principle

Dimethyl-para-phenylenediamine sulfate (DPPDS) is converted to (colorless) leuco-methylene blue in the reaction with sulfide. This substance is converted to (blue) methylene blue by the ferric ions. The amount of methylene blue is determined colorimetrically and is proportional to the sulfide in the sample assayed. The zinc acetate is used to stabilize samples by combining free sulfides as the zinc salt. Stabilization is especially important in field samples that are going to be transported or in laboratory samples that are going to be stored before assay.

Reagents

1. 2% (w/v) zinc acetate, containing 0.2 ml acetic acid per liter.
2. 0.2% (w/v) DPPDS in 2.5 M $H_2SO_4$; dissolve the DPPDS in 200 ml distilled water, very carefully mix in 200 ml concentrated $H_2SO_4$, and make up to 1 liter with $dH_2O$.
3. 1% (w/v) ferric ammonium sulfate: dissolve the salt in 50-ml distilled water in a 100-ml volumetric flask, add 2 ml concentrated $H_2SO_4$, and make up to 100 ml with water.

Procedure

1. Add 20 ml Reagent 1 to a 100-ml volumetric flask.
2. Add sample. Volume depends on sulfide content: upper limit is about 0.03 mM (i.e., 3 μmoles sulfide in the 100-ml assay).
3. Add distilled water or, in the field, sulfide-free water as available, if necessary to about 80 ml total volume in the flask.
4. Add 10 ml Reagent 2 and *mix rapidly*.
5. Immediately add 0.5 ml Reagent 3 and *mix well.*
6. Make up to 100 ml with distilled water.
7. Allow to stand at room temperature for at least 15 min and then read absorbance at 670 nm using a distilled water reagent blank. In our experience the color is stable for many hours.

Calibration and Calculation

The calibration curve is linear to about $OD_{670} = 0.8$, and 1 μmole of sulfide gives a reading of about 0.36 in a 1-cm cell. Standard curves can be prepared from sulfide solutions (e.g., using 0.1 mM sodium sulfide prepared from washed crystals of $Na_2S \cdot 9H_2O$ *freshly dissolved* in distilled water and *used immediately*). A 10-ml sample will give a reading of about 0.36 in the 100-ml assay.

Note that standard curves will be approximate because sulfide cannot be regarded as a primary standard, unless its solutions are standardized separately by iodine titration. To conserve materials in the field, the procedure can be scaled down to smaller volumes (e.g., to 25 ml or 10 ml). If a precipitate forms on assaying—for example, large volumes of field samples—move to Step 7 and then filter through glass-fiber filters to remove precipitate before reading absorbance.

## Sulfite Estimation

Two methods may be used to estimate sulfite. The former is more sensitive, but requires a spectrophotometer. The latter titration method can be used in the field if necessary and is useful as a rapid routine procedure to check for sulfite disappearance.

### *Spectroscopic Method* [11]

Principle

An exact amount of iodine is generated in the assay mixture from a solution of iodide plus iodate and acetic acid. Sulfite and sulfide or thiosulfate, but not polythionates, oxidize the iodine to iodide. By measuring the absorbance ($OD_{350}$) of a blank with no sulfite, and of a series of sulfite standards, a calibration curve of sulfite can be obtained. Note that the highest OD is the blank and the OD decreases as more sulfite is present as the iodine color is discharged by it.

$$IO_3^- + 6H^+ + 5I^- = 3I_2 + 3H_2O$$

$$SO_3^{2-} + I_2 + H_2O = SO_4^{2-} + 2H^+ + 2I^-$$

$$2S_2O_3^{2-} + I_2 = S_4O_6^{2-} + 2I^-$$

$$H_2S + I_2 = S + 2H^+ + 2I^-$$

Reagents

1. Iodate-iodide reagent: (1) Dissolve 36.1 g potassium iodide in about 150 ml distilled water in a 250-ml volumetric flask; (2) add 0.1 g sodium carbonate and allow to dissolve; (3) add 10 ml 0.025 N potassium iodate (this is prepared by dissolving 0.1338 g $KIO_3$ in 100 ml water in a volumetric flask); and (4) make up to 250 ml.
2. Buffer pH 7.0: 100 ml of 0.2 M $NaH_2PO_4$ + 59.3 ml 0.2 M NaOH.
3. 5 N acetic acid: Dilute 30 ml "glacial" acetic acid to 100 ml with distilled water (using a graduated cylinder or volumetric flask).
4. Sodium sulfite standards: (1) Make a stock of 5 mM EDTA (=1.86 g of disodium EDTA per liter); (2) dissolve 0.51 g $Na_2SO_3$ in 100 ml EDTA; and (3) dilute 1 ml of the solution in (2) to 100 ml with EDTA solution: this is 0.2 mM sulfite and contains 0.2 μmoles per ml.

Procedure

1. Use 25-ml volumetric flasks as reaction vessels.
2. Add 3.2 ml buffer to each of a series of flasks.
3. Add sample (and wash in with a little water): for sulfite the maximum amount to be present in the sample should be 1.6 μmoles (and for thiosulfate the range is 0 to 3.0 μmoles). If both ions are present their effects are additive. If samples have to be diluted, use the EDTA solution to minimize auto-oxidation.
4. Add 3 ml 5 N acetic acid and mix.
5. Immediately add 2.3 ml of the iodate-iodide reagent, make up to 25 ml with water, mix, and leave sealed with a stopper until sampling to read absorbance.
6. Read absorbance at 350 nm.

Note: Iodine is very volatile from solution, so take readings as rapidly as possible and expose solutions to the air for as little time as possible.

Calibration

In a typical calibration experiment the following data were obtained:

| ml Sulfite Solution | μmoles Sulfite | $OD_{350}$ |
|---|---|---|
| 0 | 0 | 1.764 |
| 2 | 0.4 | 1.442 |
| 4 | 0.8 | 1.094 |
| 6 | 1.2 | 0.740 |
| 8 | 1.6 | 0.396 |
| 10 | 2.0 | 0.071 |

In various calibrations, 1 μmole of sulfite gave a drop in OD of 0.85–0.91 (i.e., the negative slope of the calibration curve is 0.850–0.910 $OD_{350}$/μmole). Performing the same calibration with thiosulfate (0–10 ml of a 0.2 mM solution) gave the following data:

| ml Thiosulfate Solution | μmoles Thiosulfate | $OD_{350}$ |
|---|---|---|
| 0 | 0 | 1.764 |
| 2 | 0.4 | 1.606 |
| 4 | 0.8 | 1.416 |
| 6 | 1.2 | 1.216 |
| 8 | 1.6 | 1.010 |
| 10 | 2.0 | 0.829 |

Values for the slope obtained range between 0.456–0.475 $OD_{350}$/μmole.

If thiosulfate (or any other material likely to reduce iodine) is also present in a sample, replicate samples should be run, to one series of which is added 1.5 ml of 0.5 M formaldehyde (4 ml 40% commercial formaldehyde solution diluted to 100 ml) just before adding the acetic acid. This addition complexes the sulfite, and any reduction in OD then seen is due to materials other than sulfite. We have not encountered interference from other ions, but internal standards are obviously advisable for environmental samples.

### *Iodine Titration Method*

#### Principle

$$Na_2SO_3 + I_2 + H_2O = Na_2SO_4 + 2\ HI$$

$$2\ Na_2S_2O_3 + I_2 = Na_2S_4O_6 + 2\ NaI$$

Sulfite requires twice as much iodine as does thiosulfate on a molar basis, so 1 ml of 100 mM sulfite (or 2 ml of 100 mM thiosulfate) requires 1 ml 0.1 N or 10 ml 0.01 N iodine.

#### Reagents

1. Dilute 25 ml stock 0.1 N iodine to 250 ml (= 0.01 N).
2. Dilute 25 ml stock 0.1 M thiosulfate to 250 ml (= 0.01 M).
3. Dilute 2 ml 40% formaldehyde with 6 ml water (final concentration = 10%).
4. 10% acetic acid (10 ml glacial to 100 ml).
5. Starch solution: Prepare 1% soluble starch by making a slurry with a little water, then dissolving in hot water, and making up to 100 ml.

#### Procedure

1. 250 ml conical flask + 2 ml 10% acetic acid + 20 ml iodine: Titrate to starch blue end point with thiosulfate (about 20 ml thiosulfate).
2. 250 ml conical flask + 2 ml 10% acetic acid + 20 ml iodine: Rapidly mix in 2 ml culture sample. Titrate as in (1).
3. If (2) is less than (1), 250 ml flask + 2 ml formaldehyde + 2 ml culture sample: Mix and then add 2 ml acetic acid. Rapidly with swirling add 20 ml iodine and mix. Titrate as in (1).
4. (1) = blank. (2) = sulfite + any other reactive species such as thiosulfate. (3) = other reactive species (sulfite having been bound by formaldehyde).

#### Calculations

As iodine is not the primary standard, it is calibrated by titration with standard thiosulfate, so that 10 ml iodine will require "x" ml 0.01 M thiosulfate. Thus, x/10 ml iodine is equivalent to 1 ml 0.01 M thiosulfate (= 10 μmoles thiosulfate or 20 μmoles sulfite). If (1) = (3) then titration (1) − titration (2) = sulfite (y ml). Alternatively, titration (3) − (2) = sulfite (= y ml). y x 20 = μmoles sulfite in the 2-ml sample.

$$\text{Concentration of sulfite} = \frac{y \times 20}{2}\ \text{mM}$$

## Sulfur

### Reagents

1. Acetone solvent: 19 parts acetone:1 part water
2. 0.1 % (w/v) KCN in acetone solvent
3. 0.4% (w/v) ferric chloride in acetone solvent
4. "Flowers" of sulfur [2] dissolved in acetone to be used as the standard (sulfur does not dissolve in toluene or petroleum ether); dilutions are used of a stock solution containing 32 mg S (1 mmole) in 100 ml acetone (10 μmoles per ml stock)

### Procedure

1. Spin culture sample slowly (100–200 rpm) to remove sulfur particles but not cells from suspension.
2. Respin the supernatant to ensure complete removal of sulfur particles.
3. Dissolve the sulfur in acetone at 30°C for several hours.
4. To determine sulfur, place 5 ml sample into a 25-ml volumetric flask.
5. Add 15 ml KCN reagent. Mix and stand for 2 minutes.
6. Make up to 25 ml with acetone solvent.
7. To a 5-ml aliquot, add 5 ml ferric chloride reagent. Mix.
8. Measure absorbance at 470 nm immediately.

The lower limit of detection is about 0.2 μmole S in the 5-ml sample, and the range for the standards should be 0–5 μmoles S per assay. Thiosulfate does not interfere with this assay.

## Thiosulfate, Tetrathionate, and Polythionates: Cyanolysis Method

### Principle

Thiosulfate and the polythionates react with cyanide (either spontaneously at low or high temperatures or under copper(II) catalysis) to produce thiocyanate [7, 10]. Thiocyanate can be determined spectrophotometrically as red ferric thiocyanate, and the amount of thiosulfate and polythionates (individually and in mixture) can be estimated. Tetrathionate and thiosulfate react to produce the indicated amounts of thiocyanate:

$$S_4O_6^{2-} + CN^- \Rightarrow SCN^- + S_2O_3^{2-} + SO_4^{2-} \text{ (i.e., one } S_4O_6^{2-} \text{ produces one } SCN^-)$$

$$S_2O_3^{2-} + CN^- \Rightarrow SCN^- + SO_3^{2-}$$

(copper catalyst)

(i.e., 1 $S_4O_6^{2-}$ would give rise to 2 $SCN^-$)

The assay is of ferric thiocyanate, which has a strong absorbance at 460 nm.

### Reagents

1. 0.1 M KCN (1.63 g KCN/250 ml).
2. Buffer pH 7.4: 50 ml 0.2 M $NaH_2PO_4$ + 39 ml 0.2 N NaOH.
3. 0.1 M copper sulfate (2.5 g $CuSO_4 \cdot 5H_2O$/100 ml).
4. Ferric nitrate reagent (1.5 M in 4 N perchloric acid): Dissolve 303 g $Fe(NO_3)_3 \cdot 9H_2O$ with continuous stirring in 217.4 ml 72% perchloric acid

and 15–20 ml water. *Beware!!* This mixture gets very cold and may take some time to dissolve, even overnight. It should be done in a fume hood in a glass container (plastic will be irreversibly stained) with constant slow magnetic stirring. The fumes emanating can irreparably damage nearby equipment, such as pH electrodes.

5. Once dissolved, make up to 500 ml with water.
6. KSCN standard: 0.001 M = 0.0972 g/l (= 1 μmole/ml).
7. Thiosulfate standard: 0.001 M = 0.248 g/l.

### *Thiosulfate*

Procedure

1. Use 25-ml volumetric flasks at room temperature (but below 22°C).
2. To each flask add 4 ml buffer, pH 7.4.
3. Add sample (1–6 mmol thiosulfate ideally) and mix. In practice, cultures containing up to 20 mM thiosulfate usually have 200-μl sample assayed per flask. It is not necessary to remove the organisms from the culture sample.
4. Add 5 ml 0.1 M KCN. Rinse in with about 5 ml water. Leave for 10 minutes.
5. Add, *with constant mixing*, 1.5 ml 0.1 M copper sulfate. Leave for 5–10 min (not critical).
6. Add 3 ml ferric nitrate reagent, and make up to 25 ml final total with water.
7. Read $OD_{460}$ *at once* using cuvettes of 1-cm light path.

The blank contains buffer, water, and ferric nitrate reagent, but no thiosulfate or thiocyanate, and KCN may also be omitted if desired. The blank mixture may be kept indefinitely. Standards use 0–10 μmol KSCN per flask, which may also have copper sulfate and cyanide omitted. Indeed in the absence of thionates, adding the copper sulfate produces an interfering white precipitate. The absorbance produced by 1 mmol thiosulfate in the assay is about 0.175 O.D. unit (= A).

Where many samples are to be assayed, it is possible to preload volumetric flasks with the buffer and cyanide (Steps 2 and 4), and store at 4°C until required. Once the sample has been added to this mixture and washed in with a little water (Step 3), the flask may again be stored at 4°C until it is convenient to assay. This quality is particularly useful with field samples or those taken during weekends, holidays, or late at night.

### *Tetrathionate*

Procedure

1. Assay as for thiosulfate, but omit the copper sulfate (Step 5).
2. Where both thiosulfate and tetrathionate are present,

$$2[OD_{460}(-\ Cu)] - OD_{460}(+\ Cu) \times A = \text{tetrathionate } (\mu\text{moles in sample volume}).$$

### *Trithionate*

$$S_3O_6^{2-} + CN^- + H_2O = SO_3^{2-} + SO_4^{2-} + SCN^- + 2H^+$$

Trithionate only reacts with cyanide at high temperature, producing thiosulfate, which is converted to thiocyanate by standard copper(II) catalysis.

Procedure

1. Add the sample (0.2 ml) to the buffer plus cyanide in a 25-ml volumetric flask.
2. Incubate at 100°C for 45 minutes in a fume hood.
3. Cool, and then add 1.5 ml 0.1 M $CuSO_4$ as for thiosulfate assay.
4. Add ferric nitrate reagent.
5. Measure $OD_{460}$ immediately.

$$(OD_{460} \text{ 'hot'} - OD_{460} \text{ 'cold + Cu'}) \times A = \text{trithionate } (\mu\text{moles})$$

### *Hexathionate*

$$S_6O_6^{2-} + 5CN^- + H_2O = S_2O_3^{2-} + SO_4^{2-} + 2HCN + 3SCN^-$$

In the assay without Cu, 1 $S_6O_6^{2-}$ produces 3 $SCN^-$. Thus, $S_6O_6^{2-}$ (μmoles) = SCN/3 (μmoles)

$$S_6O_6^{2-} + 8\ CN^- + 2\ H_2O = 2\ SO_4^{2-} + 4\ HCN + 4\ SCN^-$$

$$[Cu^{2+}]$$

In the assay with Cu, 1 $S_6O_6^{2-}$ produces 4 $SCN^-$. Thus, $S_6O_6^{2-}$ (μmoles) = SCN/4 (μmoles).

Calculation

The following calculation gives the number of sulfur atoms in a pure sample of a polythionate:

$$\left[\frac{[OD - Cu]}{[OD + Cu] - [OD - Cu]}\right] + 3$$

### *Thiosulfate Assay by Iodine Titration*

$$2Na_2S_2O_3 + I_2 = Na_2S_4O_6 + 2NaI$$

Reagents

1. 0.005 N iodine: This reagent is prepared from a commercial standard volumetric stock containing 0.1 N iodine by diluting 5 ml to 100 ml. Keep sealed and in the dark to avoid loss of iodine, which is volatile from solution.
2. 1% (w/v) soluble starch solution.
3. Standard 0.01 M thiosulfate: This solution should be freshly prepared, although frozen solutions are stable indefinitely. 1 ml is equivalent to 2 ml 0.005 N iodine.

Procedure

1. In a 100-ml or 250-ml conical flask mix culture sample with 5 ml 10% (v/v) acetic acid in water.
2. Titrate with iodine (burette) to a starch-blue end point (0.5–1.0 ml starch added to flask).
3. If using 0.005 N iodine, 1 ml is equivalent to 5 μmoles thiosulfate.

## Dimethyl Sulfide (DMS)

### Procedure

1. Extract culture sample (0.2–0.5 ml) with 5 ml 2,2,4-trimethylpentane (TMP), using mechanical agitation in sealed tubes for 5–10 min.
2. Mix 2 ml of the TMP fraction with 2 ml 0.2% (w/v) iodine in TMP.
3. Measure absorbance at 300 nm.
4. In aqueous DMS solutions that have been extracted with TMP, calibration range of DMS is 0–6 mM.

## Dimethyl Disulfide (DMDS)

### Procedure

1. Extract 1 ml sample with 2 ml hexane.
2. Measure absorbance at 260 nm of the hexane (upper) phase using a hexane blank.
3. Prepare calibration curve with 0–6 mM DMDS, processed as with DMS.

## Carbon Disulfide ($CS_2$) [22]

### Principle

In this method $CS_2$ is determined spectrophotometrically as cupric complexes of N,N-bis(2-hydroxyethyl)dithiocarbamic acid. The calibration curve is linear up to 5.0 μmoles $CS_2$, where an $OD_{430}$ of 0.18 is given by 1.0 μmole $CS_2$.

### Reagents and Assay

1. To prepare the color reagent, dissolve 36 mg copper(II) acetate $H_2O$ in 100 ml diethanolamine and dilute to 1 liter with ethanol (96% v/v).
2. To prepare a standard curve, dispense 20 ml color reagent into 25-ml volumetric flasks and add 0–250 μl $CS_2$ solution (2.5 mM in 96% (v/v) ethanol).
3. Make up the mixtures to 25 ml with 96% (v/v) ethanol, and allow to stand for 15 minutes before reading the absorbance values at 430 nm against a reagent blank.
4. To assay cultures add 50–100 μl to the reaction mixture in place of the $CS_2$ solution. The blank should include 50–100 μl minimal medium.

# ENZYME ASSAYS

These assays are for enzymes of autotrophic carbon dioxide fixation and inorganic sulfur oxidation that may be found in sulfur bacteria after growth with inorganic or organic sulfur compounds (see also [8, 9]).

## Ribulose Bisphosphate Carboxylase (RUBISCO)

### Reagents

1. Tris-HCl, pH 8.0, 0.1 M (1.21 g/100 ml)
2. 0.25 M $MgCl_2$ (5.1 g/100 ml)
3. Ribulose 1,5-bisphosphate (RuBP), 12–15 mM (final concentration in assay is about 2–3 mM)

4. Triton X-100, 5 or 10% (v/v) aqueous solution
5. Cetyltrimethylammonium bromide, 0.1% (w/v) aqueous solution
6. 6 M phosphoric acid
7. $CO_2$ absorption reagent: 15 ml ethanolamine + 35 ml methoxyethanol
8. Scintillant: any proprietary mixture that is miscible with water

Procedure

1. Prepare stock solution for the $^{14}C$-assay reaction mixture as follows. The total volume is 30 ml.

| | |
|---|---|
| Tris-HCl | 23 ml |
| $MgCl_2$ | 3 ml |
| Glutathione (dissolved in 1 ml water) | 19 mg |
| $NaHCO_3$ (dissolved in 1 ml water) | 112 mg |
| Water | 1.85 ml |
| $Na_2^{14}CO_3$ (250 μCi) | 0.15 ml |

2. Store this solution frozen in aliquots of up to 5 ml in sealed tubes (e.g., with vaccine caps) that may be thawed and used individually.
3. Use the procedure either for suspensions of organisms that are pre-permeabilized with Triton X-100 or cetyl trimethylammonium bromide (CTAB) or for cell-free extracts prepared by an appropriate procedure. The whole-organism procedure can also be conducted with organisms harvested from culture onto membrane filters. In that case separate tubes are needed for each time course interval sample.

Permeabilized Cell Procedure

1. Harvest organisms from culture by centrifugation in small (4-ml) glass centrifuge tubes, and suspend in 0.3 ml of Triton X-100 or CTAB solutions to give 0.2–0.5 mg dry weight cells/tube. Incubate these cells at the standard reaction temperature (usually 30°C) for 10 min. If doing a preliminary test in which organisms may contain little RUBISCO, it is advisable to use a higher cell density.
2. Add 0.9 ml $^{14}C$-assay mixture to each tube and mix.
3. Incubate for 10 min to activate the enzyme with the $CO_2$ substrate.
4. At zero time, add 0.3 ml water (blanks = controls) or 0.3 ml RuBP solution, and mix rapidly.
5. Immediately and after intervals of (for example) 5, 10, 20, 30, and 40 minutes, remove 0.1 ml samples and mix them with 0.1 ml phosphoric acid in scintillation counter vials.
6. Allow at least 30 min for discharge of $^{14}CO_2$ from the samples after addition to the phosphoric acid. Then add scintillant and count $^{14}C$ fixed in control and RuBP-supplemented treatments.
7. Determine the $^{14}C$-content of the $^{14}C$-assay mixture by mixing 0.02 ml with 1 ml absorption reagent, adding scintillant and counting $^{14}C$ at once.

Calculation of Activity

$^{14}C$ incorporation into the cells is a measure of RUBISCO activity and is calculated as cpm/min/mg dry weight (or protein). Specific activity of the bicarbonate (cpm/nmol) allows further calculation of enzyme activity as nmol $^{14}CO_2$ fixed/min/mg protein.

Cell-Free Extract Procedure

1. Supplement six tubes (e.g., 5-ml narrow test tubes or glass or plastic centrifuge tubes) with 0.7 ml $^{14}C$-assay mixture, and make the following additions at the indicated times:

| Tube | Addition 1 (ml water*) | Addition 2 (at 10-sec intervals; ml cell-free extract) | Addition 3 (at 10-sec intervals after 10 min preincubation at assay temperature; ml) |
|---|---|---|---|
| 1 | 0.2 | 0 | 0.2 water |
| 2 | 0 | 0.2 | 0.2 water |
| 3 | 0 | 0.2 | 0.2 RuBP |
| 4 | 0.05 | 0.15 | 0.2 RuBP |
| 5 | 0.10 | 0.10 | 0.2 RuBP |
| 6 | 0.15 | 0.05 | 0.2 RuBP |

*(or buffer in which the extract was prepared)

2. After Addition 3, mix rapidly and sample 0.2 ml of each tube IMMEDIATELY. Remove further samples at predetermined intervals (e.g., 2, 4, 6, 8 minutes or 2, 5, 10, 15 minutes or longer) depending on enzyme activity.
3. Add each sample to 0.1 ml phosphoric acid in a scintillation vial. Leave these vials at room temperature for at least 30 minutes after the final sample. Then add 10 ml of the scintillant to each vial and mix.
4. Preload 3 vials with 1 ml absorption reagent, and add 10 μl $^{14}C$ mixture to each. Add scintillant (10 ml), and count radioactivity immediately, together with the experimental vials.
5. As $^{14}C$ incorporation by the extracts is a measure of RUBISCO activity, calculate it as described for the whole-cell procedure.

## Thiosulfate-Oxidizing Enzyme (Tetrathionate Synthase) [12, 20]

This enzyme is found in several thiobacilli and catalyzes the oxidation:

$$2S_2O_3^{2-} \Rightarrow S_4O_6^{2-} + 2 \text{ electrons}$$

The electrons can be accepted by a cytochrome or, in the standard assay procedure, by ferricyanide, which is reduced to ferrocyanide:

$$2S_2O_3^{2-} + 2Fe(CN)_6^{3-} \Rightarrow S_4O_6^{2-} + 2Fe(CN)_6^{4-}$$

The enzyme has been successfully assayed using different buffers (phthalate, phosphate, and acetate). Activity with ferricyanide increases as the pH is decreased. The maximum rate is obtained at pH 4.5–5.0, but below pH 4.5 there is a chemical reaction between thiosulfate and ferricyanide.

Reagents

1. Phthalate buffer pH 5.0 (0.3 M): 6.1 g KH phthalate + 12.3 ml N NaOH made up to 100 ml with water
2. Potassium ferricyanide (0.006 M) 0.2 g/100 ml water
3. Sodium thiosulfate (0.06 M) 1.5 g/100 ml water, which should be stored frozen when not in use
4. Cell-free extract prepared by French pressure cell, lysozyme treatment, or other procedures

Procedure

1. Measure ferricyanide reduction in a spectrophotometer using continuous monitoring of absorbance on a chart recorder. Use 1-cm cuvettes at 420 nm.
2. Blank cell: 1 ml phthalate buffer + 1.75 ml water. Zero the machine with this mixture in both cuvettes.
3. Experimental cuvette: 1 ml phthalate buffer + 0.75 ml water + 0.5 ml thiosulfate + 0.5 ml ferricyanide
4. Add 0.25 ml extract to both cuvettes, and follow decrease in absorbance due to ferricyanide reduction.
5. Test different amounts of extract, and also endogenous reduction in a cuvette containing no thiosulfate.
6. Ferricyanide calibration: Read $OD_{420}$ with 0.1–0.6 ml ferricyanide (0.6–3.6 μmol) in a cuvette containing 1 ml phthalate buffer plus water to a total volume of 3 ml.
7. Use this calibration to calculate the rate of ferricyanide reduction as μmoles/min. Calculate the rate due to enzyme activity by deducting any rate observed in the presence of the extract but in the absence of thiosulfate.

## Rhodanese

$$S_2O_3^{2-} + CN^- = SCN^- + SO_3^{2-}$$

This enzyme is measured by assaying the rate of thiocyanate production. Thiocyanate is determined as ferric thiocyanate ($OD_{470}$ nm).

Reagents

1. Ferric nitrate reagent: Mix $Fe(NO_3)_3 \cdot 9H_2O$, 40 g with 16 ml of concentrated nitric acid, and make up to 250 ml with water.
2. 1.0 M Tris at pH 10.6 (6 g/100 ml) (unadjusted).
3. Tris may be replaced by phosphate buffer for assays at lower pH and may be compared with phosphate buffer at higher pH. For example: 7.21 g $Na_2HPO_4 \cdot 2H_2O$ + 1.48 g $NaH_2PO_4 \cdot 2H_2O$, made up to 100 ml gives 0.5 M phosphate, pH 7.47.
4. For assay at pH 8.38 use 0.25 ml 0.5 M phosphate pH 7.47
   pH 9.33 use 0.25 ml 0.05 M phosphate pH 7.47
   pH 10.7 use 0.25 ml 0.5 M phosphate pH 8.0
   pH 11.4 use 0.25 ml 0.5 M phosphate pH 10.6
5. 1.0 M sodium thiosulfate (2.48 g pentahydrate/10 ml).
6. 1.0 M potassium cyanide (0.65 g/10 ml).

Procedure

1. Assay in small tubes (at 30°C).
2. In a blank tube, mix 0.2 ml formalin, 2.5 ml assay mix or water, and 1.3 ml ferric reagent to a total volume of 4 ml.
3. Zero the spectrophotometer at 470 nm with this mixture.
4. Standards: 0–2 μmol KSCN in a 4-ml assay. Stock = 1 μmol/ml = 0.097 g/l. An $OD_{470}$ of 0.1 is equivalent to 96 nmol thiocyanate.
5. Assay in small tubes (5-ml capacity) containing: 0.25 ml M Tris pH 10.6, 0.05 ml M $Na_2S_2O_3$, 0.01 ml extract, and 2.14 ml water.
6. Preincubate in a water bath at 30°C after having covered tubes with Parafilm and mixed.
7. At zero time add 0.05 ml M KCN, mix rapidly, and return to 30°C.

8. Stop the reaction in the tubes after 2, 4, or 6 min or longer by adding 0.2 ml formalin and mixing rapidly.
9. To assay thiocyanate, add 1.3 ml ferric nitrate reagent. Mix and read immediately at 470 nm.
10. Measure specific activity as nmol thiocyanate formed/min/mg/protein.

## Sulfur-Oxidizing Enzyme

This assay measures thiosulfate production and oxygen uptake using Warburg manometers.

$$1/8S_8^{\circ} + O_2 + H_2O = H_2SO_3 \quad \text{(Oxygenase)}$$

$$H_2SO_3 + S = H_2S_2O_3 \quad \text{(Chemical reaction)}$$

### Procedure

1. In a final volume of 2 ml water, dissolve these components:

| | |
|---|---|
| Tris (hydroxymethyl)-methylamine/HCl, pH 7.8 | 500 μmol |
| Sulfur (BDH "Optran" grade) | 48 mg |
| 2,2′ bipyridyl | 0.2 μmol |
| Catalase | 250 mg |
| Crude extract | 0.1 ml |

2. Initiate reaction in air-filled Warburg flasks by adding 5 μmol glutathione from the side arms of the flasks.
3. Record oxygen uptake, which may take several hours for some thiobacilli.
4. Sample flask contents and assay for thiosulfate by cyanolysis (see above). Enzyme-specific activity can be expressed in terms of both rate of oxygen consumption and thiosulfate formation (μmoles/min/mg crude extract protein).

## REFERENCES


1. Bak, F. and Pfennig, N. 1987. Chemolithotrophic growth of *Desulfovibrio sulfodismutans* sp. nov. by disproportionation of inorganic sulfur compounds. Arch. Microbiol. 147:184–189.
2. Bartlett, J.K. and Skoog, D.A. 1954. Colorimetric determination of elemental sulfur. Anal. Chem. 26:1008–1011.
3. Bonch-Osmolovskaya, E.A. 1994. Bacterial sulfur reduction in hot vents. FEMS Microbiol. Rev. 15:65–77.
4. Gleen, H. and Quastel, J.H. 1953. Sulphur metabolism in soil. Appl. Microbiol. 1:70–78.
5. Guede, H., Strohl, W.R., and Larkin, J.M. 1981. Mixotrophic and heterotrophic growth of *Beggiatoa alba* in continuous culture. Arch. Microbiol. 129:357–360.
6. Huber, R. and Stetter, K.O. 1992. The order Thermoproteales. In The Prokaryotes, Vol. 1, 2nd ed. (Balows, A., Trüper, H.G., Dworkin, M., Harder, W., and Schleifer, K.-H., Eds.), pp. 677–683. Springer-Verlag, New York.
7. Kelly, D.P. and Wood, A.P. 1994. Synthesis and determination of thiosulfate and polythionates. In Methods in Enzymology—Inorganic Microbial Sulfur Metabolism, Vol. 243 (Peck, H.D., Jr. and LeGall, J., Eds.), pp. 475–501. Academic Press, Orlando.
8. Kelly, D.P. and Wood, A.P. 1994. Enzymes involved in the microbiological oxidation of thiosulfate and polythionates. In Methods in Enzymology—Inorganic Microbial Sulfur Metabolism, Vol. 243 (Peck, H.D., Jr. and LeGall, J., Eds.), pp. 501–510. Academic Press, Orlando.

9. Kelly, D.P. and Wood, A.P. 1994. Whole organism methods for inorganic sulfur oxidation by chemo- and photo-lithotrophs. In Methods in Enzymology—Inorganic Microbial Sulfur Metabolism, Vol. 243 (Peck, H.D., Jr. and LeGall, J., Eds.), pp. 510–520. Academic Press, Orlando.
10. Kelly, D.P., Chambers, L.A., and Trudinger, P.A. 1969. Cyanolysis and spectrophotometric estimation of trithionate in mixture with thiosulfate and tetrathionate. Anal. Chem. 41:898–901.
11. Koh, T. and Taniguchi, K. 1973. Spectrophotometric determination of total amounts of polythionates (tetra-, penta-, and hexa-thionate) in mixtures with thiosulfate and sulfite. Anal. Chem. 45:2018–2022.
12. Lyric, R.M. and Suzuki, I. 1970. Enzymes involved in the metabolism of thiosulfate by *Thiobacillus thioparus*. III. properties of thiosulfate-oxidizing enzyme and proposed pathway of thiosulfate oxidation. Can. J. Biochem. 48:355–363.
13. Morris, I., Glover, H.E., Kaplan, W.A., Kelly, D.P., and Weightman, A.L. 1985. Microbial activity in the Cariaco Trench. Microbios 42:133–144.
14. Nelson, D.C. and Jannasch, H.W. 1983. Chemoautotrophic growth of a marine *Beggiatoa* in sulfide-gradient cultures. Arch. Microbiol. 136:262–269.
15. Nelson, D.C., Revsbech, N.P., and Jorgensen, B.B. 1986. Microoxic-anoxic niche of *Beggiatoa* spp.: microelectrode survey of marine freshwater strains. Appl. Environ. Microbiol. 52:161–168.
16. Postgate, J.R. 1974. Media for sulphur bacteria. Lab. Pract. 15:1239–1244.
17. Postgate, J.R. 1984. The Sulfate-Reducing Bacteria, 2nd ed. Cambridge University Press, Cambridge.
18. Segerer, A.H. and Stetter, K.O. 1992. The order Sulfolobales. In The Prokaryotes, Vol. 1, 2nd ed. (Balows, A., Trüper, H.G., Dworkin, M., Harder, W., and Schleifer, K.-H., Eds.), pp. 684–701. Springer-Verlag, New York.
19. Segerer, A.H. and Stetter, K.O. 1992. The genus *Thermoplasma*. In The Prokaryotes, Vol. 1, 2nd ed. (Balows, A., Trüper, H.G., Dworkin, M., Harder, W., and Schleifer, K.-H., Eds.), pp. 712–718. Springer-Verlag, New York.
20. Trudinger, P.A. 1961. Thiosulphate oxidation and cytochromes in *Thiobacillus* X. 2. Thiosulphate-oxidizing enzyme. Biochem. J. 78:680–686.
21. Vorholt, J., Kunow, J., Stetter, K.O., and Thauer, R.K. 1995. Enzymes and coenzymes of the carbon monoxide dehydrogenase pathway for autotrophic $CO_2$ fixation in *Archaeoglobus lithotrophicus* and the lack of carbon monoxide dehydrogenase in the heterotrophic *A. profundus*. Arch. Microbiol. 163:112–118.
22. Vuik, J., Dinter, R., and van de Vos, R. 1992. Improved sample pretreatment of the carbon disulphide evolution method for the determination of dithiocarbamate residues in lettuce. J. Agric. Food Chem. 40:604–606.
23. Widdel, F. 1992. The genus *Desulfotomaculum*. In The Prokaryotes, Vol. II, 2nd ed. (Balows, A., Trüper, H.G., Dworkin, M., Harder, W., and Schleifer, K.-H., Eds.), pp. 1792–1799. Springer-Verlag, New York.
24. Widdel, F. 1992. The genus *Thermodesulfurobacterium*. In The Prokaryotes, Vol. IV, 2nd ed. (Balows, A., Trüper, H.G., Dworkin, M., Harder, W., and Schleifer, K.-H., Eds.), pp. 3390–3392. Springer-Verlag, New York.
25. Widdel, F. and Bak, F. 1992. Gram-negative mesophilic sulfate-reducing bacteria. In The Prokaryotes, Vol. IV, 2nd ed. (Balows, A., Trüper, H.G., Dworkin, M., Harder, W., and Schleifer, K.-H., Eds.), pp. 3352–3378. Springer-Verlag, New York.
26. Widdel, F. and Pfennig, N. 1984. Dissimilatory sulfate- or sulfur-reducing bacteria. In Bergey's Manual of Systematic Bacteriology, Vol. 1, 9th ed. (Kriegg, N.R. and Holt, J.G., Eds.), pp. 663–679. Williams & Wilkins, Baltimore.
27. Wood, A.P., Kelly, D.P., and Norris, P.R. 1987. Autotrophic growth of four *Sulfolobus* strains on tetrathionate and the effect of organic nutrients. Arch. Microbiol. 146:382–389.

3

JOEL KOSTKA
KENNETH H. NEALSON

# Isolation, Cultivation and Characterization of Iron- and Manganese-Reducing Bacteria

Bacterial reduction of iron and manganese has been known for many years, but until recently, dissimilatory metal-reducing bacteria—those capable of coupling metal reduction to cellular respiration—were considered extremely rare and possibly nonexistent [34, 78]. Although the redox potentials of these electron acceptors are quite high—Mn(III) is near that of nitrate, and Fe(III) is well above sulfate [46, 74, 78]—oxidized iron and manganese usually exist as solid oxides or oxyhydroxides in nature and were thus viewed as unlikely candidates to be used by bacteria for respiration. This view has dramatically changed in recent years, and bacteria capable of coupling anaerobic carbon oxidation to metal reduction have been isolated from many environments [35, 36, 55, 56]. To date, only two groups of these organisms have been studied extensively: (1) facultative anaerobes in the genus *Shewanella* (*S. putrefaciens* [44, 53, 65] and a closely related species, *S. alga* [15]) and (2) obligate anaerobes in the group *Desulfuromonas acetoxidans* [64] and the closely related *Geobacter metallireducens* [38]. The challenge of understanding the role of these bacteria in nature is great because of the complexity of biotic and abiotic chemical interactions in the surrounding medium [11, 20]. The respiratory substrates (solid oxidized metals) are difficult to measure accurately and to synthesize, and the organisms using them can be difficult to culture. Furthermore, the products of metal reduction often are bound to other components or form insoluble minerals with other chemical species characteristic of anoxic environments. Some of these complexities are illustrated in Figure 3-1, in which the cycles of manganese and iron are shown in relation to other processes in the water column and sediments.

This chapter presents several detailed methods for (1) preparing growth media and metal oxides, (2) rate determinations of metal reduction rates in the laboratory, and (3) isolating metal-reducing bacteria by enrichment culture. The development of field methods for monitoring metal-reducing activity in situ is beyond the scope of this chapter, and so the reader is referred to several recent reports for this information [16, 17, 75].

It is almost certain that only a small fraction of the bacteria capable of dissimilatory reduction of oxidized metals have been cultured and characterized, and the metabolic versatility of these microbes is just beginning to be realized. For example, marine and freshwater strains of *S. putrefaciens* have recently been shown to catalyze the rapid dissimilatory reduction of dissolved Mn(III), magnetite, and Fe-rich clays, none of which were previously known as electron acceptors for bacteria [29–31]. The metabolic versatility of this one group exemplifies the physiologic diversity of metal-reducers that is likely to occur in nature.

Metal-reducing bacteria have both significant biogeochemical implications and important applications to bioremediation. In terms of biogeochemistry, metal-reducing bacteria are able to affect the geochemistry of marine and freshwater sediments, both past and present. In recent sediments, dissimilatory metal reduction results in the removal of both organic carbon via its oxidation to $CO_2$, and of solid metal oxides via conversion to the reduced soluble forms. This removal

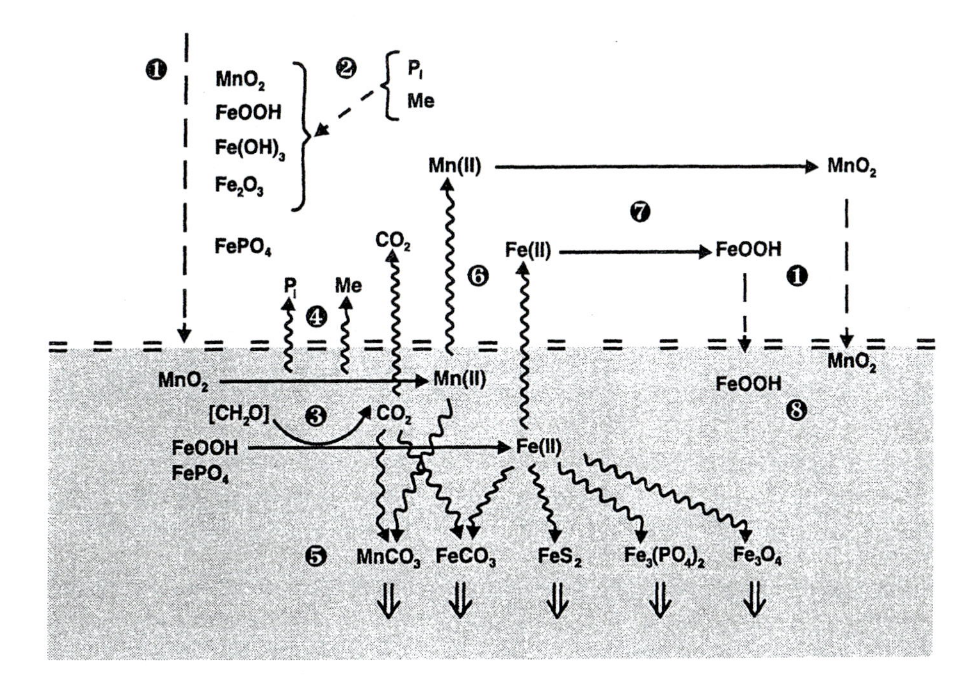


Figure 3-1 Iron and manganese cycling, particularly complexities that must be considered when studying such processes in the field (and sometimes in the laboratory). *1.* The metal oxides are solids and tend to move toward the sediments in aquatic environments, thus making gravity part of this biogeochemical cycle and moving oxidizing equivalents downwards. *2.* The metal oxides are excellent scavengers of other materials, including both organics and inorganics, and will transport other materials to the sediments with them. *3.* Because both iron and manganese have high redox potentials, they can be used by bacteria as electron acceptors for the oxidation of organic carbon, resulting in the production of reduced (soluble) metals and oxidized organic carbon. *4.* A side effect of Reaction 3 is that release and diffusion upward of adsorbed materials may occur. *5.* Reduced metals, $CO_2$, and adsorbed materials also diffuse downward, with many possible results, such as the formation of rhodochrosite ($MnCO_3$), siderite ($FeCO_3$), pyrite ($FeS_2$), vivianite ($Fe_3(PO_4)_2$), magnetite ($Fe_3O_4$), and others. *6.* Some of the reduced metals diffuse upward to the oxic zone as Mn(II) and Fe(II). *7.* Oxidation of the reduced metals occurs either biologically or abiotically to form metal oxides and oxyhydroxides, which initiates the cycle again (1). *8.* If oxidized metals are sedimented onto non-reducing surfaces, they will remain as ferromanganese layers or concretions. Metals can be removed by burial of reduced metals, such as pyrite, silicates, or carbonates. Bioturbation acts to recycle reduced metals to oxidized forms through irrigation or physical reworking. Therefore, Fe or Mn atoms may be recycled many times before permanent burial.

in turn causes the release of both nutrients via organic mineralization and of trace metals that were co-precipitated or adsorbed with metal oxides in sediments [22, 24, 25, 56, 69]. The release of such nutrients as phosphate at the sediment-water interface [24] can regenerate water column nutrients used by primary producers. Such benthic-pelagic coupling is one way that metal-reducing bacteria can contribute significantly to the flow of carbon/energy in marine and freshwater environments.

The extent to which dissimilatory metabolism affects the organic carbon oxidation in sediments is largely unknown, although recent studies in coastal marine sediments have indicated that Fe and Mn reduction can account for 50% or more of the organic carbon oxidized in some sedimentary environments [1, 2, 16, 17]. Metal-reducers are likely to compete for carbon substrates with sulfate-reducers and methanogenic bacteria depending upon the availability and reactivity of Fe and S.

Another area of possible significance involves the transformation of magnetite and clay minerals by metal-reducers. Magnetite is a mixed Fe(II)/Fe(III) oxide, and is of great importance as a contributor to sedimentary magnetism, whereas clay minerals dominate the solid phase of soils and sediments. Magnetic signatures are used as primary evidence for sea-floor spreading [68], as well as proxy indicators for global change [76]. It is now clear that dissimilatory iron-reducing bacteria can generate fine-grained magnetite as a byproduct of Fe(III) reduction [8, 43]. Under other conditions, both marine and freshwater strains of *S. putrefaciens* are able to respire millimolar amounts of Fe(III) in magnetite, resulting in its solubilization in a few days time at room temperature and near neutral pH [29]. The structural Fe(III) in clay minerals has been shown to be reduced rapidly (6–12 h) in cultures of *S. putrefaciens* MR-1 at temperatures and pHs common to soils and sediments. Structural Fe(III) reduction is thought to mediate a wealth of physicochemical changes in natural porous media [71]. Whether or not similar reactions occur in nature is not known, but if they do, they could have an important impact on the interpretation of both recent and paleoenvironments.

The potential role of metal-reducers in bioremediation seems great, but is not well understood. Some dissimilatory metal-reducers are also capable of reducing uranium, which may result in the removal of uranium from contaminated environments [42]. With *S. putrefaciens*, it is well documented that both whole cells [61] and cell extracts [62] can catalyze the anaerobic dechlorination of tetrachloromethane. *G. metallireducens* strain GS-15 has been studied more extensively for its application to bioremediation, and it is known to couple the reduction of iron to the oxidation of a variety of aromatics, including toluene, phenol, and p-cresol [37, 39]. Recently Kazumi et al. [26] studied iron-reducing enrichments for the breakdown of both chlorinated and non-chlorinated organic pollutants, and they speculate that there are novel metal-reducers in their enrichment cultures. In a general sense, the use of solid substrates like iron offers two advantages [58]: First, they can be delivered to the site of interest with no fear of diffusion until they are utilized; and second, since the ability to use solid substrates as electron acceptors is not particularly widespread, the competition for them will not be severe from indigenous populations of microbes.

## MEDIA PREPARATION

This section describes two types of culture media that are commonly used to cultivate metal reducers: (1) M1, which was developed for the cultivation of *S. putrefaciens* [54], and (2) the medium developed for the cultivation of *G. metallireducens* [38].

## M1 Medium

M1 basal medium has three main components: phosphate buffer, basal salts, and metals supplement—with the separate addition of ammonium, selenate, bicarbonate, and amino acids.

Procedure

1. To prepare 1 liter of M1, add 1.19 g $(NH_4)_2SO_4$ to a 2-liter Erlenmyer flask, followed by 15 ml of the phosphate buffer, 100 ml of basal medium, and 875 ml of distilled water.
2. Mix until all of the ammonium dissolves, and then add 0.1 ml each of a filter-sterilized 0.115 M $Na_2SeO_4$ solution and the metals supplement.
3. Adjust the pH to 7.0 with 10 N NaOH or HCl as needed.
4. For solid media only, add 15 g of agar.
5. Autoclave for 20 min and allow to cool (it may be turbid, but should slowly clear as it cools).
6. Add 10 ml of filter-sterilized 0.2 M $NaHCO_3$. The medium may be stored sterile at room temperature.
7. Before use, add amino acids, carbon substrate, and electron acceptor as desired. Ten ml of the amino acid mix is normally added, whereas carbon sources and electron acceptors are added to the levels wanted.
8. For standard growth experiments, add carbon substrates to 10–20 mM and electron acceptors at 4–100 mM.
9. To stimulate heterotrophic growth, supplement the medium with 0.01% to 0.05% of yeast extract, peptone, or casamino acids.
10. If pH variation is a problem, add HEPES [4-(2-hydroxyethyl)-1-piperazineethane-sulfonic acid] buffer at a final concentration of 10 to 50 mM before adjusting the pH of the basal medium.

### *Phosphate Buffer (in Plastic Bottle)*

| | |
|---|---|
| $KH_2PO_4$ | 30.0 g |
| $K_2HPO_4$ | 66.1 g |
| Distilled water | 800 ml |

Adjust pH to 7 with NaOH or HCl as needed. Bring final volume to 1 liter with distilled water, and store at 4°C (or freeze for long-term storage).

### *Metals Supplement*

| | |
|---|---|
| $CoSO_4 \cdot 7H_2O$ | 1.41 g |
| $Ni(NH_4)_2(SO_4)_2 \cdot 6H_2O$ | 1.98 g |
| NaCl | 0.58 g |
| Distilled water | 100 ml |

Autoclave for 20 minutes and store at 4°C.

### *Basal Salts (in Plastic Bottle)*

| | |
|---|---|
| Sterile trace elements solution (see below) | 10 ml |
| Distilled water | 800 ml |
| $MgSO_4 \cdot 7H_2O$ | 2.00 g |
| $CaCl_2 \cdot 2H_2O$ | 0.57 g |

| | |
|---|---|
| EDTA, disodium salt | 0.20 g |
| $FeSO_4 \cdot 7H_2O$ | 0.012 g |

Store at 4°C.

### *Trace Elements (Sterile, in Glass Bottle)*

| | |
|---|---|
| $H_3BO_3$ | 2.8 g |
| $ZnSO_4 \cdot 7H_2O$ | 0.24 g |
| $Na_2MoO_4 \cdot 2H_2O$ | 0.75 g |
| $CuSO_4 \cdot 5H_2O$ | 0.042 g |
| $MnSO_4 \cdot H_2O$ | 0.17 g |
| Distilled water | 1000 ml |

Filter-sterilize and store in glass bottle at 4°C.

### *Mixed Amino Acids, pH 7*

Add 0.2 g each of L-arginine, L-serine, and L-glutamic acid to 100 ml of distilled water. Autoclave for 20 minutes, and store in a sterile glass bottle at 4°C.

### *Carbon Substrates*

Carbon substrates are prepared as filter-sterilized solutions of 1 M, with pH adjusted to 7.0. They are stored in glass jars.

### *Electron Acceptors*

Electron acceptors are prepared as described (pp. 63–67), stored in glass jars as sterile suspensions.

## GS-15 Medium for *G. metallireducens*

This medium is prepared using strict anaerobic technique [38].

Procedure

1. In 1 liter of distilled water, dissolve 2.5 g $NaHCO_3$, 1.5 g $NH_4Cl$, 0.6 g $NaH_2PO_4$, and 0.1 g KCl.
2. Add acetate or the electron donor of choice to 20–50 mM.
3. Add 10 ml each of the vitamin mix and mineral mix described below.
4. Split the media into tubes, add the electron acceptor (prepared as described below), and sparge with 80% $N_2$ and 20% $CO_2$ for at least 6 minutes, with the last minute with the stopper in place.
5. Fill the tubes to no more than half-full, and adjust the final pH to 6.8–7.0.
6. Seal the tubes under a stream of $N_2/CO_2$ and autoclave for 20 min.
7. The pH will increase during the growth of metal-reducers in culture. To buffer the pH, add 10–50 mM HEPES before sparging.

### *Vitamin Mix (per Liter of Distilled Water)*

| | |
|---|---|
| Biotin | 2 mg |
| Folic acid | 2 mg |
| Pyridoxine HCl | 10 mg |
| Riboflavin | 5 mg |

| | |
|---|---|
| Thiamine | 5 mg |
| Nicotinic acid | 5 mg |
| Pantothenic acid | 5 mg |
| Vitamin B12 | 0.1 mg |
| p-Aminobenzoic acid | 5 mg |
| Thioctic acid | 5 mg |

Store at 4°C.

### *Mineral Mix (per Liter of Distilled Water)*

| | |
|---|---|
| Nitrilotriacetic acid | 1.5 g |
| $MgSO_4$ | 3.0 g |
| $MnSO_4$ | 0.5 g |
| NaCl | 1.0 g |
| $FeSO_4$ | 0.1 g |
| $CaCl_2$ | 0.1 g |
| $CoCl_2$ | 0.1 g |
| ZnCl | 0.13 g |
| $CuSO_4$ | 0.01 g |
| $AlK(SO_4)_2$ | 0.01 g |
| $H_3BO_3$ | 0.01 g |
| $Na_2MoO_4$ | 0.025 g |
| $NiCl_2 \cdot 6H_2O$ | 0.024 g |
| $Na_2WO_4$ | 0.025 g |

## PREPARATION OF THE OXIDIZED METALS

Fundamental to the success of isolating iron- and manganese-reducing bacteria is the proper preparation of the oxidized metals. Iron exists in nature as the reduced ferrous [Fe(II)] state and the oxidized ferric [Fe(III)] state. Fe(II) is thermodynamically unstable at neutral pH values and is rapidly oxidized to Fe(III), which then hydrolyzes to form a variety of hydroxides and oxides. Mn is found in three different states—the soluble Mn(II) form and the oxidized Mn(III) and Mn(IV) forms—which also hydrolyze to a variety of oxides and oxyhydroxides in water. Metal oxides are colloidal or solid and present a variety of experimental problems that have plagued workers in the past. In general, as the crystallinity of the oxide increases, its reactivity goes down, probably due to decreased surface area, among other factors [7, 50]. Thus, the preparation of fresh hydrous metal oxides is crucial to the successful study of these bacteria.

### Soluble Forms of the Oxidized Metals

Soluble Fe(III) can be prepared and used for laboratory experiments by the use of chelating agents, such as citrate, ethylenediaminetetraacetic acid (EDTA), or nitrilotriacetic acid (NTA). In contrast to the insoluble metal oxides, these soluble complexes are not surface limited for activity. Both citrate [38, 40, 54, 55] and NTA [3–5] have been used for iron reduction studies.

Citrate binds tightly to Fe(III) over a range of pH and may be purchased as an inexpensive Fe(III) salt. Fe(III) citrate is added to the medium as an autoclaved stock solution.

Procedure

1. Dissolve Fe(III) citrate to 0.5 M in 500 ml distilled water and bring the pH to 7 with 10 N NaOH.
2. Adjust the pH of the stock as the Fe(III) salt dissolves in water, which may take several hours of continuous stirring.
3. Once dissolved and pH adjusted, autoclave the solution for 20 min.

Previous studies have used final concentrations ranging from 2 to 50 mM. We recommend using 4 to 10 mM Fe(III) citrate for physiologic studies.

NTA is not readily available as an Fe(III) salt, but may also be added as an autoclaved stock solution. Dissolve equimolar amounts of NTA and $FeCl_3 \cdot 6H_2O$ in distilled water. Adjust the pH and prepare in the same manner for ferric citrate.

Mn(IV) forms no such chelates, but Mn(III) can be prepared in soluble form using pyrophosphate or other organic chelates and is sufficiently stable that it can be used for studies of bacterial reduction [30]. Though chelated dissolved Mn(III) has been studied in soils [21, 45] and in fungal cultures for years [63], until recently, no data were available on the bacterial reduction of complexed, oxidized Mn. Kostka et al. [30] showed that Mn(III)-pyrophosphate complexes were stable in culture media at a pH characteristic of aquatic environments and that these complexes could be easily reduced by *S. putrefaciens*. Mn(III) complexed to other ligands, such as citrate, malate, pyruvate, and tiron, was also reduced. The dissimilatory reduction of dissolved Mn(III) by *S. putrefaciens* strain MR-1 was shown to be coupled to energy generation, carbon metabolism, and protein synthesis [30].

Procedure: Preparation of Mn(III)-Pyrophosphate Complex

1. Dissolve sodium pyrophosphate in the culture medium to a final concentration of 40 mM, and adjust the pH to 8.2 with concentrated HCl.
2. Stir rapidly to dissolve, and while stirring, add manganese(III) acetate dihydrate to a final concentration of 10 mM and adjust the pH to 8.0.
3. Add carbon substrate and filter-sterilize.
4. Store the medium + Mn(III)-pyrophosphate at room temperature in the dark.
5. *Important note*: Omit EDTA and HEPES from the culture medium as both can cause abiotic reduction of Mn(III) complexes [30].

## Insoluble Fe or Mn Oxide Minerals

Though complexed or chelated metals have been observed in some geochemical studies, oxidized metals are predominantly stabilized in sediments as solid minerals, such as oxides, oxyhydroxides, phosphates, or clays. Many previous microbiologic studies of metal-reducing bacteria have used pulverized and sieved rock minerals. However, drying or aging has a pronounced effect on the surface chemistry of minerals [49, 52, 59], and minerals that are dried and stored are likely to have a very different chemistry from those found in the sedimentary environments from which metal-reducing bacteria have been isolated [13, 49, 52, 54]. To mimic more closely the production of minerals in natural environments, we suggest synthesizing minerals and keeping the suspension wet throughout preparation. In general, it is best to synthesize the mineral immediately before use, especially with the more reactive (less crystalline) forms.

For the preparation of metal oxides, we have consulted several sources including our own procedures. We present methods for preparing the most common minerals found in nature and for those that have been used in most microbiologic studies. Generally, the synthesis of oxides involves an initial precipitation step at high pH, followed by an aging step at a range of temperatures depending on the oxide of interest. Other important considerations are washing and storage. The synthesized oxide should be washed carefully to remove the salts used during preparation. Centrifugation is a rapid method for desalting the sample, although more gentle dialysis methods may be preferred to ensure that finer grains are preserved [6, 66].

## Ferrihydrite [$Fe(OH)_3$]

Ferrihydrite is the least crystalline of the Fe(III) oxyhydroxide minerals, has the highest surface area per volume, and is considered the most reactive. This mineral is often observed as the initial precipitate formed after the rapid hydrolysis of Fe(III) solutions, and preparations can be achieved either by heating an acid Fe(III) solution at 80°C or by rapidly raising the pH with alkali at room temperature. We describe a method from Schwertmann and Cornell [66] that uses the latter method.

### Procedure

1. Dissolve 40 g $Fe(NO_3)_3 \cdot 9H_2O$ in 500 ml distilled water, approximately 330 ml KOH to bring the pH to 7.8.
2. Add the last 20 ml of KOH dropwise with constant monitoring of the pH. Stir vigorously while dissolving and adjusting the pH.
3. Because this method produces a suspension of approximately 10 g ferrihydrite per liter, use dialysis to remove the electrolytes. The suspension may also be centrifuged and washed six times to remove salts.
4. Prepare the ferrihydrite on the same day it is to be used, since goethite or hematite may form from ferrihydrite with aging [67].

## Goethite [FeOOH]

For synthesizing goethite, generally considered to be the most stable form of Fe(III) oxide [67], the method of Atkinson et al. [6] is suggested.

### Procedure

1. Add 200 ml of 2.5 N KOH to 50 g of $Fe(NO_3)_3 \cdot 9H_2O$ in 825 ml of double-distilled water.
2. Adjust the pH of the suspension to 12, cover to avoid evaporation, and age for 24 hours in a 60°C oven. This preparation produces a suspension of 20 to 30 g of ferric oxide per liter.
3. Dialyze or wash the suspension several times to remove salts.
4. Store the resulting goethite in polyethylene bottles or centrifuge tubes for days in the refrigerator (4°C) or for months at −20°C in a freezer.
5. As with all Fe oxides, prepare the goethite just before (or within a few days of) use.

## Hematite [$Fe_2O_3$]

In order to obtain hematite in the absence of other Fe minerals, we recommend preparation by "forced hydrolysis" as described by Schwertmann and Cornell [66].

Procedure

1. Begin by heating 2 liters of 0.002 M $HNO_3$ in a Duran flask in an oven.
2. When the temperature reaches 98°C, remove the reaction vessel from the oven and add 16.6 g unhydrolyzed crystals of $Fe(NO_3)_3 \cdot 9H_2O$ (0.02 M Fe) while vigorously stirring.
3. After closing the Duran flask, immediately return the solution to the oven and incubate at 98°C for 7 days.
4. Desalt this precipitate by dialysis or by centrifugation (six times).

## Magnetite [$Fe_3O_4$]

The preparation of magnetite requires heating under anaerobic conditions. We use the method of Schwertmann and Cornell [66], which requires a 500-ml separatory funnel mounted on a 2-liter vacuum flask with a side port.

Procedure

1. Begin by deaerating 560 ml of deionized water with nitrogen.
2. Add 80 g of $FeSO_4 \cdot 7H_2O$ while the solution is sparged with nitrogen.
3. At this point, place the flask, with the separatory funnel in place, into a water bath and heat to 90°C.
4. While heating, dissolve 6.46 g $KNO_3$ and 44.9 g KOH in 240 ml deionized water, place in a separatory funnel, and sparge with nitrogen for at least 20 minutes.
5. Within approximately 5 minutes, add the basic solution from the separatory funnel dropwise into the Fe solution at 90°C and mix well.
6. Continue heating the solution for another 30–60 min, cool overnight, and wash the black precipitate.

After termination of the reaction, no further protection against air is necessary. Contrary to the popular belief that crystalline Fe(III) minerals are stable and of equal activity for years, it has been recently reported that magnetite suspensions are decreased in reactivity after only 11 days [27].

## Colloidal Mn(IV)

In low-salt environments, Mn(IV) is relatively stable as a colloidal form [60]. It can be prepared as described in this section and then used for experiments, but it suffers from being quite variable, depending on medium composition. It is, however, very reactive, and might be the oxide of choice in some cases. When working at higher salt concentrations, this colloidal form coagulates rapidly and forms more solid manganese oxides.

Procedure

1. Dissolve $Na_2S_2O_3$ and $KMnO_4$ to 0.1 M in separate solutions, and filter-sterilize these stocks, using sterile glassware.
2. To prepare the colloid, aseptically pipette 2 ml of each solution to 100 ml of culture media. The colloid should form almost immediately and will coagulate with increasing salt and Mn concentration.

## Vernadite [$MnO_2$]

The Mn oxide mineral most often used in microbiologic studies is $MnO_2$ or vernadite [11–13, 53]. The preparation described here is a slight modification of methods developed earlier [7, 47, 49]. Manganese dioxide is formed from the oxidation of manganous ion by permanganate according to the following reaction:

$$3MnCl_2 + 2KMnO_4 + 2H_2O \rightarrow 5MnO_2 + 4H^+ + 2K^+ + 6Cl^-$$

Procedure

1. Dissolve 5.55 g $MnCl_2 \cdot 4H_2O$ in 27.8 ml of distilled water.
2. In a separate 250-ml Erlenmyer flask, dissolve 2.96 g $KMnO_4$ in 74.1 ml distilled water.
3. While stirring continuously, bring $KMnO_4$ to 90°C on a hot plate.
4. With the solution at 90°C, add 3.71 ml 5 N NaOH, and then slowly add the $MnCl_2$ solution. A black precipitate should form immediately.
5. Turn off the hot plate, stir the suspension for about 30 min, and cool to room temperature.
6. Wash the suspension several times with distilled water, and allow to sit overnight before using.

## Fe-Rich Clays

Clay minerals have not been synthesized for microbiologic studies to date, but have been prepared from mined material, such as natural Fe-rich smectites or montmorillonites [71]. These minerals are fractionated and dialyzed before use. The dialyzed suspension is added to the culture medium and the mixture is autoclaved. The final concentration of the suspension used in culture media has been 1–2 g/l.

Only a few studies of the microbial reduction of Fe in clays have been carried out, and most have focused on the indirect reduction of Fe in clays by heterotrophic soil bacteria [73, 77]. Recently, it was shown that *S. putrefaciens* MR-1 can catalyze the rapid reduction of structural Fe(III) in the clay mineral, smectite [31]. The reduction of the iron in clays occurred at temperatures and pHs common to soils and sediments and was coupled to energy generation and carbon metabolism in MR-1.

## Mineral Characterization

Another aspect of this work is the characterization of the mineralogy and oxidation state of the particles used. Without this information, it is difficult to compare various studies. The primary means by which minerals are identified in culture studies is X-ray diffraction measurements, which rely on comparisons of d-spacings that correspond to bond angles in the mineral structure. A good source for d-spacing comparisons of sedimentary minerals is found in Murray [50] for Fe, and Burns and Burns [14] for Mn. Often X-ray diffraction methods are supported by transmission electron micrograph (TEM) characterization of mineral particles. Iron particles, in particular, have distinctive shapes and therefore may be identified by TEM. For example, ferrihydrite particles are amorphous, goethite particles are generally acicular, and magnetite particles are often cubic.

Minerals may also be characterized by their stoichiometry in acid extracts. For Fe oxide minerals, a 6 N HCl extraction has been used; the oxidation state of Fe is determined using a colorimetric reagent, such as ferrozine [70]. For Mn oxides, the iodometric titration method of Murray et al. [51] is suggested. Using these analytic methods, the total ratio of equivalents of oxidized metal to oxygen may be determined.

The oxidation state of Fe in clay minerals may be determined in culture samples as described in Stucki and Anderson [72] and modified by Komadel and Stucki [28].

Procedure

1. Centrifuge the culture sample to give 1 g or less of sample, and perform the extraction on the pellet under a red lamp (in an otherwise dark room) to avoid photoreduction of the iron.
2. Resuspend the pellet in 12 mol of 3.6 N $H_2SO_4$, taking care to wash down any sample clinging to the walls of the tube.
3. Add 2 ml of 10% 1, 10-phenanthroline monohydrate (in 95% ethanol), followed by 1 ml of 48% hydrofluoric acid, and place the tube into a boiling water bath for 30 min. **Caution should be exercised, as these reagents are extremely caustic!**
4. After the tube is cooled, add 10 ml of 5% $H_3BO_3$. Make known dilutions with distilled water if necessary.
5. Measure absorbance (510 nm) in the dark for the determination of the Fe(II) in the sample.
6. Then filter the sample, and measure total iron [Fe(II) + Fe(III)] using atomic absorption spectrometry. Calculate Fe(III) as the difference between the Fe(II) and total Fe.
7. Also use Mossbauer spectroscopy to measure the oxidation state of Fe in clays as described in Komadel and Stucki [28].

Another important physical property of oxidized metal particles reduced by bacteria is their surface area. Though few microbiologic studies to date have normalized reduction rates to empirically determine surface areas [4, 13], this property is thought to limit respiration and growth on oxidized metals. Particle surface area may be determined using the BET (Brunauer, Emmet and Teller) method of $N_2$ gas adsorption [10]. Future studies should employ this method more often in order to elucidate the relationship between particle surface and bacterial metal reduction.

## MONITORING METAL REDUCTION IN CULTURE

Metal reduction is commonly assessed by measuring the accumulation of soluble metal in the culture medium using flame atomic absorption spectrometry (AAS) or uv/vis spectrophotometry. As with other measurements discussed in this chapter, a wide range of techniques have been implemented in cultures of metal-reducers. The more representative methods for Fe and Mn reduction are presented in this section.

### Fe Measurements

Iron is often measured by the colorimetric reagent, ferrozine, using variations of the method of Stookey [70]. Ferrozine binds both Fe(II) and Fe(III), but a colored complex is produced only with Fe(II). Fe(II) may thus be determined by extracting the sample in 0.5 M HCl and then adding the extract to a ferrozine solution buffered with HEPES [4-(2-hydroxyethyl)-1-piperazineethane-sulfonic acid] buffer. Color development is pH dependent [70], so the addition of buffer is critical.

Procedure

1. To prepare the ferrozine reagent, dissolve 0.02% ferrozine in 1 liter of 50 mM HEPES buffer at a pH of 7.0.
2. Add 5 ml of reagent to 25-ml scintillation vials.
3. Pipette 0.05–0.5 ml of culture sample into 0.5 M HCl.
4. Wait 30 min, and pipette 0.05–0.2 ml of the acid extract (any larger volumes will exceed the buffering capacity of the ferrozine) into the 5 ml of ferrozine reagent.
5. Let sit for 10–15 min, filter through a 0.2-μm filter, and measure the absorbance on a spectrophotometer at a wavelength of 562 nm.

For Fe, culture samples are most often extracted whole in HCl, and their metal content is determined as the total amount of Fe reduced by the bacteria. The Fe measurement should be standardized using an Fe(II) salt, such as ferrous ammonium sulfate. It is best to generate the standard curve in the culture medium and to check for the effects of any additions to the medium with extra standard measurements.

Reduced Fe may also be determined by measuring the total amount of Fe in the dissolved phase of AAS or by inductively coupled plasma spectrometry (ICP). Collect a culture sample, filter through a 0.2-μm filter, and add trace metal clean $HNO_3$ to pH 1. Culture samples may then be stored at 4°C for months before analysis. The assumption for AAS analysis is that the dissolved phase is completely reduced.

The amount of oxidized Fe in a culture sample that contains Fe minerals may be determined after extraction of the sample in a strong acid solution.

Procedure

1. When using Fe oxide minerals, add 0.1 ml of culture sample to 5–10 ml of 6 N HCl.
2. Wait until the Fe mineral is dissolved completely (usually a few hours). Note: For more rapid dissolution in HCl, place into an oven at 50–60°C.
3. Then dilute the acid extract with distilled water, keeping the pH at 1.0.
4. Filter through a 0.2-μm filter, and analyze for Fe with flame AAS as described above. This measurement should give the total amount of Fe in the sample (dissolved [ferrous] + particulate [ferric]).
5. Measure the Fe(II) in the solid phase by extracting the culture sample in parallel with 6 N HCl as above. Rather than diluting the acid extract in distilled water, however, add to buffered ferrozine and measure the absorbance on a spectrophotometer as described previously. Determine the amount of oxidized Fe by subtracting the reduced Fe from the total Fe.

The amount of dissolved versus particulate Fe is determined by filtration of the culture sample before measurement. Three phases of Fe should be considered: dissolved, adsorbed, and particulate. The dissolved and absorbed phases are likely to comprise mainly reduced Fe, whereas the particulate fraction should contain mostly oxidized Fe depending upon the pH of the culture medium. The adsorbed and particulate fractions are difficult to separate. These same considerations apply to cultures with chelated metals or solid minerals added as the electron acceptors; reduced metal may absorb to the bacterial cell and therefore enter the particulate phase.

## Mn Measurements

Similar methods may be used for the measurement of reduced Mn in culture samples. In fact, most previous culture studies have measured the total Mn in the dissolved phase of culture samples using flame AAS [13, 53, 54].

Several colorimetric methods are available for Mn as well [18, 19, 23]. We have used the formaldoxime method developed by Goto et al. [23] and modified by Brewer and Spencer [9] to measure Mn in culture samples. Because this method uses hydroxylamine, which reduces Mn(IV) and Mn(III) to Mn(II) it measures the total Mn (all oxidation states) in the sample. Therefore, samples must be filtered before adding the reagent to measure the reduced Mn in the dissolved phase as with the AAS method.

### Procedure

1. To prepare the formaldoxime reagent, dissolve 20.0 g of hydroxylamine hydrochloride in 450 ml of distilled water.
2. Add 10 ml of formaldehyde solution (37%), and make up to 500 ml with distilled water.
3. Filter 1 ml of culture sample through a 0.2-μm filter.
4. Add 0.05–0.5 ml of filtered culture sample to 5.0 ml of distilled water.
5. Then add 0.5 ml of formaldoxime reagent and 0.03 ml of glacial $NH_4OH$.
6. Wait 5–10 min, measure the absorbance on a spectrophotometer at a wavelength of 450 nm.
7. Optimum color development occurs at a pH of 8.8 to 8.9 so experiment with the culture medium by adding 0.5 ml formaldoxime and then various amounts of $NH_4OH$ until the pH is optimal.

Since much Mn(II) is likely to be absorbed onto the surface of $MnO_2$ particles, the particulate material should be washed and the Mn in the wash solution added to the total amount of Mn reduced. A $CuSO_4$ solution (10 mM, pH = 4.0) can be used to desorb Mn(II), as the Cu(II) will exchange with the Mn(II) on the surface of the particles [12]. This wash step should be included with Mn measurements by AAS or colorimetric methods.

The oxidation state of Mn in culture samples may be monitored by redox titration as with chelated Mn(III) [30]. The iodometric titration used according to Murray et al. [51] is suitable for both the dissolved and particulate phase.

Though it has yet to be tested for the quantitative determination of bacterial Mn reduction, the leuco-berbelin blue-I (LBB) reagent has been used to detect and isolate both Mn-reducing and Mn-oxidizing bacteria [32]. A recent paper by Lee and Tebo [33] produced a standard curve for the measurement of Mn(IV) with the LBB reagent. This reagent should be explored further for its use in determining quantitative rates of Mn reduction.

## ENRICHMENT, ESTIMATION OF ABUNDANCE, AND ISOLATION OF METAL-REDUCING BACTERIA

A simple enrichment culture method can be used to demonstrate microbial manganese reduction potential, to estimate the relative reducing potential of microbes that are present, and to act as a source for the isolation of such organisms for further study [54, 57]. Although this method is valuable for the demonstration of the phenomenon of manganese reduction and as a supply of material for pure culture isolation, it suffers, as do all enrichment methods, from selectivity caused by medium composition, which favors the isolation of similar organisms from different environments. However, for teaching and for obtaining an estimate of environmental activity, it is an acceptable method. The method employs a soft agar base, which keeps the metal oxides in suspension and allows one to easily visualize the metal reduction as it occurs in the tubes (Fig. 3-2). Although enrichments can also be carried out in liquid suspension, it is not possible to score them easily because the metal oxides collect on the bottom of the vials. It is also possible, using microelectrodes, to examine the microenvironments around zones of reduction using this soft agar method, something that is not easily done with liquid suspensions.

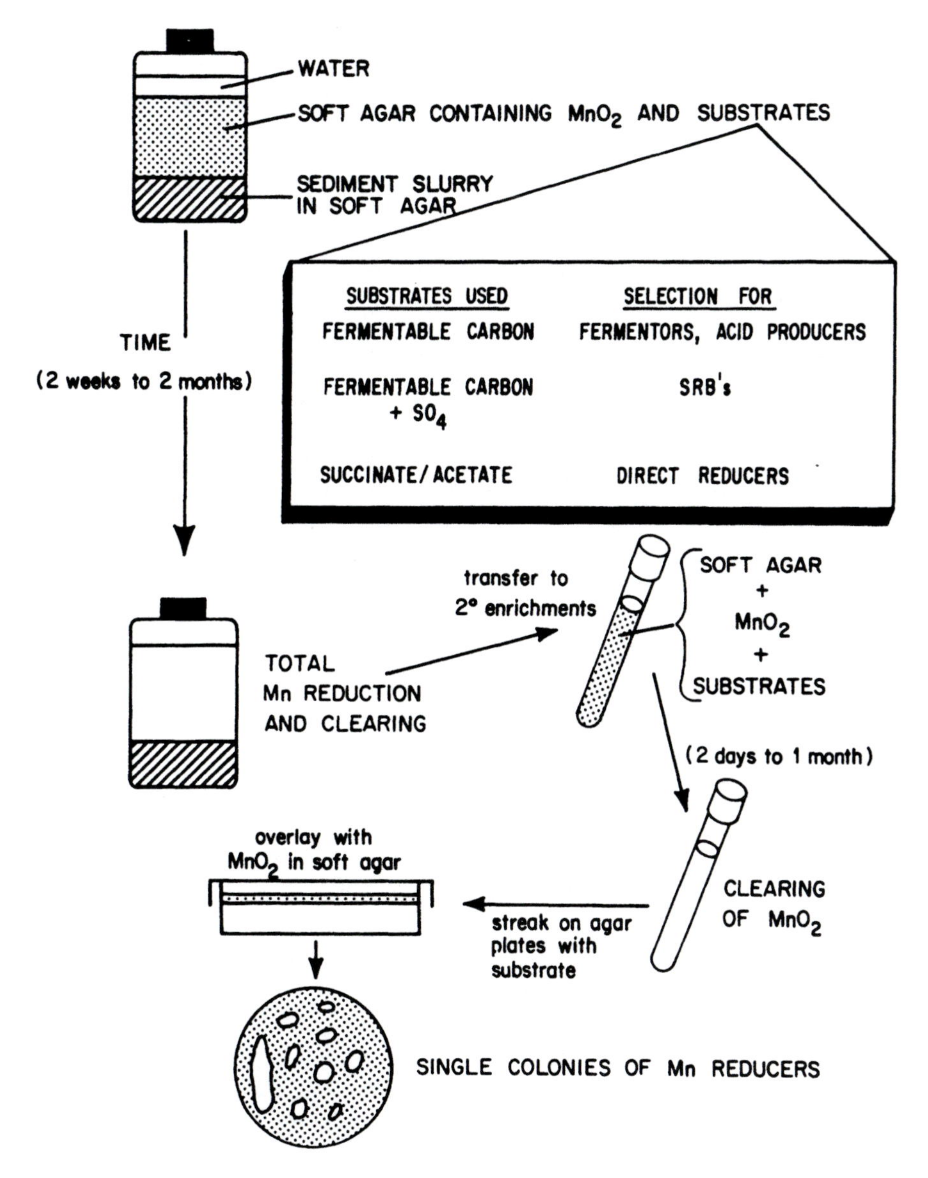


Figure 3-2 Enrichment method for both the indication of relative amounts of metal-reducing activity (see Fig. 3–3) and for the isolation of pure cultures of metal-reducers. This method should be used as a guideline, and further experimentation should be carried out to improve the enrichments. We routinely make the following additions to enrichments: (1) a series with peptone or yeast extract added to stimulate more fastidious organisms, (2) controls with molybdate added to inhibit sulfate reducers, (3) controls with mercuric chloride to inhibit biological growth and metal reduction, and (4) a series of dilutions to estimate total number of metal reducers.

## Sample Acquisition

Metal-reducers are often abundant in water columns at redox interfaces or in anoxic sediments. In stratified water environments, like the Black Sea, samples can be collected using Niskin bottles and transferred directly to degassed anaerobic enrichment media. Dilutions are made in sterile, degassed seawater medium and then added to the enrichment tubes. Metal oxides, carbon sources, inhibitors, and the like are added to the enrichment media as described below.

Sediment samples are taken by core samplers and extruded with a core extruder (or sampled from the side using sterile syringes). For obligate anaerobes, extrusion or subsampling is done under a nitrogen atmosphere, usually using a disposable glove bag (see Chapter 10 on sampling methods). Samples are placed in an anaerobic hood or glove bag and either used directly in enrichment cultures as described below or diluted into sterile anoxic lake water or seawater and used in the enrichment cultures.

## Enrichment Cultures

Water samples can be mixed directly with an equal volume of 1.5% agar (containing appropriate nutrients and metal oxides) to give a soft agar slurry. The agar is held at 45°C in a water bath, and upon mixing, the material is placed immediately into vials, where cooling and setting of the agar occur rapidly. The mixing can be done directly into anaerobic vials, using a syringe to add the sample. The samples are gently mixed and allowed to uniformly suspend the metal oxides. For sediment samples, 1 g of suspension or environmental sample is mixed with 9 ml of sterile anoxic lake (or sea) water containing 50 mM HEPES buffer at a pH of 7.4. This mixture is then treated in the same way that the water samples above were treated, except that transfers are made via pipettes rather than syringes because of the suspended solids in the samples.

## Carbon Substrates

If respiratory metal-reducers are to be isolated, it is essential to use a nonfermentable substrate, such as lactate, acetate, succinate, or glycerol (usually at a final concentration of 20 mM). If glucose is added, for example, metal reduction often occurs, but the pH of the environment is often rendered so low that this reduction is due to acid catalysis rather than direct reduction. In any case, the respiratory metal-reducers can be easily outcompeted by the glucose-fermenting bacteria.

## Metal Oxides

Metal oxides are prepared as described above and used as suspensions in the soft agar. Soluble forms of iron can be used, but it is impossible to visualize zones of reduction because of the diffusibility of the metal chelates.

## Poisons or Inhibitors

Poisons or inhibitors are added to the enrichment mixture at the appropriate concentration in solid form just before mixing the agar and sample. Mercuric chloride is added as a general poison at a final concentration of 0.15% (by weight). Sodium molybdate is added at a final concentration of 25 mM. Especially in marine samples, sodium molybdate can be used to inhibit the abundant sulfate-reducers, which, with lactate, may dominate the metal chemistry and reduction via the production of sulfide.

To this suspension is added an equivolume of 1.5% agar to give a final concentration of 0.1% tryptone and 0.05% yeast extract. The agar is premelted and held at 45°C, added to the water sample, gently mixed to suspend the cells and metal oxide, and placed in a cool water bath to harden quickly. Any kind of bottle can be used, but bottles that are easy to manipulate and store, especially for field or shipboard work, are recommended. We have routinely used either sterilized scintillation vials or anaerobic tubes with crimp tops. When scintillation vials are used, the surface is covered with a thin layer of mineral oil to maintain anaerobic conditions and retard fungal growth. The vials are incubated at the desired temperature and examined at intervals of 1–2 days for 1 week and at weekly intervals for 1 month.

## Data Analysis

Vials are scored at daily intervals for up to 2 weeks and then scored at weekly intervals until no further changes are noted. Although the scoring is somewhat qualitative, it is rather easy to judge samples that are reduced at different levels, and we have adopted a method of estimating reduction from 0% (no reduction) to 100% reduction at intervals of approximately 20%. Thus, 1 (+) indicates 20% reduction, etc. Such data can be used to show relative levels of reduction in samples as a function of time, thus giving a good representation of the zones where such organisms are abundant or active. An example of such data analysis (reduction of undiluted samples as a function of time) is shown in Figure 3-3. These data are from the Black Sea [56], where after 1 month, the entire suboxic zone showed complete reduction of $MnO_2$. However, after 1 week, zones of re-

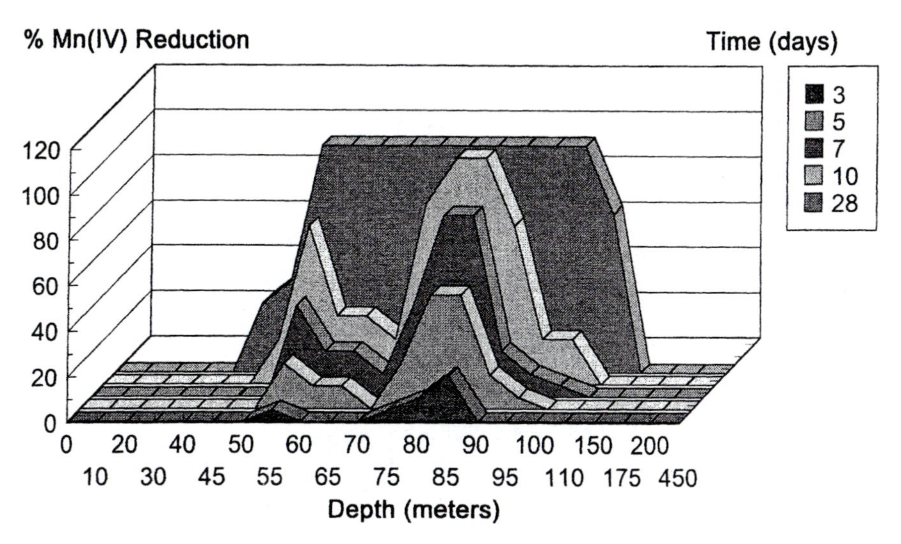


Figure 3-3 Black Sea Enrichment data [56], in which samples were obtained by pumping at intervals of 5 meters and enrichments established at each interval. These enrichments were scored visually for manganese reduction and the values plotted. Early in the process, zones of potential activity of manganese reduction were apparent. Such methods may be very valuable for field studies of stable stratified environments, where detailed activity measurements could subsequently be focused on these zones of activity. If the enrichments are allowed to proceed too long before scoring, as shown by the points after 1 month, this information is lost as all samples are reduced to near 100%. These studies would be improved with the addition of dilutions, as discussed in Figure 3-2.

duction were clearly visible in enrichment cultures from 50 meters depth (denitrification zone) and 85 meters depth (Mn reduction zone) of the water column. These data were consistent with direct plating results, which showed very high numbers of metal-reducing bacteria at both of these interfaces and very low numbers in other zones [56].

This approach can be combined with a most probable number (MPN)-like dilution method, in which a series of dilutions are set up and scored for reduction at various times. Using such an approach, it is possible to estimate the relative numbers of metal-reducers present in different sedimentary environments.

### Isolation of Pure Cultures

From enrichments it is possible to isolate pure cultures of metal-reducers. Using the media described here and the metal oxides, one can either continue enrichments or try to isolate metal-reducers directly on plates, using either solid or soft agar.

### Soft Agar Overlay Cultures

It is easy to prepare soft agar overlay cultures using either solid or colloidal $MnO_2$. An agar basal medium (25 ml) containing all needed nutrients is first poured and allowed to harden. Then, 4 ml of soft agar (0.8% by weight) containing nutrients and metal oxides is mixed with 1 ml of sample from the enrichment culture (or with a dilution of the enrichment culture sample). This mixture is poured over the surface of the base agar to give a thin layer of soft agar containing metal oxide and bacteria. Because respiratory metal-reducers require contact with the metal oxides, they do not form discrete colonies on such a medium, but instead form zones of clearing in the top agar due to their requirement for surface contact. Alternatively, the top agar medium can be made with Fe(III) citrate as the electron acceptor, and the substrate will diffuse to the bacterial colonies, which will result in zones of clearing of the iron citrate around the colonies.

## GROWTH IN THE LABORATORY: GENERAL CONSIDERATIONS

Several important variables need to be controlled or measured during experiments with metal-reducing bacteria: pH, oxygen level, electron acceptors, and electron donors. For electron acceptors, the methods of preparation will dictate much about how the experiments are performed. For example, if complexed soluble electron acceptors are used, it is easy to filter the cells and count using epifluorescence or even to follow optical density. When solid forms are utilized, counting by epifluorescence microscopy is the method of choice, although because the cells are attached to the solid oxides, obtaining good growth data is time-consuming and the data are of limited precision. Another method of possible utility involves a direct measurement of total cell carbon, using a CHN (carbon, hydrogen, nitrogen) analyzer system [48]. This system converts all organic carbon to $CO_2$, and measures the total $CO_2$ of filtered cells. Using this method, cells bound to metal oxides or clay particles can be monitored accurately in terms of total cell carbon [48].

Although it is always tempting to add high levels of carbon substrate to stimulate rapid reduction, we have had poor success using glucose or other fermentable substrates in initial enrichments. Other organisms (fermentative microbes) almost always outcompete the metal-reducers. Another important consideration is that hydrogen is an effective electron donor for metal respiration by both the *Shewa-*

*nella* and *Geobacter* groups [38, 40]. Therefore, when studying carbon metabolism in an anaerobic chamber containing a hydrogen atmosphere, hydrogen utilization needs to be corrected for or considered.

As iron or manganese proceeds in culture, protons are consumed and pH rises, so the pH must be buffered or measured or both. Organic buffers such as HEPES are often used, depending on the pH range desired, and are usually adequate for controlling pH with minimal effects on metal chemistry. Phosphate and bicarbonate buffers have also been used, although the possibility of interaction of these buffers with the metals must always be considered.

Molecular oxygen is a strong inhibitor dissimilatory metal reduction is *S. putrefaciens* [53] and is lethal to strict anaerobes such as *G. metallireducens* [41]. Thus, it must be controlled and maintained at very low levels in laboratory experiments. For *S. putrefaciens*, no special precautions are necessary, as it is tolerant of oxygen, and any standard method for obtaining anaerobic conditions can be used. We prefer to use an anaerobic chamber for routine studies because of the ease of manipulation. However, since most anaerobic chambers use a hydrogen-palladium catalyst for oxygen removal, this method is not useful if hydrogen-free growth is required. For these cases, we use stoppered tubes or serum vials with the headspace purged with inert gas (nitrogen or argon).

## CONCLUSION

This field is in its infancy, and although specific media and instructions are presented in this chapter, perhaps the best tool that can be taken to the field is some ingenuity and an open mind. It is of great importance to use freshly prepared hydrous metal oxides wherever feasible, and we strongly recommend this approach. As for the details of specific growth media, the methods presented here work well for a few organisms, but the use of other approaches may well pay large dividends. Therefore, we strongly encourage the use of new media and culture methods. In studies of metal-reducing bacteria, the field may well change as rapidly as new methods and approaches are utilized.

## REFERENCES


1. Aller, R.C. 1890. Bioturbation and manganese cycling in hemipelagic sediments. Phil Trans. R. Soc. Lond. A331:51–68.
2. Aller, R.C., Aller, J., Blair, N., Mackin, J., and Rude, P. 1991. Biogeochemical processes in Amazon shelf sediments. Oceanography, April, 27–32.
3. Arnold, R., DiChristina, T., and Hoffman, M. 1986. Inhibitor studies of dissimilative Fe(III) reduction by *Pseudomonas* sp. 200 ("*Pseudomonas ferrireductans*"). Appl. Environ. Microbiol. 52:281–289.
4. Arnold, R., DiChristina, T., and Hoffman, M. 1988. Reductive dissolution of Fe(III) oxides by *Pseudomonas* sp. 200. Biotechnol. Bioengineer. 32:1081–1096.
5. Arnold, R., Hoffman, M., DiChristina, T., and Picardal, F. 1990. Regulation of dissimilative Fe(III) reduction in *Shewanella putrefaciens*. Appl. Environ. Microbiol. 56: 2811–2817.
6. Atkinson, R.J., Posner, A.M., and Quirk, J.P. 1967. Adsorption of potential-determining ions at the ferric oxide-aqueous electrolyte interface. J. Phys. Chem. 71:550–558.
7. Balistrieri, L.S. and Murray, J.W. 1982. The surface chemistry of $MnO_2$ in major ion sea water. Geochim. Cosmochim. Acta. 46:1041–1052.
8. Bell, P.E., Mills, A.L., and Herman, J.S. 1987. Biogeochemical conditions favoring magnetite formation during anaerobic iron reduction. Appl. Environ. Microbiol. 53: 2610–2616.

9. Brewer, P.G. and Spencer, D.W. 1971. Colorimetric determination of manganese in anoxic waters. Limnol. Oceanogr. 16:107–110.
10. Brunaurer, S., Emmett, P.H., and Teller, E. 1938. Adsorption of gases in multimolecular layers. J. Amer. Chem. Soc. 60:309–319.
11. Burdige, D.J. 1993. The biogeochemistry of manganese and iron reduction in marine sediments. Earth-Sci. Rev. 35:249–284.
12. Burdige, D.J. and Nealson, K.H. 1986. Chemical and microbiological studies of sulfide-mediated manganese reduction. Geomicrobiol. J. 4:361–387.
13. Burdige, D.J., Dhakar, S.P., and Nealson, K.H. 1992. Effects of manganese oxide mineralogy on microbial and chemical manganese reduction. Geomicrobiol. J. 10:27–48.
14. Burns, R.G. and Burns, V.M. 1979. Manganese minerals. In Marine Minerals (Burns, R.G., Ed.), pp. 1–46. Mineralogical Society of America, Washington, D.C.
15. Caccavo, F., Blakemore, R.P., and Lovley, D.R. 1992. A hydrogen-oxidizing, Fe(III)-reducing microorganism from the Great Bay estuary, New Hampshire. Appl. Environ. Microbiol. 58:3211–3216.
16. Canfield, D.E., Jorgensen, B.B., Fossing, H., Glud, R., Gundersen, J., Ramsing, N.B., Thamdrup, B., Hansen, J.W., Nielsen, L.B., and Hall, P.O.J. 1993. Pathways of organic carbon oxidation in three continental margin sediments. Mar. Geol. 133:27–40.
17. Canfield, D.E., Thamdrup, B., and Hansen, J.W. 1993. The anaerobic degradation of organic matter in Danish coastal sediments: iron reduction, manganese reduction, and sulfate reduction. Geochim. Cosmochim. Acta. 57:3867–3885.
18. Chiswell, B. 1993. A new method for the evaluation of the oxidizing equivalents of manganese in surface freshwaters. Talanta 40:533–540.
19. Chiswell, B. and Mokhtar, M.B. 1986. The speciation of manganese in freshwaters. Talanta 33:669–677.
20. Davison, W. 1993. Iron and manganese in lake. Earth-Sci. Rev. 34:119–163.
21. Dion, H.G. and Mann, P.J.G. 1946. Three-valent manganese in soils. J. Agric. Sci. 36: 239.
22. Godtfredsen, K.L. and Stone, A.T. 1994. Solubilization of manganese dioxide-bound copper by naturally occurring organic compounds. Environ. Sci. Technol. 28:1450–1458.
23. Goto, K., Komatsu, T., and Furukawa, T. 1962. Rapid colorimetric determination of manganese in waters containing iron. Anal. Chem. Acta. 27:331–334.
24. Ingall, E. and Jahnke, R. 1994. Evidence for enhanced phosphorous regeneration from marine sediments overlain by oxygen depleted water. Geochim. Cosmochim. Acta. 58: 2571–2576.
25. Jenne, E.A. 1977. Trace element sorption by sediments and soil sites and processes. In Symposium on Molybdenum in the Environment (Chappel, W. and Petersen, K., Eds.), pp. 425–453. Marcel Dekker, New York.
26. Kazumi, J., Haggblom, M.M., and Young, L.Y. 1995. Degradation of monochlorinated and nonchlorinated aromatic compounds under iron-reducing conditions. Appl. Environ. Microbiol. 61:4069–4073.
27. Klausen, J., Troeber, S.P., Haderlein, S.B., and Schwarzenbach, R.P. 1995. Reduction of substituted nitrobenzenes by Fe(II) in aqueous mineral suspensions. Env. Sci. Technol. 29:2396–2404.
28. Komadel, P. and Stucki, J.W. 1988. Quantitative assay of minerals for Fe(II) and Fe(III) using 1,10-phenanthroline: III. A rapid photochemical method. Clays Clay Min. 36:379–381.
29. Kostka, J.E. and Nealson, K.H. 1995. Dissolution and reduction of magnetite by bacteria. Envir. Sci. Technol. 29:2535–2540.
30. Kostka, J.E., Luther, G.W., and Nealson, K.H. 1995. Chemical and biological reduction of Mn(III)-pyrophosphate complexes: potential importance of dissolved Mn(III) as an environmental oxidant. Geochim. Cosmochim. Acta. 59:885–894.
31. Kostka, J.E., Nealson, K.H., Wu, J., and Stucki, J.W. 1996. Reduction of the structural Fe(III) in smectite by a pure culture of the Fe-reducing bacterium. *Shewanella putrefaciens* strain MR-1. Clays Clay Min. 44:522–529.
32. Krumbein, W.E. and Altmann, H.J. 1973. A new method for the detection and enumeration of manganese oxidizing and reducing microorganisms. Helgol. Wiss. Meeresunters. 25:347–356.

33. Lee, Y. and Tebo, B.M. 1994. Cobalt(II) oxidation by the marine Mn(II) oxidizing *Bacillus* sp. strain SG-1. Appl. Environ. Microbiol. 60:2949–2957.
34. Lovley, D.R. 1987. Organic matter remineralization with the reduction of ferric iron: a review. Geomicrobiol. J. 5:375–399.
35. Lovley, D.R. 1991. Dissimilatory Fe and Mn reduction. Microbiol. Rev. 55:259–287.
36. Lovley, D.R. 1993. Dissimilatory metal reduction. Annu. Rev. Microbiol. 47:263–290.
37. Lovley, D. and Lonergan, D.J. 1990. Anaerobic oxidation of toluene, phenol, and *p*-cresol by the dissimilatory iron reducing organism, GS-15. Appl. Environ. Microbiol. 56:1858–1864.
38. Lovley, D.R. and Phillips, E.J. 1988. Novel mode of microbial energy metabolism: organic carbon oxidation coupled to dissimilatory reduction of iron or manganese. Appl. Environ. Microbiol. 51:683–689.
39. Lovley, D.R., Baedecker, M., Lonergan, D., Cozzarelli, I., Phillips, E., and Siegel, D. 1989. Oxidation of aromatic contaminants coupled to microbial iron reduction. Nature 339:297–299.
40. Lovley, D.R., Phillips, E.J., and Lonergan, D.J. 1989. Hydrogen and formate oxidation coupled to dissimilatory reduction of iron or manganese by *Alteromonas putrefaciens*. Appl. Environ. Microbiol. 55:700–706.
41. Lovley, D.R., Giovannoni, S.J., White, D.C., Champine, J.E., and Phillips, E. 1993. *Geobacter metallireducens* gen. nov. sp. nov., a microorganism capable of coupling the complete oxidation of organic compounds to the reduction of iron and other metals. Arch. Microbiol. 159:336–344.
42. Lovley, D.R., Roden, E., Phillips, E.J., and Woodward, J.C. 1993. Enzymatic iron and uranium reduction by sulfate reducing bacteria. Mar. Geol. 113:41–53.
43. Lovley, D.R., Stolz, J.F., Nord, G.L., and Phillips, E.J. 1987. Anaerobic production of magnetite by a dissimilatory iron-reducing microorganism. Nature 330:252–254.
44. MacDonell, M.T. and Colwell, R.R. 1985. Phylogeny of *Vibrionaceae*, and recommendation for two new genera, *Listonella and Shewanella*. Syst. Appl. Microbiol. 6: 171–182.
45. Mann, J.G. and Quastel, J.H. 1946. Manganese metabolism in soils. Nature 158: 154–156.
46. Morgan, J.J. 1967. Chemical equilibria and kinetic properties of manganese in natural waters. In Principles and Applications of Water Chemistry (Faust, S.D. and Hunter, J.V., Eds.), pp. 561–626. Wiley, New York.
47. Morgan, J.J. and Stumm, W. 1964. Colloid chemical properties of manganese dioxide. J. Colloid. Interface Sci. 19:347–359.
48. Moser, D.P., Brozowski, J.R., and Nealson, K.H. 1996. Elemental analysis for biomass determination in the presence of insoluble substrates. J. Microbiol. Methods 26:271–278.
49. Murray, J.W. 1974. The surface chemistry of hydrous manganese dioxide. J. Colloid. Interface Sci. 46:357–371.
50. Murray, J.W. 1979. Iron oxides. In Marine Minerals (Burns, R.G., Ed.), pp. 47–98. Mineralogical Society of America, Washington, D.C.
51. Murray, J.W., Balistrieri, L.S., and Paul, B. 1984. The oxidation state of manganese in marine sediments and ferromanganese nodules. Geochim. Cosmochim. Acta. 48: 1237–1247.
52. Murray, J.W., Dillard, J.G., Giovanoli, R., Moers, H., and Stumm, W. 1985. Oxidation of Mn(II): initial mineralogy, oxidation and aging. Geochim. Cosmochim. Acta. 49: 463–470.
53. Myers, C. and Nealson, K.H. 1988. Bacterial manganese reduction and growth with manganese oxide as the sole electron acceptor. Science 240:1319–1321.
54. Myers, C.R. and Nealson, K.H. 1988. Microbial reduction of manganese oxides: interactions with iron and sulfur. Geochim. Cosmochim. Acta. 52:751–756.
55. Nealson, K.H. and Myers, C.R. 1992. Microbial reduction of manganese and iron: new approaches to carbon cycling. Appl. Environ. Microbiol. 58:439–443.
56. Nealson, K.H. and Saffarini, D.A. 1995. Iron and manganese in anaerobic respiration: environmental significance, physiology, and regulation. Annu. Rev. Microbiol. 48: 311–343.

57. Nealson, K.H., Myers, C.R., and Wimpee, B. 1991. Isolation and identification of manganese reducing bacteria, and estimates of microbial manganese reducing potential in the Black Sea. Deep Sea Res. 38:S907–S920.
58. Nealson, K.H., Saffarini, D.A., and Moser, D. Anaerobic respiration of *Shewanella putrefaciens*: potential use of solid electron acceptors for pollutant removal and bioremediation in anoxic environments. In Bioremediation of Contaminants in Soils and Sediments (DeLuca, M., Ed.). Rutgers University Press, New Brunswick, NJ (submitted for publication).
59. Parks, G.A. 1965. The isoelectric points of solid oxides, solid hydroxides, and aqueous hydrous complex systems. Chem. Rev. 65:177–198.
60. Perez-Benito, J.F., Brillas, E., and Pomplana, R. 1989. Identification of a soluble form of colloidal Mn(IV). Inorg. Chem. 28:390–392.
61. Petrovskis, E.A., Vogel, T.M., and Adriaens, P. 1994. Effects of electron acceptors and donors on transformation of tetrachloromethane by *Shewanella putrefaciens* MR-1. FEMS Microbiol. Lett. 121:357–364.
62. Picardel, F.W., Arnold, R.G., Couch, H., Little, A.M., and Smith, M.E. 1993. Involvement of cytochromes in the anaerobic biotransformation of tetrachloromethane by *Shewanella putrefaciens* 200. Appl. Environ. Microbiol. 59:3763–3770.
63. Popp, J.L., Kalyanaraman, B., and Kirk, T.K. 1990. Lignin peroxidase oxidation of Mn(II) in the presence of veratryl alcohol, malonic or oxalic acid and oxygen. Biochemistry 29:10475–10480.
64. Roden, E.E. and Lovley, D.R. 1993. Dissimilatory Fe(III) reduction by the marine microorganism, *Desulfuromonas acetoxidans*. Appl. Environ. Microbiol. 59:734–742.
65. Semple, K. and Westlake, D.W.S. 1987. Characterization of iron-reducing *Alteromonas putrefaciens* strains from oil field fluids. Can. J. Microbiol. 33:366–371.
66. Schwertmann, U. and Cornell, R.M. 1991. Iron Oxides in the Laboratory: Preparation and Characterization. Weinheim, New York.
67. Schwertmann, U. and Fitzpatrick, R.W. 1992. Iron minerals in surface environments. In Biomineralization Processes of Fe and Mn (Skinner, H.C.W. and Fitzpatrick, R.W. Eds.), pp. 1–7. Catena Verlag.
68. Shau, Y., Peacor, D.R., and Essene, E.J. 1993. Formation of magnetic single-domain magnetite in ocean ridge basalts with implications for sea floor magnetism. Science 261:343–346.
69. Singh, S.K. and Subramanian, V. 1984. Hydrous Fe and Mn oxides—scavengers of heavy metals in the aquatic environment. Crit. Rev. Environ. Contr. 14:33–90.
70. Stookey, L.L. 1970. Ferrozine—a new spectrophotometric reagent for iron. Anal. Chem. 42:779–781.
71. Stucki, J.W. 1988. Structural iron in smectites. In Iron in Soils and Clay Minerals (Stucki, J.W., Goodman, B.A., and Schwertmann, U., Eds.), pp. 625–675. D. Reidel, Norwell, MA.
72. Stucki, J.W. and Anderson, W.L. 1981. The quantitative assay of minerals for Fe(II) and Fe(III) using 1,10 phenanthroline: sources of variability. Soil Sci. Soc. Am. J. 45: 633–637.
73. Stucki, J.W., Komadel, P., and Wilkinson, H.T. 1987. Microbial reduction of structural Fe(III) in smectites. Soil Sci. Soc. Am. J. 51:1663–1665.
74. Stumm, W. and Morgan, J.J. 1981. Aquatic Chemistry, 2nd ed. Wiley-Interscience, New York.
75. Thamdrup, B., Glud, R.N., and Hansen, J.W. 1994. Manganese oxidation and in situ manganese fluxes from a coastal sediment. Geochim. Cosmochim. Acta. 58:2577–2585.
76. Thouveny, N., deBeaulieu, J.-L., and Bonifay, E. 1994. Climate variations in Europe over the past 140 Kyr deduced from rock magnetism. Nature 371:503–506.
77. Wu, J., Roth, C.B., and Low, P.F. 1988. Biological reduction of structural Fe in sodium-nontronite. Soil Sci. Soc. Am. J. 52:295–296.
78. Zhender, A.J.B. and Stumm, W. 1988. Geochemistry and biogeochemistry of anaerobic habitats. In Biology of Anaerobic Microorganisms (Zehnder, A.J.B., Ed.), pp.1–38. John Wiley & Sons, New York.

4

MICHEL ARAGNO

# The Aerobic, Hydrogen-Oxidizing (Knallgas) Bacteria

## DEFINITION

The hydrogen-oxidizing bacteria, often called the *knallgas bacteria*, are a physiologic group of aerobic, chemolithoautotrophic bacteria. Although they comprise bacteria from several different taxonomic units, they are all able to recover part of the reducing power and of the energy resulting from the oxidation of molecular hydrogen (Fig. 4-1). This recovery of energy and reducing power allows them an autotrophic metabolism: They derive most, if not all, of their cellular carbon from $CO_2$. Some of the electrons resulting from the oxidation of the donor reduce $NAD^+$; this reduction occurs through reverse electron flow, which implies respiratory chain components and ATP. A direct reduction of $NAD^+$ through a $H_2$:NAD oxidoreductase ($H_2$ dehydrogenase) has so far been described in only three species of hydrogen-oxidizing bacteria. The remaining electrons enter the respiratory chain and thus allow ATP synthesis through oxidative phosphorylation. The final acceptor is $O_2$, which in some cases can be replaced by nitrate (denitrification). $CO_2$ fixation occurs in general through the reductive ribulose-bisphosphate cycle. In *Hydrogenobacter* and *Aquifex*, it occurs through the reductive tricarboxylic cycle [114]. Hydrogen-oxidizing bacteria have been the subjects of several reviews [5, 7, 8, 20, 64, 94, 100, 102–104] and a monograph [130].

Except for a few species, the knallgas bacteria so far studied are facultative chemolithoautotrophs; they can also grow under completely heterotrophic conditions with one of a variety of organic substrates as both electron donor and carbon source. Most hydrogen bacteria show a higher growth rate on organic substrates than under autotrophic conditions. For heterotrophic growth, the basic enzymes and metabolic pathways of most knallgas bacteria studied so far are quite similar to those of other aerobic heterotrophic bacteria. Only the combination of the two faculties—to accept hydrogen as electron donor and to use carbon dioxide as the sole carbon source—is biologically unique to this physiologic group.

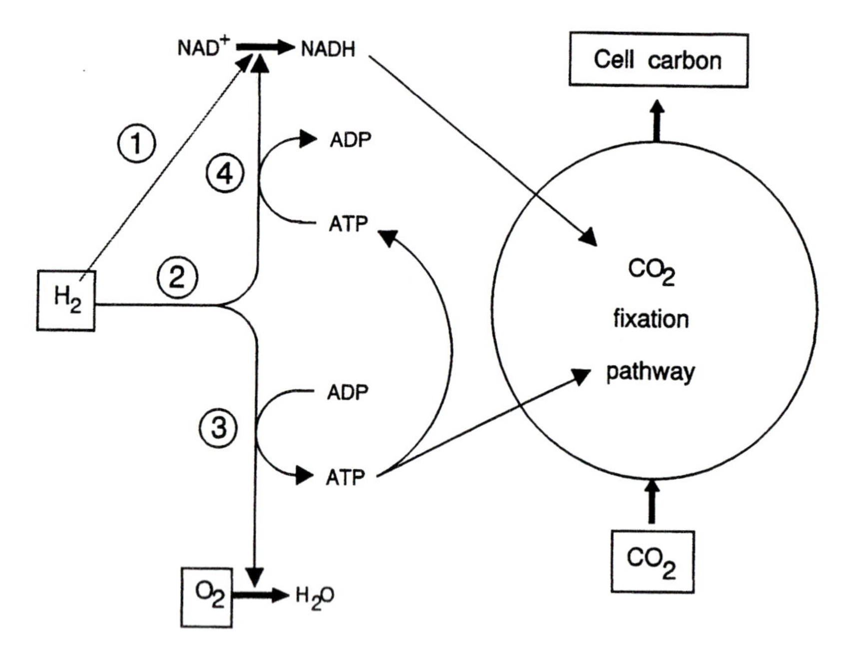


Figure 4-1 Principle of the aerobic, hydrogen-oxidizing chemolithoautotrophic metabolism of knallgas bacteria. *1*. Cytoplasmic; $NAD^+$ reducing hydrogenase (hydrogen dehydrogenase, hydrogen:$NAD^+$ oxidoreductase; C 1.12.1.1). *2*. Membrane-bound, not $NAD^+$ reducing hydrogenase. *3*. Respiratory chain. *4*. Reverse, ATP dependent electron transport.

## DIVERSITY OF KNALLGAS BACTERIA

Knallgas bacteria were once considered to form a taxonomic group and were grouped within the special genus *Hydrogenomonas*. After a large number of strains had been studied and described, it became clear that this genus was heterogeneous and that in many cases species should be moved to preexisting genera of heterotrophic bacteria, such as *Pseudomonas*, *Alcaligenes*, *Aquaspirillum*, and so on. The genus *Hydrogenomonas* was thus rejected [31]. Morphologic studies [9, 22, 123], as well as the isolation of hydrogen-oxidizing bacteria clearly belonging to other taxonomic groups, confirmed this diversity.

At present (Table 4-1), gram-positive and gram-negative hydrogen bacteria are known. There are sporeformers and non-sporeformers; mesophiles and thermophiles; immotile, polarly, or peritrichously flagellated organisms; cocci; rods; spirilla; and filamentous forms. One group of thermophilic hydrogen bacteria (comprising the genera *Hydrogenobacter* and *Aquifex*) forms the deepest branching so far described in the Domain Bacteria [54, 88, 116]. The other types belong to the gram-positive and Proteobacteria branches.

So far, all the bacteria isolated that can grow with carbon monoxide as a sole carbon and energy source have also turned out to be knallgas bacteria [66, 129, 130]. They are able to grow as well or even faster with carbon dioxide and hydrogen as with carbon monoxide.

Several knallgas bacteria possess another metabolic alternative. They can also use reduced sulfur compounds (e.g., elemental sulfur or thiosulfate) as electron donors [18, 44], and even grow mixotrophically, using both types of electron donors simultaneously.

Most dinitrogen-fixing bacteria contain a membrane-bound hydrogenase that recycles the molecular hydrogen produced in the nitrogenase reaction, so they

Table 4-1 Distinctive Characteristics of Mesophilic Knallgas Bacteria

| | Motility | Gram stain | Pigment | Shape | D-Galactose | D-Fructose | D-Xylose | L-Rhamnose | Sucrose | Lactose | Trehalose | Cellobiose | Mannitol | Adipate | 2-Oxoglutarate | Denitrification | $N_2$-fixation | Gelatin Hydrolysis |
|---|---|---|---|---|---|---|---|---|---|---|---|---|---|---|---|---|---|---|
| Gram-negative species | | | | | | | | | | | | | | | | | | |
| *Acidovorax delafieldii* | + | – | – | r | + | + | – | – | – | – | – | – | + | + | + | – | – | d |
| *Acidovorax facilis* | + | – | – | r | + | + | – | – | – | – | – | – | + | – | – | – | – | + |
| *Alcaligenes xylosoxydans* | + | – | – | r | | d | + | – | – | – | – | – | – | + | – | – | – | – |
| *[Alcaligenes] eutrophus* | + | – | – | r | – | + | – | – | – | – | – | – | – | + | + | + | – | – |
| *[Alcaligenes] latus* | + | – | – | br | – | + | – | – | + | – | d | – | – | – | + | – | + | + |
| *Ancylobacter aquaticus* | – | – | – | cu | | | | – | – | – | – | | | – | | + | – | |
| *[Aquaspirillum] autotrophicum* | + | – | – | s | – | – | – | – | – | – | – | – | – | | + | – | – | – |
| *Azospirillum lipoferum* | + | – | – | s | d | + | + | – | – | – | | – | + | | | + | + | – |
| *Bradyrhizobium japonicum* | | – | – | r | + | + | + | – | – | – | – | – | + | | | + | + | – |
| *Derxia gummosa* | | – | | | + | + | | d | d | d | d | | + | | | d | + | |
| *Hydrogenophaga flava* | + | – | y | r | + | + | – | + | + | | + | + | + | – | – | – | – | – |
| *Hydrogenophaga palleronii* | + | – | y | r | – | – | – | – | – | – | – | – | – | d | – | – | – | – |
| *Hydrogenophaga pseudoflava* | + | – | y | r | + | + | + | d | + | d | + | + | + | + | d | + | d | – |
| *Hydrogenophaga taeniospiralis* | + | – | y | r | + | + | + | + | + | – | – | + | + | – | – | + | – | + |
| *Paracoccus denitrificans* | – | – | – | co | d | + | – | d | + | – | + | – | + | – | + | + | – | – |
| [*Pseudomonas*] *carboxyhydrogena* | + | – | – | cu | – | d | – | | – | – | – | – | – | | | – | – | – |
| [*Pseudomonas*] *hydrogenovora* | + | – | ? | r | + | + | | + | d | + | – | + | – | | | – | – | + |
| [*Pseudomonas*] *saccharophila* | + | – | – | r | + | + | d | – | + | – | d | + | – | + | – | – | – | + |
| *Variovorax paradoxus* | d | – | y | r | + | + | + | d | – | – | – | d | + | + | + | – | – | – |
| *Xanthobacter agilis* | + | –v | y | r | – | – | – | – | – | – | – | – | – | + | – | – | + | – |
| *Xanthobacter autotrophicus* | d | –v | y | pl | – | d | – | – | d | – | – | d | – | d | – | – | + | – |
| *Xanthobacter flavus* | d | –v | y | pl | – | d | – | – | – | – | d | – | – | + | + | – | + | – |
| Gram-positive species | | | | | | | | | | | | | | | | | | |
| *Anthrobacter* sp. | – | + | – | pl | – | + | + | – | + | | + | – | | | | – | – | + |
| *Amycolata autotrophica* | – | + | ? | f | | + | + | | + | – | | + | | | | – | – | – |
| *Mycobacterium gordonae* | – | + | yo | cr | – | sl | – | – | – | – | – | – | | | | sl | – | – |
| *Mycobacterium* sp. | – | + | y | cr | + | 1 | d | d | – | – | d | – | + | | | – | – | – |
| *"Nocardia opaca"* | – | + | p | f | + | + | d | d | + | d | | d | + | | | – | – | – |

*Abbreviations:* +, positive; –, negative; d, variable; v, gram variable; y, yellow; yo, yellow-orange; p, pink; r, rods; br, broad rods; s, spirilla; co, cocci; cu, curved cells; pl, pleomorphic rods; f, filamentous; cr, coryneform; sl, slow growth.

already possess part of the enzyme system required for lithoautotrophic life. The ability to fix $CO_2$ and to grow as a knallgas bacterium was discovered in many strains of previously described heterotrophic dinitrogen fixers (for detailed references see [10]).

## HABITATS AND ECOLOGY

Knallgas bacteria can be isolated from a great variety of habitats, soils, freshwater, superficial layers of sediment, activated sludge, landfill leachate, geothermal manifestations, and the like. These habitats show great variation in the origin and concentration of their $H_2$.

The ability of most knallgas bacteria to use both organic and inorganic substrates enables them to occupy special niches and may explain their wide distribution. Hydrogen and carbon dioxide are produced in nature in ample amounts, and those microorganisms that can use them as growth substrates have a selective advantage. It is reasonable to assume, therefore, that knallgas bacteria would be favored in those habitats in which hydrogen, carbon dioxide, and oxygen are simultaneously available.

## $H_2$ Production in Natural Environments

In natural environments, $H_2$ can be produced either biologically or chemically. Several biologic reactions may result in the formation of $H_2$ concentrations allowing the growth of knallgas bacteria. Fermentative processes are one such mechanism, particularly the so-called formate-hydrogen lyase reaction that is catalyzed by numerous facultatively anaerobic bacteria:

$$HCOOH = CO_2 + 2e + 2H^+ \qquad E_0' = 0.5\ V$$

$$2e^- + 2H^+ = H_2 \qquad E_0' = -0.414\ V$$

The overall reaction is exothermic, so high $H_2$ concentrations may occur in this way. Other reactions, such as the anaerobic transformation of short chain fatty acids and alcohols to acetate, $CO_2$, and $H_2$ by "proton reducers" or fermentative reactions leading to acetate via acetyl-phosphate, are only possible under very low $H_2$ partial pressures, far below the threshold allowing growth of knallgas bacteria.

Hydrogen production during dinitrogen fixation also results in concentrations of $H_2$. The nitrogen-fixing enzyme system (nitrogenase) is "leaky": Part of the electrons activated reduce protons instead of dinitrogen and thus produce $H_2$. $H_2$ seems to be an unavoidable byproduct of dinitrogen fixation. It is not even suppressed by increased $N_2$ pressures [117]. In legume nodules, for example, about 30% of the electron flow is diverted to reduce protons [40, 41, 107]. Many $N_2$-fixing organisms are able to recycle part of the $H_2$ produced by means of an uptake hydrogenase [97], which enables the bacteria not only to recover energy but also to consume the oxygen that diffuses into the cell. This recycling of $H_2$ may serve to protect the highly oxygen-sensitive nitrogenase from inactivation by oxygen [21]. $H_2$ production during $N_2$ fixation occurs in oxic environments as well.

Several chemical mechanisms have been suggested to explain the occurrence of $H_2$ in geothermal gases, including the reaction between water and silicon radicals at high temperature [67], such as

$$Si + H_2O \rightarrow SiOH + H$$

$$H + H \rightarrow H_2$$

or the reduction of water by hot (over 800°C) ferrous rocks [30]:

$$2FeO + H_2O \rightarrow Fe_2O_3 + H_2$$

$$2Fe_3O_4 + H_2O \rightarrow 3Fe_2O_3 + H_2$$

## $H_2$ Concentrations in Natural Environments

$H_2$ is a trace gas in the atmosphere. Its partial pressure amounts to approximately 55 mPa in the troposphere, corresponding to a mixing ratio of 0.55 ppmv [23, 27, 108] and a concentration in water of 0.41 nM at 20°C. In soils, the $H_2$ partial pressure is normally much lower—less than 1 mPa [112]. These concentrations are far below the usual threshold allowing knallgas bacteria to take up hydrogen; that is, 1.3–8.3 ppmv [28]. However, soils covered with $N_2$-fixing legumes often show much higher $H_2$ concentrations and even emit $H_2$ into the troposphere during the vegetation period [24, 25]. It is therefore possible for relatively high $H_2$ concentrations to occur locally; for example, in the immediate vicinity of legume root nodules [71, 110] and perhaps also in the $N_2$-fixing rhizosphere of grasses.

In freshwater, the $H_2$ concentration is generally low. Contrary to early suppositions [70], the lowest concentrations are found in anoxic waters, whereas concentrations above the hydrogen uptake threshold of knallgas bacteria were measured in the oxic, metalimnic layers of a holomictic, eutrophic lake. These concentrations reached a maximum early in the morning and a minimum at the end of the afternoon [27]. This fluctuation could possibly be due to a nocturnal dinitrogen fixation by non-heterocystous cyanobacteria (e.g., *Oscillatoria* spp) that were abundant in the lake studied.

In some exceptional situations, such as in landfills, in tanks containing organic waste, and in other masses of decomposing organic materials, much higher concentrations of molecular hydrogen have been measured. During the transition between aerobic and anaerobic degradation, there is a fermentative phase in which $H_2$ production occurs, and mixing ratios of up to 30% may be measured [36, 37]. Such gases eventually mix with air—for instance, in a landfill's superficial layer—which could pose explosion hazards.

In composts, during the active thermogenic phase, significant $H_2$ concentrations (mixing ratios above 1%) were sometimes measured (T. Beffa, personal communication). These concentrations could be explained by the frequent transitions between oxic and anoxic conditions that occur in actively composting material.

Molecular hydrogen occurs frequently in the gas phase of geothermal manifestations. It can reach concentrations of the order of several percent v/v of the total dry gas; for example, 4.6–4.8% in *S. Federigo solfatara* in Tuscany, Italy [17, 29].

## $H_2$ Consumption in Natural Environments

Two types of $H_2$-consuming activities have been observed in natural environments: (1) a high $K_m$, high-threshold activity, similar to the activity of pure cultures of knallgas bacteria, and (2) a low $K_m$, low-threshold activity tentatively attributed to free enzymatic activities [26, 27, 108]. These activities showed distinct pH and temperature optima [109]. In geothermal springs, only a high $K_m$, high-threshold activity was detected [29].

## Occurrence of Knallgas Bacteria in Natural and Artificial Habitats

Knallgas bacteria occur even in environments where the hydrogen concentration measured in the water mass or in the soil atmosphere is lower than the threshold allowing hydrogen uptake. As they are facultatively chemolithoautotrophic versatile heterotrophs, most knallgas bacteria can live at the expense of organic compounds in soils, as can their purely heterotrophic neighbors. However, the occurrence of a direct hydrogen transfer between H-producing organisms (e.g., dinitrogen fixers) and knallgas bacteria cannot be excluded. In freshwater lakes the distribution of knallgas bacteria is irregular, only partially following the distribution of molecular hydrogen [27]. Although several species of knallgas bacteria are able to use reduced sulfur compounds as alternative electron sources [18, 44], we never observed any distinct enrichment of knallgas bacteria at the oxic-anoxic interface where $H_2S$ and $O_2$ mix, neither in a holomictic eutrophic lake (Le Loclat, canton Neuchâtel) nor in a meromictic one (Lago di Cadagno, canton Tessin). The cell concentration was always low (0–30 $ml^{-1}$). Most cells were retained by 3-mm membrane filters and thus were likely to be bound to particles or living cells. This binding occurred for 90% to 100% of the fast-growing knallgas bacteria, whereas about half of the slow-growing ones (forming pinpoint colonies, later identified as a hydrogen-oxidizing *Mycobacterium* spp) could pass through the membrane [27, 42]. No stimulation by $H_2$ of ($^3$H)methyl thymidine incorporation into bacterial DNA was noticed in the bacterioplankton of Lake Constance (Germany), which means the production of these bacteria probably plays a minor role compared to heterotrophic bacteria [68].

In soils, most probable numbers (MPN) of knallgas bacteria vary from none (in peat or acidic forest soils) to $10^4$–$10^5$ in neutral, well-amended garden soils (M. Aragno, unpublished results). A distinct enrichment of the knallgas flora and the establishment of higher $H_2$-oxidation activities were noticed in the environment of Hup root legume nodules [71, 110] and in the $N_2$-fixing rhizosphere of the common reed, *Phragmites communis* [10], and of rice [52, 86, 91]. This enrichment probably results from $H_2$ transfer between plant-associated $N_2$ fixers and the knallgas bacteria. The most-often isolated hydrogen-oxidizers in rich rhizosphere are members of the genus *Xanthobacter*, which are dinitrogen fixers themselves. In landfill leachate, we measured up to $10^6$ knallgas bacteria per ml: this high concentration can be related to the strong $H_2$ production in such environments. Flushing biofilter soils with a $H_2$:$CO_2$:$N_2$:$O_2$ mixture results in an enrichment of the hydrogen-oxidizing microflora to a maximum of $350 \times 10^6$ g$^{-1}$, with a specific hydrogenase activity comparable to that of knallgas bacteria in pure cultures [36, 37].

Unlike geothermal habitats, the contribution of individual species in natural, mesobiotic habitats so far is unclear. A wide range of different species can be isolated from most sampling sites. For example, we isolated the following species of knallgas bacteria from the water column of a small eutrophic lake in Switzerland (Le Loclat, near Neuchâtel): *Variovorax pradoxus*, *Aquaspirillum autotrophicum*, *Arthrobacter* sp., *Hydrogenophaga pseudoflava*, *Hydrogenophaga* sp. ($N_2$-fixing strains), *Mycobacterium* sp., *Acidovorax facilis*, *Xanthobacter autotrophicus*, *X. agilis*, and *X. flavus* [3, 6, 59–61, 111; M. Aragno, unpublished results]. Obviously, the ecology of individual species is determined by the existence of ecologic niches at microscale (e.g., adsorption to decomposing particles, clay particles, other cells, particularly algae and cyanobacteria, and so on).

In some cases, definite types of knallgas bacteria were found to be dominant in a given environment. All the knallgas bacteria isolated from soil around Hup root nodules by La Favre and Focht [71] were characterized as *Hydrogenophaga pseudoflava* (M. Aragno, unpublished results). Most hydrogen-oxidizing autotrophs isolated from the rhizosphere of the common reed (T. Beffa, unpublished results) and of rice [91] were $N_2$-fixing *Xanthobacter* spp. Most strains isolated from biofilters flushed with a $H_2$:$CO_2$:$N_2$:$O_2$ mixture, made up from a nitrogen-poor liceous soil [36], were $N_2$-fixing *Hydrogenophaga* spp. and *Xanthobacter* spp. (L. Dugnani and M. Aragno, unpublished results). A good correlation occurred in these biofilters between $N_2$-fixing and $H_2$-oxidizing activities (I. Wyrsch, unpublished observations).

As geothermal fluids and gases often contain significant concentrations of molecular hydrogen, hot springs and other geothermal manifestations might harbor thermophilic aerobic $H_2$-oxidizing bacteria. It is therefore surprising how few attempts were made before 1980 to isolate this type of bacteria from these environments [50], while thermophilic knallgas bacteria were found in cold non-geothermal habitats [6, 38, 77, 99]. The discovery of the highly thermophilic, obligately chemolithoautotrophic *Hydrogenobacter* [63, 69], of *B. tusciae* [17], and of the geothermal habitat of *Bacillus schlegelii* [16] recently confirmed that geothermal sites are the source of several original, highly thermophilic hydrogen-oxidizing aerobes and that these can play a role in the primary production in such environments.

Geothermal habitats include water from hot and warm springs (over 50°C), sediments from hot and warm ponds and creeks, and wet fumarolic soils (less than 100°C). Temperature and pH are major ecologic factors determining the presence or absence and the type of knallgas bacteria in a given spring.

Warm non-geothermal habitats, such as compost piles or waste masses during the transient phase between aerobic and anaerobic decomposition, are likely to harbor thermophilic knallgas bacteria because significant concentrations of hydrogen are produced by fermentation reactions [36]. Recently we isolated a variety of thermophilic knallgas bacteria from such environments, both sporeformers and non-sporeformers (M. Blanc, T. Beffa, and M. Aragno, unpublished observations).

Surprisingly, several strains of thermophilic knallgas bacteria were isolated from cold (meso and psychrobiotic) habitats. The moderately thermophilic *Pseudomonas thermophilia* were isolated from soil [77] and from drainage [39], and *Bacillus schlegelii* were isolated from lake sediments and even from glacier ice [4, 19, 99]. If moderately thermophilic non-sporeforming knallgas bacteria are likely to be found in non-geothermal environments, the presence of strict thermophiles is clearly allochthonous and due to the fact that endospores allow transport by air.

## FURTHER PERSPECTIVES IN THE ECOLOGY OF KNALLGAS BACTERIA

Knallgas bacteria can be useful tools for different approaches in microbial ecology, either fundamental or applied to environmental or biotechnologic problems. This section describes several such applications.

### Experimental Ecology

As facultative chemolithoautotrophs, knallgas bacteria are valuable tools for studies of mixotrophic behavior in the simultaneous presence of inorganic and organic energy and carbon sources. Moreover, in knallgas bacteria as in other chemolithoautotrophs, the inorganic carbon source is distinct from the inorganic energy source. This feature allows experiments to be designed with sole energy or sole carbon limitation, which is not possible with such heterotrophs as *Escherichia coli*.

Not only steady state but also transitions between trophic conditions can be studied with chemostat cultures of knallgas bacteria. This was tested with success in our laboratory using *Aquaspirillum autotrophicum* [87].

### Primary Production in Particular Habitats

In geothermal environments, such as hot springs and submarine geothermal vents, knallgas bacteria could play a role similar to sulfur-oxidizing bacteria as primary producers in aerobic conditions. So far this possibility has not been studied in depth.

#### *Knallgas Bacteria as Indicators of $H_2$ Evolution*

Production of $H_2$ is often difficult to measure in situ, such as in soils. The presence of important populations of knallgas bacteria would then be a good indicator for such production accompanying dinitrogen fixation (e.g., in the rhizosphere of grasses or around legume nodules). In other words, knallgas bacteria could serve as indirect indicators for $N_2$ fixation in soil biology.

### *Knallgas Bacteria and Organic Waste Management*

Facilitating the development of knallgas bacteria in or around decomposing waste masses would decrease the concentration of hydrogen, and so avoid the production of the explosive mixtures of $H_2$ and air that are likely to occur in such conditions, particularly in landfills in waste containers or even in composts [47].

### *Knallgas Bacteria as Agents of Bioremediation of Groundwater Contaminated by Nitrate*

Recently, Smith et al. [118] found numerous (still unidentified) denitrifying knallgas bacteria in a nitrate-contaminated aquifer in which denitrifying activity was enhanced by the addition of hydrogen or formate. They suggest therefore that such organisms may have a significant potential for in situ bioremediation of nitrate contamination in groundwater.

## Knallgas Bacteria as Tools for Energy Conversion

As they use the energy from the $H_2/O_2$ mixture to build up their biomass, the knallgas bacteria are excellent energy-to-biomass converters. During autotrophic growth of *Alcaligenes eutrophus*, cell yields of 18 to 20 g dry weight/l have been obtained [95, 96, 101]. The efficiency of energy conversion is high: Only 6 mol of $H_2$ are consumed when 12 g of carbon are transformed from $CO_2$ to organic material [$CH_2O$]. With hydrogen produced by electrolysis of water, 1 kilowatt hour is equivalent to 18.65 mol $H_2$, which amounts to about 70 g dry weight of biomass.

Early studies on the isolation of metabolic mutants indicated that *A. eutrophus* might be used to overproduce amino acids, such as valine, leucine, or isoleucine [92, 105], or D(-)-β-hydroxybutyric acid [121]. Although the yields were low, the accessibility of *A. eutrophus* to the methods of recombinant DNA technology will probably encourage further attempts to use knallgas bacteria as overproducers [2, 43, 45, 48, 56].

The thermophilic aerobic autotrophic bacteria still have a largely underexploited potential for biotechnological applications. In addition to the characteristics they share in common with the other thermophiles, they are among the fastest-growing autotrophic organisms, and they often reach high cell densities. If the cultivation of sulfo-oxidizers is often problematic because of the accumulation of sulfuric acid and the concomitant decrease in pH, it is much easier to culture knallgas bacteria which does not require the addition of important amounts of neutralizing solutions nor lead to precipitation reactions.

Although the ecology of knallgas bacteria is no more an unexplored field, the data presented here are still like a jigsaw puzzle with many detached and unrelated pieces, the complete image remaining largely hypothetical. Much more research is necessary, mainly at the microenvironmental scale, on $H_2$ production, concentration, and consumption, as well as on cell numbers, identification of knallgas bacteria, and on their relationships with the other microorganisms in their microbiocenosis. An environment particularly promising for the study of aerobic $H_2$ oxidation is the rhizosphere, not only of legume nodules but also of other plants known to harbor an active $N_2$-fixing flora. In water, the possible association of knallgas bacteria with $N_2$-fixing cyanobacteria or $H_2$-producing phototrophs should be studied more closely. Finally, almost nothing is known about knallgas bacteria in marine environments.

## HYDROGEN IN NATURAL ENVIRONMENTS

### Sampling

It is often difficult to obtain good samples of gases (e.g., from geothermal fumaroles) without significant air contamination. However, in geothermal bubbling ponds, we succeeded in obtaining samples with less than 10% air contamination by using the simple device shown in Figure 4-2.

Procedure

1. Fill 100-ml serum bottles (b) with boiled water, and use 1–2 ml of the carrier gas for gas chromatography (GC) analysis to avoid an explosion due to expansion.
2. Dip a funnel (c) in the bubbling pond, leaving just the outlet emerging.
3. The tube from the funnel (c) should be made of natural rubber, PVC, or neoprene, but not of silicone as it is highly permeable to gases. First, purge it completely with bubbling gas and then connect it to the serum bottle through a syringe needle, with another needle connecting the bottle to a syringe with two valves (a).
4. Holding the bottle neck down, pump the syringe slowly, until the gas has almost completely replaced the water in the bottle and then withdraw the needles.

For hydrogen dissolved in liquid samples, it is best to extract the gas with air. Due to its low solubility in water (16.3 ml/l at 30°C), 97% to 99% of the dissolved

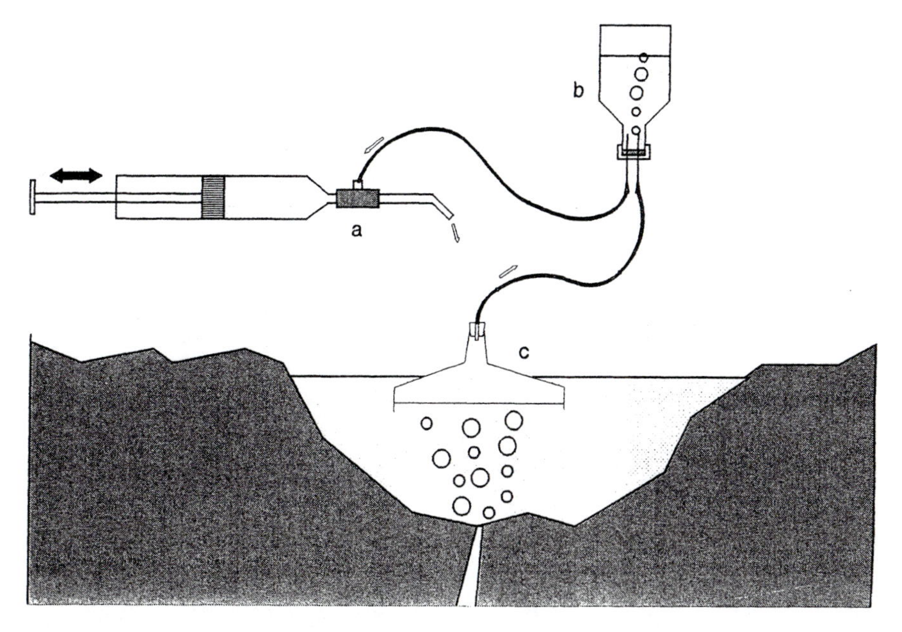


Figure 4-2 Sampling system for collecting bubbling gases in geothermal ponds. *a*. Two-valve syringe; *b*. 100 ml serum bottle; *c*. funnel (15–25 cm in diameter).

hydrogen will diffuse in the air phase at equilibrium if equal water/air volumes are used.

Procedure

1. Use a two-valve gas flask (Fig. 4-3) with a lateral sampling port. The volume is determined exactly, and a mark (line) delimitates the half-volume.
2. Fill the bottle completely with the sample, and add a concentrated $HgCl_2$ solution to a final concentration of 50 mg $ml^{-1}$, sufficient for complete inhibition of $H_2$ consumption.
3. Store the flasks in the dark and transport them to the laboratory for analysis of $H_2$ within 2 h.
4. Then replace half of the water in the flask with $H_2$-free synthetic air.
5. After opening both valves, disconnect the gas bottle and shake vigorously in order to equilibrate both phases. The hydrogen concentration in the gas phase (e.g., in ml air) will then be 97% to 99% of the concentration in the sample, depending on the temperature.

## Analysis of Hydrogen

Semiquantitative determinations of the $H_2$ content of the ground gases can be made using Draeger tubes for 0.5% $H_2$ (Draegerwerk AG, Lübeck, Germany), enabling direct field detection and estimation of concentrations above 0.5% $H_2$ v/v. Gases bubbling through small ponds are best suited for such analyses.

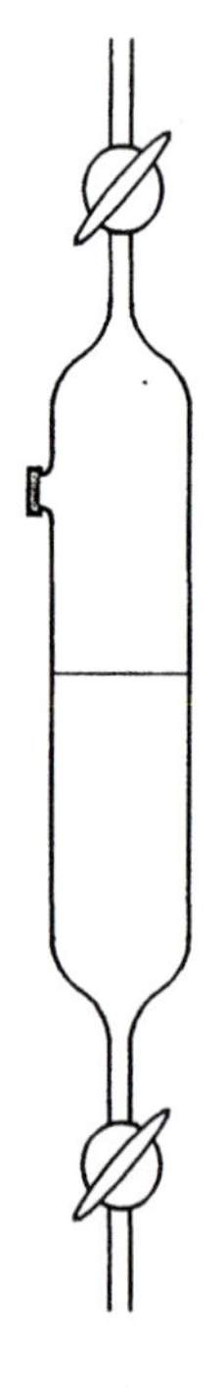

Figure 4-3 Gas flask for sampling gases dissolved in liquid samples.

Procedure

1. Fit a funnel (minimum of 600 ml) with a tube of sufficient length and dip it into the pond water beside the bubbling zone. The tube should be made of natural rubber, PVC, or neoprene, but not of silicone as it is highly permeable to gases.
2. Fill 25% of the funnel volume with air: The reaction in the tube is the oxyhydrogen reaction, and the gas analyzed must contain at least 5% $O_2$.
3. Then transfer the funnel to the bubbling one and fill it with gas. The gas will then be pumped through the tube. The length of the pink zone indicates the $H_2$ concentration.
4. *Follow the manufacturer's instructions exactly.*

For laboratory analyses, GC with a thermal conductivity detector gives good results at concentrations above 0.1% (v/v). It is then preferable to use argon or nitrogen as carrier gas, rather than helium.

For measurement of hydrogen in most natural environments (air, waters, soils), a much higher sensitivity is required. Seiler et al. [113] developed a detector system based on a heated HgO column (RGD2, Reduced Gases Trace Detector, is sold by Trace Analytical, Menlo Park, California). When a reductive gas passes through the column, HgO is reduced to Hg vapor, which is then detected. The lower limit of sensitivity of such a hydrogen analyzer is 2.5 nl $H_2 \cdot l^{-1}$ water (0.1 nM $H_2$).

## CULTURE METHODS

### Culture Conditions

#### *Media*

Knallgas bacteria have simple growth requirements. They use ammonium, nitrate, or urea as nitrogen sources and require potassium, magnesium, calcium, phosphate, sulfate, and iron as nickel. Maintenance of the pH is critical; it is kept constant either by the high phosphate concentration present in most of the recommended media or by a $CO_2$-bicarbonate buffer system; under the latter conditions, an excess of phosphate is unnecessary. A requirement for nickel salt during autotrophic growth has been shown for *Alcaligenes eutrophus* [12, 119] and for hydrogenase synthesis [45]. The base mineral medium [8, 106] is used routinely.

Procedure: Preparation of Basal Mineral Medium

1. To avoid precipitation, prepare three solutions, sterilize them separately by autoclaving and mix after cooling:

*Solution I*

| | |
|---|---|
| $Na_2HPO_4 \cdot 12H_2O$ | 9 g |
| $KH_2PO_4$ | 1.5 g |
| $NH_4Cl$ | 1.0 g |
| $MgSO_4 \cdot 7H_2O$ | 0.2 g |
| Trace elements solution | 1 ml |
| Distilled water | 1000 ml |

*Trace Element Solution*

| | |
|---|---|
| $ZnSO_4 \cdot 7H_2O$ | 100 mg |
| $MnCl_2 \cdot 4H_2O$ | 30 mg |
| $H_3BO_3$ | 300 mg |
| $CoCl_2 \cdot 6H_2O$ | 200 mg |
| $CuCl_2 \cdot 2H_2O$ | 10 mg |
| $NiCl_2 \cdot 6H_2O$ | 20 mg |
| $Na_2MoO_4 \cdot 2H_2O$ | 30 mg |
| Distilled water | 1000 ml |

*Solution II*

| | |
|---|---|
| Ferric ammonium citrate | 50 mg |
| $CaCl_2 \cdot 2H_2O$ | 100 mg |
| Distilled water | 100 ml |

*Solution III*

| | |
|---|---|
| $NaHCO_3$ | 5 g |
| Distilled water | 5 g |

2. After sterilization and cooling, mix solutions I, II, and III in these proportions: I: 1000 ml, II: 10 ml, and III: 10 ml.
3. To prepare solid medium, supplement the liquid medium with 1.7% (wt/vol) purified agar.
4. For cultivation under heterotrophic conditions, use the same medium—without solution III and supplemented by a suitable carbon source; the carbon sources are added to a final concentration of 0.2% to 0.5% or even 1.0% for sugars or 0.1% to 0.2% for other organic compounds.

For special requirements, different compositions may be used. For example, for pure cultures of the acidophilic, thermophilic *B. tusciae*, it is preferable to use a low-phosphate (less than 10 mM) medium amended with a non-metabolizable compound with good buffering capacities around pH 4.5. For the type strain, L(+)-tartrate at a concentration of 50 mM and pH 4.5 has been used successfully. For marine strains of knallgas bacteria, the NaCl-containing media described by Nishihara et al. [81, 82] can be used.

Many knallgas bacteria are able to grow in complex media, such as yeast extract or peptone nutrient broth. However, some strains or species (e.g., *Xanthobacter agilis*) grow poorly on nutrient broth even if a suitable carbon source is added.

## Handling of Gases

### *Safety in Handling Hydrogen*

CAUTION! Before using hydrogen, the researcher should be aware of the dangers of handling this gas. Normal hydrogen [78] is a flammable gas, and mixtures with air or oxygen are highly explosive. The lower and upper flammability limits in air at 20°C, 1 bar, are 4% $H_2$ and 74.5% $H_2$, respectively; the corresponding limits in oxygen are 4% $H_2$ and 94% $H_2$, respectively. This means that mixtures of 95% $H_2$ and 5% $O_2$ are completely safe. Mixtures of 65% $H_2$ + 25% $O_2$ + 10% $CO_2$ are highly explosive, but can be handled safely when precautions are taken.

Hydrogen is a nontoxic, odorless, and colorless gas. Its solubility in water amounts to 16.3 ml gas/liter water at 30°C (Bunsen coefficient). It has a low density (molecular weight 2.016 g/mol).

When handling flammable gases, take the following precautions:

Do not leave open flames (Bunsen burner, cigarette), a hot platinum wire, or any spark-generating apparatus in the room where you are handling the gases.

Put a warning on the door of the laboratory, and label the gas containers.

Take care when handling gas containers, and avoid scratches or cracks.

When evacuating the container, cover it with wire netting, and wear safety glasses when handling evacuated containers.

If you must put the container filled with an explosive gas mixture into a refrigerator or an incubator, be sure that the apparatus is protected against sparks. Place the electrical thermostatic device in a tightly closed box, completely isolated from the incubation chamber. Be sure that the pressure in your container is low enough at room temperature and that it does not increase too much at high temperatures.

Be careful when opening the container and manipulating its contents! Even if a desiccator containing an explosive gas mixture has been standing open for several days, there can still be a flammable gas mixture inside. Remove the gas by rinsing with nitrogen or air.

### *Composition of the Gas Mixture*

The oxygen partial pressure in the gas atmosphere may be kept as low as 5 kPa in order to avoid explosive gas mixtures. Since many knallgas bacteria are microaerophilic, the oxygen content in the gas mixture should be lower than its atmospheric content. An oxygen partial pressure of 5 kPa in the gas mixture, corresponding to about 1.8–2.0 mg of dissolved oxygen per liter of medium, is often favorable. A lower concentration may be necessary to start cultures of highly oxygen-sensitive bacteria, such as *Hydrogenobacter* spp. or dinitrogen-fixing organisms. The concentration of other components is less critical, and a mixture containing 5 kPa oxygen, 10 kPa carbon dioxide, and 85 kPa hydrogen generally gives good results.

In liquid cultures with increasing cell densities, the concentration of dissolved oxygen becomes limiting when the above-mentioned mixtures are used. Higher oxygen concentrations in the gas phase must be used, therefore, and the optimum gas supply is provided by a mixture that contains the gas components at a ratio at which they are consumed by the cells [105].

During exponential growth of knallgas bacteria, the gases are usually consumed at the following ratios:

$$2H_2 + O_2 \rightarrow 2H_2O \tag{1}$$

$$2H_2 + CO_2 \rightarrow (CH_2O) + H_2O \tag{2}$$

Equation 1 represents the oxy-hydrogen reaction that generates energy; Equation 2 represents the reductive fixation of $CO_2$ to the level of cell material. The ratio in which these gas components are consumed varies from species to species.

### *Incubating Cultures under Gas Mixtures*

Procedure

1. For incubating Petri dishes or small flasks under a given gas mixture, usually use closed vessels (desiccators, anaerobic jars, or Witt's vessels).
2. First evacuate the vessels, and then successively introduce the gas components into the vessel.

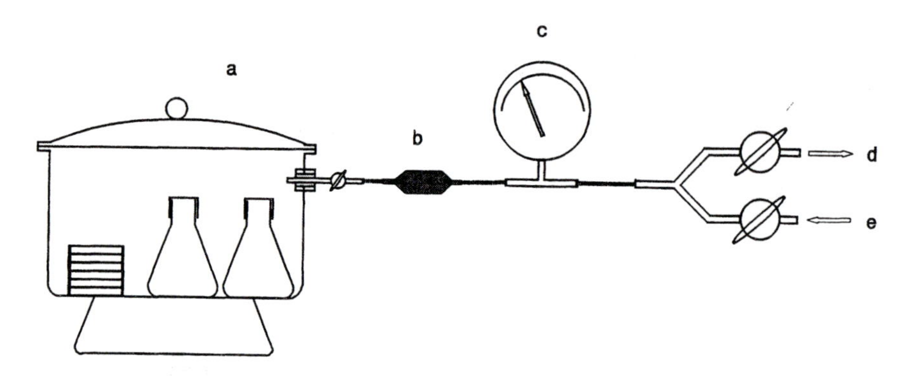


Figure 4-4 Device for evacuating dessicator and anaerobic jars and refilling with gas mixtures. *a*. Desiccator; *b*. sterile filter; *c*. culture vessel manometer; *d*. vacuum pump; *e*. gas tank(s).

3. Measure the partial pressures of the gases in the vessels by means of a manometric device (Fig. 4-4). Normally, the vessels are filled with 5 kPa $O_2$ (or less, see above), 10 kPa $CO_2$, and 75 kPa $H_2$. The 10 kPa underpressure helps keep the vessel's cover on tight.
4. For thermophiles, reduce the hydrogen partial pressure to 30–40 kPa $H_2$ (all partial pressures are measured at normal laboratory temperature). This initial underpressure of 45–55 kPa is necessary to avoid overpressure when the vessel is heated to an incubation temperature over 60°C.

During routine laboratory operation, Petri dishes are incubated upside down to avoid condensation of water vapor on the lid. However, because the air dissolved in the agar causes the agar to fall down during evacuation, Petri dishes must not be placed upside down in this case. Without agitation, liquid media should be

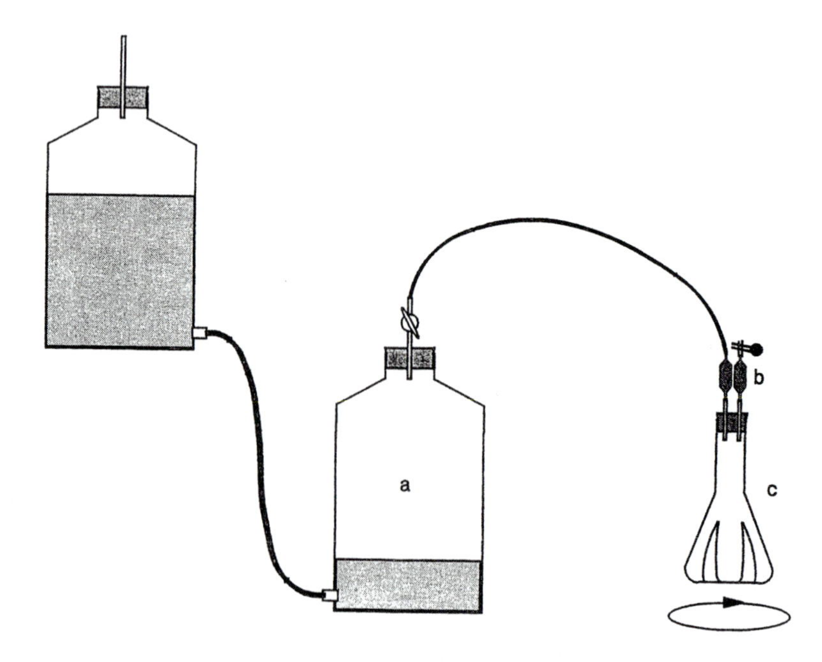


Figure 4-5 Gasometer system for growing knallgas bacteria in liquid culture in agitated vessels under a gas atmosphere. *a*. Gas reservoir; *b*. sterile filters; *c*. culture vessel (here, in Erlenmeyer with blown-in cones).

dispensed in shallow layers (e.g., 10 ml in 50-ml flasks or 20 ml in 100-ml flasks). Fernbach vessels allow the incubation of volumes up to 200 ml.

For larger volumes, a better exchange between the gas and liquid phases is achieved using an Erlenmeyer flask with blown-in "cones." It is then simplest to connect the flask to a gasometer (gas container system; see Fig. 4-5) and to put it on a rotary shaker. The gas is sucked into the culture vessel as it is consumed by the bacteria. Under these conditions, if the gas mixture does not contain the gas components in the ideal ratio, the gas present in excess accumulates in the gas phase of the culture vessel and cell growth is impaired. A closed circuit prevents the accumulation of excess gas. The gas is pumped from the gasometer into the culture vessel and then back to the gasometer. With this system, the exponential growth rates can be maintained for a longer period, resulting in higher densities and greater yields (Fig. 4-6).

Growth in fermentors requires a continuous flow system. The gases are taken from steel containers and lead via reducing wells, flowmeters, and a mixing chamber to the culture vessel. If a continuous flow system with its additional equipment is available, very simple culture vessels (Kluyver flasks or washing bottles) with appropriate spargers may be used instead of a fermentor. The gas absorption rate in these vessels is not extremely high, but is sufficient for yields of up to 5 or 8 g cell dry weight per liter.

Continuous cultures under hydrogen limitation are feasible, but difficult to control and to model. Indeed, hydrogen limitation normally occurs when only a fraction of the hydrogen in the input gas phase has been transferred to the liquid phase and then to the biomass. In other words, no equilibrium between the gas phase and the liquid phase is reached, and the hydrogen concentration in the output gas is not related to its concentration in the liquid phase. The fraction actually consumed depends on the gas-liquid contact surface and on the duration of this contact and therefore on the dimensions of the gas bubbles. Gas bubbles dimension and duration of gas-liquid contrast depends not only on the agitation system and its intensity but also on the viscosity of the culture, which can vary

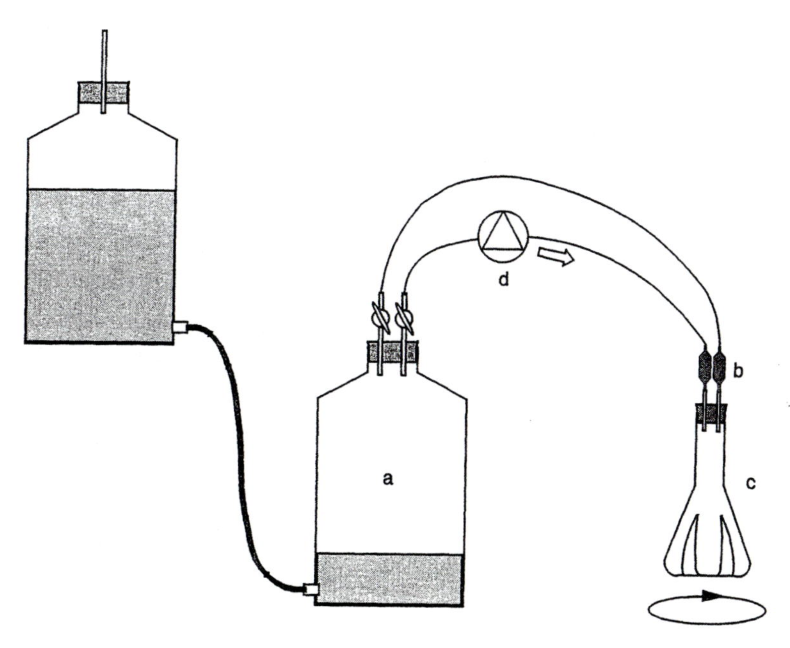


Figure 4-6 Closed-circuit system for gassing cultures of knallgas bacteria. *a–c*. See Fig. 4-4; *d*. gas pump.

according to cell density and possible secretions in the medium. Therefore, working at low cell densities and with a non-slime-producing organism is a prerequisite for such studies. Low cell densities allow a better gas transfer as well. Then, keeping the gas flow and agitation speed strictly constant and considering the limiting hydrogen concentration in the culture (liquid phase) to be negligible, one could consider an apparent feeder concentration of hydrogen under steady-state conditions to be

$$S_f = \frac{(\%\ H_2\ \text{input} - \%\ H_2\ \text{output}) \times \text{gas flow}}{\text{medium low}}$$

## Enrichment and Isolation

Designing an elective medium for enrichment of knallgas bacteria is simple, as knallgas bacteria are the only organisms able to grow aerobically in the dark with $CO_2$ as the sole carbon source and with $H_2$ as the energy and electron source. Therefore, a mineral medium, such as the one described earlier, under a gas mixture containing $H_2$, $CO_2$, and $O_2$, is appropriate. However, these enrichment conditions suppose that the desired bacteria have several other properties as well: a tolerance to phosphate concentration, an ability for assimilatory sulfate reduction, and independence of nutritional supplements, such as vitamins, amino acids, and other growth factors. Most "classical" knallgas bacteria have been isolated under these or similar conditions.

By modifying the enrichment conditions, special strains or species have been isolated. In the absence of combined nitrogen and in the presence of gaseous $N_2$, nitrogen-fixing strains of *X. autotrophicus* [124] and of other dinitrogen-fixing knallgas bacteria [76] were obtained. Nokhal and Schlegel [84] isolated strains related to *Paracoccus denitrificans* under autotrophic conditions by using a nitrate-containing medium under an oxygen-free atmosphere containing $H_2 + CO_2$. Since strains of *X. flavus* isolated from aquatic habitats proved to be biotin-dependent [3, 60] enrichments and purification in the presence of biotin resulted in the isolation of other biotin-requiring stains that could be attributed to *X. flavus* (B. Jenni and M. Aragno, unpublished results). At elevated temperatures, thermophilic knallgas bacteria were isolated, such as *Hydrogenobacter* spp. and *Bacillus schlegelli* at neutral pH and *B. tusciae* at a moderately acidic pH [5]. Two main types of enrichment/isolation procedures can be used: the direct-plating method (direct isolation on solid media) or enrichment in liquid culture.

### *Direct Isolation on Solid Media*

The direct plating of samples taken from natural habitats on selective solid media results in the growth of a great variety of different bacterial strains. All the bacteria able to grow on the surface of the selective medium will produce colonies. Because the bacteria contained in the inoculum are spread over the surface and are well separated, even the slow growers have a chance to grow and form colonies. In this way not only the predominant organisms but even those belonging to the minority populations can be isolated.

#### Procedure

1. Disperse about 1 g of the solid or semisolid inoculum, such as soil or mud, in 10 ml saline solution (0.9% NaCl in water).

2. Mix 1 ml of this inoculum suspension or of its serial dilutions in Petri dishes with 20 ml of the mineral agar medium, melted and cooled to 45°C. Be sure that the agar surface has been well dried before it is inoculated.
3. Allow the agar to solidify and incubate under the appropriate gas mixture for each subgroup of knallgas bacteria, as described in Cultivation and Maintenance of Knallgas Bacteria. The suspension medium must be soaked into the agar either in a desiccator or in a draught of dry air (laminar flow cabinet) before the samples are exposed to the gas atmosphere in a closed container and incubated.
4. Alternatively, spread small volumes (0.02–0.1 ml) of the diluted samples on the solidified agar medium. A turntable and a triangular spatula are usually used for spreading.

For liquid samples from ponds and lakes with low numbers of knallgas bacteria, the membrane filter method permits direct isolation. A certain volume of the sample (varying between 2 and 500 ml, depending on the expected cell concentration and on the size of the filter apparatus) is filtered through a membrane with pore diameters of 0.2–0.45 μm. The membrane is then deposited on the mineral agar and incubated as described earlier in this chapter for each subgroup. In order to avoid spreading of colonies, the agar must be well dried before placing the filter on top of the agar.

### *Enrichment Cultures in Liquid Media*

In contrast to the direct isolation of hydrogen-oxidizing bacteria on solid media, the organisms of any sample introduced into a liquid medium compete with each other, and the most rapidly growing bacteria in the sample are selected. This can lead to the selection of the best "r-strategists" in the population and not necessarily of the most abundant and relevant types. r-strategists are dominant in high nutrient and high energy-flux systems and are characterized by high exponential growth rates in unlimited conditions. It corresponds more or less to Winogradski's "zymogenous" strategists, or to "opportunistic" strategists. This is in opposition to "K-strategists" (please respect r in low case and K in upper case) which dominate in low nutrient and low energy fluxes systems, and are characterized essentially by high affinities for the limiting substrate. This fits more or less with Winogradski's "autochtonous" strategists. However, procedures using liquid media allow the addition of a heavy inoculum and thus the selection of rare bacteria. This procedure is particularly important for the enrichment of fast-growing organisms that form heavy suspensions, which are best suited for industrial applications.

During the initial incubation period, the bacteria being selected will predominate. However, in the course of time the microorganisms (amoebae, mycobacteria, bdellovibrios) that prey on hydrogen-oxidizing bacteria are favored; heterotrophs also grow at the expense of metabolites secreted by the autotrophs. After a few days the bacteria selected initially may well have been overwhelmed and may even disappear completely. Therefore, primary liquid enrichment cultures with heavy inocula should be transferred after short time intervals (1–5 days).

For ecologic and taxonomic purposes, to isolate the predominant organism(s) in a given habitat, the following dilution procedure should be applied. At the same time, it provides an MPN estimation for the actual population of knallgas bacteria in the sample.

Procedure: Preparation of a $10^{-1}$ Suspension-Dilution from Solid Samples

1. To relate the number of bacteria to a standard value, measure the water content of the sample.

2. Suspend a sample weight corresponding to 2 g dry matter in 20 ml sterile suspension medium: either tap water, saline (0.9% NaCl), or another physiologic mineral solution.
3. Either mix the suspension thoroughly in a sterile mortar, treat by a vortex mixer, or lightly sonicate it.

Procedure: Preparation of Serial Dilutions of the Sample

1. After mixing on a vortex, transfer 1 ml of the liquid sample or of the $10^{-1}$ suspension of the solid sample into a 18-mm test tube containing 9 ml of the sterile suspension medium.
2. After mixing thoroughly (vortex), transfer 1 ml of the diluted suspension to another tube of suspension medium, and so on. For water samples, dilution steps to $10^{-4}$ are generally sufficient. For soil samples, dilutions to $10^{-6}$–$10^{-7}$ and up to $10^{-10}$ for presumably enriched samples, such as soil from biofilters (see below), are recommended.

Procedure: Enrichment Cultures

1. Prepare selective culture media in 18-mm test tubes (4 ml/tube) or in 25- to 50-ml pyrex flasks.
2. Inoculate three to five tubes (or more) for each dilution with 1 ml of suspension.
3. Place the tubes or the bottles in a desiccator or Witt's vessel, wedged with paper or foam rubber, under the proper gas mixture (see above) and if desired on a rotary shaker at the proper temperature.
4. Allow sufficient time for growth (usually 1–2 weeks).
5. Count the number of positive tubes at each dilution for MPN determination [55], and use the last positive dilution for isolation.

### *Other Enrichment Procedures*

Human imagination is the only limit to designing enrichment and isolation procedures. Consider these two approaches:

1. Inoculating chemostat cultures with a natural sample results through competition to a selection based on the highest affinity for the limiting substrate and not on the highest growth rate in unrestricted conditions.
2. Flushing soil columns with a gas mixture provides conditions in which attachment to solid particles and nutrient limitations dictated by soil compositions are dominating parameters. Such columns are used for the enrichment of $N_2$-fixing knallgas bacteria.

## Enrichment/Isolation of Physiologic Subgroups of Knallgas Bacteria

Several subgroups of knallgas bacteria are characterized by particular metabolic properties, some of which can be used for specific enrichment. The special features that have so far been discovered in knallgas bacteria are the ability (1) to fix dinitrogen, (2) to use nitrate as an electron acceptor under autotrophic conditions, (3) to grow on carbon monoxide or on propane and other hydrocarbons, (4) to form heat-resistant spores, and (5) to grow as thermophiles at elevated temperatures. Other special features that can be taken into consideration are a reduced or increased tolerance to certain environmental conditions, such as the partial pressure of oxygen, the concentration of phosphate or magnesium, and the presence of antibiotics or toxic substances.

Although microbes are found everywhere and the environment selects particular types, the choice of inoculum may affect the results of the selective culture, as shown by procedures that allow isolation of dinitrogen-fixing knallgas bacteria. For example, when an enrichment procedure using a nitrogen-free medium with a heavy inoculum was applied, only strains of *Xanthobacter autotrophicus* [125] were isolated [124]. Using different enrichment and isolation procedures, other authors were able to isolate other species of *Xanthobacter* [59, 60], as well as $N_2$-fixing knallgas bacteria belonging to other genera [76]. By using an enrichment procedure in soil columns, Dugnani et al. [36] were able to isolate, along with *Xanthobacter* spp. dinitrogen-fixing strains related to *Hydrogenophaga pseudoflava* [62], as well as another group of dinitrogen-fixing knallgas bacteria probably related to *Hydrogenophaga* (M. Aragno, unpublished results). The enrichment of dinitrogen-fixing knallgas bacteria under these conditions can be explained by the nitrogen limitation in the soil, whereas energy (as $H_2 + O_2$) and carbon (as $CO_2$) were provided continuously. The MPN of $3.5 \times 10^8$ knallgas bacteria per g soil was reached under these conditions.

Procedure (adapted from [36])

1. Fill a glass cylinder, 10 cm in diameter and closed at both ends with silicone rubber (Fig. 4-7), with 1–2 cm silica gravel (or acrylic cotton) and 25–30 cm soil. In our experience, the best results were obtained with an alluvial, siliceous soil.
2. Flush the column upward with a gas mixture containing 5% $O_2$, 30% $N_2$, 10% $H_2$, and 55% $CO_2$, at a rate of 250–500 ml/h over 2 to 4 weeks. After this time, hydrogen usually has disappeared from the gas escaping the col-

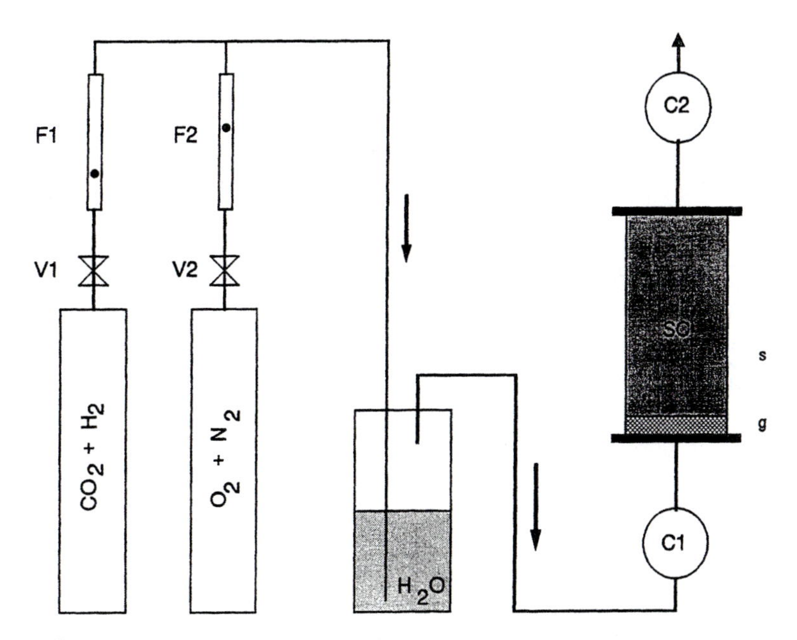


Figure 4-7 Soil column system for enrichment of $N_2$-fixing knallgas bacteria. *V1* and *V2*. Precision valves; *F1* and *F2*. flowmeters; *C1* and *C2*. gas counters (may be omitted for simpler operations); *SC*. soil column; *g*. silica gravel; *s*. soil. (Adapted from Dugnani et al. [36]).

umn, being completely oxidized by the knallgas bacteria when developed spontaneously on the soil particles.

3. Then take samples of soil from different levels of the column and use for isolation, according to the serial dilution enrichment technique.

## Obtaining Pure Cultures of Knallgas Bacteria

### *Mesophiles*

It is generally easy to obtain pure cultures from cultures and colonies that have been repeatedly transferred under primary selective conditions or from cultures inoculated with the last positive dilution of a sample. Streaking on mineral agar plates incubated under autotrophic conditions must be repeated several times. The number of transfers required is dependent on several factors: the number and kind of accompanying bacteria, the consistency of the colonies, and the amount of slime production. A short alkaline treatment [124] or enzyme treatment may help dissolve the slime. If this treatment fails, a colony should be inoculated into liquid medium and the suspension then restreaked on agar. In every case a final confirmative test must be made to ascertain the purity of the culture.

### *Thermophiles*

When isolating thermophilic knallgas bacteria, it is often difficult to obtain well-separated single colonies on solidified medium. Often, the cultures spread on the entire medium surface. Increasing the agar concentration and air drying the surface before incubation often favor the formation of single colonies.

*Hydrogenobacter* spp. showed no growth at all on agar plates when cultivated autotrophically with $H_2$ as the sole electron and energy source. However, Alfredsson et al. [1] found that Icelandic *Hydrogenobacter* strains grew well on agar media supplemented with thiosulfate (3.92 g/l $Na_2S_2O_3 \times 5H_2O$) under air + 10 kPa $CO_2$, and even better under air + 10 kPa $H_2$ + 10 kPa $CO_2$, but not with $H_2$ as the only electron and energy source. Ishii et al. [58] achieved colony formation by *H. thermophilus* on media containing 10 mg/l $CuSO_4$ and solidified with GELRITE, an agar-like bacterial polysaccharide (Merck & Co, Darmstadt, Germany).

If it is not possible to isolate single colonies by streaking on solidified media, serial dilution of the last enrichment culture may be done. To obtain pure cultures, streaking must be repeated several times using the last positive tube. Purity can be checked by microscopic observation confirming morphologic homogeneity and by the presence of a single inflexion point in the DNA denaturation curves.

It is often difficult to separate *Hydrogenobacter* spp. and *Bacillus schlegelli* when both are present in an enrichment culture. Separation can be achieved by raising the enrichment temperature to 77°C, which inhibits *B. schlegelli*, or by boiling the sample for 5 minutes, which would preserve *B. schlegelii* spores but would kill the non-sporeforming *Hydrogenobacter* cells.

## Maintenance

### *Maintenance on Solid Media*

Some strains of knallgas bacteria have kept their original autotrophic properties in spite of countless transfers under heterotrophic conditions. Other species, however, easily lose their ability to grow autotrophically. Why the autotrophic capabilities are lost during non-selective subculture is not known in most cases. For *Nocardia opaca* lb, in which the information for lithoautotrophic characters is not an integral part of the bacterial chromosome but is localized on a plasmid, these characters are lost at a frequency of 0.4% over 15 generations.

In a transconjugant of *N. erythropolis* that received the lithoautotrophic characters by conjugation, the frequency of loss was still higher (20% over 15 generations [93]). For this reason, it is better to maintain knallgas bacteria under autotrophic conditions. Furthermore, autotrophic cultures maintain their viability for longer periods than heterotrophic ones. At 4°C most strains stay alive for 6 months or longer. The longevity of the agar slant cultures may be increased by covering the inoculated agar with paraffin oil.

#### *Lyophilization*

Most of the knallgas bacteria survived conventional lyophilization in skim milk very badly. Because complete desiccation of bacteria during freeze-drying results in death or mutations due to the breakdown of the water crystal lattice of the DNA, compounds that bind water should be added. Therefore, the skim milk was supplemented with glutamate (5%), *meso*-inositol (5%), or honey (10%). Most knallgas bacteria, especially sensitive strains, survived lyophilization in these media. Even 5% glutamate alone protected the cells [74].

In any case, the conditions for the preservation of knallgas bacteria should be studied for each strain or species before irreplaceable agar slant cultures are discarded. For lyophilization, autotrophically grown knallgas bacteria cultures are preferable to heterotrophically grown ones.

## IDENTIFICATION

Knowledge of the distribution of individual species in natural environments is an important chapter in ecology, although often underestimated in recent literature. Individual species harbor particular sets of properties that make them much more accurate ecological bioindicators than a whole physiological group harboring a common function (like the knallgas bacteria); for example, the distribution of individual species of thermophilic knallgas bacteria in hot springs (Fig. 4-8). Individual species occupy distinct niches, as shown in a temperature versus pH plot.

In this section, we present two sets of tests. The first one enables the recognition of knallgas bacteria as such; that is, it differentiates them from oligocarbophilic heterotrophs that may exhibit modest growth under autotrophic conditions (especially on solid media), which may be interpreted erroneously as autotrophic growth. The second one is a rapid, primary identification procedure using an identification key and a species/characters matrix and based on a limited choice of significant characters.

### Confirmative Tests for $H_2$ Chemolithoautotrophy

To recognize a bacterium as a hydrogen-oxidizing autotroph, the most suitable and simple tests measure the ability to grow autotrophically in an atmosphere consisting of hydrogen, oxygen, and $CO_2$ and the presence of hydrogenase activity in cells grown in mineral medium in a hydrogen, oxygen, and $CO_2$ gas mixture.

#### *Testing Autotrophic Growth*

Procedure

1. Prepare Petri dishes or agar slants with the solidified basal mineral medium (1.7% agar); for each strain prepare two plates or agar slants.

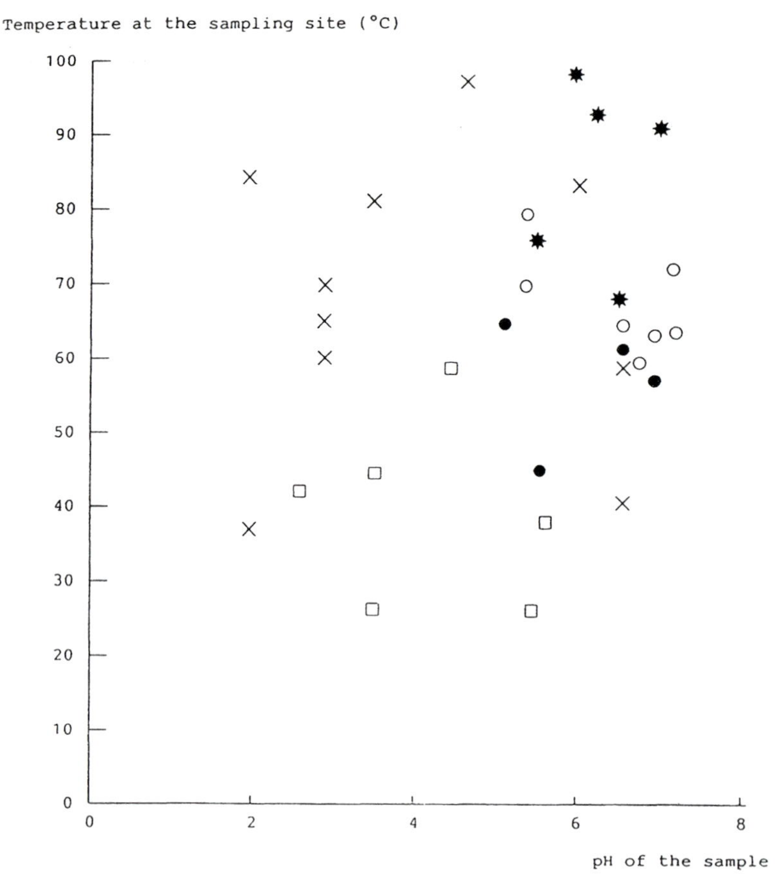


Figure 4-8 Distribution of different types of thermophilic, hyrogen-oxidizing bacteria from Tuscany (Italy) and S. Miguel (Acores) in a pH versus temperature plot. Each point represents a single sample. * *Hydrogenobacter*, homology group 1; ● *Hydrogenobacter*, homology group 2; ○ *Bacillus schlegelli*; □ *Bacillus tusciae*; × no bacteria found (from [5]).

2. Incubate one under an atmosphere of 5% $O_2$, 10% $CO_2$, and 85% $H_2$, and the other under air or preferably under 5% $O_2$, 10% $CO_2$, and 85% $N_2$.
3. After 1 week or longer, compare growth.

The test may be performed equally well in liquid media and under agitation. However, remember that at least some hydrogen-oxidizing bacterial strains do not tolerate submerged cultivation or agitation.

### *Qualitative Spot Test for Hydrogenase Activity* [10]

The occurrence of a hydrogenase able to transfer electrons from molecular hydrogen to cellular acceptors (NAD or respiratory chain components) is characteristic of all the hydrogen bacteria. There are several tests for hydrogenase activity, ranging from a simple qualitative spot test to more elaborate quantitative measurements on whole cells or on cell-free extracts, using either methylene blue or NAD as potential acceptors. This section presents a simple, qualitative spot test; a more thorough description of hydrogenase assays is given elsewhere [10].

The irreversible reduction of triphenyltetrazolium (TTC) salt to water-insoluble red triphenylformazan is used to differentiate between colonies of cells containing a hydrogenase or not. Whereas hydrogenase-free cells, when incubated under a hydrogen atmosphere, only slightly reduce TTC, the hydrogenase-containing colonies become intensely colored. This test may sometimes give uncertain results, since either TTC diffusion to the cells is restricted by abundant slime or a limited TTC reduction occurs from endogenous electron donor. Then, if applicable, a control without hydrogen should be performed in parallel.

Procedure

1. Deposit a sterile membrane filter with a pore diameter of 0.20–0.45 mm on a mineral agar plate.
2. Transfer the strains to be tested by spotting them on the membrane. If the colonies do not spread, up to 10 strains may be tested on one membrane.
3. Incubate them under 5 kPa $O_2$ + 10 kPa $CO_2$ + 75 kPa $H_2$.
4. After growth has occurred, deposit the membranes on filter paper pads soaked with a 0.1% (wt/vol) freshly prepared solution of TTC.
5. Incubate 10 min under air at room temperature in the dark. Note the appearance and intensity of a red coloration of the colonies.
6. Then incubate 10 min under 80 kPa $H_2$ + 10 kPa $O_2$ at room temperature. The strains that do not stain after incubation under air and that become reddish-brown after incubation under $H_2$ have hydrogenase activity.

This method may also be applied directly to colonies grown after direct isolation from water on membrane filters (see above). The cells are still alive after the test and may be transplanted further.

## Rapid, Primary Identification of Knallgas Bacteria

The unambiguous identification of any bacterial strain is, as a rule, a lengthy procedure. For small, well-defined groups, such as the hydrogen-oxidizing bacteria, a limited program of simple tests facilitates the determination of newly isolated strains. We present here a procedure, with a key and a matrix suited for the rapid identification of hydrogen-oxidizing bacteria. Using this key, one has to keep in mind the following facts:

The key and the matrix are based upon well-described species or strains published or known to the author at the time of writing this chapter. Species or strains that have been described only incompletely, particularly many "historical" species, such as *Hydrogenomonas pantotropha* and *H. vitrea* were not taken into account due to the loss of the type strains.

Likely, many types of hydrogen-oxidizing bacteria are not yet known or described. It is possible, therefore that strains, species, and genera will be discovered that do not fit the key and the matrix.

The key uses features that are constant in all well-studied strains of a species. Many species, however, have been described on the basis of only one or a few strains. As a consequence, the variability of features in such a species is not well established.

The matrix comprises the mesophilic species only.

The key and the matrix (Tables 4-1 to 4-3) are meant only as an orientation; identification must be completed by referring to a more detailed description of the species (see [8] for mesophilic species published before 1990).

## Other Approaches to Identification

The development of chemotaxonomic and molecular methods in bacterial taxonomy has allowed immense progress in this discipline in the last 20 years. Yet,

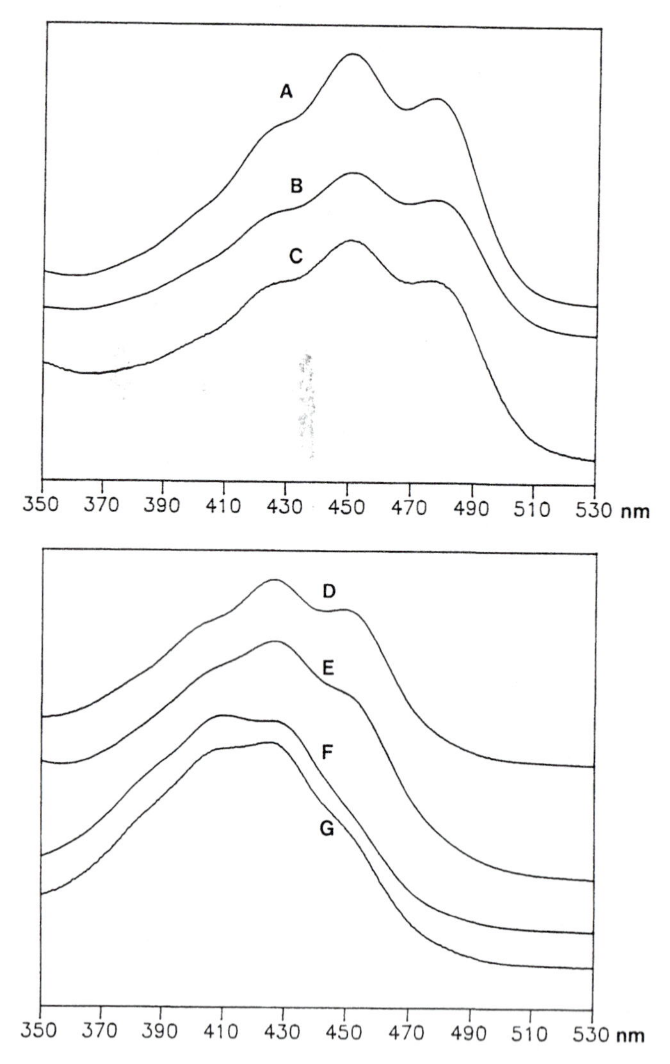


Figure 4-9 Absorption spectra of yellow-pigmented, gram-negative knallgas bacteria. *Top*: *A. Xanthobacter autotrophicus, B. X. agilis; C. X. flavus. Bottom*: *D. Hydrogenophaga pseudoflava; E. Alcaligenes paradoxus; F. H. palleronii; G. H. taeniospiralis.*

these methods do not always prove satisfactory for ecologic purposes. With the example of the hydrogen-oxidizing bacteria in mind, we suggest the following reasons for the limitations of these methods:

Knallgas bacteria are a functional entity, *not* a taxon. They share common properties, but occupy dispersed positions in the phylogeny of Bacteria. Even the genes coding for the enzymes characteristic of their typical functions (hydrogenase and key enzymes for $CO_2$ fixation) are not unique to these organisms. Uptake hydrogenases are found in many non-$H_2$-oxidizing bacteria, such as heterotropic dinitrogen fixers, Enterobacteraceae, and so on. $CO_2$ fixation occurs in other autotrophs, as well as in several methylotrophs, particularly in formate utilizers. A DNA sequence signature of this functional group is therefore probably impossible to find.

The development of probes for the recognition of individual species is a powerful tool. However, probes cannot substitute for descriptions based on a wide collection of phenotypic characters, because the list of these char-

**Table 4-2** Key to the Identification of Hydrogen-Oxidizing Bacteria

| | | |
|---|---|---|
| 1– | Thermophilic: growth at 5°C, no growth at 27°C | 24 |
| – | Mesophilic: no growth at 50°C, growth at 27°C | 2 |
| 2– | Gram negative or variable | 3 |
| – | Gram positive | 20 |
| 3– | Coccoid cells, even in young cultures, occurring frequently in pairs or short chains Anaerobic growth with nitrate formation of $N_2$ in heterotrophic conditions | *Paracoccus denitrificans* [14,31] |
| – | Cells sharply curved, gas-vacuolated | *Ancylobacter aquaticus* [85,90] |
| – | Spiral cells | *Aquaspirillum autotrophicum* [6] |
| – | Not as above | 4 |
| 4– | Obligately lithoautotrophic; no growth under air on organic substrates and/or complex media | 5 |
| – | Facultatively lithoautotroph; growth under air at the expense of a more or less wide spectrum of organic carbon sources | 6 |
| 5– | Halophilic, found in marine environments G + C content about 44% | *Hydrogenovibrio marinus* [83] |
| – | Not halophilic. Uses $H_2S$ and PbS as alternative electron donors | *Thiobacillus plumbophilus* [35] |
| 6– | White or cream-colored colonies; no pigment in acetone: methanol extract | 7 |
| – | Colonies more or less yellow; pigment soluble in acetone: methanol extract, showing peaks between 735 and 480 nm | 14 |
| 7– | Cells more than 1 microns broad | *Alcaligenes latus* [87] |
| – | Cells less than 0.8 microns broad | 8 |
| 8– | Growth on L-rhamnose and lactose | *Pseudomonas hydrogenovora* [57] |
| – | Not as above | 9 |
| 9– | Starch hydrolyzed, growth on sucrose and cellobiose | *Pseudomonas saccharophila* [34] |
| – | Not as above | 10 |
| 10– | Growth on D-galactose and mannitol | *Acidovorax facilis* [98,127] |
| – | Not as above | 11 |
| 11– | Growth on D-xylose; no growth on p-hydroxybenzoate | *Alcaligenes xylosoxidans* [65] |
| – | Not as above | 12 |
| 12– | Growth on D-fructose; anaerobic growth with nitrate and dinitrogen evolution under heterotrophic conditions | *Alcaligenes eutrophus* [31] |
| – | Not as above | 13 |
| 13– | Dinitrogen is fixed; scant growth on nutrient broth | Non-pigmented strains of *Xanthobacter agilis* [59] |
| – | Dinitrogen not fixed | *Oligotropha carboxidovorans* [79] |
| 14– | Pigment in acetone:methanol with absorption maxima at 426, 451, and 478 nm: genus *Xanthobacter* | 15 |
| – | Not as above | 17 |
| 15– | Very motile rods, not pleomorphic; scant growth on nutrient broth | *Xanthobacter agilis* [59] |
| – | Non-motile or rarely pleomorphic rods: coryneform in young, heterotrophic cultures; coccoid in older cultures and under autotrophic conditions | 16 |
| 16– | Biotin required or simulatory | *Xanthobacter flavus* [75] |
| – | Biotin not required | *Xanthobacter autotrophicus* [13,125] |
| 17– | Pigment in acetone:methanol with absorption maxima at 399 and 420 nm and shoulders at 376 and 455 nm; growth on D-xylose, not on lactose, saccharose, or trehalose (Fig. 4–9) | *Variovorax paradoxus* [31,128] |
| – | Pigment in acetone:methanol with absorption maxima at 408 and 431 nm and shoulders at 384 and 455 nm; no growth on sugars (except D-glucose) | *Hydrogenophaga palleronii* [32,126] |
| – | Pigment in acetone:methanol not as above; growth on sucrose, cellobiose, D-galactose, mannitol, D-fructose | 18 |
| 18– | Gelatin hydrolyzed; growth with D-xylose, not with trehalose | *Hydrogenophaga taeniospiralis* [72,126] |
| – | Gelatin not hydrolyzed; pigment in acetone:menthanol with absorption peak at 430 nm and shoulders at 404 and 455 nm; growth with trehalose | 19 |
| 19– | Growth with D-xylose; anaerobic growth with nitrate and dinitrogen production under heterotrophic conditions | *Hydrogenophaga pseudoflava* [11,126] |
| – | No growth with D-xylose; not denitrifying | *Hydrogenophaga flava* [80,126] |

(continued)

Table 4-2 (*continued*)

| | | |
|---|---|---|
| 20– | Acid-fast | 21 |
| – | Not acid-fast | 22 |
| 21– | Growth on D-galactose and mannitol; generally unpigmented in the dark | *Mycobacterium* sp. ref. strain PP7 (M. Aragno, unpublished results) |
| – | No growth on D-galactose and mannitol; pigmented in the dark (scotochromogenic) | *Mycobacterium gordonae* [15] |
| 22– | Pleomorphic rods, not branching; formation of coccoid cells in old cultures | *Arthrobacter* sp. ref. strain 11/x [13,22] |
| – | Formation of a branched mycelium, at least in young cultures, which then fragments into rods or conidia | 23 |
| 23– | Formation of an aerial mycelium; colonies snow-white at the upper face, and yellowish at the underside | *Amycolata autotrophica* [73,120] |
| – | No aerial mycelium formed; colonies rough, cream, gray, or pink | "*Nocardia opaca*" [33,122] |
| 24– | Endospore formers, gram positive to variable | 25 |
| – | Non-endospore formers | 26 |
| 25– | Growth at pH 7.0, no growth at pH 4.0; endospores terminal, spherical, deforming the sporangium; %(G + C) 66–68 | *Bacillus schlegelii* [99] |
| – | Growth at pH 4.0, no growth at pH 7.0; endospores subterminal to terminal, oval, deforming the sporangium; %(G + C) 57–58 | *Bacillus tusciae* [17] |
| 26– | Gram-positive, filamentous, obligately autotrophic | *Streptomyces thermoautotrophicus* [46] |
| – | Gram-negative | 27 |
| 27– | Growth at 70°C; %(G + C) 37–46 | 28 |
| – | No growth at 70°C; growth at 50°C | 31 |
| 28– | Acidophilic, optimum pH 3.0 to 4.0 | *Hydrogenobacter acidophilus* [115] |
| – | Neutrophilic, growth at pH 7.0 | 29 |
| 29– | Halophilic, growth in 0.5 NaCl | 30 |
| – | Not halophilic, no growth at 0.5 M NaCl | *Hydrogenobacter thermophilis* [63], as well as several genospecies almost indistinguishable phenotypically, but showing low DNA homology with *H. thermophilus* [5,16]. *Calderobacterium hydrogenophilum* [69] should be included among these genospecies of *Hydrogenobacter*. |
| 30– | Growth at 85°–90°C | *Aquifex pyrophilus* [54] |
| – | No growth at and above 80°C | *Hydrogenobacter halophilus* [82] |
| 31– | Three species of facultative knallgas bacteria, all moderate thermophiles, described on the basis of one to two strains isolated; so far they have not been validly described: | |
| – | No carotenoid pigment formed; colonies light brown | "*Pseudomonas thermophila*" [39,77] |
| – | Presence of carotenoid pigment; colonies dull yellow | |
| – | Motile | "*Pseudomonas hydrogenothermophila*" [51] |
| – | Non-motile | "*Flavobacterium autothermophilum*" [51] |

Adapted from [5,8].

acters is the functional identity card of the organism, needed by the ecologist. Moreover, several species comprise both hydrogen-oxidizing and non-hydrogen-oxidizing strains, such as *Alcaligenes eutrophus* [61]. In most cases this is related to the location on a plasmid of some or all of the genes coding for lithoautotrophy.

Like other groups of microorganisms, it is highly probable that a number of species of knallgas bacteria are at present unknown and that the study of the new potential habitats will reveal new taxa. Even if organisms that have not been cultivated so far can be recognized in natural environments by molecular methods and even be classified phylogenetically, nothing is revealed of their functional capabilities by these means. We will still have to turn to the good, old, and painful cultural methods to get this information.

However, clearly chemotaxonomic and molecular techniques will equip microbial ecology with increasingly important new tools, in **addition** (but not as a substitute) to the classical approaches. Whatever the sophistication of the molec-

Table 4-3 Identification Methods for the Characters in Tables 4-1 and 4-2

**MORPHOLOGY AND MOTILITY.** The cell morphology and motility must be observed in young, actively growing cultures; use phase-contrast microscopy. Many bacteria are motile only on certain substrates, during special growth phases, and after growth at special temperatures (e.g., cells of *Hydrogenophaga pseudoflava* are immotile after growth on succinate, but strongly motile after growth on gluconate). For most strains, gluconate peptone medium either in liquid form or solidified by 0.5% agar is well suited for motility tests.

**GRAM STAIN.** Young cultures must be used for the gram stain, possibly after 24 h of incubation. The staining method of Harrigan and McCance [53] has been used successfully, but other methods might also be applied.

**ACID-FAST STAIN.** Use cells from gluconate-peptone medium for the acid-fast strain. The staining methods of Harrigan and McCance [53] and of Gilstrap et al. [49] have been applied successfully. Better results were obtained using 20% (v/v) $H_2SO_4$ instead of acidic alcohol for discoloration [42].

**PIGMENTATION.** The pigmentation of the colonies is generally easy to recognize in cultures growing on mineral agar supplemented by a suitable carbon source. In doubtful cases, collect cells from several Petri dishes or cell suspensions, and treat the paste of cells with acetone:methanol (3:1); then analyze the filtered extract for absorption peaks and shoulders between 340 and 500 nm. If any pigmentation is observed, the spectrum has a good value for species identification (Fig. 4-9) and should in all cases be established.

**GELATIN HYDROLYSIS.** Use a conventional procedure (e.g., liquefaction of nutrient gelatin or the absence of gelatin around colonies on gelatin agar) as revealed by a precipitation reaction.

**DENITRIFICATION.** To judge denitrification by the evolution of gaseous nitrogen, use the Durham method. The medium contains nutrient broth (DIFCO), 8 g; sodium succinate, 2 g; $KNO_3$, 1 g; and water to 1000 ml. Pipette medium into 16-mm diameter test tubes containing an inverted Durham tube. Sterilize the tube at 120°C for 15 min; after cooling, inoculate the medium, and incubate the cultures at 30°C (thermophilic strains at 50° or 60°C) for at least 1 week. Denitrification results in the accumulation of gas in the Durham tubes.

**NITROGEN FIXATION.** Good growth in the basal mineral medium without combined nitrogen is indicative of nitrogen fixation. Use the basal mineral medium without the addition of ammonium chloride; this salt may be replaced by an equivalent amount of sodium chloride. Incubate the liquid and solid media for 1 week or more under an atmosphere of 5 kPa air, 5 kPa $CO_2$, 10 kPa hydrogen, and 80 kPa nitrogen. Make a control with cultures growing in the same nitrogen-free medium under an atmosphere of 10 kPa oxygen, 5 kPa $CO_2$, and 85 kPa hydrogen; no nitrogen fixation should occur under the latter conditions due to the lack of nitrogen in the gas mixture and to the repressive effect of oxygen. A nitrogen-fixing strain will grow better in a gas mixture containing nitrogen but low in oxygen than in the control. The most reliable procedure for detecting nitrogen-fixing activity is the acetylene reduction test [89]. (See also Chapter 1.)

**UTILIZATION OF CARBON SOURCES.** The ability to grow on and use various carbon and energy sources can be determined on solid media. Add sugars (0.2%) or other substrates (0.1%). Some substrates (e.g., phenol, formate, oxalate) may be inhibitory at a concentration of 0.1% and must be used at lower concentrations. Spot inocula of different strains (template or sectorial streaked) on the agar surface; a plate without carbon source serves as a control. Incubate the plates for 1 and 2 weeks; in case of doubt, repeat the test using static or agitated liquid media. Substrates, such as formate or oxalate, which give low yields and may be inhibitory at too high concentrations, should be tested in agitated liquid cultures of sufficient volume to allow an evaluation of the turbidity compared to a substrate-free control.

*Note*: The proposed identification key and matrix (see Tables 4-1 and 4-2) require growth assays on D-galactose, D-fructose, D-xylose, L-rhamnose, sucrose, lactose, trehalose, cellobiose, and mannitol.

ular methods used, a limiting factor in microbial ecology will probably always be the difficulties in performing correct in situ analyses and sampling and in applying reliable enumeration, enrichment, and isolation methods.

*Acknowledgments* Part of the documentation presented here originated from our collaboration with Prof. H. G. Schlegel and from the papers published together. Prof. Schlegel is the former director of the Gottingen Microbiology Institute. Under his direction, this school educated many general, open-minded microbiologists, who today are active on both sides of the Atlantic. With some delay we would like to dedicate this paper to Hans Schlegel on the occasion of his 70th birthday.

The research undertaken in the author's laboratory and presented here benefited from grants from the Swiss National Science foundation. We are grateful to Catherine Fischer for revising the manuscript and to Ralf Conrad, Louis Egger, and Trello Beffa for their helpful information.

## REFERENCES

1. Alfredsson, G.A., Ingason, A., and Kristjansson, J.K. 1986. Growth of thermophilic, obligately autotrophic hydrogen-oxidizing bacteria on thiosulfate. Lett. Appl. Microbiol. 2:21–24.
2. Anderson, K., Tait, R.C., and King, W.R. 1981. Plasmids required for utilization of molecular hydrogen by *Alcaligenes eutrophus*. Arch. Microbiol. 129:384–390.
3. Aragno, M. 1975. Mise en évidence d'hydrogénobactéries corynéformes auxo-hétérotrophes pour la biotine dans l'eau d'un lac eutrophe. Ann. Microbiol. 126A: 539–542.
4. Aragno, M. 1978. Enrichment, isolation and primarily characterization of a thermophilic endospore-forming hydrogen bacterium. FEMS Microbiol. Lett. 3:13–15.
5. Aragno, M. 1991. Thermophilic aerobic, hydrogen-oxidizing (knallgas) bacteria. In The Prokaryotes, A Handbook on Biology of Bacteria, Vol. 4, 2nd ed. (Balows, A., Trüper, H.G., Dworkin, M., Harder, W., and Schleifer, K.H., Eds.), pp. 3917–3933. Springer Verlag, New York.
6. Aragno, M. and Schlegel, H.G. 1978. *Aquaspirillum autotrophicum*, a new species of hydrogen-oxidizing, facultatively autotrophic bacteria. Int. J. Syst. Bacteriol. 28: 112–116.
7. Aragno, M. and Schlegel, H.G. 1981. The hydrogen-oxidizing bacteria. In The Prokaryotes, A Handbook on Habitats, Isolation and Identification of Bacteria (Starr, M.P., et al., Eds.), pp. 865–893. Springer Verlag, Berlin.
8. Aragno, M. and Schlegel, H.G. 1991. The hydrogen-oxidizing (knallgas) bacteria. In The Prokaryotes, A Handbook on Biology of Bacteria, Vol. 1, 2nd ed. (Balows, A., Trüper, H.G., Dworkin, M., Harder, W., and Schleifer, K.H., Eds.), pp. 344–384. Springer Verlag, New York.
9. Aragno, M., Walther-Mauruschat, A., and Schlegel, H.G. 1977. Micromorphology of Gram-negative hydrogen bacteria. I. Cell morphology and flagellation. Arch. Microbiol. 114:93–100.
10. Arreguit, M. 1988. Dinitrogen fixation in the rhizosphere of the Common Reed (*Phragmites communis L.*). Diploma Work in biology, University of Neuchâtel.
11. Auling, G., Reh, M., Lee, C.M., and Schlegel, H.G. 1978. *Pseudomonas pseudoflava*, a new species of hydrogen-oxidizing bacteria: its differentiation from *Pseudomonas flava* and other yellow-pigmented, Gram-negative hydrogen-oxidizing species. Int. J. Syst. Bacteriol. 28:82–95.
12. Bartha, R. and Ordal, E.J. 1965. Nickel-dependent chemolithotrophic growth of two *Hydrogenomonas* strains. J. Bacteriol. 89:1015–1019.
13. Baumgarten, J., Reh, M., and Schlegel, H.G. 1974. Taxonomic studies on some Gram-positive coryneform hydrogen bacteria. Arch. Microbiol. 100:207–217.
14. Beijerinck, M. and Minkman, D.C.J. 1910. Bildung und Verbrauch von Stickoxydul durch Bakterien. Zentralbl. Bakteriol. Infektionskr. Hyg. Abt. 2, 25:30–63.
15. Bojalil, L.F., Cerbon, J., and Trujillo, A. 1962. Adansonian classification of mycobacteria. J. Gen. Microbiol. 28:333–346.
16. Bonjour, F. 1988. Hydrogénobactéries des milieux géothermaux et volcaniques: taxonomie et écologie. PhD thesis, University of Neuchâtel (Switzerland).
17. Bonjour, F. and Aragno, M. 1984. *Bacillus tusciae*, a new species of thermoacidophilic, facultatively chemolithoautotrophic hydrogen-oxidizing sporeformer from a geothermal area. Arch. Microbiol. 139:397–401.

18. Bonjour, F. and Aragno, M. 1986. Growth of thermophilic, obligatorily chemolithoautotrophic hydrogen oxidizing bacteria related to *Hydrogenobacter* with thiosulfate and elemental sulfur as electron and energy source. FEMS Microbiol. Lett. 35:11–15.
19. Bonjour, F., Graber, A., and Aragno, M. 1988. Isolation of *Bacillus schlegelli*, a thermophilic hydrogen-oxidizing aerobic autotroph from geothermal and nongeothermal environments. Microb. Ecol. 16:331–338.
20. Bowien, B. and Schlegel, H.G. 1981. Physiology and biochemistry of aerobic hydrogen oxidizing bacteria. Ann. Rev. Microbiol. 35:405–452.
21. Brotonegoro, S. 1974. Nitrogen fixation and nitrogenase activity in *Azotobacter chroococcum*. Ph.D. Thesis. University of Wageningen, The Netherlands.
22. Canevascini, G. and Eberhardt, U. 1975. Chemolithotrophic growth and regulation of hydrogenase formation in the cornyneform hydrogen bacterium strain 11/x. Arch. Microbiol. 103:283–291.
23. Conrad, R. 1988. Biogeochemistry and ecophysiology of atmospheric CO and $H_2$. Adv. Microb. Ecol. 10:231–283.
24. Conrad, R. and Seiler, W. 1979. Field measurement of hydrogen evolution by nitrogen-fixing legumes. Soil. Biol. Biochem. II:689–690.
25. Conrad, R. and Seiler, W. 1980. The role of microbiological processes for the atmosphere hydrogen cycle. Forum. Microbiol. 3:219–225.
26. Conrad, R. and Seiler, W. 1981. Decomposition of atmospheric hydrogen by soil microorganisms and soil enzymes. Soil Biol. Biochem. 13:43–49.
27. Conrad, R., Aragno, M., and Seiler, W. 1983. Production and consumption of hydrogen in a eutrophic lake. Appl. Environ. Microbiol. 45:502–510.
28. Conrad, R., Aragno, M., and Seiler, W. 1983. The inability of hydrogen bacteria to utilize atmospheric hydrogen is due to threshold and affinity for hydrogen. FEMS Microbiol. Lett. 18:207–210.
29. Conrad, R., Bonjour, F., and Aragno, M. 1985. Aerobic and anaerobic microbial consumption of hydrogen in geothermal spring water. FEMS Microbiol. Lett. 29: 201–205.
30. d'Amore, F. and Nuti, S. 1977. Note on the chemistry of geothermal gases. Geothermica 6:39–45.
31. Davis, D.H., Doudoroff, M., Stanier, R.Y., and Mandel, M. 1969. Proposed to reject the genus *Hydrogenomonas*: taxonomic implications. Int. J. Syst. Bacteriol. 19: 375–390.
32. Davis, D.H., Stanier, R.Y., Doudoroff, M., and Mandel, M. 1970. Taxonomic studies on some Gram-negative polarly flagellated "hydrogen bacteria" and related species. Arch. Mikrobiol. 70:1–13.
33. den Dooren de Jong, L.E. 1927. Ueber protaminophase Bakterien. Zentralbl. Bakteriol. Parasitenkd. Infektionskr. Hyg. Abt. 71:193–232.
34. Doudoroff, M. 1940. The oxidative assimilation of sugars and related substances *Pseudomonas saccharophila* with a contribution to the problem of the direct respiration of the di- and polysaccharides. Enzymologia 9:59–72.
35. Drobner, E., Huber, H., Rachel, R., and Stetter, K.-O. 1992. *Thiobacillus plumbophilus* new species, a novel galena and hydrogen oxidizer. Arch. Microbiol. 157: 213–217.
36. Dugnani, L., Wyrsch, I., Gandolla, M., and Aragno, M. 1986. Biological oxidation of hydrogen in soils flushed with a mixture of $H_2$, $CO_2$, and $N_2$. FEMS Microbiol. Ecol. 38:347–351.
37. Dugnani, L., Wyrsch, I., Gandolla, M., and Aragno, M. 1987. Ossidazione biologica nel terreno di copertura di uno scarico controllato dei gas maleodoranti prodotti nelle prime fasi di degradazione dei rifiuti. Rifiuti Solidi 1:163–168.
38. Emnova, E.E. and Romanova, A.K. 1977. Hydrogenase activity of the thermophilic hydrogen-oxidizing bacterium *Pseudomonas thermophilus* (sic). Microbiology 46: 505–509.
39. Emnova, E.E. and Zavarzin, G.A. 1977. Additional characteristics of a thermophilic hydrogen bacteria *Hydrogenomonas thermophilus*. Microbiology 46:321–324.
40. Evans, H.J. and Barber, L.E. 1977. Biological nitrogen fixation for food and fiber production. Science 197:332–339.

41. Evans, H.J., Ruiz-Argüeso, T., Jennings, N., and Hanus, J. 1977. Energy coupling efficiency of symbiotic nitrogen fixation. In Genetic Engineering for Nitrogen Fixation (Hollander, A., Ed.), pp. 333–354. Plenum Press, New York.
42. Fischer, C. 1988. Etude d'hydrogénobactéries acidorésistantes. Diploma Work at the University of Neuchâtel (Switzerland).
43. Friedrich, B. and Hogrefe, C. 1984. Genetics of lithoautotrophic metabolism in *Alcaligenes eutrophus*. In Microbial Growth On Cl Compounds (Crawford, R.L. and Hanson, R.S., Eds.), pp. 244–247. ASM Press, Washington, DC.
44. Friedrich, C.G. and Mitrenga, G. 1981. Oxidation of thiosulfate by *Paracoccus denitrificans* and other hydrogen bacteria. FMS. Microbiol. Lett. 10:209–212.
45. Friedrich, B., Hogrefe, C., and Schlegel, H.G. 1981. Naturally occurring genetic transfer of hydrogen-oxidizing ability between strains of *Alcaligenes eutrophus*. J. Bacteriol. 147:198–205.
46. Gadkari, D., Schricker, K., Acker, G., Kroppenstedt, R.M., and Meyer, O. 1990. *Streptomyces thermoautotrophicus*, a thermophillic carbon monoxide and hydrogen-oxidizing obligate chemolithoautotroph. Appl. Environ. Microbiol. 56:3727–3734.
47. Gandolla, M. and Aragno, M. 1992. The importance of microbiology in waste management. Experientia 48:362–366.
48. Gerstenberg, C., Friedrich, B., and Schlegel, H.G. 1982. Physical evidence for plasmids in autotrophic, especially hydrogen-oxidizing bacteria. Arch. Microbiol. 133: 90–96.
49. Gilstrap, M., Kleyn, J.G., and Nester, E.W. 1983. Experiments in Microbiology, 2nd ed. Sunders College Publishing, Philadelphia.
50. Goto, E., Kodama, T., and Minoda, Y. 1977. Isolation and culture conditions of thermophilic hydrogen bacteria. Agric. Biol. Chem. 41:685–690.
51. Goto, E., Kodama, T., and Minoda, Y. 1978. Growth and taxonomy of thermophilic hydrogen bacteria. Agric. Biol. Chem. 42:405–408.
52. Hai, W., Wang, Y., You, C., and Thou, H. 1993. Studies on the associative diazotrophs in rice rhizosphere. Acta Microbiol. Sin. 33:79–85.
53. Harrigan, W.F. and McCance, M. 1966. Laboratory Methods in Microbiology. Academic Press, New York.
54. Huber, R., Wilharm, T., Huber, D., Trincone, A., Burggraf, S., König, H., Rachel, R., Rockinger, J., Fricke, H., and Stetter, K.-O. 1992. *Aquifex pyrophilus* new genus new species represents a novel group of marine hyperthermophilic hydrogen-oxidizing bacteria. Syst. Appl. Microbiol. 15:340–351.
55. Hurley, M.A. and Roscoe, M. 1983. Automated statistical analysis of microbial enumeration by dilution series. J. Appl. Bacteriol. 55:159–164.
56. Husemann, M., Klintworth, R., Butcher, V., Salnikov, J., Weissenborn, C., and Bowien, B. 1988. Chromosomally and plasmid-encoded gene clusters for $CO_2$ fixation (cfx genes) in *Alcaligenes eutrophus*. Mol. Gen. Genet. 214:112–120.
57. Igarashi, Y., Kodama, T., and Minoda, Y. 1980. Identification and physiological characterization of analytic hydrogen bacterium *Pseudomonas hydrogenovora*. Agric. Biol. Chem. 44:1277–1280.
58. Ishii, M., Igarashi, Y., and Kodama, T. 1987. Colony formation of *Hydrogenobacter thermophilus* on a plate solidified with gelrite. Agric. Biol. Chem. 51:3139–3142.
59. Jenni, B. and Aragno, M. 1987. *Xanthobacter agilis* nov. sp., a motile dinitrogen-fixing hydrogen-oxidizing bacterium. Syst. Appl. Microbiol. 9:254–257.
60. Jenni, B., Aragno, M., and Wiegel, J.K.W. 1987. Numerical analysis and DNA-DNA hybridization studies on *Xanthobacter* and emendation of *Xanthobacter flavus*. Syst. Appl. Microbiol. 9:247–253.
61. Jenni, B., Realini, L., Aragno, M., and Tamer, A.U. 1988. Taxonomy of non-hydrogen-lithotrophic oxalate-oxidizing bacteria related to *Alcaligenes eutrophus*. Syst. Appl. Microbiol. 10:126–133.
62. Jenni, B., Isch, C., and Aragno, M. 1989. Nitrogen fixation by new strains of *Pseudomonas pseudoflava* and related bacteria. J. Gen. Microbiol. 135:461–467.
63. Kawasumi, T., Igarashi, Y., Komama, T., and Minoda, Y. 1984. *Hydrogenobacter thermophillus* n.gen. n. spec., an extremely thermophilic aerobic hydrogen-oxidizing bacterium. Int. Syst. Bacteriol. 34:5–10.

64. Kelly, D.P. 1971. Autotrophy: concepts of lithotrophic bacteria and their organic metabolism. Annu. Rev. Microbiol. 5:177–210.
65. Kiredjian, M., Holmes, B., Kersters, K., Guilvout, I., and DeLey, J. 1986. *Alcaligenes piechaudii*, a new species from human clinical specimens and the environment. Int. J. Syst. Bacteriol. 36:282–287.
66. Kistner, A. 1953. On a bacterium oxidizing carbon monoxide. Proc. Koninklijke Nederlandse Akademie van Wetenschappen C56:443–450.
67. Kita, I., Matsuo, S., and Wakita, H. 1982. $H_2$ generation by reaction between $H_2O$ and crushed rock: an experimental study on $H_2$ degassing from the active fault zone. J. Geophys. Res. 87:10789–10795.
68. Kraffzik, B. and Conrad, R. 1991. Thymidine incorporation into lake water bacterioplankton and pure cultures of chemolithotrophic carbon monoxide, hydrogen and methanotrophic bacteria. FEMS Microbiol. Ecol. 86:7–14.
69. Kryukov, V.R., Savel'eva, N.D., and Isheva, M.A. 1983. *Calderobacterium hydrogenophilum* n.gen. n.spec., an extremely thermophilic hydrogen bacterium and its hydrogenase activity. Microbiology 52:611–618.
70. Kuznetsov, S.I. 1959. Die Rolle der Mikroorganismen im Stoffkreislauf der Seen, VEB Deutscher Verlag für Wissenschaften, Berlin.
71. LaFavre, J.S. and Focht, D.D. 1983. Conservation in soil of $H_2$ liberated from $N_2$ fixation by Hup nodules. Appl. Environ. Microbiol. 46:304–311.
72. Lalucat, J., Pares, R., and Schlegel, H.G. 1982. *Pseudomonas taeniospiralis* sp. nov., an R-body-containing hydrogen bacterium. Int. J. Syst. Bacteriol. 32:332–338.
73. Lechevalier, M.P., Prauser, H., Labeda, D.P., and Ruan, J.S. 1986. Two new genera of nocardioform actinomycetes: *Amycolata* gen. nov. and *Amycolatopsis* gen. nov. Int. J. Syst. Bacteriol. 36:29–37.
73a. Marchiani, M., Aragno, M. 1995. A modified method for extraction of low molecular weight (LMW) RNAs from thermophilic bacteria. J. Microbiol. Meth. 21: 217–223.
74. Malik, K.A. 1976. Preservation of knallgas bacteria. In Abstracts of the Fifth International Fermentation Symposium Berlin, Session 9.08 (Dellweg, H., Ed.), p. 180. Verlag Versuchs- und Lehranstalt für Spiritusfabrikation und Fermentationstechnologie, Berlin.
75. Malik, K.A. and Claus, D. 1979. *Xanthobacter flavus*, a new species of nitrogen-fixing hydrogen bacteria. Int. J. Syst. Bacteriol. 29:283–287.
76. Malik, K.A. and Schlegel, H.G. 1980. Enrichment and isolation of new nitrogen-fixing hydrogen bacteria. FEMS Microbiol. Lett. 8:101–104.
77. McGee, J.M., Brown, L.R., and Tischer, R.G. 1967. A high temperature hydrogen-oxidizing bacterium—*Hydrogenomonas thermophilus* (sic) n. sp. Nature 214:715–716.
78. Medard, L., Angibaud, G., Barbe, Bergeon, J.J., Bryselbout, J., Creuse, R., Deflers, J., Feiche, C., Goardou, T., Laloe, M., David, M., Ducoux, C., and Koenig, J. In Gas Encyclopedia (Allamagny, P., Ed.), Elsevier, Amsterdam.
79. Meyer, O., Stackebrandt, E., and Auling, G. 1993. Reclassification of ubiquinone Q-10 containing carboxidotrophic bacteria: transfer of "(*Pseudomonas*) *carboxydovorans*" OM5-T to *Oligotropha*, gen. nov., as *Oligotropha carboxidovorans*, comb. nov., transfer of "(Alcaligenes) carboxydus" DSM 1086-T to *Carbophilus*, gen. nov., as *Carbophilus carboxidus*, comb. nov., transfer of "(*Pseudomonas*) *compransoris*" DSM 1231-T to *Zavarzinia* gen. nov., as *Zavarzinia compransoris*, comb. nov., and amended descriptions of the new genera. Syst. Appl. Microbiol. 16:390–395.
80. Niklewski, B. 1910. Ueber die Wasserstoffoxydation durch Mikroorganismen. Jahrb. f. wissenschaftl. allg. Mikrobiol. 3:251–264.
81. Nishihara, H., Igarashi, Y., and Kodama, T. 1989. Isolation of an obligately chemolithoautotrophic, halophilic and aerobic hydrogen-oxidizing bacterium from marine environment. Arch. Microbiol. 152:39–43.
82. Nishihara, H., Igarashi, Y., and Kodama, T. 1990. A new isolate of *Hydrogenobacter*, an obligately chemolithoautotrophic, thermophilic, halophilic and aerobic hydrogen-oxidizing bacterium from seaside saline hot springs. Arch. Microbiol. 153:294–298.

83. Nishihara, H., Igarashi, Y., and Kodama, T. 1991. *Hydrogenovibrio marinus* new genus new species, a marine obligately chemolithoautotrophic hydrogen-oxidizing bacterium. Int. J. Syst. Bacteriol. 41:130–133.
84. Nokhal, T.H. and Schlegel, H.G. 1983. Taxonomic study of *Paracoccus denitrificans*. Int. J. Syst. Bacteriol. 33:26–37.
85. Orskov, J. 1928. Berschreibung einer neuen mikroben, *Microccyclus aquaticus* mit eigentümlicher Morphologie. Zentralbl. Bakteriol. Parasitenkd. Infektionskr. Hyg. Abt. 1, 107:180.
86. Oyaizu-Mazuchi, Y. and Komagata, K. 1988. Isolation of free-living nitrogen-fixing bacteria from the rhizosphere of rice. J. Gen. Appl. Microbiol. 34:127–164.
87. Pagni, M., Egger, L., and Aragno, M. 1995. The relationship between kinetics of substrate-limited transitions and steady-state growth in continuous cultures of *Aquaspirillum autotrophicum* limited by pyruvate. Antonie van Leeuwenhoek 68: 181–189.
88. Palleroni, N.J. and Palleroni, A.V. 1978. *Alcaligenes latus*, a new species of hydrogen-utilizing bacteria. Int. J. Syst. Bacteriol. 28:416–424.
89. Postgate, J.R. 1972. The acetylene reduction test for nitrogen fixation. In Methods in Microbiology, Vol. 6B (Norris, J.R. and Ribbons, D.W., Eds.), pp. 343–356. Academic Press, New York.
90. Raj, H.D. 1989. Oligotrophic methylotrophs: *Ancylobacter* (basonym "Microcyclus" frskov) Raj gen. nov. CRC Crit. Rev. Microbiol. 17:89–106.
91. Reding, H.K., Hartel, P.G., and Wiegel, J. 1991. *Xanthobacter* as a rhizosphere organism of rice. Plant Soil 137:1–12.
92. Reh, M. and Schlegel, H.G. 1969. Die Biosynthese von Isoleucin und Valin in Hydrogenomonas H16. Arch. Mikrobiol. 67:110–127.
93. Reh, M. and Schlegel, H.G. 1975. Chemolithoautotrophie als eine übertragbare, autonome Eigenschaft von Nocardia opaca lb. Nachrichten Akad. Wiss. Göttingen, Math.-Phys. Kl. 2 12:207–216.
94. Repaske, R. 1966. Characteristics of hydrogen bacteria. Biotechnol. Bioengin. 8: 217–135.
95. Repaske, R. and Mayer, R. 1976. Dense autotrophic cultures of *Alcaligenes eutrophus*. Appl. Environ. Microbiol. 32:595–597.
96. Repaske, R. and Repaske, A.C. 1976. Quantitative requirements for exponential growth of *Alcaligenes eutrophus*. Appl. Environ. Microbiol. 32:585–591.
97. Robson, R.L. and Postgate, J.R. 1980. Oxygen and hydrogen in biological nitrogen fixation. Annu. Rev. Microbiol. 34:183–207.
98. Schatz, A. and Bovell, C. Jr. 1952. Growth and hydrogenase activity of new bacterium, *Hydrogenomonas facilis*. J. Bacteriol. 63:87–98.
99. Schenk, A. and Aragno, M. 1979. *Bacillus schlegelii*, a new species of thermophillic, facultatively chemolithoautotrophic bacterium oxidizing molecular hydrogen. J. Gen. Microbiol. 115:333–342.
100. Schlegel, H.G. 1966. Physiology and biochemistry of Knallgas bacteria. Adv. Comp. Physiol. Biochem. 2:185–236.
101. Schlegel, H.G. 1969. From electricity via water electrolysis to food. In Fermentation Advances (Perlman, D., Ed.), pp. 807–832. Academic Press, New York.
102. Schlegel, H.G. 1976. The physiology of hydrogen bacteria. The fifth A.J. Kluyver memorial lecture. Antonie van Leeuwenhoek 42:181–201.
103. Schlegel, H.G. 1989. Aerobic hydrogen-oxidizing (Knallgas) bacteria. In Autotrophic Bacteria (Schlegel, H.G. and Bowien, B., Eds.), pp. 305–329. Science Technology Publishing, Madison, WI.
104. Schlegel, H.G. and Eberhardt, U. 1972. Regulatory phenomena in the metabolism of Knallgas bacteria. Adv. Microb. Physiol. 7:205–242.
105. Schlegel, H.G. and Lafferty, R.M. 1971. Novel energy and carbon sources. A: The production of biomass from hydrogen and carbon dioxide. Adv. Biochem. Engin. 1:143–168.
106. Schlegel, H.G., Kaltwasser, H., and Gottschalk, G. 1961. Ein Submersverfahren zur Kultur wasserstoffoxydierender Bakterien: Wachstumsphysiologische Untersuchungen. Arch. Mikrobiol. 38:209–222.

107. Schubert, K.R. and Evans, H.J. 1976. Hydrogen evolution: a major factor affecting the efficiency of nitrogen fixation in nodulated symbionts. Proc. Nat. Acad. Sci. USA 73:1207–1211.
108. Schuler, S. and Conrad, R. 1990. Soils contain two different activities for oxidation of hydrogen. FEMS Microbiol. Ecol. 73:77–84.
109. Schuler, S. and Conrad, R. 1991. Hydrogen oxidation activities in soil as influenced by pH, temperature, moisture and season. Biol. Fertil. Soils 12:127–130.
110. Schuler, S. and Conrad, R. 1991. Hydrogen oxidation in soil following rhizobial hydrogen production due to nitrogen fixation by a *Vica faba* and *Rhizobium leguminosarum* symbioses. Biol. Fertil. Soils 11:190–195.
111. Schweizer, C. and Aragno, M. 1975. Etude des hydrogénobactéries dans un petit lac (le Loclat, ou lac de Saint-Blaise). Bull. Soc. Neuchâtel. Sci. Nat. 98:79–87.
112. Seiler, W. 1978. The influence of the biosphere and the atmospheric CO and $H_2$ cycles. In Environmental Biogeochemistry and Geomicrobiology (Krumbein, W.E., Ed.), pp. 773–810. Ann Arbor Sciences Publishers, Ann Arbor, MI.
113. Seiler, W., Giehl, H., and Roggendorf, P. 1980. Detection of carbon monoxide and hydrogen by conversion of mercury oxide to mercury vapor. Atmos. Technol. 12: 40–45.
114. Shiba, H., Kawasumi, T., Igarashi, Y., Kodama, T., and Minoda, Y. 1985. The carbon dioxide assimilation via the reductive tricarboxylic acid cycle in an obligately autotrophic aerobic hydrogen-oxidizing bacterium, *Hydrogenobacter thermophilus*. Arch. Microbiol. 141:198–203.
115. Shima, S. and Suzuki, K.I. 1993. *Hydrogenobacter acidophilus* sp. nov., a thermophilic, aerobic, hydrogen-oxidizing bacterium requiring elemental sulfur for growth. Int. J. Syst. Bacteriol. 43:703–708.
116. Shima, S., Yanagi, M., and Saiki, H. 1994. The phylogenetic position of *Hydrogenobacter acidophilus* based on 16S rRNA sequence analysis. FEMS Microbiol. Lett. 119:119–122.
117. Simpson, F.B. and Burris R.H. 1984. A nitrogen pressure of 50 atmosphere does not prevent evolution of hydrogen by nitrogenase. Science 224:1203–1215.
118. Smith, R.L., Ceazan, M.L., and Brooks, M.H. 1994. Autotrophic, hydrogen-oxidizing, denitrifying bacteria in groundwater: potential agents for bioremediation of nitrate contamination. Appl. Environ. Microbiol. 60:1949–1955.
119. Tabillion, R. and Kaltwasser, H. 1977. Energieabhängige $^{63}$Ni-Aufnahme bei *Alcaligenes eutrophus* Stamm H1 und H16. Arch. Microbiol. 113:145–151.
120. Takamiya, A. and Tubaki, K. 1956. A new form of *Streptomyces* capable of growing autotrophically. Arch. Mikrobiol. 25:58–64.
121. Vollbrecht, D. and Schlegel, H.G. 1979. Excretion of metabolites by hydrogen bacteria. III. D(-)-3-hydroxybutanoate. Eur. Appl. Microbiol. Biotechnol. 7:259–266.
122. Waksman, S.A. and Henrici, A.T. 1948. Family II: *Actinomycetaceae* Buchanan. In Bergey's Manual of Determinative Bacteriology, 6th ed. (Breed, R.S., Murray, E.G.D., and Hitchens, P.A., Eds.), pp. 892–928. Baillère, Tindall & Cox, London.
123. Walther-Mauruschat, A., Aragno, M., Schlegel, H.G. 1977. Micromorphology of Gram-negative hydrogen bacteria. II. Cell envelope, membranes and cytoplasmic inclusions. Arch. Microbiol. 114:101–110.
124. Wiegel, J. and Schlegel, H.G. 1976. Enrichment and isolation of nitrogen-fixing hydrogen bacteria. Arch. Microbiol. 107:139–142.
125. Wiegel, J., Wilke, D., Baumgarten, J., Opitz, R., and Schlegel, H. 1978. Transfer of the nitrogen-fixing hydrogen bacterium *Corynebacterium autotrophicum* Baumgarten et al. to *Xanthobacter* gen. nov. Int. J. Syst. Bacteriol. 28:573–581.
126. Willems, A., Busse, J., Goor, M., Pot, B., Falsen, E., Jantzen, E., Hoste, B., Gillis, M., Kersters, K., Auling, G., and De Ley, J. 1989. *Hydrogenophaga*, a new genus of hydrogenoxidizing bacteria that includes *Hydrogenophaga flava* comb. nov. (formerly *Pseudomonas flava*), *Hydrogenophaga palleronii* (formerly *Pseudomonas palleronii*), *Hydrogenophaga pseudoflava* (formerly *Pseudomonas pseudofava* and "*Pseudomonas carboxydoflava*"), and *Hydrogenophaga taeniospiralis* (formerly *Pseudomonas taeniospiralis*). Int. J. Syst. Bacteriol. 39:319–333.
127. Willems, A., Falsen, E., Pot., E., Jantzen, E., Hoste, B., Vandamme, P., Gillis, M., Kersters, K., Auling, G., and Deley, J. 1990. *Acidovorax*, a new genus for *Pseu-*

*domonas facilis*, *Pseudomonas delafieldii* E Falsen (EF) group 13, EF group 16, and several clinical isolates, with the species *Acidovorax facilis* comb. nov., *Acidovorax delafieldii* comb. nov. and *Acidovorax temperans* sp. nov. Int. J. Syst. Bacteriol. 40:384–398.

128. Willems, A., DeLey, J., Gillis, M., and Kersters, K. 1991. *Comamonadaceae*, a new family encompassing the *Acidovorans* ribosomal RNA complex including *Variovorax paradoxus* new genus new combination for *Alcaligenes paradoxus* Davis 1969. Int. J. Syst. Bacteriol. 41:445–450.
129. Zavarzin, G.A. 1976. Hydrogen and carboxydobacteria belonging to the microflora of dispersal Mikrobiologiya 45:20–22 (in Russian, with English translation).
130. Zavarzin, G.A. 1978. Hydrogen Bacteria and Carboxydobacteria. Nauka, Moscow (in Russian).

5

STEPHEN H. ZINDER

# Methanogens

Methanogens are responsible for $CH_4$ production in a wide variety of anaerobic habitats, including marine and freshwater sediments, marshes and swamps, flooded soils, certain groundwater aquifers, bogs, geothermal habitats, animal gastrointestinal tracts, and anaerobic bioreactors. They convert a relatively narrow range of simple substrates to $CH_4$ (Table 5-1). Therefore, in habitats with complex organic substrates, they must interact with other anaerobes that catabolize complex substrates into simpler methanogenic ones.

A few electron donors can be used by methanogenes to reduce $CO_2$ to $CH_4$ (Tables 5-1 and 5-2). Nearly all methanogens can use $H_2$, and this reaction is considered important in almost all methanogenic habitats, often accounting for one-third of the methane produced in many freshwater sediments and anaerobic bioreactors and nearly all of the methane produced in certain gastrointestinal habitats, including the animal rumen [40]. The ability of methanogens to scavenge $H_2$ in natural habitats allows certain reactions, especially oxidation of substrates to acetate, to become more thermodynamically favorable, and this process has been termed *interspecies hydrogen transfer* [37]. Most methanogens that use $H_2$ can also use formate as an electron donor for $CO_2$ reduction, and some evidence has been obtained that formate may be more important than $H_2$ as an electron donor for $CO_2$ reduction in some habitats [5, 34]. More recently, it has been found that some methanogens can use certain short-chain alcohols for $CO_2$ reduction, although the importance of this reaction in natural habitats in unclear.

Acetate decarboxylation is carried out by only two methanogenic genera (see Table 5-2), but has been shown to be responsible for approximately two-thirds of the methane produced in most freshwater sediments and in anaerobic bioreactors [40]. Methanol fermentation is carried out by a limited number of methylotrophic methanogens and may be important in habitats where it is released from methylated sugars. Methylamines are products of the metabolism of N-methylated compounds, including the important lipid constituent choline and the osmoprotectants, glycine-betaine and trimethylamine N-oxide. Since the latter two compounds are often accumulated in high amounts in tissues of organisms growing at high salin-

Table 5-1 Methanogenic Reactions

| Reactants | Products | $\Delta G^{\circ\prime}$ (kJ/mol $CH_4$) | Symbol Table 2 |
|---|---|---|---|
| $4H_2 + HCO_3^- + H^+$ | $CH_4 + 3H_2O$ | −135 | H |
| $4HCO_2^- + H^+ + H_2O$ | $CH_4 + 3HCO_3^-$ | −145 | F |
| $2CH_3CH_2OH + HCO_3^-$ (a) | $2CH_3COO^- + H^+ + CH_4 + H_2O$ | −116 | A |
| $CH_3COO^- + H_2O$ | $CH_4 + HCO_3^-$ | −31 | Ac |
| $4CH_3OH$ | $3CH_4 + HCO_3^- + H_2O + H^+$ | −105 | M |
| $4(CH_3)_3\text{-}NH^+ + 9H_2O$ (b) | $9CH_4 + 3HCO_3^- + 4NH_4^- + 3H^+$ | −76 | MA |
| $2(CH_3)_2\text{-}S + 3H_2O$ (c) | $3CH_4 + HCO_3^- + 2H_2S + H^+$ | −49 | MeS |
| $CH_3OH + H_2$ | $CH_4 + H_2O$ | −113 | H/M |

$\Delta G^{\circ\prime}$ values from Thauer et al. [33].

ities, methylamines can be important methane precursors in marine and high-salinity environments [27]. Methylated sulfides can be derived from the metabolism of methionine or the osmoprotectant propiothetin or from the methylation of sulfide by acetogens and can play an important role in the sulfur cycle in many freshwater and marine habitats.

Methanogens are relatively slow-growing strict anaerobes, and our understanding of methanogen diversity increased greatly in the last two decades with the development of improved anaerobic techniques for their culture, which are described in greater detail in this chapter. The methanogens are charter members of the domain Archaea (Archaebacteria), and represent several orders within that domain. The hierarchy of their taxonomy is based essentially on 16S rRNA sequencing [6] and is shown to the genus level in Table 5-2. Our understanding of the biochemistry of methane production has also greatly increased in the last two decades, and several novel enzymes, co-enzymes, and biochemical reactions have been discovered in methanogens [13].

Because of their important role in the carbon cycle and their practical significance, the methanogens and their ecology have been studied fairly intensively. Although the methanogens are relatively difficult to study in pure culture, they possess unique properties that make studying their ecology relatively simple compared with other organisms, such as glucose-fermenting heterotrophs. Among those properties are (1) the obligatory production of a unique product ($CH_4$) from a relatively narrow range of substrates; (2) the presence of unique biochemical markers, such as high concentrations of co-factor $F_{420}$, which makes many of them autofluorescent when viewed with a fluorescence microscope; and (3) the presence of the enzyme, methyl co-enzyme M methylreductase, which can be specifically inhibited by the co-enzyme M analog, bromoethane sulfonic acid. That methanogens are part of the Archaea domain makes them resistant to many common antibiotics, including inhibitors of peptidoglycan and protein synthesis. Thus, there are many useful tools that can help us understand the numbers and activity of methanogens in natural habitats.

This chapter focuses on relatively simple cultural methods used to enumerate, isolate, identify, and grow methanogens since these approaches are of general utility. Activity measurement techniques, microscopic observation techniques, immunologic, and molecular biologic techniques are described in less detail, but with references to direct the researcher to appropriate sources. Ideally, several methods can be used to provide a more complete picture of the number of methanogens present and their activity. A detailed summary of more general issues in methanogen ecology was recently published by the author [40] in a monograph [13] with excellent descriptions of methanogen taxonomy, cell structure, biochemistry, and genetics.

**Table 5-2** Taxonomy of Methanogens

| Order | Family | Genus | Morphology | Substrates[a] | Comments[b] | Typical Species[c] |
|---|---|---|---|---|---|---|
| *Methanobacteriales* | *Methanobacteriaceae* | *Methanbacterium* | Rod | H, F*, A* | Thermophilic species may form a new genus | *formicicum* (1535)<br>*thermoautotrophicum* (1053) |
| | | *Methanobrevibacter* | Short rod | H, F | | *smithii* (861) |
| | | *Methanosphaera* | Coccus | H/M | | *stadtmanii* (3091) |
| | *Methanothermaceae* | *Methanothermus* | Rod | H | All known species are extreme thermophiles | *fervidus* (2088) |
| *Methanococcales* | *Methanococcaceae* | *Methanococcus* | Cocci | H, F* | All known strains are marine; thermophilic hyperthermophilic species may form new | *maripludis* (2067)<br>*thermolithotrophicus* (2095) |
| *Methanomicrobiales* | *Methanomicrobiaceae* | *Methanomicrobium* | Short rod | H, F*, A* | | *jannashii* (2661)<br>*mobile* (1539) |
| | | *Methanogenium* | Irregular coccus | H, F, A* | Marine | *cariaci* (1497) |
| | | *Methanoculleus* | Irregular coccus | H, F, A* | One species thermophilic; was previously classified as *Methanogenium* | *marisnegri* (1498) |
| | | *Methanoplanus* | Plate | H, F | | *limicola* (2297) |
| | *Methanocorpusculaceae* | *Methanocorpusculum* | Irregular coccus | H, F, A* | | *parvum* (3823) |
| | *Methanospirillaceae* | *Methanospirillum* | Long spirals | H, F, A | | *hungateii* (864) |
| *Methanosarcinales* | *Methanosarcinacae* | *Methanosarcina* | Packets or irregular cocci | H*, M, MA, Ac | One species thermophilic; most species factultatively marine | *barkeri* (800)<br>*thermophila* (1825) |
| | | *Methanolobus* | Irregular coccus | M, MA, MeS | All species marine | *tindarius* (2278) |
| | | *Methanococcoides* | Irregular coccus | M, MA | Marine | *methylutans* (2657) |
| | | *Methanohalophilus* | Irregular coccus | M, MA | Moderate halophile | *halophilus* (3094) |
| | | *Methanohalobium* | Irregular coccus | M, MA | Halophile, moderate thermophile | *evestugatyn* (3721) |
| | *Methanoseataceae* | *Methanosaeta* (*Methanothrix*) | Sheathed filament | Ac | One species thermophilic | *concillii* (3671) (*soehngenii*) |
| *Methano-pyrales* | *Methanopyraceae* | *Methanopyrus* | Rod | H | Hyperthermophlilic | *thermophila* (6194)<br>*kandleri* (6324) |

*Denotes that only some species use this substrate.

[a]Substrate legend: H, $H_2/CO_2$; F, formate; A, alcohols (other than methanol); M, methanol; H/M, $H_2$/methanol; MA, methylamines; Ac, Acetate.

[b]Unless stated otherwise, it is assumed that all species are mesophiles and freshwater. Thermophiles have temperature optima near 60°C, extreme thermophiles near 80°C, and hyperthermophiles ≥100°C.

[c]Deutche Sammlung von Mikroorganism (DSM) number of the type species are in parentheses.

## CULTURAL METHODS FOR STUDYING METHANOGENS

The first major improvements in methods used to culture methanogens were the anaerobic techniques developed by R.E. Hungate in the 1940s. One should still consult his classic review of these techniques [16], as well as a more recent review [7] for discussions of the general principles of anaerobiosis and of habitat simulation in culture media. The basic technique involves boiling the medium to drive off $O_2$, and then performing all subsequent manipulations under a blanket of an inert gas, usually either $N_2$ or $CO_2$ or a mixture thereof. Reducing agents, such as cysteine ($E^{0\prime} = -210$ mv) and $Na_2S$ ($E^{0\prime} = -270$ mv), were used, and the dye resazurin ($E^{0\prime} = -51$ mv for the pink/clear transition) was used as an oxidation-reduction indicator. The original Hungate tubes used simple rubber stoppers that fit into a tooled neck. The trick to the technique was to remove a gassing needle from the culture vessel while simultaneously pushing the rubber stopper into the tooled opening of the vessel. A major disadvantage of the Hungate tube was that, if the gas pressure in the tube built up to any appreciable level, the stopper would pop out, so that the tubes needed to be held in a special press during autoclaving to hold the stoppers down. In addition, for cultures that produced large amounts of gas, there was always the danger of finding the culture with a popped stopper. In general, the Hungate technique was laborious and difficult to master, and only a few pure cultures of methanogens were obtained using it.

A quantum leap in the ability to culture methanogens was achieved when the methods of Balch et al. [1] were developed in the late 1970s. One major improvement was replacement of the butyl-rubber stoppers used by Hungate with thick butyl-rubber serum-septa held in place by aluminum crimps so that the seals were pressure-tight. These stoppers could be used with specially made culture tubes with crimp-top, or with standard serum bottles. Methanogens growing on $H_2/CO_2$ could be pressurized with several atm of these gases, allowing much greater growth. The other major improvement was the use of the relatively inexpensive Coy (Coy Laboratory Products, Grass Lake, MI) anaerobic chambers for many of the anaerobic manipulations that had previously been done on the benchtop under a blanket of gas. Anaerobic gloveboxes have allowed researchers to dispense culture medium, harvest cells, purify enzymes, and plate out cells in the absence of oxygen. Anaerobic chambers are usually not anaerobic enough to allow growth of methanogens, so Petri plates are typically incubated in air-tight chambers [1]. Most researchers use a combination of benchtop and glovebox techniques for the handling of methanogen cultures, and each anaerobe laboratory has its own set of techniques and lore. However, although these newer techniques have made culture of methanogens tractable, even experienced researchers have unexplained lapses in the ability to grow them.

The techniques outlined in this chapter are relatively simple and have worked well in our laboratory for both ecologic and pure culture studies of methanogens. The methods described use the minimum amount of sophisticated anaerobic equipment possible, so that laboratories lacking anaerobic chambers and gassing manifolds plumbed with copper tubing and expensive valves can perform some simple studies on methanogens. They are also as simple as possible. For example, rather than boiling the medium to remove $O_2$, the medium is simply flushed with $N_2$, which is much less laborious, but allows adequate anaerobiosis when used in concert with the other techniques described.

### Preparation of Culture Medium

The formulation of the growth medium we use for culturing methanogens is described in this section. We have used this medium and its variants (marine salts, etc.) to grow a variety of methanogens, and all methanogens, as well as many other anaerobes we have studied, grow in this medium. The reducing agent is $Na_2S$, which allows less growth of contaminants in mixed culture studies than does cysteine, which can be used as an energy source by some fermentative heterotrophs. However, some anaerobes find $Na_2S$ toxic. The redox indicator is resazurin, which changes from purple to pink to clear when reduced. The buffer system is $HCO_3^-/CO_2$, which is often the buffer system found in natural habitats for methanogens and ensures that $CO_2$ requirements of methanogens are met. Chloride salts are used instead of sulfate salts to discourage the growth of sulfate-reducing bacteria. To culture methanogens and other anaerobes from natural habitats, we also suggested adding a vitamin solution. Other more complex nutrient amendments, such as rumen fluid or yeast extract, may increase the probability of culturing methanogens from a given habitat, but at the cost of supporting many other organisms. A discussion of methanogen nutrition is given by Jarrell and Kalmokoff [18].

Most studies use 18 × 150 mm crimp-top tubes (volume ≈27 ml), and so the methods described involve these tubes. For studies involving larger amounts of cells, 100-ml serum bottles (actual volume = 120 ml) or 1-liter bottles can be used [1]. For the growth of methanogens on aqueous substrates, 10 ml of medium is used, whereas $H_2/CO_2$, 5 ml is used so that there is a larger headspace/liquid volume, allowing better gas transfer to the liquid and more substrate present per ml culture.

The minimum requirement for gases is one tank each of prepurified 100% $N_2$ and mixtures of 70% $N_2$/30% $CO_2$ and 80% $H_2$/20% $CO_2$. The $N_2$ can usually be obtained locally and inexpensively, but the quality of such products can vary. If problems are observed, such as poor reduction of the growth medium, one may have to obtain a better-quality gas from a gas supplier, such as Matheson or Linde. The gas can be scrubbed over hot copper coils [16] or by passing through a solution of Ti(III) citrate [38], but with good-quality gas, this step should not be necessary.

A simple anaerobic gassing manifold can be made using butyl-rubber tubing, and glass T-type connectors, and other materials (see Appendix 5-A). It is important to use screw-type hose clamps for connections using rubber tubing, since any leaks will cause air to come into the system because of counterflow, even under positive pressure. Plastic pinch clamps can be used as on/off valves and to provide crude flow control. There should be at least five outlets (rubber hoses) to the manifold, and they should be able to accept Luer-hub needles. Either specialized barbed Luer-Lok adapters can be inserted into the end of the rubber tubing, or one can simply cut a 1-ml disposable syringe in half and fasten it to the end with a hose clamp. Suitable gassing needles are 17-gauge × 3½ inch and should be bent slightly near the hub so that they can easily rest on the lip of vials and tubes. If medium is to be dispensed on the benchtop, the minimum requirement is one outlet for gassing the vessel containing the medium, four others so that four tubes or vials can be gassed at once, and one permanently sterile outlet for aseptic gassing. This gassing manifold should be attached to the $N_2$ and $N_2/CO_2$ tanks together. A gas pressure of 5 pounds per square inch (psi) is sufficient. For temporary aseptic gassing, the 17-gauge needles can be replaced with a sterile Acrodisc or another membrane filter apparatus fitted with a sterile disposable 23-gauge × 1-inch needle. The needles can be replaced after several uses when they become dull, and the filters can be used indefinitely until there is reason to believe that their sterile bottom sides have become contaminated. At least one more outlet in the manifold should be present for permanently sterile gassing. In this outlet, a wad of cotton is placed upstream of the hub to filter microorganisms from the gas flow. This assembly should be autoclaved intact, and the 3½-inch needle can be subsequently sterilized by flaming. The $H_2/CO_2$ tank can have a single tube attached to its regulator terminating in a Luer-hub connector and membrane filter apparatus and can be used for providing overpressure of $H_2/CO_2$.

Syringes and needles replace pipettes for transferring liquids in anaerobic technique. **Note**: It is essential that all personnel using syringes and needles, especially involving natural samples, are adequately immunized against tetanus. In most cases, sterile disposable plastic syringes and needles can be used (see Appendix 5-A). However, when transferring organisms, a disposable glass syringe is preferable to a plastic one, since some oxygen is dissolved in the plastic. The stoppers on tubes or vials should be swabbed with 95% ethanol and flamed before making aseptic additions. Since the syringes have a dead volume of approximately 0.1 ml of air, it is important to flush them with $N_2$ just before use; when the transfers need to be aseptic (i.e., when transferring sterile solutions), this gas should be sterile. A convenient method for making gas sterile is to prepare sterile 150 × 18 mm standard culture tubes covered with plastic caps. Immediately after removing a cap, the permanently sterilized gassing needle should be inserted to flush the tube with sterile $N_2$. One can then insert the syringe into the tube and flush it several times rapidly (the gas flow should be high enough so that no air is pulled into the tube on the upstroke) to remove air.

Procedure

1. Using the recipe provided below, make an appropriate volume of the medium in a flask with a constricted neck, such as an Erlenmeyer flask or a boiling flask.
2. Bubble the medium vigorously with $N_2$ for 45 minutes.
3. If the medium is to be dispensed inside an anaerobic glovebox, seal it with a stopper that is secured in place (wire can be used for this purpose) so that it does not pop out during evacuation of the airlock. Dispense the medium into tubes or vials with a Repipet-type dispenser or with a syringe fitted with a 14-gauge cannula. Seal the tubes with stoppers and crimps, and then take them out of the glovebox for subsequent autoclaving.
4. If the medium is dispensed on the benchtop, after bubbling, raise the gassing needle so that it rests on the lip of the vessel. Begin flushing the first 4 tubes in a 40-tube rack with $N_2$ using the gassing manifold.
5. Dispense the medium with a syringe (10 ml, glass or plastic) fitted with a 14-gauge cannula. Before dispensing the first sample, flush the syringe several times with $N_2$ using one of the tubes. Dispense 10 ml of medium per tube or 5 ml if $H_2/CO_2$ is the methanogenic substrate.
6. After dispensing the medium, place a rubber stopper partway into the neck, keeping the needle in place. The needle can then be removed while pushing down on the stopper. If it is removed in a smooth motion, no air should enter the culture vessel. Place the gassing needle in the next tube one row back in the rack.
7. When all of the tubes have been stoppered, crimp the stopper in place.
8. Prepare anaerobic stock solutions (see below) at this point as needed.
9. Sterilize the medium and stock solutions in an autoclave for at least 15 min for tubes and 30 min for vials, and allow the medium to cool.
10. Flush the headspaces of culture tubes with sterile $N_2/CO_2$ using the sterile Acrodisc-type membrane filter assemblies. Before flushing, lightly swab all stoppers with alcohol and flame them. Put the needle on the membrane filter assembly through the stopper before an exit needle to ensure positive pressure. Four tubes can be gassed simultaneously. After 1 min of flushing,

remove the gassing needle and place it in the next tube down the line, followed after a few seconds by the exit needle, thereby allowing the vessel to re-equilibrate to 1 atm pressure.

11. Next swab the stoppers with alcohol again, and add to each tube, using a sterile $N_2$-flushed disposable plastic syringe, 0.1 ml of a sterile anaerobic 10% w/v $NaHCO_3$ followed by 0.1 ml of 5% v/v $Na_2S \cdot 9H_2O$. The stoppers do not need to be swabbed with alcohol again between additions unless they are touched by a non-sterile surface, such as fingers. The resazurin in the medium should turn from purple to pink to clear within 5 min if the medium is prepared correctly. If there is a pink meniscus at the top of the liquid layer or if the medium turns pink upon shaking the tube, then the headspace of the culture is contaminated with $O_2$, the tubes should be discarded, and the anaerobic techniques need to be examined. Other nutrient amendments can also be added at this point (see below).
12. Finally, add the methanogenic substrate, either 0.1 ml of 4 M sodium acetate solution for acetotrophic methanogens, 0.1 ml of 2 M methanol or methylamines for methylotrophic methanogens, or 2 atm overpressure of $H_2/CO_2$ for hydrogenotrophic methanogens. *Note*: This last component should be added after inoculation because it is difficult to add the inoculum to highly pressurized tubes.

## Growth Medium Formulation

### *Basal Medium*

| | |
|---|---|
| $NH_4Cl$ | 0.5 g/l |
| $K_2HPO_4$ | 0.4 g/l |
| $MgCl_2 \cdot 2H_2O$ | 0.1 g/l |
| $CaCl_2 \cdot 2H_2O$ | 0.05 g/l |
| Trace metal solution | 10 ml |
| Resazurin solution (stock solution 1 g/l) | 1 ml |

Adjust pH to 6.8 with 1 M HCl. For marine medium, also add 30 g/l NaCl and 5 g/l $\cdot$ $MgCl_2 \cdot 2H_2O$.

### *Trace Metal Solution*

| | *g/l* |
|---|---|
| Nitrilotriacetic acid (disodium) | 4.5 |
| $FeCl_2 \cdot 7H_2O$ | 0.4 |
| $MnCl_2 \cdot H_2O$ | 0.1 |
| $CoCl_2 \cdot 6H_2O$ | 0.17 |
| $ZnCl_2$ | 0.1 |
| $H_3BO_3$ | 0.19 |
| $NiCl_2 \cdot 6H_2O$ | 0.02 |
| $Na_2MoO_4 \cdot 2H_2O$ | 0.01 |

Adjust pH to 7 with 10 N KOH.

### *100X Stock Solutions* (Conveniently Prepared in 50-ml Amounts in 120-ml Serum Vials)

| | |
|---|---|
| 5% w/v $Na_2S \cdot 9H_2O$ | 200 mM |
| 10% w/v $NaHCO_3$ | 1.2 M |

| | |
|---|---|
| Na acetate | 4 M |
| Na formate | 5 M |
| 10% v/v methanol | 2.4 M |
| Na trimethylamine | 1 M |

Notes

1. Wearing rubber gloves, rinse crystals with distilled water and pat dry before weighing. Add 200 mM $Na_2S \cdot 9H_2O$ to distilled water that has been previously bubbled for at least 10 min, and seal immediately.
2. Bubble distilled water for at least 10 min before adding methanol, and seal immediately.
3. Bubble 1 M NaOH for at least 10 min before adding trimethylamine, and seal immediately.

### *Vitamin Solution (100X)* (Modified from Balch et al. [1])

| | *mg/l* |
|---|---|
| Biotin | 2 |
| Folic acid | 2 |
| Pyridoxine hydrochloride | 10 |
| Thiamine hydrochloride | 5 |
| Riboflavin | 5 |
| Nicotinic acid | 5 |
| Calcium pantothenate | 5 |
| Vitamin $B_{12}$ | 2 |
| p-Aminobenzoic acid | 5 |
| Lipoic acid | 5 |

The vitamin solution can be anaerobically filter-sterilized by bubbling with $N_2$, pouring into an upright 50-ml syringe with the plunger removed, a 3½-inch needle gassing the barrel, and fitted with a 25 mm × 0.2 μm Acrodisc filter and needle. The plunger can be inserted while removing the needle, similar to the Hungate technique for inserting a stopper. The solution can then be dispensed through the needle (using a second exit needle to relieve the pressure) into a sterile anaerobic serum vial.

### *Other Nutrient Amendments to Consider Adding*

10% v/v clarified rumen fluid or anaerobic digestor sludge supernatant
2 g/l each of yeast extract and casein hydrolysate
1 mM sodium acetate
2 mg/l of $Na_2WO_4 \cdot 2H_2O$ and $Na_2SeO_3$

## Most Probable Number Determination

The most probable number (MPN) determination is the simplest viable counting method of determining methanogen numbers. In this method, a set of serial (usually 10-fold) dilutions are made from a sample into replicate vessels (usually culture tubes) containing liquid growth medium, and the number of those replicates that are positive for a property (e.g., growth or methanogenesis) is scored for each dilution after incubation. At dilutions in which some of the samples remain negative, the distribution of viable units follows a Poisson distribution, and the number of cells in the original sample can be estimated using tables or computer programs based on probability theory [8, 20]. Typically three or five replicate tubes are inoculated per dilution, with the higher number allowing more precision. However, MPN determinations are imprecise by nature, essentially providing an order-of-magnitude estimate of viable numbers.

Colony counts in agar roll tubes are more precise than MPN determinations, but are more laborious and involve more perturbation of the sample, as it is placed in molten agar at 45°C and then chilled rapidly to 0°C. Spread-plate counts are difficult to perform for methanogens because the methanogens on the surface of the agar are afforded little protection from exposures to $O_2$, which greatly affects their reliability [19].

The most serious limitations of the MPN technique for methanogens are those typical of any viable counting technique. One is that the growth medium may not support all of the methanogens present because of lack of a required nutrient, the presence of a compound toxic to the desired organisms, or poor physiochemical simulation of the organism's habitat. Another problem is inadequate dispersal of aggregated organisms or organisms attached to surfaces so that a viable unit represents more than one cell. A related problem is that some methanogens are not present in the environment as single cells, but instead are colonial, such as *Methanosarcina*, which can form clumps with greater than 1000 cells, and *Methanothrix*, which forms long filaments.

Thus, one must interpret viable counts with caution; having said that, viable count techniques have been fairly successful for methanogens compared with other organisms. The general procedure outlined in this section gave us MPN counts in a thermophilic anaerobic bioreactor that were high enough to account for the rates of methanogenesis obtained [42] and were comparable to microscopic counts [S. Zinder, unpublished]. By manipulating the electron donor and acceptor, nearly any anaerobic physiologic group (for example, benzoate-oxidizing sulfate-reducers) can be enumerated using these techniques. It is convenient to make a set of master dilutions in basal medium that can then be used as inoculum for several different sets of MPN dilutions.

Procedure

1. Prepare a set of master dilution tubes. These tubes can contain basal medium to which only $N_2/CO_2$, bicarbonate, and $Na_2S$ or other reductant have been added. For accurate 10-fold dilutions, these tubes should have a final liquid volume of 9 ml. If a large number of dilutions are going to be performed, 45 ml of medium in 120-ml serum vials or even 90 ml in 160-ml vials can be used, although this method is for tubes.
2. Add a 1-ml sample to the first dilution tube. If this sample contains a large amount of solids that would clog a syringe needle, such as a sediment or digestor sludge, it may be easier to add it by removing the stopper, gassing the tube with $N_2/CO_2$, and using a spatula or syringe fitted with a cannula or with no needle at all. If the culture was gassed for a long time when open, it may be necessary to add more sulfide to maintain reduced conditions.
3. Homogenize the sample well with a vortex mixer. In some studies more stringent homogenization conditions may be required, and one can use a tissue homogenizer, for example, under anaerobic conditions either inside an anaerobic chamber or on the benchtop under a stream of anaerobic gas.
4. Add 1 ml of this initial sample (which can be called $10^0$, because it contains 1 ml of sample) to the next dilution tube using a disposable glass syringe to maximize viability.

5. Make serial dilutions past the point at which it is expected that no microorganisms of the given type will be detected. MPN's greater than $10^{10}$/ml have been detected for fermentative microorganisms in sewage sludge, and habitats that make fairly large amounts of methane can have more than $10^8$ $H_2$-utilizing methanogens/ml. It is better to err on the side of having too many dilutions. It is imperative to use a fresh syringe for each dilution, or carryover of organisms in the syringe may give the mistaken impression that there are unreasonably high numbers of microorganisms in the sample, a common mistake made by beginners.
6. From these master dilutions, make dilutions into tubes containing substrates that enrich for the desired group, such as $H_2/CO_2$, methanol, or acetate for methanogens.
7. Then incubate the tubes. The dilutions will turn positive sequentially until a dilution is finally reached at which no tubes are positive. This process can take a long time in the case of methanogens and other slow-growing anaerobes. If $10^6$ organisms are required to detect their presence by turbidity or methanogenesis, then a single organism must go through approximately 20 doublings ($2^{20} \approx 1.05 \times 10^6$) to be detected. Thus, if an organism has a doubling time of 1 hour, it will take approximately 20 hours for the culture from a single organism to be positive, and if it has a doubling time of 1 day, it will take 20 days. Some anaerobes, such as acetate-using methanogens and some fatty-acid-oxidizing syntrophic bacteria, have doubling times of several days [40]. If it is anticipated that long incubation times are needed, it is imperative to use thick black butyl-rubber stoppers, rather than the thinner gray butyl-rubber ones.
8. Score tubes as positive either on the basis of visible growth, typically indicated by turbidity, or on the basis of product formation, such as methane. It is best to compare these findings with those for a control set of tubes inoculated with medium lacking the substrate in question. For example, one may detect low amounts of methane in tubes receiving methanol or acetate as the major methanogenic substrate, but this methane may have been produced by $H_2$-utilizing methanogens that used $H_2$ produced by organisms fermenting the nutrient supplements, such as yeast extract. Thus, the amount of growth or methanogenesis should be significantly greater than the negative controls lacking the substrate.
9. Consult tables [8] for the MPN values and their confidence limits. The confidence limits depend on the actual pattern obtained since, for serial dilutions of a five-tube MPN, the pattern of positive tubes 5,2,0 is more likely than 4,0,1, which is reflected in the confidence limits.
10. Examine the positive dilutions microscopically, and compare their populations with those original samples using phase-contrast and epifluorescence microscopy to determine whether there is a correspondence between populations in the MPN tubes and those in the sample.

## Isolation of Methanogens from Enrichments

It is often desirable to have an isolate of an organism from the habitat in question to study its physiology and to obtain its 16S rRNA sequence. Isolating methanogens, especially slow-growing or poorly characterized ones, can require considerable skill and patience. This section presents only some of the general principles underlying techniques used to isolate methanogens; the reader is directed to seek more detailed descriptions of these procedures.

The classic method for isolation of methanogens and other stringent anaerobes is the agar roll tube technique adapted to anaerobic conditions by Hungate [16].

It is essentially a pour-plate technique in which a suitable dilution of the sample containing 10–100 colony-forming units (CFUs) is added to a tube containing molten agar medium kept at 45°C. For a 18 × 150 mm crimp-top tube, 7 ml of agar medium (1.5% to 2.2% agar) gives a good solid agar layer.

Procedure

1. Rotate the tube rapidly in ice water either by hand or by using a specially designed rolling apparatus available from Bellco (cat. #7790-44125) until the agar has solidified as a thin layer along the side of the tube.
2. Incubate the tube upright until colonies appear in the agar layer.
3. Examine these colonies using a dissecting microscope, always holding the tubes at an angle such that the fluid at the bottom of the tube does not flow onto the agar.
4. Use a marking pen to identify colonies that are desirable to pick.

The simplest way to pick these colonies is to unstopper the tube and work rapidly. One can pick the colonies with a sterile Pasteur pipette that has a 90° bend about 1/4 inch from the tip (this can be done before autoclaving) and transfer them to a liquid growth tube or dilution tube for further purification as described by Hungate [16]. Using a Pasteur pipette requires opening the receiving tube and gassing aseptically, and the manipulations are best carried out by two people. However, it is much simpler to flush a sterile syringe and needle with $N_2$ and to pick the colony by stabbing it with the needle and then transferring it through the stopper to liquid growth medium for growth or further dilution. Although the medium can be boiled to melt the agar and then dispensed hot, it is also convenient to add 0.15 g of agar to each tube before dispensing liquid medium into it. The tubes are shaken well before and after autoclaving. Additions can be made to molten agar as prewarmed solutions to prevent solidification.

If one has an anaerobic chamber, it is possible to use standard pour-plate or streak-plate methods to isolate methanogens. However, the ambient oxygen partial pressures in glovebox atmospheres are apparently too high to allow the growth of methanogens, so that special chambers (typical Gas-Pak plastic anaerobic jars will not suffice) must be used to incubate methanogens [1, 19]. A simpler method for streaking anaerobic cultures is to use the flat upright bottles described by Hermann et al. [15]. These bottles accept a serum stopper that is secured with a screw cap.

Procedure

1. Add 0.15–0.2 g of agar to the bottles before dispensing liquid medium (10 ml).
2. After autoclaving, add prewarmed sterile anaerobic solutions before the agar is allowed to solidify in the specially designed well on one side of the bottle.
3. Open the bottle on the benchtop and gas with sterile $N_2$ or inside an anaerobic chamber for streaking.
4. Streak a culture across the rectangular agar surface either as one large back-and-forth streak or as three successive streaks on top of each other as described by Hermann et al. [15].
5. Incubate the bottles either upright or laying on their sides with the agar sides up.

Since methanogens are of the Archaea domain, they are resistant to many but not all antibiotics that act on eubacterial cell wall or protein synthesis [3]. This resistance can be used as an advantage to eliminate eubacterial contaminants when isolating methanogens, and the penicillin family and cycloserine have been used with success. A particularly useful antibiotic is vancomycin, which did not inhibit the growth of *Methanosaeta* (*Methanothrix*), whereas penicillins did [17].

It is also essential to check the purity carefully of the methanogen culture, since many false claims have been made about methanogens based on impure cultures. It is philosophically impossible to prove a culture is pure; one can only prove that it is impure. However, careful microscopic examination and use of a variety of contaminant-checking media are essential. At the minimum these media should include the growth media used for the methanogen in which the methanogenic substrate has been replaced by fairly low concentrations (less than 1 g/l) of various amendments supporting the growth of contaminants, including yeast extract, sugars, and lactate and sulfate. Sulfate-reducing bacteria can be tenacious contaminants, as they use the oxidation products of sulfide, which inevitably form due to contamination with trace amounts of $O_2$, as electron acceptors.

### Methods to Identify Isolated Methanogens

Methanogens can often be identified to the genus level and sometimes to the species level simply on the basis of their morphology, substrates utilized, and other physiologic traits, such as optimal growth temperature. A long mesophilic motile spirillum that uses hydrogen or formate is *Methanospirillum hungatei* (Table 5-2), whereas a packet-forming organism that uses methanol or acetate at 55°C is *Methanosarcina thermophila*. In many cases, however, the only sure way to identify cultures to the species level is to sequence their 16S rDNA, and if you believe you have a new genus or species, extensive characterization should be done [4].

## METHODS TO QUANTIFY METHANOGENIC ACTIVITY

### Methanogenic Potential

It is relatively easy to measure the ability of a microbial community (or culture) to convert a substrate to methane (see Appendix 5-B). That substrate may be a direct methane precursor, such as $H_2/CO_2$, formate, acetate, or methanol, or it could be one that must be initially metabolized by other trophic groups, such as cellulose or benzoate. Simply measure the rate and amount of methane produced in samples receiving the substrate versus those same quantities in ones not receiving it. The amount of $CH_4$ expected to be produced can be estimated from the Buswell equation:

$$C_nH_aO_b + [n - a/4 - b/2]H_2O \rightarrow [n/2 - a/8 + b/4]CO_2 + [n/2 + a/8 - b/4]CH_4$$

For nitrogenous compounds, $NH_2$ can be considered to be the equivalent of OH and will require an extra mole of $H_2O$ for metabolism to $CH_4$ and $CO_2$. One can measure either $CH_4$ production using a gas chromatograph or measure the pressure increase using a pressure transducer or pressure gauge [31]. To use pressure increase as a measure of methanogenesis, one must take into account the high solubility of $CO_2$ in the liquid phase and the $CO_2/HCO_3^-$ equilibrium, and it would be wise to verify that the expected fraction of the partial pressure is indeed $CH_4$.

For the differential to be statistically significant in methanogenesis between samples receiving additions and unamended controls, the amount of additional methane produced must be of a similar order of magnitude or greater than the endogenous methanogenesis in the system. Usually the addition of millimolar quantities of substrates is required. This is not a problem for many substrates, but some, such as chlorinated aromatics, may be toxic at fairly low concentrations. Toxicity should be suspected when rates of methanogenesis are below those in unamended controls. To accommodate substrates that must be added at low concentrations, the amount of endogenous methanogenesis can be reduced by diluting the sample in medium [31] or by preincubating the sample for a period before substrate addition sufficient to allow most of endogenous methanogenesis to run its course. Another consideration is to ensure that there is enough buffering capacity in the system to prevent acidification. For example, 2 moles of acetic acid are produced from 1 mole of glucose, and they are often produced more rapidly than they can be consumed. Thus, there must be sufficient buffering capacity to counteract acetic acid production.

$CH_4$ is quantified most readily by gas chromatography (GC). Thermal conductivity detector (TCD) gas chromatographs are inexpensive and can detect $\leq 10^{-3}$ atm $CH_4$ when using helium as the flow gas. Flame ionization detector (FID) gas chromatographs are more expensive and can detect $\leq 10^{-6}$ atm $CH_4$, as well as other organics. For thermal conductivity analysis, one can use a 6-foot × 1/8-inch stainless steel column packed with 80/100 mesh Spherocarb operated at 140°C with a He flow of 30 ml/min. For FID analysis one can use a 6-foot × 1/8-inch stainless steel column packed with 80/100 mesh Poropak Q operated at 40°C with a carrier gas flow rate of 30 ml/min. Numerous other GC columns, including high-resolution capillary columns, are suitable to various applications involving the measurement of $CH_4$ and other gases.

A 0.1–0.5 ml sample is taken from the headspace using a pressure-tight syringe, allowing the researcher to remove a defined portion of the total headspace volume even if the gas pressure in the headspace is greater than 1 atm. A 0.5 ml Pressure Lok syringe with a side-port needle to prevent coring of GC septa provides excellent accuracy and precision and has nearly zero dead volume (the extra volume due to the needle and the needle housing), allowing for greater accuracy when assaying samples under pressure. For aseptic sampling of pure cultures, the needles for these syringes can be resterilized in small plastic-capped tubes, but are expensive to buy in the numbers needed. A more economical solution for aseptic sampling is to use a 1-ml Glaspak syringe fitted with a Mininert Luer-tip syringe valve, which can accept sterile disposable needles. Monoject needles are supplied in containers that can be resterilized and therefore can be used several times until they become dull. Syringes with disposable Luer-hub needles are much less accurate and have a dead volume $\geq 0.05$ ml.

A calibration curve can be obtained using $CH_4$ standards prepared by injecting a known amount of pure $CH_4$ into vials or tubes with known volumes and containing a few glass beads to disperse the added gas by shaking. For example, a 0.01 atm standard can be prepared by injecting 1.2 ml of methane into a 120-ml serum vial. (A correction for ambient barometric pressure should be made for measurements done at high altitude.) A $10^{-4}$ atm standard can be obtained by injecting 1.2 ml of the 0.01 atm sample into a 120-ml vial. The number of calibration points in the curve depends on the precision of the measurements. For gas chromatographs outfitted with integrators, peak areas can be measured. For those with output to an analog chart recorder, the peak height can be measured in millimeters using a ruler and multiplied by the attenuation to obtain an absolute peak height. If the calibration curves are linear (as most are for either peak height or area), then one need only inject a single standard to recalibrate the instrument during subsequent measurements.

Using the standard curve, one can readily calculate the $CH_4$ partial pressure in a sample. To obtain the volume at 1 atm pressure of $CH_4$ in a sample, multiply the partial pressure by the headspace volume, so that in a tube with a 17-ml headspace, if the methane measured is 0.1 atm, the volume of $CH_4$ at standard pressure is 1.7 ml. This value can then be converted into a more convenient molar value using the ideal gas equation, which states that at 25°C, 1 mole of gas at 1 atm pressure occupies approximately 24 liters (and 1 μmole occupies 24 μl). Thus, the 1.7 ml of $CH_4$ in this example represents approximatley 71 μmoles (=1.7 ml ÷ 0.024 ml/μmol) present in the headspace. A small amount of $CH_4$ is dissolved in the liquid phase. The dimensionless Henry's law constant at 25°C for methane is approximately 0.03, meaning that in a crimp-sealed tube with a 17-ml headspace and 10 ml liquid volume, an additional 1.8% of the $CH_4$ found in the headspace (≈1.3 μmoles) will be dissolved in the liquid phase, so that the total amount of $CH_4$ in the tube was approximately 72 μmoles. This value can then be compared to the methane production in samples not receiving additions, and the net $CH_4$ produced can be compared to the theoretical amount expected for the addition made.

Inhibitors have been useful in understanding the role of methanogens in anaerobic processes in anaerobic habitats. Chloroform was used in early studies, and in micromolar concentrations it inhibits methanogenesis. A more specific inhibitor is bromo-ethane sulfonic acid (BES or BESA), which is an analog of co-enzyme M. Co-enzyme M is central to the final step of methanogenesis, the methylreductase reaction, and has not been found in any non-methanogenic organism to date [1]. Although methylreductase is sensitive to micromolar concentrations of BES, millimolar concentrations are often required for rapid inhibition of whole cells [41]. It is also important to note that BES can be biodegraded in long-term incubations [2]. Finally, BES may not be 100% specific for methanogenesis. For example, reductive dechlorination of tetrachloroethene was inhibited by BES [9], even though evidence indicated that methanogens were not involved in the process. In other studies, eubacteria have been inhibited by antibiotics, with vancomycin being the most popular, but these antibiotics are not metabolic poisons acting in the short term, but rather are growth inhibitors with unpredictable effects on mixed populations.

## Radioisotope Techniques

Measurement of conversion of $^{14}C$-labeled $CH_4$-precursors to $CH_4$ is necessary to determine the contribution of that precursor to overall methanogenesis in a natural sample. $^{14}CH_4$ is difficult to measure because it is volatile and not easily trapped. One useful method is the tandem gas chromatograph-gas proportional counter method first described by Nelson and Zeikus [26]. In this method the effluent from a thermal conductivity detector is directed into a gas proportional counter that counts $^{14}CH_4$ as it passes by a detector, giving a peak on a chart recorder that, together with the $CH_4$ peak from the gas chromatograph, can provide the specific activity of the $^{14}CH_4$ in a single measurement. This method is convenient, but sensitivity is low compared with scintillation counting, there is a tendency toward quenching at high methane partial pressures [39], and the instruments are expensive and not widely available.

A simpler method is introducing a gas sample into the flow of a gas train in which it passes through a alkaline solution to trap $^{14}CO_2$, followed by passage over hot copper oxide in an oxidation furnace to convert $^{14}CH_4$ to $^{14}CO_2$, followed by a second alkaline trap for the $^{14}CO_2$ produced. This method is sensitive, but laborious.

In an even simpler method for measuring $^{14}CH_4$, a gas sample is injected into a scintillation vial sealed with a Teflon-coated silicone stopper held in place with a cap with a hole drilled in it [39]. The vial is nearly filled with scintillation fluid so that almost all of the $CH_4$ is dissolved in the fluid ($CH_4$ is more soluble in organic solvents than water). $^{14}CO_2$ can be removed from the gas sample by first injecting the sample into a serum vial nearly full of 1 N NaOH before transferring it to the scintillation vial. This method is convenient, sensitive, and inexpensive.

To determine the rate of methanogenesis from a potential methane precursor, one must determine the pool size (i.e., concentration) of that precursor and its rate of turnover. If the system is at steady state, such that a precursor has a constant concentration, the disappearance of the label from the precursor pool will be first-order and one can determine a turnover time (i.e., a half-life). One can also measure the instantaneous rate over short time periods, which may be more useful in systems in which pool size is not constant, such as an intermittently fed bioreactor in which acetate concentrations change over the feeding cycle. The pool size for $CO_2/HCO_3^-$ is typically rather large and unchanging. One can simply add a large amount of $H^{14}CO_3^-$ and measure its short-term conversion to $^{14}CH_4$ to determine the contribution of $^{14}CO_2$ reduction—the electron donor can be $H_2$, formate, or alcohols (Table 5-1)—to methanogenesis. Acetate concentrations tend to be in the millimolar range in anaerobic bioreactors and the micromolar range in aquatic sediments. Acetate can be measured conveniently by high-performance liquid chromatography [12] or gas chromatography. Acetate turnover has been measured using either steady-state [22, 32] or short-term methods [42]. Methylamines can be important methane precursors in marine systems, are usually present in micromolar concentrations, and can be measured by gas chromatography, as can methanol [27]. However, the concentrations of methylamines and methanol are often too low for quantifying their contribution to methanogenesis, either because they are below the limit of detection or because adding $^{14}C$-labeled substrates carrier-free greatly increases their pool size.

## MICROSCOPIC METHODS TO STUDY METHANOGENS IN NATURAL HABITATS

### Visible and Fluorescence Microscopy

Some methanogens, such as *Methanosarcina*, *Methanothrix*, and *Methanospirillum*, have distinct morphologies and can be tentatively identified in natural samples using a phase-contrast microscope, although such identification should always be corroborated using other methods. For example, *Sarcina ventriculi* might be mistaken for *Methanosarcina*.

Epifluorescence microscopy is a method useful in studying methanogen ecology. Many methanogens contain high concentrations of the electron carrier co-factor $F_{420}$ [14], so that their cells autofluoresce blue-green (near 480 nm) when illuminated with light near 420 nm using an epifluorescence microscope [10]. Although filter combinations for acridine orange can give some fluorescence of methanogens, often their excitation wavelength is too long to provide good co-factor $F_{420}$ autofluorescence.

Co-factor $F_{420}$ has been detected in eubacteria, but not in high enough concentrations to autofluoresce. However, $F_{420}$ levels vary widely in methanogens, and growth conditions may affect those levels [14]. For example, *Methanosaeta* usually does not have high enough $F_{420}$ concentrations to autofluoresce [17, 43], so that non-fluorescent cells in samples are not necessarily non-methanogens. The only non-methanogenic microorganism found to show co-factor $F_{420}$ autofluorescence is the hyperthermophilic sulfate-reducing Archaeon, *Archaeoglobus fulgidus* [6]. This technique has facilitated qualitative identification of methanogens in natural samples, including those with endosymbiotic and ectosymbiotic relationships with protozoa [21, 35]. Fluorescence microscopy can also provide information useful to interpretation of viable counts, such as the morphology of predominant methanogenic species, clump size in *Methanosarcina*, and the presence of methanogens in aggregated forms (i.e., associated with surfaces or with protozoa). Extracting natural samples and then measuring $F_{420}$ concentrations have been suggested as a means for estimating methanogenic biomass, but the great variation in $F_{420}$ content in different methanogens [14] makes this method untenable.

## Immunomicroscopy

The use of antibodies to study methanogen diversity and ecology has been spearheaded by E. Conway de Macario and colleagues, who developed a panel of antibodies against a wide variety of methanogens and some other anaerobes and determined their cross-reactivities [23]. In general, antibody specificity correlates with phylogenetic relationship. More recently, they have used these antibodies to directly identify and enumerate methanogenic populations in anaerobic bioreactors using fluorescent immunomicroscopy [25, 36].

The limitations on the use of antibodies to study methanogen ecology center around their specificity, which can be too low and therefore may cross-react with organisms other than the one desired or too high and essentially give information about the ecology of individual stereotypes of the same species. An example of low specificity in an instance in which antibodies against *Methanosarcina thermophila* TM-1 cross-reacted with *Methanothrix*-like cells in one bioreactor sample [24]. For this reason, bulk immunologic methods, such as enzyme-linked immunoassay (ELISA), should not be used to study natural habitats, since microscopy will at least ensure that the reacting organisms have the expected morphology. Finally, this method does not identify organisms that are not reactive to the panel of antibody probes, so that novel undescribed methanogens might be missed. Despite these limitations, immunologic analysis is a valuable tool to study methanogen ecology. Immunogold-labeling of electron microscope specimens has also been used to identify methanogens and other anaerobes in mixed microbial populations [11, 30].

## Molecular Biologic Methods to Identify Methanogens in Natural Communities

Molecular biologic methods are a powerful addition to the arsenal of tools available to the microbial ecologist, and their utility and application to microbial ecology are described elsewhere in this book. A recent study applied the methods of 16S rRNA technology to studying methanogen ecology in natural habitats. Raskin et al. [28, 29] devised methanogen 16S rRNA probes of various specificities (Table 5-3) and determined their optimal conditions for hybridization. For example, probe MS821 reacted only with members of the genus *Methanosarcina*, whereas probe MS1414 reacted with essentially all of the methylotrophic methanogens and MSMX860 reacted with these plus the acetate-utilizing methanogenic genus *Methanosaeta* (i.e., all of the *Methanosarcinales*). These probes were first tested against a slot-blot panel of 16S rRNAs from various methanogens and other Archaea. It

**Table 5-3** Oligonucleotide Probes Devised for Detection of Methanogens

| Probe | Sequences (5′–3′) | Target site (E. coli numbering) | Td (°C) | Reacts with (see Table 5-2) |
|---|---|---|---|---|
| MC1109 | GCAACATAGGGCACGGGTCT | 1128–1109 | 55 | *Methanococcales* (except *M. jannaschii*) |
| MB310 | CTTGTCTCAGGTTCCATCTCCG | 331–310 | 57 | *Methanobacteriaceae* |
| MB1174 | TACCGTCGTCCACTCCTTCCTC | 1195–1174 | 62 | *Methanobacteriaceae* |
| MG1200 | CGGATAATTCGGGGCATGCTG | 1200–1200 | 53 | *Methanogenium, Methanomicrobium, Methanoplanus, Methanospirillum* |
| MSMX860 | GGCTCGCTTCACGGCTTCCCT | 880–860 | 60 | *Methanosarcinales* |
| MS1414 | CTCACCCATACCTCACTCGGG | 1434–1414 | 38 | *Methanosarcinaceae* |
| MS821 | CGCCATGCCTGACACCTAGCGAGC | 844–821 | 60 | *Methanosarcina* |
| MX825 | TCGCACCGTGGCCGACACCTAGC | 847–825 | 59 | *Methanosaeta* (poor reaction with thermophiles) |
| ARC915 | GTGCTCCCCCGCCAATTCCT | 934–915 | 56 | Universal archael probe |
| ARC344 | TCGCGCCTGCTGCICCCCGT | 363–344 | 54 | Universal archael probe |

was found that the probes reacted as expected with rRNA from target methanogens and showed essentially no cross-reactivity with other organisms.

Raskin et al. [28] went on to apply these probes to 16S rRNA extracted from various laboratory-scale and full-scale anaerobic bioreactors and obtained results in fairly good agreement with expectations. Of particular importance, they demonstrated that the probes "nested" (i.e., the sum of responses to probes for non-overlapping subgroups roughly equaled that of a more universal probe encompassing them). For example, the sum of the responses to MC1109, MB310, MG1200, and MSMX860 was approximately equal to ARC915, indicating that these methanogen probes had reacted with essentially all the Archaea present in significant numbers. A large discrepancy would be a clue that Archaea are present that have not been detected by the extant probes. The fluorescent staining of large *Methanosarcina*-like clusters using rhodamine-labeled MS821 was also demonstrated in a bioreactor. These studies represent a promising first application of molecular biology techniques to the study of methanogen ecology. The power of these methods is greatest when complemented by other methods described in this chapter to provide a more complete description of the methanogens and their activities.

*Acknowledgments* I would like to thank all of my colleagues, especially technician Timothy Anguish, for their assistance in developing the techniques used in our laboratory. I would also like to acknowledge the support of the U. S. Department of Energy for their support of my ecologic and physiologic studies of methanogens.

## REFERENCES

1. Balch, W.E., Fox, G.E., Magrum, L.J., Woese, C.R., and Wolfe, R.S. 1979. Methanogens: reevaluation of a unique biological group. Microbiol. Rev. 43: 260–296.
2. Belay, N. and Daniels, L. 1987. Production of ethane, ethylene, and acetylene from halogenated hydrocarbons by methanogenic bacteria. Appl. Environ. Microbiol. 53: 1604–1610.
3. Bock, A. and Kandler, O. 1985. Antibiotic sensitivity of Archaebacteria. In The Bacteria: A Treatise On Structure and Function. Vol. VIII: Archaebacteria (Woese, C.R. and Wolfe, R.S., Eds.), pp. 525–544. Academic Press, New York.
4. Boone, D.R. and Whitman, W.B. 1988. Proposal of minimal standards for describing new taxa of methanogenic bacteria. Int. J. Syst. Bacteriol. 38: 212–219.
5. Boone, D.R., Johnson, R.L., and Liu, Y. 1989. Diffusion of the interspecies electron carriers $H_2$ and formate in methanogenic ecosystems and its implications in the measurement of $K_m$ for $H_2$ or formate uptake. Appl. Environ. Microbiol. 55: 1735–1741.
6. Boone, D.R., Whitman, W.B., and Rouvire, P. 1993. Diversity and taxonomy of methanogens. In Methanogenesis (Ferry, J.G., Ed.), pp. 36–80. Chapman & Hall, New York.
7. Breznak, J.A. and Costilow, R.N. 1994. Physiochemical factors in growth. In Methods for General and Molecular Bacteriology (Gerhardt, P., Ed.), pp. 137–154. American Society for Microbiology, Washington, DC.
8. de Man, J.C. 1975. The probability of most probable numbers. Eur. J. Appl. Microbiol. 1: 67–78.
9. DiStefano, T.D., Gossett, J.M., and Zinder, S.H. 1992. Hydrogen as an electron donor for the dechlorination of tetrachloroethene by an anaerobic mixed culture. Appl. Environ. Microbiol. 58: 3622–3629.
10. Doddema, H.J. and Vogels, G.D. 1978. Improved identification of methogenic bacteria by fluorescence microscopy. Appl. Environ. Microbiol. 36: 752–754.

11. Dubourgier, H.C., Archer, D.B., Algagnac, G., and Prensier, G. 1988. Structure and metabolism of methanogenic microbial conglomerates. In Anaerobic Digestion 1988 (Hall, E.R. and Hobson, P.N., Eds.), pp. 13–23. Pergamon Press, Oxford.
12. Ehrlich, G.G., Goerlitz, D.F., Bourell, J.H., Eisen, G.V., and Godsy, E.M. 1981. Liquid chromatographic procedure for fermentation product analysis in the identification of anaerobic bacteria. Appl. Environ. Microbiol. 42: 878–885.
13. Ferry, J.G. 1993. Methanogenesis. Chapman and Hall, New York.
14. Gorris, L.G.M. and Van der Drift, C. 1994. Cofactor contents of methanogenic bacteria reviewed. Biofactors 4: 139–145.
15. Hermann, M., Noll, K.M., and Wolfe, R.S. 1986. Improved agar bottle plate for isolation of methanogens or other anaerobes in a defined gas atmosphere. Appl. Environ. Microbiol. 51: 1124–1126.
16. Hungate, R.E. 1967. A roll tube method for cultivation of strict anaerobes. In Methods in Microbiology. Vol. 2B (Norris, J.R. and Ribbons, D.W., Eds.), pp. 117–132. Academic Press, New York.
17. Huser, A.A., Wuhrman, K., and Zehnder, A.J.B. 1982. *Methanothrix soehngenii* gen. nov. sp. nov., a new acetotrophic non-hydrogen-oxidizing methane bacterium. Arch. Microbiol. 132: 1–9.
18. Jarrell, K.F. and Kalmokoff, M.L. 1988. Nutritional requirements of the methanogenic archaebacteria. Can. J. Microbiol. 34: 557–576.
19. Jones, W.J., Whitman, W.B., Fields, R.D., and Wolfe, R.S. 1983. Growth and plating efficiency of methanococci on agar media. Appl. Environ. Microbiol. 46: 220–226.
20. Koch, A.L. 1994. Growth measurement. In Methods for General and Molecular Bacteriology (Gerhardt, P., Ed.), pp. 448–477. American Society for Microbiology, Washington, DC.
21. Lee, M.J., Schreurs, P.J., Messer, A.C., and Zinder, S.H. 1987. Association of methanogenic bacteria with flagellated protozoa from a termite hindgut. Curr. Microbiol. 15: 337–341.
22. Lovley, D.J. and Klug, M.J. 1982. Intermediary metabolism of organic matter in the sediments of a eutrophic lake. Appl. Environ. Microbiol. 43: 552–560.
23. Macario, A.J.L. and Conway de Macario, E. 1983. Antigenic fingerprinting of methanogenic bacteria with polyclonal antibody probes. Syst. Appl. Microbiol. 4: 451–458.
24. Macario, A.J.L., Conway de Marcario, E., Ney, U., Schoberth, S., and Sahm, H. 1989. Shifts in methanogenic subpopulations measured with antigenic probes in a fixed-bed loop anaerobic bioreactor. Appl. Environ. Microbiol. 55: 1996–2001.
25. Macario, A.J.L., Visser, F.A., van Lier, J.B., and Conway de Macario, E. 1991. Topography of methanogenic subpopulations in a microbial consortium adapting to thermophilic conditions. J. Gen. Microbiol. 137: 2179–2189.
26. Nelson, D.R. and Zeikus, J.G. 1974. Rapid method for the radioisotopic analysis of gaseous end products of anaerobic metabolism. Appl. Microbiol. 28: 258–261.
27. Oremland, R.S. and Polcin, S. 1982. Methanogenesis and sulfate reduction: competitive and noncompetitive substrates in estuarine sediments. Appl. Environ. Microbiol. 44: 1270–1276.
28. Raskin, L., Poulsen, L.K., Noguera, D.R., Rittman, B.E., and Stahl, D.A. 1994. Quantification of methanogenic groups in anaerobic biological reactors by oligonucleotide probe hybridization. Appl. Environ. Microbiol. 60: 1241–1248.
29. Raskin, L., Stromley, J.M., Rittman, B.E., and Stahl, D.A. 1994. Group-specific 16S rRNA hybridization probes to describe natural communities of methanogens. Appl. Environ. Microbiol. 60: 1232–1240.
30. Robinson, R.W. and Erdos, G.W. 1985. Immuno-electron microscopic identification of *Methanosarcina* spp. in anaerobic digestor fluid. Can. J. Microbiol. 31: 839–844.
31. Shelton, D.R. and Tiedje, J.M. 1984. General method for determining anaerobic biodegradation potential. Appl. Environ. Microbiol. 47: 850–857.
32. Smith, P.H. and Mah, R.A. 1966. Kinetics of acetate metabolism during sludge digestion. Appl. Micribiol. 14: 368–371.
33. Thauer, R.K., Jungermann, K., and Decker, K. 1977. Energy conservation in chemotrophic anaerobic bacteria. Bacteriol. Rev. 41: 100–180.

34. Thiele, J.H. and Zeikus, J.G. 1988. Control of interspecies electron flow during anaerobic digestion: significance of formate transfer versus hydrogen transfer during syntrophic methanogenesis in flocs. Appl. Environ. Microbiol. 54: 20–29.
35. van Bruggen, J.J.A., Stumm, C.K., and Vogels, G.D. 1983. Symbiosis of methanogenic bacteria and spropelic protozoa. Arch. Microbiol. 136: 89–95.
36. Visser, F.A., van Lier, J.B., Macario, V., and Conway de Marcario, E. 1991. Diversity and population dynamics of methanogenic bacteria in a granular consortium. Appl. Environ. Microbiol. 57: 1728–1734.
37. Wolin, M.J. and Miller, T.L. 1982. Interspecies hydrogen transfer: 15 years later. ASM News 48: 561–565.
38. Zehnder, A.J.B. and Wuhrmann, K. 1976. Titanium (III) as a nontoxic oxidation reduction buffering system for the culture of obligate anaerobes. Science 194: 1165–1166.
39. Zehnder, A.J.B., Huser, B., and Brock, T.D. 1979. Measuring radioactive methane with the liquid scintillation counter. Appl. Environ. Microbiol. 37: 897–899.
40. Zinder, S.H. 1993. Physiological ecology of methanogens. In Methanogenesis (Ferry, J.G., Ed.), pp. 128–206. Chapman & Hall, New York.
41. Zinder, S.H., Anguish, T., and Cardwell, S.C. 1984. Selective inhibition by 2-bromoethanesulfonate of methanogenesis from acetate in a thermophilic anaerobic digestor. Appl. Environ. Microbiol. 47: 1343–1345.
42. Zinder, S.H., Cardwell, S.C., Anguish, T., Lee, M., and Koch, M. 1984. Methanogenesis in a thermophilic (58°C) anaerobic digestor: *Methanothrix* sp. as an important aceticlastic methanogen. Appl. Environ. Microbiol. 47: 796–807.
43. Zinder, S.H., Anguish, T., and Lobo, A.L. 1987. Isolation and characterization of a thermophilic acetotrophic strain of *Methanothrix*. Arch. Microbiol. 146: 315–322.

# Appendix 5-A

# Supplies for Culturing Methanogens

Table 1 Supplies for Constructing an Anaerobic Gassing Manifold

| Item | Supplier | Catalog Number | Comments |
|---|---|---|---|
| $N_2$ Gas | Matheson | Prepurified grade | Locally supplied $N_2$ may be adequate |
| 70% $N_2$/30% $CO_2$ | Matheson | Certified grade | |
| 80% $H_2$/20% $CO_2$ | Matheson | Certified grade | For growing hydrogen-utilizing methanogens |
| Two-stage gas regulator | Matheson | 3104-580 for $N_2$ and 3104-350 for $N_2$/CO, for $H_2$/CO | Provides better pressure control than single-stage regulators |
| Butyl tubing | Fisher | 14-1688 | Latex or Tygon can be adequate, but do not give as good performance |
| Glass tees, 1/4 inch, and hose clamp, 1/2 inch | Fisher | 15-328A | For securing connections; available at hardware stores |
| Pinch clamp, polypropylene | Fisher | 05-869 | |
| Needle tubing connector, Luer-hub | Sigma | C3399 | Provides a good connection between tubing and gassing needles; can be replaced with the front half of a 1-ml Glaspak syringe |
| Gassing needles, 17-gauge × 3½ inch | Sigma | N 9010 | |
| Acrodisc filters, 25 mm × 0.2 5 m | VWR | 28143-310 | Can be reused for filtering gases as long as sterility is ensured |

Table 2 General Supplies Needed for Anaerobic Culture

| Item | Supplier | Catalog Number | Comments |
|---|---|---|---|
| Disposable syringes, Glaspak, 1 ml | VWR | BD5292 | Used for inoculating cultures |
| Disposable syringes, plastic | Fisher | 14-823-2F | 1 ml is most useful for 10 ml cultures; 10-ml syringes useful for dispensing medium into tubes |
| Disposable needle, 25-gauge × 5/8 inch, Monoject | Krackler | 24-1-250313 | Useful for piercing gray butyl stoppers |
| Disposable needle, 23-gauge × 1 inch, Monoject | Krackler | 24-1-250255 | Useful for piercing thicker black butyl stoppers |
| Pipetting needle, 14-gauge × 4 inch | Fisher | 14-825-16K | Useful for dispensing medium |
| Black butyl stoppers, 13 × 20 mm | Bellco Geo Microbial Technologies | 1313<br>2048-11800 | Survive more punctures; to be used for pressurized cultures, especially those using $H_2/CO_2$, and for long-term incubations |
| Gray butyl stoppers, 13 × 20 mm flanged | Fisher | 06-447J | Flanged stoppers provide better performance than slotted ones |
| Aluminum crimp-top tube, 18 × 150 mm | Bellco | 2048-18150 | |
| Aluminum crimps, 20 mm | Fisher | 06-406-14B | For low-pressure applications, tear-off seals (06-406-15) can be used |
| Crimper, 20 mm | Fisher | 10-319-490 | |
| Decrimper | Supelco | 3-3283 | Can also use long-nose pliers |
| Serum vials, 100 ml | Fisher | 06-406J | Actual volume is 120 ml |
| Anaerobic streak bottles | Bellco | 2535-50020 | Used with agar medium for streaking out anaerobic cultures |

# Appendix 5-B

# Materials Needed for Methane Analysis (excluding Gas Chromatograph)

Table 1 Materials for Methane Analysis

| Item | Supplier | Catalog Number | Comments |
|---|---|---|---|
| Pressure-Lok syringe, Series A-2, 0.5 ml | Supelco | 2-2272 | |
| Needles, side port | Supelco | 2-2289 | Less septum coring than beveled point needles |
| Syringe valve, Teflon, Luer-tip | Supelco | 2-2285 | Fits on conventional Luer-tip syringes, such as a 1-ml Glaspak syringe; when Teflon stretches, can be secured to syringe tip with Teflon tape |
| Sterile, disposable needle, 25-gauge × 5/8 inch, Monoject | Krackler | 24-1-250313 | Useful for aseptic sampling of gray butyl stoppers; can be reautoclaved |
| Sterile, disposable needle, 23-gauge × 1 inch, Monoject | Krackler | 24-1-250255 | Usefor for aseptic sampling or gray butyl stoppers; can be reautoclaved |
| Gas standard, 99% $CH_4$ | Supelco | 2-2562 | |
| Syringe adapter for gas standard | Supelco | 6-09-010 | |
| 6-foot × 1/8-inch SS column packed with 80/100 Spherocarb | Alltech | 5683 PC | For TCD GC applications |
| 6-foot × 1/8-inch SS column packed with 80/100 Poropak Q | Alltech | 2701PC | For use with flame ionization detector GC applications |

## SOURCES FOR CULTURES OF METHANOGENS

Oregon Collection Of Methanogens
Oregon Graduate Institute
P.O. Box 91000
Portland, Oregon 97291, (503) 690-1146,

Web address: http://www.ese.ogi.edu/ese_docs/bugs/ocm.html
Deutsche Sammlung von Mikroorganismen (German Collection of Microorganisms)
Mascheroder Weg 1 B
D-3300 Braunchweig, Germany, 49 (0531) 61870

The American Type Culture Collection also can supply cultures of methanogens, but these two organizations have more extensive collections and greater expertise in handling them.

## SUPPLIERS LISTED

| | |
|---|---|
| Alltech Associates, Inc., 2051 Waukegan Rd., Deerfield, IL 60015 | (800) 255-8324 |
| Bellco Glass, Inc., PO Box B, 340 Edrudo Rd., Vineland, NJ 08360 | (800) 257-7043 |
| Coy Laboratory Products, 14500 Coy Dr., Grass Lake, MI 49240 | (313) 475-2200 |
| Fisher Scientific Co., 711 Forbes Ave., Pittsburgh, PA 15219 | (800) 766-7000 |
| Geo-Microbial Technologies, PO Box 132, E. Main St., Ochelata, OK 74051 | (918) 535-2281 |
| Krackler Scientific, Inc., PO Box 1849, Albany, NY 12201 | (800) 334-7725 |
| Matheson Gas Products, Inc., 959 Rt. 46E, Parsippany, NJ | (973) 257-1100 |
| Sigma Chemical Co, 3050 Spruce St., St. Louis, MO 63178 | (800) 325-3010 |
| Supelco, Inc., Supelco Park, Bellefonte, PA 16823 | (800) 247-6628 |
| VWR Scientific, 1310 Goshen Pkwy., West Chester, PA 19380 | (800) 932-5000 |

6

RICHARD S. HANSON

# Ecology of Methylotrophic Bacteria

## ECOLOGY OF METHANOTROPHIC MICROORGANISMS

Microorganisms that grow on one-carbon compounds that are more reduced than carbon dioxide and assimilate formaldehyde as a major source of carbon are referred to as *methylotrophs*. These bacteria are ubiquitous: They are found in freshwater rivers and lakes, in marine environments, in soils, on plant surfaces, and as air-borne microbes [29, 34]. The descriptor, *methanotroph*, is reserved for those gram-negative methylotrophs that use methane as a source of carbon and energy [2, 28]. Methanotrophic bacteria are usually obligate and do not use multicarbon compounds as sources of carbon or energy [2, 28]. Microbes that use methanol are very diverse and include a number of yeast belonging to six different genera and three mycelial fungi, a restricted group of gram-positive endospore-forming bacteria, and a very diverse group of gram-negative bacteria [2, 34].

Methanotrophs are also ubiquitous and occur in abundance at oxic-anoxic interfaces in aquatic environments, in soils, on the surfaces of aquatic plants, in or on the roots of several plants, and as endosymbionts of mussels [11, 12, 25, 27, 29]. Methane produced by methanogens in the sediments and hypolimnia of eutrophic freshwater lakes is oxidized by methanotrophs as it diffuses to the oxygen-containing metalimnia [25, 27]. Similarly, methane is produced in anoxic soils and diffuses upward where it is oxidized when oxygen is present [4, 21, 27, 47]. Methanotrophs play important roles in the oxidation of methane in peat and wetlands [4, 47]. Some floating aquatic plants, such as Lemna minor (little duckweed), that are abundant on the surfaces of water in wetlands contain methanotrophs as epiphytes and may be important in the oxidation of methane before it escapes to the atmosphere [27]. Geothermally produced methane and methane in deposits below the ocean floor serve as the carbon and energy source of endosymbionts and free-living methanotrophs that are at the base of marine food chains [11, 12].

Methane absorbs 20 to 30 times more infrared irradiation mole for mole than carbon dioxide and has been increasing in concentration at the rate of 1% per year for the past 200 years [21]. Therefore, it is considered an important radiatively active compound that may contribute significantly to global warming. As tundra soils and marsh sediments dry, they become aerobic sinks for atmospheric methane, whereas wet tundra soils and water-saturated sediments are anoxic and are net methane sources [4, 21, 26, 47]. The oxidation of atmospheric methane may increase as these soils dry during global warming and reverse the trend toward increased rates of warming.

Some methanotrophic bacteria fortuitously oxidize a large number of organic compounds including priority pollutants, such as the halogenated low-molecular weight compounds: trichloroethylene, dichloroethane, dichloroethylene, vinyl chloride, chloroform, and dichloromethane [1, 15, 22, 26]. These bacteria may prove to be useful for the bioremediation of soils and groundwaters contaminated with these compounds [26]. The methanotrophs that oxidize these chemicals produce a soluble methane monooxygenase (sMMO) during copper-limited growth and oxidize them at high rates only when copper is limited [15, 26]. A limited number of methanotrophic bacteria are genetically competent to produce sMMO, whereas all methanotrophs produce a membrane-bound or particulate form of MMO (pMMO) when copper is present in the growth medium at concentrations above 0.9 mg per gram of dry weight of bacterial cells [15, 44].

Methane is also oxidized in anaerobic marine environments [3], soda (alkaline) lakes [31], and in sediments of eutrophic freshwater lakes [37]. The bacteria that oxidize methane in the absence of oxygen remain to be characterized. It has been suggested that some enrichment cultures that oxidize methane use sulfate anaerobically as a terminal electron acceptor and require acetate or lactate as carbon sources [32, 37]. In other environmental samples where anaerobic methane oxidation has been detected, the terminal electron acceptor remains to be identified.

Methanol-utilizing bacteria are abundant on several plant surfaces and may have a role in protecting the plants from invasion by pathogenic bacteria [30]. Most of these bacteria are pink-pigmented facultative methylotrophs that grow with a variety of multicarbon sources of carbon and energy. A few methanol-utilizing bacteria are obligate methylotrophs, and others are obligate or restricted facultative methylotrophs that use a limited range of organic compounds. Some bacteria oxidize one-carbon substrates, such as methanol, to carbon dioxide and assimilate the carbon dioxide via the Calvin cycle. These bacteria are known as pseudomethylotrophs [34].

Hyphomicrobia, prosthecate gram-negative bacteria that multiply by budding, are methylotrophs that can employ methanol as an energy source for anaerobic respiration with nitrate as an electron acceptor [40]. These bacteria are employed for denitrification in sewage treatment plant effluents with methanol as an electron donor [13]. They are also able to mineralize a variety of pollutants, including aromatic hydrocarbons, dimethyl sulfoxide, methyl chloride, and a variety of alcohols [40].

Methylamines, dichloromethane, and dimethylsulfide also serve as carbon and energy sources for methylotrophs [2, 34]. Most, but not all, bacteria that use methylamines and some that use dichloromethane also grow with methanol as a sole source of carbon and energy.

## TAXONOMIC, PHYSIOLOGIC, AND PHYLOGENETIC COMPARISONS OF METHYLOTROPHIC BACTERIA

Methylotrophic bacteria are unique because the pathways used for the synthesis of adenosine triphosphate (ATP) and for the synthesis of intermediates that serve as precursors of biosynthetic pathways [2, 28, 34] are found only in methylotrophs. All methylotrophic bacteria use one of two pathways for the assimilation of formaldehyde [2, 49]. The gram-negative methane-utilizing bacteria that employ the ribulose monophosphate pathway as the major pathway for formaldehyde fixation are referred to as Group I methanotrophs [49]. It has been proposed that all these methanotrophs be included in the family *Methylococcaceae* [8]. The members of this family are found in three clusters within the gamma-subdivision of the Proteobacteria [8, 49]. The Group I methanotrophs all have disc-shaped bundles of intracytoplasmic membranes and contain DNA G+C values of 51–59 mol%. Optimal growth occurs between 25 and 35°C, and does not occur at all above 40°C. Three genera of Group I methanotrophs have been proposed: *Methylomonas*, *Methylobacter* and another group (genus) identified by phenotypic analysis that has not yet been named [8, 28].

The genus name *Methylococcus* has been proposed for the moderately thermophilic methanotrophs, which also contain disc-shaped bundles of intracytoplasmic membranes, use the ribulose monophosphate pathway for formaldehyde fixation, and follow a functional Calvin cycle that allows incorporation of carbon dioxide for macromolecular synthesis in the presence of methane, although autotrophic growth does not occur in the absence of methane [49]. These bacteria also contain low levels of the enzymes of the serine pathway for formaldehyde fixation. Optimal growth temperatures are between 37 and 50°C. The moles % G+C content of their DNA is 59% to 66%. Members of this genus are capable of fixing atmospheric nitrogen and possess SMMO. A separate group designation (Group X) has been suggested for these unique methanotrophs [49], although it has also been proposed that these bacteria that are related to the gamma-subdivision Proteobacteria should also be included in the family *Methylococcaceae* [8].

Those methanotrophs that use the serine pathway for formaldehyde assimilation are referred to as Group II methanotrophs and form a phylogenetically coherent cluster within the alpha-subdivision of the Proteobacteria [9]. They possess DNA with a moles % G+C content of 62% to 67%, and many are capable of fixing dinitrogen. These bacteria are also distinguished from the Group I and Group X methanotrophs by the morphology of their intracytoplasmic membranes, phospholipid fatty acid profiles, pathways utilized for nitrogen assimilation, types of resting cells formed, and several other characteristics (Table 6-1) [8, 28, 49]. Two genera, *Methylosinus* and *Methylocystis*, have been proposed for Group II methanotrophs based on cell morphology, cyst formation, and several other phenotypic and genotypic characteristics [8].

The methylotrophic bacteria that utilize one-carbon substrates other than methane are very diverse and do not form a single phylogenetically coherent group [9]. However, those obligately aerobic bacteria that use methanol and employ the serine pathway for formaldehyde fixation are found within the alpha-subdivision of the Proteobacteria, whereas the gram-negative bacteria that employ the ribulose monophosphate pathway are more closely related to other Proteobacteria in the beta-subdivision. This phylogenetic separation of methylotrophic bacteria should be viewed cautiously because they are so diverse, and 16S rRNAs have not been sequenced from a sufficient number of diverse bacteria that utilize one-carbon compounds other than methane. In Figure 6-1, the bacteria in Group Ib are those aerobic gram-negative bacteria that utilize methanol but not methane, and employ the ribulose monophosphate pathway for formaldehyde assimilation. These bacteria are obligate or restricted facultative methylotrophs. Bacteria in Group IIa are those that employ the serine pathway for formaldehyde assimilation and grow with methanol but not methane as a sole source of carbon and energy. These bacteria are facultative methylotrophs that employ a wide variety of multicarbon compounds as carbon and energy sources. Strain ER2 is a facultative methylotroph

Table 6-1 Some Characteristics of Methanotrophs in Different Groups

| Characteristics | Group I | Group II | Group X |
|---|---|---|---|
| Carbon assimilation pathways | | | |
| Ribulose monophosphate pathway | + | − | + |
| Serine pathway-hydroxypyruvate reductase | − | + | + |
| Calvin cycle | − | − | + |
| Presence of a complete tricarboxylic acid cycle and α-ketoglutarate dehydrogenase | − | + | − |
| Intracytoplasmic membrane arrangement | | | |
| Bundles of vesicular discs | + | − | + |
| Paired membranes aligned to periphery of cells | − | + | − |
| Moles % G + C content of DNA | 51–59 | 62–67 | 59–66 |
| Predominant phospholipid fatty acid chain length | 16 | 18 | 16 |
| Cell shape | Short rods | Rod or pear-shaped | Cocci |
| Formation of rosettes | − | + | − |
| Isocitrate dehydrogenase electron acceptor | | | |
| NAD+ dependent only | − | − | + |
| NADP+ dependent only | − | + | − |
| NADP+ and NAD+ dependent | + | − | − |
| Proposed genera | *Methylomonas* *Methylobacter* | *Methylosinus* *Methylocystis* | *Methylococcus* |

that degrades carbofuran and uses methylamine produced as a result of carbofuran degradation, but does not utilize methanol. Although it is found within the alpha-subdivision of the Proteobacteria, it is not closely related to other methylotrophs that utilize methanol. The pseudomethylotrophs, such as *Paracoccus denitrificans*, that do not assimilate formaldehyde are not closely related to other serine pathway methylotrophs. The bacteria of the genus *Hyphomicrobium* are prosthecate bacteria that are also facultative, serine pathway methylotrophs (see above). They are also found within the alpha-subdivision of the Proteobacteria, but are not closely related to other facultative serine pathway methylotrophs [9].

Therefore, microbes that use reduced one-carbon compounds are clearly important in global carbon cycles and in food chains that involve both free-living and symbiotic bacteria and some fungi. The methylotrophic bacteria are a diverse group that have in common the ability to assimilate formaldehyde as a major carbon source. The gram-negative methylotrophs are phylogenetically related to the Proteobacteria and can be separated into groups that utilize methane and those methylotrophs that use one-carbon compounds more oxidized than methane as sources of carbon and energy. The methanotrophs and gram-negative methylotrophs that utilize the serine pathway for formaldehyde assimilation form separate clusters within the alpha-subdivision of the Proteobacteria. The methanotrophic bacteria that employ the ribulose monophosphate (RuMP) pathway as the major route for formaldehyde formation form three clusters within the gamma-subdivision of the Proteobacteria, and the RuMP pathway methylotrophs that do not utilize methane are found within the beta-subdivision. The similarities among physiologic groups and phylogenetic groupings has made it possible to develop specific probes capable of detecting 16S rRNAs of each physiologic group and to indirectly characterize uncultured bacteria in different environments. The availability of signature probes and gene probes that allow detection of specific groups of methylotrophs has been shown to be useful for ecologic studies and the identification of bacteria that degrade some toxic chemicals.

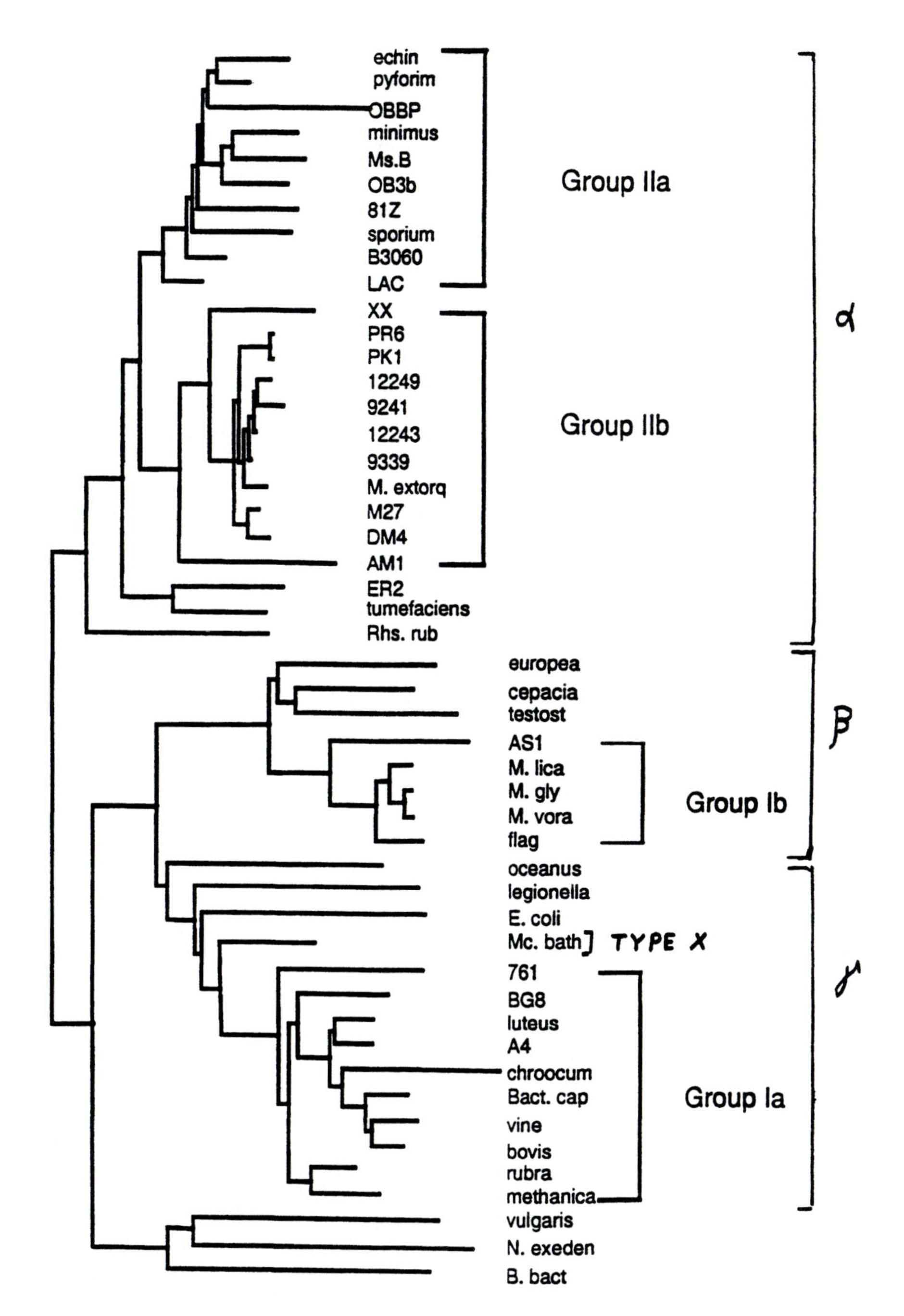


Figure 6-1 Phylogenetic relationships between some methylotrophic bacteria and other proteobacteria.

## MEASUREMENTS OF METHANE OXIDATION IN ENVIRONMENTAL SAMPLES

Mason canning jars (30 oz size) make excellent, inexpensive containers for measuring methane consumption by soils and plant material when they are equipped with screw-on lids through which Swagelock fittings (1/4-inch tube opening) and 3/8-inch Teflon septa have been placed [48]. A hole (1/2 inch in diameter) is drilled through the lid of a Mason jar, and the threaded bottom part of the Swagelock R fitting (Swagelock part no. B-400-1-OR) with a 1-inch washer over the threads is inserted upward through the hole. Another washer is placed over the threads on the upper surface of the lid and upper part of the Swagelock fitting is screwed onto the bottom part of the fitting. One-quarter-inch septa are placed in the fitting above and below the lid to provide a double gas-tight seal. This method does not allow accurate measurements of the rates of methane oxidation, but it can be employed to follow the course of enrichments and is useful for inexpensive demonstrations of methane oxidations by soil samples in teaching laboratories.

## Measurements of Methane Oxidation in Soil Samples and in situ Enrichments for Methanotrophic Bacteria in Soils

### Procedure

1. Add 100 g of soil or plant material to the Mason jars equipped with Swagelock fittings and rubber septa. It is not necessary to add water or nutrients. Wet soil tends to consume methane more slowly than moist soil. However, the enrichment for methane-oxidizing organisms can be optimized by the addition of water until the water content is equal to 25% of the weight of the dried soil.
2. Add 50 ml of methane to the closed jars through the septa using a disposable syringe. Incubate the jars at room temperature. Open the jars daily, and change the gas phase.
3. Measure methane consumption using a gas chromatograph equipped with a flame ionization detector and a Porapak Q column. Measurements of methane in the headspace at 0.5- to 2-h intervals, depending on the rate of methane consumption, provide good estimates of the rates of methane consumption.

### Notes

Total gas consumption can also be measured routinely with a U-tube manometer fitted with rubber tubing and a syringe needle. Insertion of the needle through the septa in the Mason jars permits measurements of the vacuum created by methane and oxygen consumption. A control jar without added methane can be used to roughly determine the amount of oxygen consumed in the absence of methane.

When using soil samples that oxidize methane at the time they are removed from the field, soil gas consumption increases to a maximum of 0.2 to 20 μmol methane consumed per hour per gram of dry soil after 10 to 14 days of enrichment with an atmosphere containing 2% to 5% (vol/vol) of methane. The maximum rate of methane oxidation depends on the type of the soil and nutrients available in the soil. This method can also be used to measure methane consumption by soil samples taken in the field. Methane oxidation by water samples and enrichments in Mason jars or smaller serum vials can be used successfully if magnetic bars are placed in the jars and the samples are placed on magnetic stirrers or if they are shaken to facilitate gas transfer to the liquid.

## Measurements of Methane Oxidation in Water and Soil Samples with $^{14}C$ Methane

### Procedure

1. Evacuate 16-ml serum vials closed with Teflon-lined septa and crimp-tops using a vacuum pump, with vacuum tubing attached to a 24-gauge syringe

needle. Insert the needle through the septa, and evacuate the vials as completely as possible with a vacuum pump.

2. Add 5 ml of water samples through the septa of each of five replicate vials with a disposable syringe.
3. With a gas-tight syringe, add 0.1 ml (10 microcuries) of $^{14}C$-$CH_4$, prepared from $^{14}C$-bicarbonate as described by Daniels and Zeikus [16]. *Note*: Methane prepared using some methanogens contains trace amounts of $^{14}C$-carbon monoxide that may be oxidized to $^{14}C$-$CO_2$ under aerobic or anaerobic conditions. It is important that the amount of $^{14}C$-$CO_2$ collected is greater than 5% of the total radioactivity of the methane carbon added in order to draw accurate conclusions about the rates of methane oxidation and amounts of methane oxidized. Alternatively, commercially prepared $^{14}C$-$CH_4$ of high purity may be purchased from one of several dealers in radiochemicals.
4. Add 0.5 ml of methane to each of the vials, and equilibrate the pressure in the vials with the atmosphere by inserting a 24-gauge needle through the septum of each vial. Immediately remove samples to measure the specific activity of the methane in the headspace, as described in Step 13 below.
5. Add acetylene, a potent inhibitor of methane oxidation, to one vial to give a final concentration of 8.8 millimolar in the gas phase. This vial serves as a control. Alternatively, add 0.5 ml of 1 M NaOH to one vial to prevent methane oxidation.
6. Place the vials in a water bath shaker at the same temperature as the environment from which the samples are taken. Continue incubation with vigorous shaking for 2–5 h in the dark.
7. Add 0.5 ml of 1 M solution of NaOH to stop the reactions and trap the carbon dioxide.
8. After 10 minutes, flush the vials with nitrogen in an approved fume hood by slowly bubbling nitrogen gas through the liquid for approximately 10 min. This procedure strips radioactive methane from the aqueous phase while the radioactive carbonate produced by the oxidation of methane and radioactive cell material remain in the water.
9. Remove an aliquot of the liquid sample in the vials for determination of its radioactivity.
10. Acidify the solution in the vials with 1 ml of 10 N HCl, and purge again with nitrogen to remove the carbon dioxide.
11. Remove another sample for determination of the radioactivity in the cell material, and dispose of the remainder of the contents of the vials as prescribed by radiation safety requirements.
12. Subtract the radioactivity in samples from acetylene-inhibited vials from the radioactivity detected in samples removed from the experimental vials. The amount of carbon dioxide produced is calculated by dividing the difference in the radioactivity in the samples taken before and after acidification of the samples by the specific activity of the methane (dpm methane/$\mu$mol methane added to each vial).
13. Determine the specific activity of the methane in the headspace as follows. Remove 10 $\mu$l of the headspace through the septum of the experimental vial with a gas-tight syringe. Inject the sample into a 1-ml serum vial that is filled completely with a toluene-based scintillation solvent and closed with a Teflon membrane and crimp-top. Place this vial in an empty standard scintillation vial, and determine the radioactivity of the sample. Determine the counting efficiency by adding a $^{14}C$-standard to the same scintillation solvent in a serum vial of the same size. Methane is soluble in toluene-

based scintillation solvents, but is not totally soluble in many other scintillation solvents.

14. Remove another 10 μl of the headspace to measure the methane concentration in the gas phase by gas chromatography. Calculate the specific activity by dividing the dpm of methane by the amount of methane (micromoles) present in a given volume of the headspace.

Notes

This method is a very sensitive means for detecting the conversion of methane to carbon dioxide and cell material when low numbers of methanotrophs are present in samples or when the rates of methane oxidation are too low to measure by other methods. However, precautions should be observed. With modifications, the method can be used to measure the kinetics of methane oxidation by water or soil samples and to determine $K_s$ and $V_{max}$ values for methane consumption in natural samples [48]. The effects of temperature, pH, and other parameters on methane oxidation rates can also be measured. Soil samples can be used in place of water samples. When soil samples are employed, the soil is weighed and added to the vials before closure and evacuation.

These methods are very useful for field work, particularly for aquatic samples. Several evacuated vials and a closed container containing $^{14}$C-methane of the desired specific activity can be transported to a field site. Water samples are injected into the evacuated vials followed by injection of the radioactive methane and equilibration of the vials to atmospheric pressure. The samples can be incubated in water at the site to provide the proper temperature and light conditions. After the samples are incubated for 2–5 hours, they can be killed by injection of NaOH and processed further after returning to the laboratory. This method is the only technique available for assessing the end products of methane oxidation—carbon dioxide, cell material, and soluble organic metabolites.

## Other Methods for Measuring Methane Oxidation

Methane oxidation by soil samples can be measured more conveniently in Mason jars as described earlier if a proper gas chromatograph with gas sampling loops and good sensitivity is available. Before determining the rates of methane oxidation, a period of time—usually 30 to 60 minutes—is required for equilibration of the methane with soil; the exact time required can be determined using soil that has been autoclaved. The kinetics of methane oxidation can be determined by adding different amounts of methane to the headspace and following the rate of methane consumption at different methane concentrations as the methane is consumed. This method is best employed for measurement of the potential of a soil to oxidize methane at $V_{max}$ concentrations in the gas phase [48]. Sectioned soil cores can be used to measure the depth and distribution of methane-oxidizing bacteria.

The populations of methanotrophs in samples can be approximated by assuming that these bacteria oxidize methane at the same rates in the environment as they do in pure cultures. The rate of methane oxidation by pure cultures of methanotrophs is approximately 500 nmoles methane oxidized per mg dry weight of bacterial cells per minute at $V_{max}$ concentrations of methane.

In situ measurements of methane oxidation in soils can also be done using static chambers placed over soils [48]. The design and use of flux chambers of approximately 0.25 $m^3$ placed over soils have been described by Whalen and Reeburgh [46]. Methane concentrations in the chambers were monitored by removing samples of the atmosphere from within the chambers at intervals with a gas-tight syringe and measuring the methane concentrations by gas chromatography. A sensitive gas chromatograph and gas sampling loop are required to measure the oxidation of methane at ambient (atmospheric) concentrations in field studies. These methods are useful for measurements of net production or consumption of methane, but do not give direct measurements of either process if they occur simultaneously. Methylfluoride ($CH_3F$) at concentrations of 1.2% in the gas phase is believed to specifically inhibit methane oxidation [36]. Its addition to the headspace of flux chambers or atmospheres can be used to estimate the effects of methane oxidation on methane emissions from soils. The differences in rates of methane production in the absence and presence of methylfluoride have been used to determine the contribution of methanotrophs to soil methane budgets [46].

Methane oxidation by pure cultures of methanotrophs can be measured most conveniently by harvesting cells from the growth medium by centrifugation, suspending the cell pellet (1 mg dry weight per ml) in nitrate mineral salts (NMS) media, and measuring methane consumption in an oxygen electrode.

## ISOLATION OF PURE CULTURES OF METHYLOTROPHIC BACTERIA FROM NATURAL SAMPLES AND THEIR CHARACTERIZATION

### Isolation of Methanotrophic Bacteria

No one medium, temperature, or atmosphere is optimal for enrichment of all methanotrophs because this group of microorganisms is so diverse. Different media and modifications of these media are normally employed for the isolation of methanotrophs. The most useful is the nitrate mineral salts (NMS) medium.

Procedure

1. Inoculate soil or other environmental samples (1 g or 1 ml per 100 ml of enrichment medium) into an NMS medium, and incubate the enrichments at the desired temperature under an atmosphere containing 50% air and 50% methane; shake slowly. There are several possible means for preparing the appropriate gaseous atmospheres. For enrichments in serum vials, remove part of the atmosphere with a gas-tight syringe and inject the proper amount of methane into the vials to complete the atmosphere or inject methane under pressure into an air-filled serum vial using a disposable syringe. For enrichments in flasks, evacuate half of the air atmosphere through a black stopper fitted with a glass tube and rubber hose that is plugged with a small inverted test tube. Fill the vessel with methane to 1 atm using an in-line manometer or vacuum gauge to monitor the gas pressure in the flasks.
2. Transfer cultures to fresh media (10% inocula) when growth is evident ($A_{550nm}$ = 0.2), and incubate again until turbidity is approximately $A_{550}$ = 0.2.
3. Prepare dilutions of the enrichment cultures, and spread samples onto solid media containing 1.5% (wt/vol) agar, 100 mg/l of cyclohexamide, and 10 mg/l of amphotericin B. These two antibiotics are added to inhibit the growth of fungi that commonly contaminate cultures of methanotrophic bacteria.
4. Incubate agar plates inside gas-tight containers (desiccators or anaerobic jars fitted with tubing connectors). Fit vacuum tubing onto the containers, and use a gas filter (membrane filter or glass wool filter) in the vacuum line to

remove fungi and other microbes that enter with the gas used to fill the containers. Evacuate the containers as completely as possible with a vacuum pump or water aspirator. A manometer or vacuum gauge can be placed in the vacuum line to monitor the gas pressure as each gas is added to the chambers.

5. Fill the container with air and methane from a tank to give the desired atmosphere. Most methanotrophs can be isolated using an atmosphere containing equal parts of methane and air. The addition of 5% to 10% carbon dioxide stimulates the growth of most methanotrophs. The addition of carbon dioxide can be accomplished most conveniently by purchasing a mixture of 90% methane and 10% carbon dioxide from a supplier, such as Air Products Corporation.
6. Inspect the plates under a dissecting microscope at intervals of 2 to 3 days. When colonies are just visible (about 1 mm in diameter), restreak them onto the same agar used for the initial isolation and isolate pure cultures. The first colonies of mesophilic methanotrophs typically appear in 1 to 2 weeks.
7. Test each colony for growth on agar with and without methane in the atmosphere. Some colonies that appear on agar plates are bacteria that do not utilize methane, but grow on contaminants in the agar or degrade agar.
8. Test colonies for purity by streaking onto nutrient agar, Luria broth agar, or another rich medium. Obligate methanotrophic bacteria do not grow on these media.
9. Store pure cultures that grow in the presence but not in the absence of methane on the surfaces of agar slants in anaerobic tubes closed with black rubber stoppers. Add a methane-air atmosphere of the appropriate composition, or incubate cultures streaked on agar slants in cotton-plugged culture tubes, in jars, or in dessicators containing methane-air atmospheres. Pure cultures should be transferred monthly or stored at −80°C after growth in broth in a growth medium containing 20% glycerol. Methanotrophic bacteria do not survive lyophilization well.

Notes

It is important to use methane of the highest purity when isolating methanotrophs (Air Products Corporation is one possible source). Some methane sources, including natural gas, used in laboratories contain impurities that inhibit the growth of methanotrophs.

Modifications of the enrichment and isolation procedures can be employed to isolate different groups of methanotrophs. Type II methanotrophs capable of fixing elemental nitrogen can be isolated by omitting the nitrate or ammonia from the growth media and reducing the air to less than 20% of the final volume of the atmosphere. The atmosphere used for enrichments for nitrogen-fixing methanotrophs normally contains (vol/vol) 20% air, 5.0% carbon dioxide, 25% nitrogen, and 50% methane.

Marine methanotrophs typically require 10–40 g/l of NaCl for growth. The growth of some marine methanotrophs has been reported to be inhibited by light, and none are known to be stimulated by light. Therefore, it is advisable to incubate all cultures in the dark. Marine methanotrophs isolated by Lees et al. [33] would not grow on normal solid media, although they grew well in liquid media. They were isolated as pure colonies using "sloppy agar" containing 0.4% (wt/vol) Noble agar.

Moderately thermophilic methanotrophic bacteria, such as *Methylococcus capsulatus*, can be enriched for and isolated at 45–50°C. Psychrophilic methanotrophs can be isolated at 4°C. Some isolated at 4°C appear to be true psychrophiles.

Much of the global methane oxidation occurs at temperatures below those that have normally been used for the isolation of methylotrophic bacteria. Methane oxidation in tundra soils and after fall turnover in eutrophic freshwater lakes occurs between 0–10°C. Rapid methane oxidation occurs in dry soils, including those in landfill covers, in which the water activity is less than that required for the growth of known methanotrophs. Because heterotrophic bacteria rapidly outgrow methanotrophs in media supplemented with complex nutrients, most pure cultures of methanotrophs have been isolated using defined media without organic supplements. These isolations may have missed significant populations of methanotrophs that require vitamins or other growth supplements. The growth of some Type I methanotrophs is also greatly stimulated by sugars (e.g., glucose [52]). However, the inclusion of sugars in enrichment or isolation media permits the growth of heterotrophic bacteria that tend to dominate the cultures. To enhance the variety of the isolates selected, isolate methanotrophs in media containing vitamin supplements and choose colonies that grow slowly without organic supplements and grow more rapidly with glucose or peptones in the growth medium.

The inclusion of 250 g or more per liter of polyethylene glycol (PEG) 500 into enrichment media (to reduce the water activity of the media) allows the growth of methanotrophs that appear different from previously characterized isolates. Known isolates do not grow in the presence of these concentrations of PEG (R. Hanson, unpublished results).

During their growth on methane, several methanotrophic bacteria excrete methanol and organic waste products that are toxic to pure cultures. It is often difficult to obtain pure cultures free of consorts that consume these waste products. A few facultative methanotrophic bacteria that grow on succinate, fatty acids, sugars, and rich media as well as methane have been described (*M. organophilum* strain XX; ATCC 27886). The purity of these cultures must be rigorously established [23, 38].

## Characterization of Methanotrophic Bacteria

Colony morphology and pigmentation of methylotrophic bacteria vary widely [50]. Colonies of different methanotrophic bacteria are white, tan, pink, orange, deep red to purple, yellow, or colorless and transparent. Cells can be coccoid, rod shaped, pear shaped, or vibroid and arranged as diplococci, rosettes, or single cells. Mature cultures of most Group II methanotrophs contain exospores or cysts, whereas mature cultures of Group I methanotrophs contain desiccation-resistant cysts or lack a differentiated state. The presence of exospores and cysts can be verified in microcolonies by phase-contrast microscopy after growth of bacteria on media solidified with 16% (wt/vol) Noble agar for 24–48 h. *Azotobacter*-like cysts can be detected when present in colonies from 14-day-old cultures when the cells are stained with the Ziehl-Nielsen stain [8]. Photographs of cells of methanotrophic bacteria and their resting stages are available [50]. Capsules can be observed by phase-contrast microscopy after staining with 2.5% (wt/vol) nigrosine. The presence of poly-β-hydroxybutyrate inclusions is determined chemically, and polyphosphate granules are detected by the method of Duguid et al. [18]. Other features that are useful for the characterization of methanotrophs include growth at pH 4.0–6.6, and at temperatures from 4°C to 60°C and tolerance to NaCl from 0.5% to 5.0% (wt/vol), to KCN (0.01%), to $NaN_3$(0.01%), and to sodium dodecyl sulfate (0.01%). The abilities of methanotrophs to use a variety of carbon sources as determined by stimulation of growth in the presence of methane, as well as growth with the carbon sources in the absence of methane, are often employed for determining their relationships to know isolates. Growth with atmospheric nitrogen, nitrate, ammonia, methylamine, and a variety of other amino acids and organic nitrogen sources described by Bowman et al. [8] is also important for the phenotypic characterization of methanotrophs, as are assays for the enzymes employed for assimilation of ammonia. Different groups of methanotrophs differ significantly in the moles % guanosine plus cytosine present in DNA extracted from cells.

Methanotrophs have been divided into three groups that differ in the pathways used for formaldehyde assimilation and the fine structure of intracytoplasmic membranes observed by transmission electron microscopy of thin sections of the bacteria. The presence of the ribulose monophosphate (RuMP) pathway can be determined by assays for two unique enzymes—hexulose-phosphate synthase and hexulose-phosphate isomerase as described by Bowman et al. [8]. Group II methanotrophs contain the serine pathway for formaldehyde fixation and lack enzymes of the hexulose-phosphate pathway. The presence of the serine pathway for formaldehyde assimilation can be determined by assays for serine hydroxymethyltransferase or hydroxypyruvate reductase. Group X methanotrophs contain enzymes of the RuMP pathway, lesser amounts of enzymes of the serine pathway, and ribulose-1,5-diphosphate carboxylase. The enzymes of these pathways can be assayed as described by Bowman et al. [8].

It is not practical to determine the presence or absence of all of these characteristics for a new isolate unless the bacterium is obviously different from known isolates, has a unique environmental role, or possesses physiologic properties of commercial or environmental importance. It may be more practical to determine the phylogenetic relationship of an isolate to known methanotrophs by sequencing 16S rRNA isolated from pure cultures or environmental samples.

## Identification of Methanotrophic Bacteria That Produce Soluble Methane Monooxygenase sMMO and Measurements of sMMO Using the Naphthalene Oxidation Assay

To determine the potential of methanotrophs to oxidize halogenated chemicals, it is often important to identify the presence of methanotrophic bacteria in bioreactors and in environmental samples that produce sMMO [10]. Only those methanotrophs that produce sMMO oxidize naphthalene to naphthols [10, 26]. Naphthols are readily detected as colored products produced by their reaction with tetrazolitized o-dianisidine.

### Procedure: Detection of sMMO in Colonies of Bacteria on Agar [39]

1. Replicate colonies of bacteria grown on NMS agar media to agar media prepared without copper in the trace elements and supplemented with 5.6 mg/l of $FeSO_4 \cdot 7H_2O$ dissolved in 1 mM $H_2SO_4$.
2. Add a 1:50 dilution of a 1% naphthalene dissolved in 20% Pluronic L-62 surfactant (BASF Corp.) to a 1% low-melting agarose solution in 25 mM MOPS buffer, pH 7.2. Pour 3 ml of the melted agarose solution cooled to 50°C over the replicate plates.
3. When the agarose overlayer has solidified, incubate the plates at 30°C with an atmosphere of 50% oxygen for 1 h.
4. Add 1 ml of a fresh solution of 1% Fast Blue BN (tetrazolitized o-dianisidine, Sigma Chem. Co.) to cover the plates. Colonies expressing sMMO will turn dark red in color within 5 min at room temperature. Stabilize the color by adding 1.0 ml of glacial acetic acid.

Procedure: Measurements sMMO in Broth Cultures of Methanotrophic Bacteria [10]

1. Grow cultures of bacteria in copper-free NMS media supplemented with 5.6 mg of $FeSO_4$ dissolved in 1 mM $H_2SO_4$ with an atmosphere of 50% methane and 50% air.
2. Add 10 ml of culture (diluted to $A_{600} = 0.2$ with copper-free NMS media) to 58-ml serum bottles sealed with rubber stoppers; evacuate the vials and backfill them with air to remove methane.
3. Transfer 1-ml aliquots of the diluted cultures to 10-ml screw-cap test tubes, add sodium formate to give a final concentration of 25 mM, and add 1 ml of a filtered solution saturated with naphthalene.
4. Incubate the closed tubes on a rotary shaker at 200 rpm for 1–3 h.
5. Add 0.1 ml of a freshly prepared 1% solution of Fast Blue BN (Sigma Chemical Co., St. Louis, MO).
6. Stabilize the color of the reaction products after 1 min by adding 0.4 ml of glacial acetic acid.
7. Measure the absorbance (540 nm) of the solution after centrifugation at 5,000 × g for 5 min. The extinction coefficient of the diazo-dye product formed by reaction of o-dianisidine with naphthols is 38,000 $M^{-1}$ $cm^{-1}$.

## Enrichments of Methylotrophic Bacteria That Do Not Use Methane

Because of the great diversity of methylotrophic bacteria that do not use methane, no one method will permit enrichment and isolation of all bacteria in this group. Perhaps the most ubiquitous group of methylotrophic bacteria are the pink pigmented facultative methylotrophs (PPFMs) [38]. These bacteria are facultative methylotrophs that employ the serine pathway for formaldehyde assimilation. They can be enriched and isolated using the NMS medium described for the isolation of methanotrophs when it is supplemented with 0.1% to 0.2% (vol/vol) of methanol. Cyclohexamide is often added to prevent the growth of fungi.

Procedure: Isolation of PPFM Bacteria from the Surfaces of Leaves

1. Press leaves of white clover or soybean plants on the surface of NMS-methanol agar containing 100 mg/l of cyclohexamide.
2. Remove the leaves after 5 to 10 min.
3. Seal the plates by wrapping the edges with parafilm and incubate them at 30°C. Pink pigmented colonies should appear in 1 week or less. These colonies contain rod-shaped, oxidase-positive, methanol-oxidizing bacteria that also grow on a variety of other carbon and energy sources.

### *Isolation of Pure Cultures of PPFM Bacteria*

PPFMs that use the serine pathway for formaldehyde assimilation can be isolated by enrichment from plant leaves with relative ease. They may exist in large numbers (more than $10^4$ per gram of tissue) in extracellular symbiotic or mutualistic relationships with plant tissues. They have been encountered on the surfaces of more than 50 species of plants and were found to be most abundant near the margins of the abaxial surfaces of leaves. They are located below the cutin layer of leaves [14]. When leaves are pressed onto the NMS agar supplemented with 0.5% methanol, red-colored colonies form an outline of the leaves. Closely related PPFMs can be isolated from freshwaters, the air, soil, bronchial tracts of humans, and groundwaters.

Procedure

1. Remove fresh leaves from plants 1 h or less before isolation attempts, or store leaves frozen.
2. Aseptically trim stems from leaves, and rinse the leaves in sterile distilled water.
3. Grind the leaves in a mortar and pestle with superbrite glass beads (grade 100, 3M Co., Minneapolis, MN).
4. Dilute portions of the leaf extracts in the NMS medium, and plate dilutions onto NMS agar media containing 100 mg/l of cyclohexamide and 0.5% vol/vol of methanol.

### *Isolation of Other Methanol-Utilizing Bacteria*

A wide variety of other gram-negative methylotrophs can be isolated from natural sources using NMS-methanol media or a mineral salts medium described by MacLennon et al. [35]. To increase the variety of methylotrophs, enrichments should be carried out with 0.05% to 0.2% (vol/vol) methanol. Some methylotrophs are sensitive to methanol concentrations above 0.05%. Variations in enrichment and isolation temperatures from 4–50°C and at pH values from 4.0–9.0 yield many different cultures of serine pathway methylotrophs and methylotrophs that employ the RuMP pathway for formaldehyde fixation. The inclusion of a vitamin mixture in the isolation and enrichment media increases the diversity of the methylotrophs isolated. The methylotrophs that employ the RuMP pathway for formaldehyde fixation are usually obligate or restricted facultative methylotrophs that use a limited number of multiple carbon compounds. An exception to this general rule are the gram-positive methylotrophs that employ the RuMP pathway but are true facultative methylotrophs.

## Isolation of Gram-Positive, Endospore-Forming Facultatively Methylotrophic Bacteria

Gram-positive, endospore-forming methylotrophs are readily isolated from decaying plant material, such as compost or dead leaves of deciduous trees, after enrichment at 50°C in a mineral salts medium supplemented with 1% (vol/vol) of methanol as described by Schendel et al. [41]. To enrich for a variety of isolates, the medium should be supplemented with biotin and folic acid at 20 mg/l and vitamin $B_{12}$ at 1 mg/l or with 0.5 g/l of yeast extract. Some of the bacteria belonging to the genus *Methylobacillus* require vitamins; others do not. All of the representatives of this genus that have been isolated are moderate thermophiles. It is possible that different enrichment procedures will yield other gram-positive methylotrophs that do not form endospores and species that grow at lower temperatures.

Procedure

1. Add 1g of decaying plant material to 100 ml of minerals vitamin (MV) media, and incubate at 50°C until turbidity is apparent (1–3 days). Transfer a 2% inoculum to a second flask, and continue the enrichment for a total of three or more transfers. Incubate the final enrichment at 50°C for 2 days and at 37°C for another 3 days. Some of these bacteria sporulate at 37°C, but fail to form endospores at 50°C.
2. Examine the enrichment cultures by phase-contrast microscopy. Short gram-positive rods should be present. Endospores should be apparent in some cells or occur as free spores.

3. Heat-shock a sample of the culture at 80°C for 30 min (or at 100°C for 10 min) to kill vegetative cells, and spread aliquots onto the surfaces of MV agar plates.
4. Determine the nutritional requirements of all pure cultures obtained by streaking them on trypticase soy agar to isolate pure clones. The gram-positive methylotrophs are facultative methylotrophs that grow well on rich media and use mannitol and a limited number of sugars as carbon and energy sources. Determine the ability of isolates to grow in MV media in the presence or absence of biotin and vitamin $B_{12}$. Determine their abilities to use mannitol, glucose, methylamine, succinate, acetate, and other carbon and energy sources in the absence of the methanol.

### Isolation of Hyphomicrobia That Use Methanol and Employ Nitrate as a Terminal Electron Acceptor

Dimorphic, prosthecate Hyphomicrobia that respire anaerobically with nitrate as a terminal electron acceptor and utilize methanol as a carbon and energy source can be isolated using a mineral salts medium [40] containing 0.5% methanol and 0.5% potassium nitrate. These bacteria occur in a variety of soils, sediments, and aquatic environments. They are also capable of heterotrophic or methylotrophic growth with oxygen as a terminal electron acceptor and occur in relatively low numbers in waters low in organic matter or as consorts of methanotrophic bacteria that excrete methanol. They are often co-enriched with methanotrophic bacteria.

#### Procedure

1. Add a sample of water, sewage, or soil to the mineral salts medium (medium 337) containing nitrate and methanol in a filled, tightly screw-capped tube. Incubate at 25–30°C until turbidity and bubbles of nitrogen appear. Growth often occurs as a film attached to glass surfaces or as pellicles in 3 days to 3 weeks. Phase-contrast microscopy should reveal motile swarmer cells and sessile cells with prosthecae that reproduce by budding.
2. Streak samples of the enrichments onto the same medium containing 1.5% (wt/vol) of agar, and incubate anaerobically under an atmosphere of nitrogen for the isolation of single colonies.

## DETECTION OF METHYLOTROPHIC BACTERIA IN ENVIRONMENTAL SAMPLES USING GENE PROBES AND PHYLOGENETIC PROBES

Methylotrophic bacteria are phylogenetically diverse. It has been possible to select 16S rRNA sequences that differentiate methanotrophs from methylotrophs that do not utilize methane within the serine and RuMP pathways [9]. Probes complementary to these sequences can be employed to detect 16S rRNAs isolated from bacteria in these groups. These probes and gene probes described in this section have been employed to detect and characterize methylotrophs in environmental samples and bioreactors without the need for culturing the bacteria. Other probes have been employed to localize methanotrophic endosymbionts in tissues of mussels and tubeworms [17].

## Isolation of Nucleic Acids from Environmental Samples

### Procedure: Isolation of RNA from Water Samples

1. Concentrate water from lakes and ponds or groundwater using a tangential flow filtration system until a 10- to 500-liter sample has been concentrated to 1–2 liters. The amount of source water that is required to produce sufficient biomass is dependent on the productivity of the source.
2. Further concentrate the bacteria by centrifugation at 10,000 × g for 15 min, wash the sedimented cell material once in 1 M NaCl, and suspend it in 50 mM acetate buffer, pH 5.1.
3. Isolate RNA as described for the isolation of RNA from bacterial cells in Chapter 12.

### Procedure: Isolation of RNA from Soil Samples, Plant Material, and Sediment Samples

1. In a Waring blender, briefly homogenize 50 g of a sample in 200 ml of an ice-cold homogenization solution—0.5 g/l $KH_2PO_4$, 0.4 g/l $MgCl_2 \cdot 6H_2O$, 0.4 g/l NaCl, 0.4 g/l $NH_4Cl$, 0.05 g/l $CaCl_2 \cdot 2H_2O$, and 0.2 M sodium ascorbate. Add 15 g/l of acid-washed polyvinylpolypyrrolidone and 10 g of Chelex-100, and homogenize the suspension (three 1-min cycles with cooling on ice between cycles).
2. Centrifuge the cells at 650 × g for 5 min at 4°C to remove cell debris and the Chelex resin; repeat the extraction of the soil sediment twice by adding 200 ml of the homogenization solution.
3. Centrifuge the combined supernatants at 10,000 × g for 2 min at 4°C, and suspend the cells in 1 M NaCl. Centrifuge again at 10,000 × g for 10 min. Wash the cells once with ice-cold 1 M NaCl, and suspend them in 50 mM sodium acetate and 10 mM EDTA, pH 5.1.
4. After extraction of RNA with acidic phenol and in a bead beater and purification of the RNA as described in Chapter 12, load solutions containing 5 μg of RNA in 20 μl of a freshly prepared solution of 0.01 M disodium phosphate, pH 6.8, containing 8% (wt/vol) glyoxal into wells in a 1.2% (wt/vol) agarose gel (11 × 4 cm.). RNA from pure cultures should be used at concentrations of 0.1–0.5 μg per well.
5. Separate RNA fractions at 75 volts for 3.5 h, and transfer them by Northern blotting to Magnagraph membranes (Micron Separations, Inc.).
6. Air-dry the Northern blots, and bake them in a vacuum oven for 1 hour at 80°C.
7. Perform hybridization reactions with oligonucleotide probes to detect 16S rRNAs with sequences complementary to the probes as follows: Use 20 ml of hybridization and prehybridization solutions per 100 $cm^2$ of membrane surface area.
8. Prehybridize blots in 0.9 M NaCl, 50 mM sodium phosphate, pH 7.2, 50 mM EDTA, 0.5% (wt/vol) sodium dodecyl sulfate (SDS), 10 X Denhardt's solution at 45°C for probe F, and 47°C for all others in Table 6-2 for 2 h.
9. Replace the prehybridization solution with hybridization solution—0.9 M sodium phosphate, pH 7.2, 1.0% SDS, 1% bovine serum albumin, 1 mM EDTA, pH 8.0. Add labeled probe (10–50 ng/ml) and incubate at 45°C for probe F and 45°C for all other probes listed in Table 6-2 for 6 h.
10. Wash the blots for 30 min at room temperature with a mixture of 40 mM sodium phosphate, pH 7.2, 1 mM EDTA, 0.5% SDS, and 0.5% bovine serum albumin.

**Table 6-2** Oligonucleotide Probes Employed to Detect Methylotrophic Bacteria Belonging to Different Groups

| Description | Probe Sequence | Target Sequence Position | Wash Temp (°C) |
|---|---|---|---|
| A. Probes for detection of 16S rRNAs from methane-utilizing bacteria in Group Ia (RuMP pathway methanotrophs) | 5′-CTCCGCTATCTCTAACAGATT-3′ | 1004–1024 | 52 |
| | 5′-GATTCTCTGGATGTCAAGGG-3′ | 988–1007 | 52 |
| B. Probes for detection of endosymbionts of marine mussels and tubeworms [17] | 5′-CCGCCACTAAACCTGTATATAGG-3′ | 838–859 | 46 |
| | 5′-GTAGGGCATATGCGGTATTAGCATGGG-3′ | 163–187 | 46 |
| C. A probe for detection of 16S rRNAs from methylotrophic bacteria in Group Ib (RuMP pathway methylotrophs that do not oxidize methane) | 5′-ATGCATCTCTGCTTCGTT-3′ | 1013–1020 | 52 |
| D. A probe for detection of 16S rRNAs from all RuMP pathway methylotrophs | 5′-GGTCCGAAGATCCCCCGCTT-3′ | 197–216 | 52 |
| E. A probe for detection of 16S rRNAs from methane utilizing bacteria in Group IIa (serine pathway methanotrophs) | 5′-CCATACCGGACATGTCAAAAGC-3′ | 987–1008 | 50 |
| F. A probe for detection of 16S rRNAs from methylotrophic bacteria in Group IIa (serine pathway methylotrophs that do not use methane) | 5′-GGTAACATGCCATGTCCAG-3′ | 990–1008 | 50 |
| G. A probe for detection of 16S rRNAs from all serine pathway methylotrophs | 5′-CCCTGAGTTATTCCGAAC-3′ | 142–159 | 50 |

**Note**: The target sequence positions refer to the position of the sequences complementary to each probe on the *E. coli* 16S rRNA molecule.

11. Replace the first wash solution with the stringent wash solution—40 mM sodium phosphate, pH 7.2, 1% SDS, and 1 mM EDTA.
12. Incubate in the stringent wash solution at the temperatures indicated in Table 6-2 for 15 min.
13. Repeat this stringent wash and prepare the blots for detection of hybridization reactions as described in Chapter 13.

### Notes

The two probes for Group Ia methanotrophs have slightly different specificities, and both should be used. A eubacterial probe should be used to detect ribosomal RNAs of non-methylotrophic bacteria. The hybridization temperature is 40°C, and the final wash temperature is 37°C for the eubacterial probe (5′-ACCGCTTGTGCGGGCCC-3′) [4]. RNA samples from methylotrophic bacteria belonging to each group and RNA samples from non-methylotrophic bacteria should be added to each blot to confirm that hybridization reactions occur with the specificity predicted.

The use of agarose gel electrophoresis to separate RNAs from contaminating material in extracts of bacteria from soil and water is necessary to obtain good hybridization with oligonucleotide probes. The probes most recently employed by the author's group have been 3′ end-labeled with digoxygenin-11-ddUTP according to the protocols provided by Boehringer Mannheim Biochemicals (Indianapolis, IN). Positive hybridization reactions were detected and quantified using a chemiluminescence detection system (Genius system) as described in the user's guide for filter hybridization provided with the reagents purchased from Boehringer Mannheim Biochemicals. Oligonucleotide probes end-labeled with $^{32}PO_4$ prepared as described in Chapter 13 can also be used to detect hybridization reactions.

## Extraction of DNA from Environmental Samples and Separation of DNA from Humic Acids and Other Materials That Interfere with PCR Amplification and Restriction Analysis

### Procedure

1. Extract DNA from cells in a bead beater using the alkaline phenol extraction or other methods that produce less shearing of the DNA, as described in Chapter 12. DNA from aquatic samples, groundwaters, and some sediments can be blotted onto nylon membranes for hybridization reactions or amplified by polymerase chain reaction (PCR) for further analysis. Some soils and plant materials contain organic materials, including humic acids, that interfere with PCR amplification and hydrolysis by restriction endonucleases. This procedure has been used with good success to purify DNA of these inhibitory materials.
2. Load 1 to 5 μg of DNA into an agarose gel containing 2% wt/vol of polyvinylpyrrolidone and 1.25% wt/vol of agarose [51].
3. Locate DNA in the gels by exposure to ultraviolet light after staining with ethidium bromide.
4. Isolate the DNA from agarose gels onto NA45 anion membrane strips (Schleicher & Schuell, Keene, NH) as described in the product literature.

### Notes

The isolated DNA can be treated with restriction endonucleases, subjected to electrophoresis, and Southern blotted to nylon membranes to determine the sizes of restriction fragments that hybridize to different gene probes (see Chapter 13). The lengths of hybridizing fragments in DNAs from different bacteria can be used to compare unknown bacteria in environmental samples or isolates with hybridizing fragment lengths from known bacteria. Alternatively, large DNA fragments prepared by gentle lysis of cells and treatment of lysates with infrequently cutting restriction endonucleases can be separated by pulsed-field gel electrophoresis. These approaches are especially useful for determining whether bacteria isolated from a sample are identical to uncultured bacteria in a sample [1, 43].

The cloned methanol dehydrogenase *moxF* from *Methylobacterium organophilum* strain XX hybridizes with DNA from serine pathway methylotrophic bacteria that have DNA with a moles % G+C of 52% to 57% (Groups IIa and IIb; Fig. 6-1). The conditions for hybridization with a 2.5-kb cloned fragment are that the hybridization was carried out at 65°C in 2 X SSC, and the blots were washed at 65°C as described by Southern [42]. Dot blots or Southern blots of DNA fragments can also be hybridized to cloned DNA fragments encoding protein subunits or the sMMO hydroxylase component in order to determine if Group IIa methanotrophs capable of expressing this enzyme are present in an environmental sample [7]. *E. coli* clones carrying these genes are available from the author. The DNA purified by these methods is also suitable for PCR amplification ligation with appropriate vectors and cloning into *E. coli* in order to isolate clones carrying plasmids with 16S rDNA inserts, as described in Chapter 13.

The clones carrying 16S rDNA from all methanotrophic bacteria isolated at this time can be identified by colony hybridization with probes specific for Group Ia or Group IIa methanotrophs in Table 6-2, with the exception of some strains of genus *Methylomonas*. New probes will be necessary to ensure that all PCR-amplified 16S rDNA sequences from all methanotrophs can be detected by hybridization with probes. Clones containing inserts encoding 16S rDNA from some methylotrophs that do not utilize methane can be identified on colony blots by hybridization with oligonucleotide probes complementary to 16S rDNA from methylotrophs in Groups Ib and IIb. The plasmids carrying 16S rDNA inserts that hybridize to the probes can then be isolated and the inserts sequenced in order to relate the uncultured methanotrophs in a sample to previously isolated methanotrophic bacteria from which 16S rRNAs have been isolated and sequenced.

As an alternative to these procedures, cells can be embedded in agarose and lysed. The DNA released from the cells can be treated with infrequently cutting restriction endonucleases, and the fragments can be separated by pulsed-field gel electrophoresis. After separation, the DNA fragments can be transferred to membranes by Southern blotting [42] and hybridized with methylotroph-specific gene probes to compare gene fragment sizes from environmental sources with those from known cultures [43].

## ECOLOGIC STUDIES THAT HAVE EMPLOYED GENE AND PHYLOGENETIC PROBES

### Endosymbionts of Mussels

The mussels (Family *Mytildae*) collected from sediments where methane and oxygen occur near cold gas seeps in the Gulf of Mexico contain bacterial symbionts in their gill tissue [12]. Ultrastructural studies using transmission electron microscopy revealed coccoid to rod-shaped gram-negative bacterial cells with stacked intracytoplasmic membranes similar to those observed in Group I methanotrophs [12]. Attempts to culture the methanotrophs were unsuccessful. Methane utilization was demonstrated in fresh tissues of three species of mussels from different sites. Methanol dehydrogenase present only in methylotrophs and hexulose phosphate synthase present only in Group I methylotrophs were detected in tissue extracts. Hydroxypyruvate reductase and serine glyoxylate aminotransferase, enzymes of the serine pathway present in Group II methanotrophs, were not detected. The 16S rDNA from a symbiont of a mussel from the Louisiana continental slope was amplified by PCR, and the DNA was cloned. Phylogenetic analysis of the sequences of the cloned 16S rDNAs indicated that the endosymbiont was related to but differed from the known free-living Group I methanotrophs. It is found between the clusters of Group I and Group X methanotrophs.

Two probes, complementary to the unique, highly variable regions of the endosymbiont 16S rRNA, were identified and synthesized. The symbiont-specific probes bound to structures in the gill tissues that were identical in size, shape, location, and distribution to the symbionts observed by electron microscopy [17].

Taken together these and other published observations [17] provide strong support for the existence of a previously uncharacterized Group I methanotrophic bacterial endosymbiont that provides organic carbon required for the nutrition and energy of the mussel.

Similar methanotrophic endosymbionts have been observed in Pogonophoran tube worms [17], and in one case the coexistence of methanotrophic and chemoautotrophic endosymbionts in mussel tissue of a mussel has been reported [19].

## Methanotrophic Bacteria in Freshwater Lakes and Aquifers

Groundwater in aquifer contaminated with trichloroethylene and tetrachloroethylene (TCE; U.S. Department of Energy Savannah River site near Aiken GA) was collected from 13 monitoring wells. Twenty-five different cultures of methanotrophs were isolated by enrichment culture followed by single colony isolation [7]. Characterization of the cultures by phospholipid fatty acid analysis and phenotypic testing revealed that most strains were members of the genus *Methylosinus* and produced sMMO when grown in copper-limited media. One culture was characterized as a member of the genus *Methylocystis*, which grew and produced sMMO in groundwater samples. DNA extracted from core samples hybridized to the sMMO-B gene probe. This study supported the hypothesis that indigenous microbes in aquifers were capable of producing sMMO and could be used for the removal of chlorinated aliphatic hydrocarbons when methane and air were added to the subsurface waters. The Group II methanotrophs appeared to be well adapted to oligotrophic conditions found in the aquifers, as evidenced by their growth in groundwater exposed to methane. Further evidence has suggested that the methanotrophs increased by at least 4 orders of magnitude after injection of methane into the groundwater.

Alvarez-Cohen et al. [1] isolated a consortium of bacteria from an aquifer that rapidly degraded chloroform and trichloroethylene. Curved rod-shaped bacteria in the consortium hybridized to a probe for Group II methylotrophic bacteria when fixed to microscope slides, and RNA isolated from the consortium hybridized to the same probe, but not a probe specific for Group I methanotrophs. A pure culture of methanotrophic bacterium was isolated from the consortium. Both cells in the consortium and pure culture oxidized TCE and naphthalene, indicating that they contained sMMO. DNA from both pure culture and consortium hybridized with the probe for the sMMO-B gene; were digested with *Ase*I, an infrequently cutting restriction enzyme; and were separated by pulse-field gel electrophoresis. The sizes of the restriction fragments that hybridized to the sMMO-B gene probe were indistinguishable, but differed from the sizes of hybridizing fragments from other methanotrophs. This evidence indicated that the pure culture isolate was identical to the methanotroph in the consortium. The sequence of 16S rRNA from the pure culture was most closely related to, but was slightly different from the sequences of 16S rRNAs from Group II methanotrophs of the genus *Methylosinus* [1, 29]. The phylogenetic relationship of this isolate to other Group II methanotrophs is illustrated in Figure 6-1. The isolate is identified as LAC in this figure.

In eutrophic lakes that are thermally stratified in the summer, methane oxidation occurs at the interface (metalimnion) between the anaerobic hypolimnion and the oxygenated epilimnion. Methane-oxidizing activity was located by measuring the oxidation of radioactive methane in water samples taken from different depths [29]. Water samples were also concentrated using a tangential flow filtration device, and the biomass was used to characterize the methanotrophic bacteria present in the samples. Extracts of the biomass from the metalimnion of Lake Mendota, Madison,Wisconsin, contained hexulose phosphate synthase activity, but hydroxypyruvate reductase was not detected. RNA isolated from the biomass and fixed to nylon filters hybridized to probes specific for Group I methylotrophic bacteria, but not to probes complementary to 16S rRNA sequences of serine pathway methylotrophs. The largest amounts of RNA that hybridized to the Group I methylotroph probe were found in metalimnion samples where peak rates of methane oxidation occurred [29].

## Methanotrophic Bacteria in Terrestrial Environments

Vecherskaya et al. [45] have used fluorescent antibodies prepared against a variety of methanotrophs to detect bacteria present in low-temperature tundra bog soils that were shown to oxidize methane using radioactively labeled methane. Their results indicated that 40% to 74% of the methanotrophs detected in samples from five different sites were related to species of Group I methanotrophs, whereas 24% to 43% of the methanotrophs were related to species of Group II methanotrophs. The total numbers of methanotrophs varied from $0.1–23 \times 10^6$ per gram of soil and represented 1% to 23% of the total bacterial numbers determined by staining with fluorescein isothiocyanate and counting bacteria by fluorescence microscopy.

Brusseau et al. [9] have enriched tundra soil by adding methane to the atmospheres above soil samples in Mason jars. After 1 month the rates of methane oxidation in the soils increased from 0.1 to 5.0 micromoles of methane oxidized per hour per gram of soil. Soil samples that did not oxidize methane without enrichment did not do so after enrichment. RNA was isolated from the enriched soil samples and was fixed to nylon membranes and hybridized with oligonucleotide probes specific for Group I and Group II methanotrophs. Group II methanotrophs were dominant in these methane-enriched soils.

Several other studies of the ecology of methanotrophic bacteria have been reviewed [5, 27, 32], and readers are encouraged to consult these reviews for further information.

## Unresolved Questions About the Ecology of Methylotrophic Bacteria

There are several questions regarding the ecology of methanotrophic bacteria and their roles in the global carbon cycles that remain to be answered. It is often stated that only a small fraction of the bacteria that exist in natural environments have been cultivated and studied in pure culture, and this is probably true of methanotrophs as well. We do not know whether only members of certain phylogenetic groups have been isolated while unrelated methanotrophs have not. Probes for the detection of genes and rRNAs of many of the environmentally significant methane-utilizers may not be available. Although endosymbiotic relationships of methanotrophs with marine mussels and tube worms have been described, there may be many other symbiotic relationships with freshwater plants and animals that have not. What are the relationships between methanotrophs and the plants with which they are associated [26, 32]? Do different methanotrophs exist in soils that are exposed only to atmospheric methane at low concentrations compared with soils exposed to higher methane concentrations? Bender and Conrad [6] recently published results that indicated that some soils exposed to atmospheric methane have half saturation constants ($K_m$ values) of 20–45 ppmv (30–63 nM methane in the vapor phase), whereas pure cultures of methanotrophic bacteria or soils naturally exposed to higher methane concentrations or that had been enriched by exposure to higher methane concentrations have much higher $K_m$ values. Does this observation indicate that different bacteria with different affinity constants for methane dominate in environments exposed to widely different methane concentrations? Can one isolate methanotrophs with high-affinity uptake systems for methane? Which methanotrophs oxidize methane at low temperatures (below 10°C)? Many of the environments where methane oxidation occurs, including marine environments and tundra soils, remain below 10°C most or all of the time. Is methane oxidized slowly by mesophiles, or are uncharacterized psychrophilic methanotrophs responsible?

## GROWTH MEDIA

### Modified Nitrate Mineral Salts Medium (NMS)

Phosphate solution (autoclave separately and add 20 ml per liter of sterile media after cooling). Add 3.6 g $Na_2HPO_4$ and 1.4 g of $KH_2PO_4$ per liter of stock solution. Adjust pH to 6.8.

Trace elements solution (add 10 ml per liter of media). Add per liter of stock solution:

| | |
|---|---|
| Tetrasodium EDTA | 100 mg |
| $ZnSO_4 \cdot 7H_2O$ | 7 mg |
| $MnCl_2 \cdot 4H_2O$ | 3 mg |
| $H_3BO_3$ | 30 mg |
| $CoCl_2 \cdot 6H_2O$ | 20 mg |
| $CuCl_2 \cdot 2H_2O$ | 1 mg |
| $NiCl_2 \cdot 6H_2O$ | 2 mg |
| $Na_2MoO_4 \cdot 2H_2O$ | 3 mg |

Ferrous sulfate solutions (autoclave separately, and aseptically add 10 ml per liter of media). Add 0.05 g of $FeSO_4 \cdot 7H_2O$ per 100 ml of stock solution. Dissolve in 100 ml of $10^{-3}$ M $H_2SO_4$.

Magnesium sulfate solution (add 10 ml per liter of media). Add 2 g $MgSO_4 \cdot 7H_2O$ per 100 ml of stock solution.

Calcium chloride solution (add 10 ml per liter of media). Add 0.2 g of $CaCl_2 \cdot 2H_2O$ per 100 ml of stock solution.

When methanol is used as a substrate, add 0.1% to 0.2% (vol/vol) methanol to sterile medium after cooling or after temperature of agar is 55°C. When it is added, add agar at 1.5% (wt/vol). Use Noble agar for the growth of copper-limited cultures.

Procedure: Preparation of Media

1. Dissolve 1 g of potassium nitrate in distilled water; add the trace elements, ferrous sulfate, magnesium sulfate, and calcium chloride solutions. Autoclave.
2. Cool the medium, add the phosphate solution, and store until use.
3. Prepare water supplies for copper-free media passing distilled water through a column of Chelex-100 (BioRad Corp.). Glassware for cultures grown with low levels of copper should be acid-washed and rinsed with Chelex-treated water.

## Minerals Vitamin (MV) Medium for Isolation of Gram-Positive Methylotrophs

Phosphate-ammonia stock solution (per liter):

| | |
|---|---|
| $K_2HPO_4$ | 38 g |
| $NaH_2PO_4$ | 28 g |
| $(NH_4)_2SO_4$ | 36 g |

Adjust pH to 7.0 with HCl.

Trace elements solution stock solution (autoclave separately and add 1 ml per liter). All figures are per liter of solution. Dissolve and store in $10^{-3}$ M $H_2SO_4$.

| | |
|---|---|
| $FeCl_2$ | 2.0 g |
| $CaCl_2 \cdot 2H_2O$ | 5.3 g |
| $MnSO_4 \cdot H_2O$ | 200 mg |
| H3$BO_4$ | 30 mg |
| $CuSO_4 \cdot 5H_2O$ | 40 mg |
| $ZnSO_4 \cdot 7H_2O$ | 200 mg |
| $Na_2MoO_4 \cdot 2H_2O$ | 47 mg |
| $CoCl_2 \cdot 6H_2O$ | 40 mg |

Magnesium sulfate solution (autoclave separately and add 1 ml per liter): 246.5 g/liter of $MgSO_4 \cdot 7H_2O$

Biotin, vitamin $B_{12}$ solution (filter-sterilize and store in the dark at 4°C. Add 0.1 ml per liter of sterile media when desired).

| | |
|---|---|
| Biotin | 200 mg/liter |
| Vitamin $B_{12}$ | 10 mg/liter |

Procedure: Preparation of Media

1. Add 100 ml of the phosphate-ammonia solution to 1 liter of distilled water and autoclave.
2. After cooling add the trace elements, magnesium sulfate, and vitamins solutions.
3. Add 1.0% (vol/vol) of methanol.
4. Add 1.5% (wt/vol) of agar for preparation of the solid media.

## Medium 337 for the Isolation of Hyphomicrobia

Procedure

1. Add per liter of distilled water:

| | |
|---|---|
| $KH_2PO_4$ | 1.35 g |
| $Na_2HPO_4$ | 2.13 g |

| | |
|---|---|
| $(NH_4)_2SO_4$ | 0.5 g |
| $KNO_3$ | 5.0 g |
| $MgSO_4 \cdot 7H_2O$ | 0.2 g |
| $CaCl_2 \cdot 2H_2O$ | 5.0 mg |
| $FeSO_4 \cdot 7H_2O$ | 2.0 mg |
| $MnSO_4 \cdot 7H_2O$ | 0.5 mg |

2. Autoclave, cool, and add 5 ml of methanol per liter.
3. Add 21 ml of the following vitamin mixture per liter: biotin (10 mg), niacin (35 mg), thiamine dichloride (30 mg), p-aminobenzoic acid (20 mg), calcium pantothenate (10 mg), and vitamin $B_{12}$ (5 mg) dissolved in 100 ml of distilled water.
4. Filter-sterilize, and store refrigerated.

## REFERENCES


1. Alvarez-Cohen, L., McCarty, P.L., Boulygina, E., Hanson, R.S., Brusseau, G., and Tsien, H.C. 1992. Characterization of a methane-utilizing bacterium from a bacterial consortium that rapidly degrades trichloroethylene and chloroform. Appl. Environ. Microbiol. 58: 1886–1893.
2. Anthony, C. 1982. The Biochemistry of Methylotrophs. Academic Press, London.
3. Barnes, R.O. and Goldberg, E.D. 1976. Methane production and consumption in anoxic marine sediments. Geology 4: 297–300.
4. Bartlett, K.B. and Harris, R.C. 1993. Review and assessment of methane emissions from wetlands. Chemosphere 26: 261–320.
5. Bedard, C. and Knowles, R. 1989. Physiology, biochemistry and specific inhibitors of $CH_4$, $NH_4^{+1}$ and CO oxidation by methanotrophs and nitrifiers. Microbiol. Rev. 53: 68–84.
6. Bender, M. and Conrad, R. 1994. Kinetics of methane oxidation in oxic soils. Chemosphere 26: 1–4.
7. Bowman, J.P., Jiminez, L., Rosario, I., Hazen, T.C., and Sayler, G.S. 1993. Characterization of a methanotrophic bacterial community present in a trichloroethylene-contaminated groundwater site. Appl. Environ. Microbiol. 59: 2380–2387.
8. Bowman, J.P., Sly, L.I., and Nichols, P.D. 1993. Revised taxonomy of the methanotrophs: description of *Methylobacter* gen. nov., emendation of *Methylococcus*, validation of *Methylosinus* and *Methylocystis* species, and a proposal that the Family *Methylococcaceae* includes only the group 1 methanotrophs. Int. J. Syst. Bacteriol. 43: 735–740.
9. Brusseau, G.A., Bulygina, E.S., and Hanson, R.S. 1994. Phylogenetic analysis and development of probes for differentiating methylotrophic bacteria. Appl. Environ. Microbiol. 60: 626–636.
10. Brusseau, G.A., Tsien, H.C., Hanson, R.S., and Wackett, P.P. 1990. Optimization of trichloroethylene oxidation by methanotrophs and the use of a colorimetric assay to detect soluble methane monooxygenase activity. Biodegradation 1: 19–29.
11. Cavanaugh, C.M. 1993. Methanotroph-invertebrate symbiosis in the marine environment: ultrastructural, biochemical and molecular studies. In Microbial Growth on C1 Compounds (Murrell, J. C. and Kelly, D. P., Eds.), pp. 315–328. Intercept, Andover, United Kingdom.
12. Cavanaugh, C.M., Wirsen, C.O., and Jannasch, H.W. 1992. Evidence for methylotrophic endosymbionts in a hydrothermal vent mussel (*Bivalvia Mytilidae*) from the mid-Atlantic ridge. Appl. Environ. Microbiol. 58: 3799–3803.
13. Claus, G., and Kutzner, H.J. 1985. Denitrification of nitrate and nitric acid with methanol as carbon source. Appl. Microbiol. Biotechnol. 22: 378–381.
14. Corpe, W.A. and Rheem, S. 1989. Ecology of the methylotrophic bacteria on living leaf surfaces. FEMS Microbiol. Ecol. 62: 243–250.

15. Dalton, H. 1992. Methane oxidation by methanotrophs: physiological and mechanistic implications. In Methane and Methanol Utilizers (Murrell, J.D. and Dalton, H., Eds.), pp. 85–114. Plenum Press, New York.
16. Daniels, L. and Zeikus, J.G. 1983. Convenient biological preparation of pure high specific activity $^{14}C$-labeled methane. J. Labeled Comp. Radiopharmaceuti. 20: 17–24.
17. Distel, D.L. and Cavanaugh, C.M. 1994. Independent phylogenetic origins of methanotrophic and chemoautotrophic bacterial endosymbioses in marine bivalves. J. Bacteriol. 176: 1932–1938.
18. Duguid, J., Marmion, B., and Swain, R. 1975. Staining of diphtheria bacillus and volutin-containing organisms. In Medical Microbiology, 12th ed. (Cruikshank, R., Duguid, J., Marmion, B., and Swain, R., Eds.), p. 41. Churchill-Livingstone, Edinburgh.
19. Fisher, C.R., Brooks, J.M., Vodenichar, J.S., Zande, J.M., Childress, J.J., and Burke, R.A. 1993. The cooccurrence of methanotrophic and chemautotrophic sulfur-oxidizing symbionts in a deep-sea mussel. Mar. Ecol. 114: 277–289.
20. Giovannoni, S.J., DeLong, E.F., Olsen, G.J., and Pace, N.R. 1988. Phylogenetic group-specific oligonucleotide probes for identification of single microbial cells. J. Bacteriol. 170: 720–726.
21. Graedel, T.E. and Crutzen, P.J. 1989. The changing atmosphere. Sci. Am. 261: 136–143, 22: 378–381.
22. Green, H. and Dalton, H. 1989. Substrate specificity of soluble methane monooxygenase: mechanistic implications. J. Biol. Chem. 264: 17698–17703.
23. Green, P.N. and Bousfield, I.J. 1982. Toxonomic study of some Gram-negative facultatively methylotrophic bacteria. J. Gen. Microbiol. 128: 623–638.
24. Haber, C.L., Allen, L.N., Zhao, S., and Hanson, R.S. 1983. Methylotrophic bacteria: biochemical diversity and genetics. Science 221: 1147–1153.
25. Hanson, R.S. 1980. Ecology and diversity of methanotrophic organisms. Adv. Appl. Microbiol. 26: 3–39.
26. Hanson, R.S. and Brusseau, G.A. 1993. Biodegradation of low molecular weight halogenated hydrocarbons. In Degradation and Bioremediation Technologies for Toxic Chemical, (Chaudry, G.R., Ed.), pp. 277–297. Disocorides Press, Portland.
27. Hanson, R.S. and Wattenberg, E. 1991. Ecology of methane-oxidizing bacteria. In Biology of Methylotrophs (Goldberg, I. and Rokem J.S., Eds.), pp. 325–348. Butterworth Publishers. London, United Kingdom.
28. Hanson, R.S., Netrosev, A.I., and Tsuji, K. 1991. The obligate methanotrophic bacteria *Methylococcus*, *Methylosinus* and *Methylomonas*. In The Procaryotes. Vol. II (Balows, A., Truper, H.G., Dworkin, M., Harder, W., and Schliefer, K., Eds.), pp. 2350–2364. Springer-Verlag, New York.
29. Hanson, R.S., Bratina, B.J., and Brusseau, G.A. 1993. Phylogeny and ecology of methylotrophic bacteria. In Microbial Growth on C1 Compounds (Murrell, J.C. and Kelley, D.P., Eds.), pp. 285–302. Intercept Publishers, Andover, United Kingdom.
30. Hirano, S.S. and Upper, C.D. 1991. Bacterial community dynamics. In Microbial Ecology of Leaves (Andrews, J.H. and Hirano, S.S., Eds.), pp. 271–294. Springer-Verlag, New York.
31. Iverson, N., Oremland, R.S., and Klug, M.J. 1987. Big Soda Lake (Nevada). Pelagic methanogenesis and anaerobic methane oxidation. Limnol. Oceanogr. 32: 804–818.
32. King, G.M. 1992. Ecological aspects of methane oxidation, a key determinant of global methane dynamics. Adv. Microb. Ecol. 12: 431–469.
33. Lees, V., Owens, N.J.P., and Murrell, J.C. 1991. Nitrogen metabolism in marine methanotrophs. Arch. Microbiol. 157: 60–65.
34. Lidstrom, M.E. 1991. The aerobic methylotrophic bacteria. In The Prokaryotes, Vol. I. (Balows, A., Truper, H.G., Dworkin, M., Harder, W., and Schliefer, K., Eds.), pp. 431–445. Springer-Verlag, New York.
35. MacLennan, D.G., Onsby, J.C., Vasey, R.B., and Cotton, N.T. 1971. The influence of dissolved oxygen on *Pseudomonas* AMI grown on methanol in continuous culture. J. Gen. Microbiol. 69: 395–404.
36. Oremland, R.S. and Cumberland, C.W. 1992. Importance of methane-oxidizing bacteria in the methane budget as revealed by the use of a specific inhibitor. Nature 356: 421–423.

37. Panganiban, A.T., Patt, T.E., Hart, W., and Hanson, R.S. 1979. Oxidation of methane in the absence of oxygen in lake water samples. Appl. Environ. Microbiol. 37: 303–309.
38. Patt, T.E., Cole, G.C., Bland, J.A., and Hanson, R.S. 1974. Isolation and characterization of bacteria that grow on methane and organic compounds as sole sources of carbon and energy. J. Bacteriol. 120: 955–964.
39. Phelps, P.A., Agarwal, S.K., Speitel, G.E.J., and Georgiou, G. 1992. *Methylococcus trichosporium* OB3b mutants having constitutive expression of soluble methane monooxygenase in the presence of high levels of copper. Appl. Environ. Microbiol 58: 3701–3708.
40. Poindexter, J.S. 1991. Dimorphic prosthecate bacteria: The genera *Caulobacter, Asticcacaulis, Hyphomicrobium, Pedomicrobium, Hyphomonas* and *Thiodendron*. In The Prokaryotes, Vol. III. (Balows, A., Truper, H.G., Dworkin, M., Harder, W., and Schleifer, K.-H., Eds.), pp. 2177–2196. Springer-Verlag, New York.
41. Schendel, F.J., Bremmon, C.E., Guettler, M.C., Flickinger, M.C., and Hanson, R.S., 1990. L-lysine production at 50°C by mutants of a newly isolated and characterized methylotrophic *Bacillus* sp. Appl. Environ. Microbiol. 56: 963–970.
42. Southern, E.M. 1975. Detection of specific sequences among DNA fragments separated by gel electrophoresis. J. Mol. Biol. 98: 503–517.
43. Tsien, H.-C. and Hanson, R.S. 1992. A soluble methane monooxygenase gene probe for the identification of methanotrophs that rapidly oxidize trichloroethylene. Appl. Environ. Microbiol. 58: 335–345.
44. Tsien, H.C. and Hanson, R.S. 1992. Soluble methane monoxygenase component B gene probe for identification of methanotrophs that rapidly degrade trichloroethylene. Appl. Environ. Microbiol. 58: 953–960.
45. Vecherskaya, M.S., Galchenko, V.F., Sololova, E.N., and Samarkin, V.A. 1993. Activity and species composition of aerobic methanotrophic communities in tundra soils. Curr. Microbiol. 27: 181–184.
46. Whalen, S.C. and Reeburgh, W.S. 1988. A methane flux time series for tundra environments. Global Biogeochem. Cycles 2: 399–409.
47. Whalen, S.C. and Reeburgh, W.S. 1990. Consumption of atmospheric methane by tundra soils. Nature (Lond) 346: 160–162.
48. Whalen, S.C., Reeburgh, W.S., and Barber, V.A. 1993. Oxidation of methane in boreal forest soils: a comparison of seven measures. Biogeochemistry 16: 181–211.
49. Whittenbury, R. and Krieg, N. 1984. *Methylococcaceae* fam. nov. In Bergey's Manual of Determinative Bacteriology, Vol. 1. pp. 256–262. Williams & Wilkins, Baltimore.
50. Whittenbury, R., Phillips, K.C., and Wilkinson, J.F. 1970. Enrichment, isolation and some properties of methane-utilizing bacteria. J. Gen. Microbiol. 61: 205–218.
51. Young, C.C., Burghoff, R.L., Keim, L.G., Minak-Bernero, V., Lute, J.R., and Hineton, S.M. 1993. Polyvinylpyrrolidone-agarose gel electrophoresis purification of polymerase chain reaction-amplifiable DNA from soils. Appl. Environ. Microbiol. 59: 1972–1974.
52. Zhao, S. and Hanson, R.S. 1984. Isolate 761: a new Type I methanotroph that possesses a complete tricarboxylic acid cycle. Appl. Environ. Microbiol. 48: 6–15.

7

ALAN D.M. RAYNER
LYNNE BODDY

# Terrestrial Fungal Communities

Members of the fungal kingdom permeate terrestrial ecosystems as decomposers and symbionts. Collectively, they probably outweigh the animal kingdom and outstrip vascular plants in their species diversity. They integrate and regulate energy and nutrient fluxes on spatial scales ranging from micrometers to kilometers. As decomposers, they are major participants in the global carbon cycle. As parasites, they play a regulatory role in the population dynamics of their hosts. As mycorrhizal partners, they form absorptive accessories to roots, linking the activities of separate plants and underpinning primary production in forests, heath lands, and grasslands. In lichens, they clothe what might otherwise be bare parts of the planet. In summary, terrestrial fungi are fundamentally important in the complex interactions and feedbacks that govern life processes in the biosphere. Furthermore, many of these organisms are readily accessible to experimentation, and despite widespread misconceptions to the contrary, their patterns of distribution and activity in natural environments can often be observed or assayed directly.

Knowledge of patterns of distribution and activity is crucial to understanding the relationship between ecologic structure and function. Only with such knowledge is it possible to provide meaningful answers to questions about where individual fungi (and their offspring) are, what they are doing there, how they arrived, whether they will persist, and how they may change in character or occurrence.

However, a vast amount of fungal ecologic work has been done and continues to be done without taking due account either of the need or the appropriate means to locate and characterize individual population components. This seems to be due to a failure to recognize the full significance of the fact that, for all or part of their lives, the majority of terrestrial fungi exist as structured colonies. Correspondingly, efforts to quantify the natural occurrence of terrestrial fungi have focused on counting and weighing these organisms, assuming that they can, at least approximately, be treated as fully particulate or homogeneous entities. Such assumptions are invalid, and counting is wholly inappropriate since mycelia can expand indefinitely. When systems become fragmented, they can exist as separate entities. Conversely, when genetically identical mycelia meet, they merge to become a single functional individual. Mycelia are structurally and functionally heterogeneous, indefinitely proliferating systems that thrive in locally unpredictable niches by varying their developmental pattern as circumstances change. Numbers alone mean little when they are applied to entities ranging from a single viable spore to a dynamic system covering hectares. Weights alone mean little if it is not known how the biomass of the system in question varies in organization and activity through space and time. Methods for quantifying fungal biomass have been reviewed elsewhere [36].

In order to enhance our understanding, investigations of terrestrial fungal communities therefore need to go beyond counting and weighing fungi without reference to their organizational pattern. In particular, it is important to be able to construct and interpret *maps* that configure, at *an appropriate scale*, both the regional limits or boundaries of colonies and the heterogeneous disposition of structure within these boundaries. The approaches and techniques described in this chapter have these ends in view. Although many of the techniques are relatively "low-tech," do not assume that fungal ecology is way behind other branches of microbial ecology. Indeed, fungal ecology is ahead in terms of applying concepts from macroecology and in understanding community structure and dynamics.

## LOCATING AND IDENTIFYING FUNGI

### Direct Observation—At Macroscopic Scales

It is often not realized how much information can be gleaned concerning fungal distribution patterns without recourse to special preparative techniques or culturing; in other words, just by making careful observations in the field. In fact, good initial field observations can, in the long run, save enormous amounts of time and resources in fungal ecologic studies, helping in the framing of relevant questions and in devising appropriate sampling and experimental procedures. The premature application of high-resolution techniques has contributed to the profusion of uninterpretable details and lack of insights that have characterized many aspects of fungal ecology to this day. It might not seem clever to wander into an ecosystem armed with little more than a cutting or digging tool, a hand lens (maybe even a pair of binoculars to help spot those fungi not accessible at ground level!), and some knowledge of where and how to look. However, the rewards can be considerable and can embarrass the efforts of those who are equipped more ambitiously!

Of course, much depends on the scale of the systems and of the organisms being studied. It is especially important to take account of the heterogeneous and often discontinuous distribution of the resources that fulfill the growth requirements of fungi. Such resources are commonly packaged as more or less discrete units, ranging in size from soil particles to whole trees. Fungi show two distinctive kinds of behavior with respect to their pattern of occupation of these resource units [26]. Either they are non-unit-restricted—able to produce migratory mycelium that interconnects the units—or they are unit-restricted, able to proliferate between units only by means of temporally or spatially dispersed propagules. The scale of unit-restricted fungi varies with the resource supply available to them from individual units. Basically, the larger the resource pool contained within or commanded by (as in a mycorrhizal root, for example) a resource unit, the larger can be the operational scale of the fungus. Unit-restricted fungi inhabiting tree trunks can be much larger than those inhabiting leaf petioles! Non-unit-restricted fungi, in contrast, are not so limited in scale by the boundaries of resource units and so are capable of producing among the largest individuals on earth [81].

By understanding these simple scaling rules, it is possible for fungal ecologists to adjust their search and sampling patterns to correspond with particular kinds of ecosystems and ecosystem components, much as an indeterminate mycelium adjusts its foraging pattern to suit its niche [68]. The refusal to make such adjustments, which is usually the result of attempts to produce quantitatively analyzable data, is equivalent to using a microscope at only one magnification—and generally yields either too much or too little detail.

## Collecting and Identifying Sporophores

Sporophores (fruit bodies), formed on or in the ground or at the surface of resource units, have long provided the first and often the only indication of the presence of particular kinds of fungi in natural habitats. They can provide very useful information in that they allow identification at least to the species or species-complex level, and their size and pattern of distribution to some extent reflect the resource pool available to the underlying mycelium.

However, for two reasons data based on sporophore distribution need to be interpreted with considerable caution and, if possible, to be combined with other kinds of information. First, the production of sporophores is often seasonal, strongly dependent on local environmental conditions, and coupled only loosely and often nonlinearly to the activity of mycelium. The absence of sporophores does not therefore necessarily imply the absence of mycelium. Similarly, although the biomass of sporophores indicates a lower limit to the resource pool available to the mycelium that produced them, it cannot be used as evidence of upper limits to this pool. Second, with the exception of fairy ring and other fungi with directly discernible mycelial limits (see the section, Observing and Identifying Mycelia), it is not usually possible to be sure whether separate sporophores of the same species have arisen from the same or different mycelial individuals.

Where possible, sporophores should be identified and their location recorded undisturbed in the field, especially at specific field sites where continuous monitoring is intended (see the section, Systematic Sampling at Specific Field Sites). However, if field identification is not immediately possible or if an estimate of biomass, cultural studies, or genetic analysis are intended, then it will be necessary to remove the sporophores to the laboratory. This practice should be kept to the minimum necessary to address the particular issues in question. Although there is little indication that harvesting sporophores damages mycelium, it is best to conserve study systems in as undisturbed a state as possible.

### Procedure: Removing Sporophores to the Laboratory

1. For large sporophores of basidiomycetes and ascomycetes remove either whole specimens or representative samples from them using a suitable cutting or digging tool. For non-critical work (e.g., preliminary field surveys), place specimens in a collecting basket. Otherwise keep them separate from one another, and place them after removal of any extraneous material, in suitably labeled, reasonably watertight containers (to prevent desiccation while allowing some aeration), such as plastic bags, bottles, envelopes of grease-proof paper, or compartmented boxes. Keep as fresh as possible (e.g., in a cooler), especially non-durable specimens.
2. Back at the laboratory, remove small fragments of tissue (less then 1 $mm^3$) or cut sections from the sporophore, especially from the spore-producing

region, and examine microscopically. Collect spore deposits by placing the spore-producing region over a suitable surface (e.g., Petri dish bottom, piece of cellophane, glass slide—not paper) for 4–24 h and covering to prevent drying out.

3. Identify using a suitable manual or guide. It is often helpful to use more than one text; for example an accessible, well-illustrated, but incomplete text to get to the right area and a more complete text for rigorous confirmation. Identification is not always easy (!), and in critical cases, especially before publication, it may be best to have your identification confirmed by an expert. Do not let the lack of a positive identification impede you—just record the characteristics of the specimen(s) concerned and give it a provisional code or name. If it is an important component of the community you are studying, you will know its true identity in due course! Fungi that you come across only rarely, although they may be exciting, may have little relevance to the ecologic issues of concern to you (except, for example, biodiversity), so do not waste too much time on them.
4. Be acutely aware of the effect of "experience." As your skills in finding and identifying sporophores develop, so too will your search pattern and personal bias. Ask yourself whether this bias is likely to affect the issues you are addressing and the way you interpret data.
5. Collect small sporophores (including conidiophores or sporangiophores) on a portion of the material that they are growing on, and place in suitable, small, reasonably watertight containers.
6. For small sporophores, collect individual resource units or pieces and incubate them in a damp chamber (e.g., a polythene bag, a plastic box, or a Petri dish containing moist paper) for several days, allowing sporophores to emerge.
7. Identify by means of microscopic examination and the use of suitable reference texts [see 44, 45, 48, 80 for guides to the literature], noting the same advice given under Steps 3 and 4.

## Observing and Identifying Mycelia

Mycelia are commonly regarded as hidden, nondescript systems that can only be visualized and identified by isolation into pure culture or by using specialized microscopic or analytic procedures. However, many fungal mycelia inhabit relatively large domains and either have directly visible effects on the substratum in which they are growing or are directly visible themselves.

Both lichenized and, to a lesser extent, non-lichenized mycelia commonly grow over exposed surfaces of living organisms or non-living materials or structures. Such mycelia, especially those of crustose lichens, clearly comprise mosaics of discrete individuals. There distribution can be recorded directly by mapping or photography.

The migratory mycelium of non-unit-restricted fungi, especially basidiomycetes, is commonly organized into dense, mat-like structures or cable-like mycelial cords and rhizomorphs. Their structures can simply be uncovered by carefully moving aside the superficial layers of leaf litter, soil, tree bark, or the like that protect them from desiccation (Fig. 7-1). In very humid environments, migratory mycelium may be directly visible as they bridge between resource units, and some suitably specialized forms protect themselves sufficiently from desiccation to be visible in open air even in relatively dry habitats [2, 46]. Distribution patterns of all such systems can be recorded by direct mapping and photography.

Figure 7-1 Example of a directly visible migratory mycelial systems.

The identity of directly visible mycelial systems can sometimes be ascertained through their organic connection to sporophores. However, it is important to appreciate that proof of an organic connection—the demonstration that a sporophore has actually emerged from, rather than become intermixed with, a mycelium—is not always easy to obtain. In doubtful cases it is necessary to undertake confirmatory microscopic, cultural, or analytic studies. Where sporophores are not attached to the mycelium, then small samples of the latter should be excised and taken to the laboratory for examination or analysis. Often, the mycelium has clear distinguishing features when viewed microscopically, and with experience it can even be possible to achieve field identification [70, 71]. However, such identification generally requires experience gained, for example, by examining material connected to sporophores or cultures derived from sporophores. Again, do not let the lack of an immediate identification be a deterrent; the organisms producing visible migratory mycelium at a particular site are generally relatively few in number, and their identity will become obvious in time.

Often, the mycelia of fungi growing within resource units have obvious, visible effects, such as causing necrosis, decay, or characteristic staining. Moreover, the boundaries between mutually exclusive, different individuals of the same and different species are often clearly demarcated by characteristic interaction zones, which are evident as lines of different texture or hue or both in cross section (Fig.7-2) However, it is important to differentiate between "zone lines" that are due to interaction and those that have other causes [70, 71]. Zone lines with other causes include "pseudosclerotial plates" (PSPs)—melanized, crust-like aggregates of hyphae produced by a single mycelial system—and "reaction" and "barrier" zones produced at interfaces between the functional tissues of a living host and an invading fungus. Such patterns are most evident at mature community development stages in relatively bulky, durable resource units (especially in wood), but can also be discerned at smaller scales; for example on close inspection of smaller units, such as leaves and straws.

Figure 7-2 Mosaic pattern produced by different mycelial individuals of the same and different species, growing within a beech log, as revealed in cross section.

The identity of mycelia occupying distinctive regions within a resource unit may be evident by correlation with the presence of sporophores or descriptions of symptoms caused by particular fungi growing in plant or animal hosts. Otherwise, it is helpful to persuade the mycelium to grow out in some way. This growth can often be achieved by direct incubation techniques (see Collecting and Identifying Sporophores, Step 6). Emerging sporophores and mycelium can then be identified as described elsewhere. Alternatively or additionally, isolation into pure culture may be attempted.

## *Systematic Sampling at Specific Field Sites*

Studies of the field distribution of fungi have commonly been based on haphazard, idiosyncratic, and destructive methods of sampling and observation. Such methods may have some value in preliminary surveys and in characterizing the niches of particular fungi. However, they cannot provide any useful index of spatial and temporal diversity and heterogeneity in natural fungal populations and communities. For such purposes, it is necessary to make use of experimental plots or study sites that can be sampled systematically and, where appropriate, repeatedly.

A study site or experimental plot amounts to a *fixed quadrat*, a region enclosed by a static boundary. The optimal size of the quadrat depends on the kind of ecosystem and operational scale and diversity of the organisms being studied. A major problem in fungal ecology is that mycelia, as indeterminate systems, can vary enormously in spatial extent. This problem is illustrated diagrammatically in Figure 7-3.

### Procedure

1. Before setting up your sample site or plot, consider carefully what the range in spatial scale of the organisms under consideration is likely to be and support this consideration, if possible, by preliminary observations.
2. Choose the size of the site or plot simply on the number of resource units that need to be included to address the question at hand. For unit-restricted fungi, an upper limit to the size of individuals is obviously set at the bound-

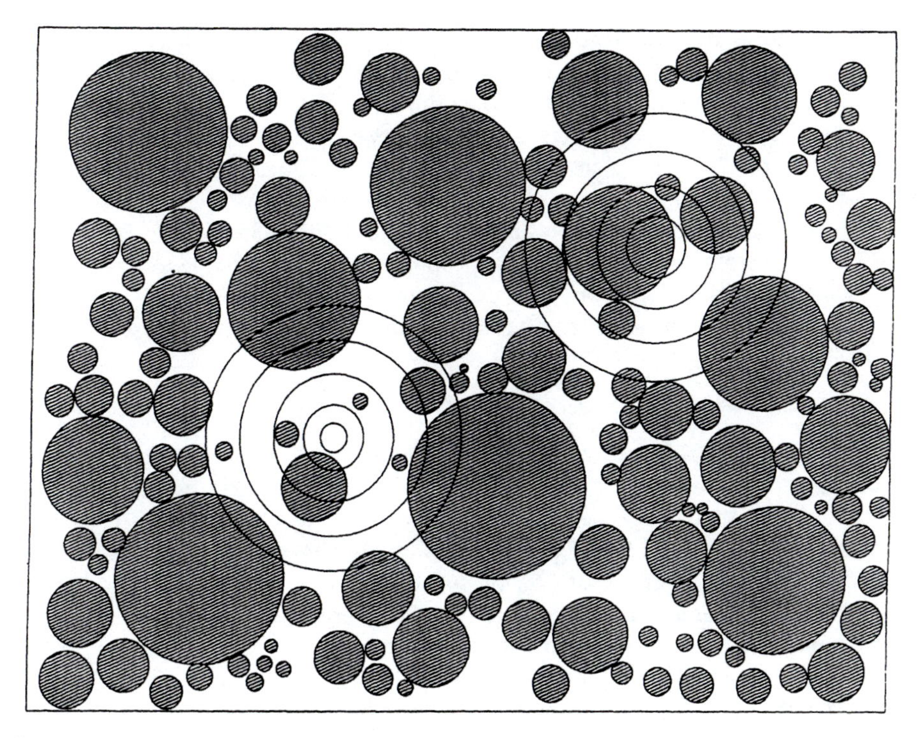

Figure 7-3 A rectangular "field" containing a set of individuals (*shaded circles*) of very varied spatial extent. Concentric circles indicate a range of quadrat sizes within which distribution patterns may be assessed. Consider the difficulties of designing a sample pattern that reflects adequately the heterogeneous distribution pattern within the set. (Courtesy of Mark Ramsdale.)

aries of resource units. For example, if inhabitants of leaf litter are the primary concern, then a quadrat size of 1 $m^2$ or less may be appropriate. Very large resource units, such as tree stumps, may themselves provide suitable quadrats.

3. In experimental plots, deliberately introduce resource units of various kinds, in appropriate arrays, within the site boundaries [25]. The benefit of studying patterns of colonization in this way lies in having a known starting point and, consequently, a full history of events in individual resource units. However, since an initial disturbance is involved, the processes revealed may not correlate well with those in natural communities.
4. For non-unit-restricted fungi, the quadrat size required to enclose the largest individuals may be very large. For some fairy ring fungi and rhizomorphic inhabitants of woody resources on the forest floor, for example, a quadrat size measurable in hectares might be needed for this purpose! However, use of such a large quadrat size would inevitably cause small individuals to be overlooked.
5. Adopt a *nested quadrat* approach as a general solution to the problem of making studies of fungal populations or community diversity over a range of spatial scales (see Fig. 7-3). This approach entails marking out a site or plot and then dividing it up into sets of progressively smaller subdomains or plots within plots. Each subdomain can then be examined as intensively as befits its size.
6. Sample with the minimal disturbance possible. Doing so is not so much of a problem if only a single "snapshot" sampling is intended, but it is best

to cause as little damage to the ecosystem as possible in all instances. In large sites, mark out paths that are consistently followed in order to minimize the effects of trampling. Mapping can be achieved by marking out some kind of grid reference system or points from which compass directions can be plotted. As much as possible, make identifications in situ. Where specimens have to be removed for identification purposes, try to find suitable reference specimens outside the plot. Replace any material that has to be removed from the site itself for inspection as close as possible to its original location. A dental mirror can be a useful aid to identification (e.g., as a means of examining the gills of agaric fungi).

## Direct Observation—At Microscopic Scales

Although much valuable information can be gained by direct macroscopic observation, light or even electron microscopy must be used for direct visualization of the location and spatial arrangement of microscopic fungal structures. The major problem with microscopic techniques is that, although microscopic sporocarps are often produced on the surface of organic substrata, mycelia typically grow within opaque resources. Thus, some disruption of the microhabitats is inevitable with microscopic techniques. Investigations usually require the removal of field material, "baits" inserted into soil or litter, or impressions or peels to the laboratory for observation, although it is possible to perform limited light microscopy in situ in the field! Kubiena [52, 53] employed an elegant technique in which a soil microscope was attached to soil profiles in the field.

Initial investigations need not be elaborate; mycelium growing on the surface or emanating from organic matter can be removed and mounted under a microscope, and soil particles or soil smears in water or agar can be examined under bright-field illumination. It is wise also to use phase-contrast microscopy, since it reveals more detail about the physiologic state of hyphae (e.g., cytoplasmic content, vacuolation, wall lysis), and hyphae that are ghost-like or invisible under bright-field illumination even when stained can be seen with phase-contrast microscopy [34, 35] Transmission and scanning electron microscopy (SEM) can be used to achieve higher resolution. Many striking photographs have been produced, but interpretation is often difficult due to artifacts, and procedures are time-consuming and costly. Thus SEM is currently not useful in routine ecologic work, though it can be invaluable for certain specific problems. SEM is also valuable in combination with electron microprobe analysis for determining elemental composition.

Revealing the three-dimensional relationships among hyphae, the substrata on/within which they are growing, and between hyphae and other organisms is of major concern when elucidating resource relationships and trying to understand the distribution patterns of fungi. The principal difficulties to be overcome are distinguishing hyphae within plant tissues, discriminating between living and dead hyphae, and identifying the species. The first two can be achieved by employing staining or labeling, and the last by using immunologic techniques or gene probes.

Procedure: Sectioning to Reveal Three-Dimensional Distribution

1. Compact, hard resources, such as wood that is only slightly decayed, by hand or with a sledge microtome for microscopic examination to reveal hyphal distribution [93, 94]. Embed more decayed samples before sectioning. Paraffin embedding is useful for serial sectioning on a rotary microtome, but since it employs alcohol dehydration it is time-consuming. Celloidin embedding works well even with very decayed wood, but takes several

months to complete [93, 94]. Embedding in polyethylene glycol is a rapid alternative [65].

2. For soils, which are generally less coherent, the three-dimensional structure must be retained both before and during sectioning. Freeze soil blocks or cores in the field by adding liquid nitrogen in situ to a region of soil that has been enclosed vertically by inserting a coring cylinder. When frozen, extract the core, take it to the laboratory, impregnate it with resin and grind with Carborundum [60] or with gelatine, and slice sections with a knife [6]. The latter technique is not suitable for minerals that lack sufficient air spaces for the penetration of gelatine.

### *Staining and Labeling*

It is often difficult to distinguish hyphae within plant tissues, and in such cases labeling the mycelium in situ or staining microscopic preparations is necessary. Considerable care must, however, be taken because various artifacts will be introduced by staining and labeling.

Procedure

1. Decolorize heavily pigmented plant material before staining; otherwise it is difficult to distinguish hyphae. Sodium chlorite/methyl salicylate is useful for leaf litter with high tannin content [47], although hydrogen peroxide causes less damage to more delicate leaves [91]. Potassium hydroxide is usually used for the clearance of cytoplasm in mycorrhizal roots [66]. Several simple methods decolorize plant material overnight using chlorine from domestic bleach or methanol [7].
2. Use the appropriate stain and procedure. Phenolic aniline blue is one of the best general-purpose stains, but there are a host of different stains and procedures suitable for different purposes [see 24, 41]. Safranin-picroaniline blue, Pianese III B, thionin-orange G, toluidine blue, and rhodamine B/methyl green give good differentiation of hyphae in wood [64, 65]. Fluorescent-labeled lectins also enhance hyphal visibility in wood and other plant tissues [58]. Trypan blue staining in lactophenol is useful for mycorrhizal fungi [66], though it is safer to use lactic acid in place of phenol. Chlorazol black E can be superior to trypan blue VA mycorrhizas [23], and lactic acid with fuchsin is useful [51].
3. Use fluorescent stains for specimens viewed against an opaque background with epi-ultraviolet or B.V. illumination [77]. Claims that they distinguish living from dead hyphae are not reliable, although fluorescein diacetate seems to more successful than most fluorescent stains in this respect [82, 83]. Use brighteners—fluorescent stilbene compounds—when stable binding and lack of toxicity are important [see 36]. Isotopic labeling and subsequent autoradiography are another means of distinguishing between metabolically active and inactive hyphae [36].

### *Immunologic Techniques and Polymerase Chain Reaction (PCR)*

The ability to identify and locate fungi in field material considerably enhances ecologic studies. Both the relatively recent immunologic and PCR (DNA polymerase chain reaction) techniques show considerable promise for recognizing mycelia in their microhabitats. Which method is better is not clear since there are many proponents for both, and perhaps it is a matter of "horses for courses."

Both involve time-consuming preliminary procedures and standardizations for each new system, and they should not be embarked on in the hope of a quick answer. However, the methods are essentially simple, and once specific antisera/monoclonal antibodies and PCR probes have been developed, their application is relatively straightforward (see Chapter 13). The immunologic methods depend on the specificity of antibodies for fungal antigens and whether the antibody is being labeled with a fluorescent compound, radioisotope, or enzyme that can be visualized or measured—the fluorescent antibody technique (FAT), radioimmunoassay (RIA), and enzyme-linked immunosorbent assay (ELISA), respectively. Early work employed polyclonal antisera, but monoclonal antibodies (MAbs) are the current choice, because of their specificity, which overcomes problems of cross-reactivity [36]. Care must, however, be taken since MAbs can exhibit assay-type specificity, and thus those selected using one approach may not function with another [42]. Mitchell [56] and Tiffin [88] outline the principles underlying immunologic techniques.

## ISOLATION OF SPORES—NON-SPECIES-SPECIFIC TECHNIQUES

### Obtaining Single Spore Isolates

A crucial step in many ecogenetic studies of fungi is obtaining strains of particular species from single spores. Such strains are invaluable as aids to identification of field isolates, in direct comparison of culture characteristics or by mating tests, and in studies of variation and gene flow in natural populations. Many, often elaborate techniques have been described for this purpose. The approach detailed in this section is among the most dependable and least demanding in terms of equipment and manipulative skill, although it takes a bit of practice.

Procedure

1. Obtain an old 100x microscope objective. Remove the lens and cement a sharp metal tube about 5 mm long in its place.
2. Obtain a 2–3 cm length of tungsten wire and attach to a suitable metal holder. Heat a layer 2–3 mm deep of sodium nitrite in a tin lid until molten. *Take care; the compound is potentially explosive—use a fume hood and use suitable protection.* Drag the wire through the sodium nitrite. If the sodium nitrite is hot enough, the wire will glow red hot and be etched to a very fine but rigid point, which can be sterilized by repeated flaming.
3. Obtain spores either from a deposit (as from a hymenomycete fruit body) or by macerating a fragment of sporophore. Suspend the spores in a drop of sterile water. Make appropriate dilutions, and spread the spore suspension over the surface of a suitable agar medium in a 9-cm Petri dish.
4. Observe microscopically at regular intervals (e.g., every 12 h). When the spores have just started to germinate, locate well-separated germlings using low power so that they are centered within the field of view; then swing the objective with the cutting tube into place and lower gently so that it marks a ring of agar around the germling. Observing all the time through a dissecting microscope, cut out the ring using the sharpened tungsten wire and transfer to mesh agar medium. Subculture when sufficient mycelium has grown out.
5. Spores do not always germinate readily! If appropriate, enhance germination by suitable amendments to the media (see the section, Using a Selective

Medium), temperature shock, or by growing a mycelium on the agar. However, there is no generally applicable rule other than using spores that have been as freshly collected as possible and that have not been kept in damp conditions.

## Isolation of Spores from the Natural Environment—Species-Specific Techniques

An important component of autecologic studies is assessment of methods of dispersal and establishment within and between populations. For this purpose, it is desirable to be able to detect viable spores of the fungal species in question in natural environments. Therefore, it is necessary to have some means of selection that eliminates spores of innumerable other fungi. There are two main approaches.

### *Using a Selective Medium*

For spores of culturable fungi, a medium containing a cocktail of inhibitors that in combination can only be tolerated by the target organism may be employed. A good example of such a medium is the one devised by Barrett [8] to isolate basidiospores of *Phaeolus schweinitzii* from soil.

Procedure

1. Obtain a pure culture of the target organism.
2. Obtain a set of antifungal compounds.
3. Incorporate each compound at various concentrations into a general-purpose medium.
4. Grow the target organism on the medium, and ascertain the inhibitor concentrations above which it is unable to grow at an appreciable rate.
5. Incorporate a mixture of inhibitors, each at concentrations just below the tolerance limits of the target organism, into the general-purpose medium.
6. Test the ability of the target organism and of a range of other fungi likely to be present in the environment being sampled to grow on the selective medium.
7. Make adjustments to concentrations, and add other inhibitors as necessary to improve the performance of the selective medium.
8. Field test the medium by exposing it to air, adding soil and the like.

### *Using "Live Bait"*

The spores of some symbiotic fungi only give rise to viable thalli if they arrive at the surface of a specific host organism, be it a plant, animal, or in the case of certain mycoparasitic fungi, another fungus. In such cases, the host organisms may be used as a kind of selective medium by exposing them in various arrays to natural spore sources.

For sexually outcrossing fungi, an unfertilized mating partner can be an ideal selective medium! By exposing unfertilized thalli in suitable arrays to natural spore sources and recording when and where sexual conjugation events occur, it is possible not only to detect viable spores but also, through subsequent genetic segregation studies, to trace the genetic lineage of these spores. In ascomycetes, conjugation events are detected most easily in those fungi that readily form ascomata (sexual sporophores) in culture, such as *Neurospora crassa*. The biallelic mating systems of these fungi allow conjugation and hence detection only of those viable spores that carry the requisite complementary allele (i.e., 50% of spores in an unbiased population). In most basidiomycetes (i.e., excluding hemibasidiomycetes), conjugation results in the formation of a distinctive "secondary mycelium" (heterokaryon, or, exceptionally, allodiploid; for further discussion see [70]). The multiallelic mating systems ensure that virtually all viable spores arriving on an unfertilized "primary mycelium" or homokaryon will be capable of mating with this mycelium. The protocol below, adopted from Adams et al. [1] and illustrated in Figure 7-4, describes a technique based on this fact. The technique may be modified to suit various purposes with different fungi.

Procedure

1. Obtain a primary mycelium of the target fungus by isolating from single basidiospores. Choose a mycelium that is readily converted to a secondary mycelium (as evidenced in many basidiomycetes by the production of clamp connections) by mating partners and, if possible, where the conversion event is marked by an obvious morphologic change (as in Fig. 7-4). The formation of clamp connections is often (but not always) diagnostic of secondary mycelia.
2. Obtain a suitably subdivided culture dish. Plastic repli dishes, 100 $cm^2$, divided into 25 compartments have proved suitable in previous studies. Partly fill each compartment with an appropriate growth medium, and inoculate each with a fragment of mycelium of the target homokaryon.
3. Allow the mycelia to grow sufficiently to occupy the full area of each compartment.
4. Expose the dish to air at the study site by removing the lid. Vary the exposure time as appropriate up to 24–48 h before replacing the lid. It may be appropriate to mount the dish on a stand above the ground to reduce invasion by invertebrates! Also place lidded dishes at the study site to act as controls.

Figure 7-4 A Repli dish in which each of 25 compartments was inoculated with a homokaryon of *Stereum hirsutum* before exposure to air-borne spores at a field site. The darker mycelium is homokaryotic. The lighter, thicker mycelium is heterokaryotic. Demarcation lines within the compartments indicate multiple spore arrivals. (Courtesy of Christian Taylor).

5. Return to the laboratory and incubate the dishes for 2–14 days at 15–25°C to allow the development of secondary mycelium to become evident. It is often possible to record fertilization events in situ. Where demarcation zones due to somatic incompatibility (recognition of non-self and subsequent delimination of individual mycelia) occur (as in Fig. 7-4), multiple basidiospore arrivals can be detected. In other cases it may be necessary to subculture from each chamber onto fresh medium in another divided dish to check for secondary mycelium formation.
6. Record the number of fertilization events, and analyze genetically and statistically as appropriate.

## ISOLATION OF CULTURABLE FUNGI—FROM MYCELIA

Fungi can be isolated into pure culture directly from pieces of mycelium/mycelial aggregates or by allowing mycelium to grow out of organic resources.

### Procedure: Isolation from Bulky Wood Substrata

1. Cut substrata to a size that is easy to handle (e.g., cut branches into 2-cm diameter discs). To ensure that the fungus obtained is the one that is wanted, it is essential to remove contaminants. Surface-sterilize bulky materials (e.g., large petioles, branches, twigs, and woody roots that are not excessively decayed) by immersion in a solution of domestic bleach containing 1% free chlorine for 5 min.
2. As an added precaution, where possible, discard the outer layers before aseptically extracting inner material and placing on artificial media. Standard agar media are often sufficient, although it is sometimes expedient to add antibiotic components or to develop a selective medium (see above). For wood, excising chips of about 2 or 3 × 3 × 3 mm with a sharp chisel and placing six or eight chips on agar is suitable.
3. Incubate agar plates under conditions that mimic as far as possible those of the natural physical environment.
4. Subculture fungi that develop by transferring a small amount (say 2 × 2 × 2 mm) of agar plus mycelium onto fresh medium. To give the target fungi a chance to develop, remove isolates that are obviously contaminants (e.g., those with very rapid growth) before they sporulate or overgrow the original isolation plates.

### Procedure: Isolation from Leaves

1. Remove attached debris by brushing with a fine brush.
2. Separate petioles from lamina, as they may contain different sets of species.
3. Cut into pieces of about 1–2 $mm^2$, and place them in batches of about 30 into sterile distilled water in 25-ml bottles.
4. Wash in repeated (about five) changes of sterile distilled water that is vigorously agitated for 5 min to remove contaminating spores and mycelia from the surface; the number of water changes and degree of agitation will vary for different materials and should be determined by preliminary examination.
5. Plate out leaf particles onto agar, and subculture as for isolations from wood.

Procedure: Isolation from Soil

1. Several techniques can be employed for making isolations from soil. In the dilution plate method, make a soil suspension and then prepare a dilution series of aliquots that are spread onto the surface of agar plates. It is time-consuming, favors sporulating fungi above all else, does not provide accurate information on numbers or fungal biomass, and its use in fungal ecologic studies is rarely justified [38, 62].
2. The soil plate method is a better alternative. Disperse less than 1 mg of soil over the bottom of a 9-cm diameter Petri dish, and cover with cool, molten agar. Incubate and subculture as for isolations from wood. Individual hyphae can be picked from soil particles and plated onto agar, but the success rate is extremely low.

Procedure: Isolation from Mycelial Cords and Rhizomorphs

1. Remove adhering debris by brushing or scraping with a scalpel.
2. Cut cords into 3–4 cm lengths, place into 10–15 ml sterile distilled water in a small bottle, and agitate for 3 min on a "Whirlymixer."
3. Repeat washing with fresh sterile distilled water five times or more as necessary. This number of times must be determined by experimentation.
4. Cut cords into 0.5–1.0 cm lengths, and plate five or six of these onto malt agar and incubate.
5. Subculture mycelial outgrowth.

Procedure: Isolation from Sporocarps (Fruit Bodies)

1. Briefly flame tough/leathery fruit bodies with or without previous brief immersion in 70% alcohol. For less substantial fruit bodies wash the stipes, as for mycelial cords.
2. Remove the outer layer of pileus or stipe to expose uncontaminated tissue.
3. Excise exposed tissue and plate five or six pieces onto agar. Take care to avoid the hymenium to avoid culture of spores.
4. Incubate and subculture as for mycelial cords.

## MOLECULAR DETECTION OF UNCULTURABLE FUNGI

Classical methods for recovering fungi from their natural communities are liable to yield a distorted picture of species composition, since some organisms are detected more readily than others. Moreover, for many groups of fungi these methods are either highly inefficient or impossible (e.g., in mycorrhizal symbionts). Even for those organisms that are culturable, comprehensive identification is often very difficult in the absence of characteristic sexual or asexual reproductive structures.

These difficulties have imposed many constraints on the development of fungal ecology and are in part responsible for the proliferation of "black-box" or process-oriented approaches to investigating fungal communities in terms of overall biomass or activity. However, it may be possible to overcome many of the difficulties by using DNA polymerase chain reaction (PCR)-based approaches [28].

Fungi need not be cultured before their detection by PCR. Moreover, the sensitivity of the techniques, a single target molecule in a complex mixture is theoretically detectable, allows them to be used on very small samples of material. There are three main approaches to identification:

1. *Sequencing and placement of the organism within a phylogenetic framework*: DNA of unknown organisms that cannot be recovered or identified conventionally is amplified, sequenced, and analyzed by comparison with existing sequence databases. Phylogenetic trees inferred from ribosomal DNA have been used most often so far to resolve relationships between fungi at a variety of taxonomic levels.
2. *Hybridization with specific DNA probes*: Probe design is still an art driven by immediate necessity. There is no single path to the design of determinative probes. Common approaches include random screening of recombinant libraries or random amplified polymorphic DNA (RAPD) fragments and comparative sequence analysis of closely related organisms for probes of desired specificity.
3. *Generation of characteristic banding patterns*: A relatively simple and rapid way of assaying DNA variation is through the use of restriction fragment or restriction fragment length polymorphism (RFLP) analysis. This approach is based on amplification of a target region, digestion with restriction enzymes, electrophoretic separation of the fragments, and side-by-side comparison of the resulting RFLP patterns. Approaches, such as DGGE, SSCP, or related methods, are based on differences in mobility of PCR-products that differ in sequence. Finally, the random amplified polymorphic DNA (RAPD) approach relies on screening DNA for variability using short primers of arbitrary sequence. The amplification products are compared by agarose gel electrophoresis.

## IDENTIFYING CULTURES AND SPORES

Many keys are available for identification of temperate fungi that are fruiting. However, not all fungi fruit readily in culture, and several methods can be used to induce fruiting in microfungi (e.g., exposure to UV light [18, 19, 37]). Basidiomycetes are more problematic; appropriate culture conditions have been devised for some, but it is questionable whether they are worthwhile since the fruit bodies produced are often abnormal. Nonetheless, these fungi can often be identified from their cultural characteristics. Since keys (including an expert system) are available for only a limited range of species [61, 75, 84] researchers often need to construct their own identification schemes. Immunologic and genetic probes are also likely to become increasingly available.

### Procedure: Identifying Mycelia in Culture

1. Collect and identify fruit bodies from the habitat concerned.
2. Isolate mycelia into culture from these fruit bodies, ensuring that enough individuals have been obtained to cover inherent cultural variability.
3. Define characteristics that can discriminate between species in the habitat. Useful vegetative characteristics have been detailed by Stalpers [84] and Rayner and Boddy [70] and include color; odor; texture; tissue-like structures; hyphal characteristics—diameter, wall thickness, clamp connections, branch angle and frequency, straightness, surface deposits, inclusions, and terminal structures; extension rates under different conditions; and enzyme tests [67, 87].
4. As some characteristics are better discriminators than others for a particular set of fungi, reduce the range of characteristics recorded accordingly. Nonetheless, many characteristics may still be necessary, especially if there are a lot of species in the habitat. Also, it may not always be possible or con-

venient to record all of the characteristics for an "unknown." Further, it is important to note that characteristics of an individual may not always be the same—some may be "turned on" by physiologic switches and environmental conditions. In these cases, some sort of computerized pattern recognition system will be useful (e.g., probabilistic methods or artificial neural networks [22, 59]).

5. Confirm identifications or discriminate among several possibilities by determining whether the isolate can mate with "tester strains" of known identity; strains are simply juxtaposed on agar media (see [70]). Principles are similar to those used for isolating spores using "live bait."

## Quantifying Extent—Mapping Regional Boundaries

The relation between coverage of domain (extent) and space-filling capacity (content) is of critical importance in assessing the ecologic roles of mycelial fungi. Assessment of extent involves identifying the regional boundaries or territorial limits within which individual mycelial systems occur and outside which they are absent. A critical consideration in delimiting regional boundaries is the degree of genetic heterogeneity that can be tolerated within these boundaries, both by the observer and by the system itself. It has become clear that in most if not all fungi, stable populations of more than two (and, in fungi other than basidiomycetes, more than one) genetically disparate kinds of nuclear genome and one kind of mitochondrial genome cannot be sustained in continuous protoplasm. This limitation is due to somatic incompatibility reactions [68]. Regional boundaries are therefore best defined, both practically and biologically, on the basis that somatic proliferation within these boundaries occurs from a single haploid, diploid, or heterokaryotic genetic source. Systems with regional boundaries that are defined in terms of such self-proliferation can be described as genetic individuals or *genets* [21].

In some cases, demarcation zones marking the regional boundaries of genets can readily be detected in the field (see Fig. 7-2). Where the distribution of mycelia cannot be visualized so readily or where confirmation or elaboration of provisional maps is necessary, then techniques that distinguish among samples on a more rigorous genetic basis are required. It is particularly important to use a technique that discriminates at the appropriate level of resolution. Although some techniques may detect fine-scale genetic variation even below the level of a genet, others only discriminate at the breeding population level or higher. In all cases, a nested quadrat sampling pattern may be necessary to take account of the potentially wide range in size of individual genets (see Fig. 7-3).

### *Direct Pairing Tests for Somatic Incompatibility*

In culturable ascomycetes and basidiomycetes, a simple test can be done by pairing isolates as described here and recording whether they are capable of complete physiologic integration or form persistent demarcation zones. Where clear-cut results cannot be obtained by this test, then molecular markers, such as isoenzyme and DNA polymorphisms, may prove useful, providing that they resolve differences at the level of the genet. In ascomycetes, the absolute inability to form a heterokaryon—heterokaryon incompatibility—is often used in population studies, commonly based on some kind of nutritional complementation [54]. This approach can provide valuable information, but at least in some cases it is more likely to resolve differences among groups of related genets, rather than among individual genets. This is because somatic rejection mechanisms commonly do not absolutely preclude heterokaryon formation. Indeed, in basidiomycetes (except hemibasidiomycetes) it is essential that they should not do so if sexual outcrossing is to be achieved [68].

Procedure

1. Obtain isolates from mycelium or fruit body tissue (not single spore or polyspore isolates) of field samples from various locations and grow on a suitable agar medium. In sexually outcrossing basidiomycetes these isolates are likely to be heterokaryons; in ascomycetes they are likely to be homokaryons.
2. Cut plugs from the growing margins of individual colonies, and place them adjacent to one another on fresh medium in various combinations. Always include some self-pairings between plugs cut from the same colony to act as controls. Incubate in darkness at 15–25°C for as long as necessary, which may be several weeks, for either integration (complete merging of colonies) and rejection (formation of some kind of demarcation zone) reactions to become evident. To allow long incubation periods, do not use ventilated dishes that dry out too quickly.
3. Record whether rejection or integration occurs. Only identify a demarcation zone as being due to rejection when it can be contrasted clearly with integration reactions in self-pairings.
4. Be aware that in some circumstances the results may not be clear cut. For example, demarcation zones can occur in self-pairings due to mutual "staling" reactions (i.e., alteration of the physico-chemical environment such that growth is prevented). It is often possible to eliminate such effects by changing the growth medium, and in general it is often worth experimenting with a range of media to find the one most suitable for the particular species being studied. Generally, rejection reactions are most distinctive on high-nutrient media under well-aerated conditions. In addition, the form of demarcation zones in different genets can vary considerably, both within and between species. They may be pigmented or not pigmented and be associated with increased or decreased production of aerial mycelia or sporophores. In pairings between homokaryons both rejection and "acceptance" reactions may occur in the same interaction. The distance between inocula can also be critical. Isolates paired too far apart (more than 1 cm) are more prone to exhibit self-inhibition reactions. It may prove economic to make multiple pairings on a single plate or to make pairings in compartmented containers, such as Repli plates (see Fig. 7-4). However, these alterations should only be done once it has been ascertained that the results obtained do not differ significantly from those obtained between single pairings in larger containers, such as 5- or 9-cm diameter Petri dishes.

## Quantifying Content—Mapping Topographic Boundaries

The concentration of biomass within the territorial limits of an individual genet is liable to be very variable. The variability stems both from (1) heterogeneities in microenvironmental conditions and resource supplies and (2) the capacity of mycelia as indeterminate systems to alter their organizational state in accord with changing circumstances and functional requirements. To quantify variations of content it is necessary to obtain some kind of image, either by direct observation or by calculation, of mycelial concentration within regional boundaries. Methods for obtaining such images, both at microscopic and macroscopic scales, have already been described. The following protocol details how such images may be analyzed and quantified using fractal geometry.

### *Image Analysis and Fractal Geometry*

Quantifying the content of mycelial systems in relation to their overall extent amounts to providing some estimate of their space-filling capacity. Density is an inappropriate measure of this capacity in irregularly heterogeneous (fractal) structures, such as mycelia, because the number of units of length, area, or volume identified varies with the scale at which observations are made: As the scale is reduced, more units become evident. Moreover, by averaging out the heterogeneities of mycelial systems, density estimates eliminate all kinds of functionally important local details of potential value in comparisons among different taxa, genets, and organizational states.

This situation can be resolved by characterizing the topographic distribution of mycelial systems in terms of their degree of irregularity or fractal (fractional) dimension [55] at different spatiotemporal locations. This measure can be obtained by relating the content or "mass" ($N$) of material in a portion of a system to its extent, the radius of the field within which it is contained ($r$), according to the formula:

$$N(r) = kr^D$$

where $k$ is a constant and $D$ is the dimension. For homogeneous structures, $D$ is the integer, but for fractal structures it is fractional and can be found as the slope of the graph of $\ln N(r)$ against $\ln r$. The higher the $D$ value, the more thoroughly the structure permeates space. There are several ways by which fractal dimension can be determined; in practice computerized image analysis is essential to achieve the large number of operations necessary. Image analysis can also be used simply for mapping two- or three-dimensional patterns of mycelial domains (e.g., maps of genets within wood) and for estimating mycelial biomass of observable mycelial systems.

Procedure

1. Obtain quantitative information on the total length of mycelium in two-dimensional systems (e.g., mycelial cords on the surface of soil) non-destructively by photographing the systems and then measuring with a planimeter (map-measurer); however, this process is laborious and prone to error. Capturing photogenic images on a framestore, and subsequent analysis using image-processing software, is more accurate and flexible [17]. Analysis is achieved by performing a large number of simple computations on the pixels (picture elements). Total length and area of mycelium can be estimated by counting the pixels; biomass can be estimated by calibrating against the actual weight of mycelium removed from systems whose images have been analyzed.
2. Obtain the fractal dimension of mycelia excluding their borders by determining the number of pixels representing mycelium within concentric circles with at least five different radii [16]. Determine fractal dimension from a double logarithmic plot of $N(r)$ against $r$, the slope corresponding to $D$ [55].
3. Determining fractal dimension using concentric circles has the drawback that the mycelial margin (i.e., the "search front") is excluded from analysis. The box-counting method allows estimation of the fractal dimension of the whole structure and enables discrimination between structures that are entirely fractal (*mass fractal*) and those that are only fractal at their boundaries (*surface fractal*). Estimate the fractal dimension of a set of pixels that make up the mycelial image by overlaying the image with grids of square boxes of different size, spanning at least a decade, and, for each box size, counting

the number of boxes intersected by the image. The box-counting fractal dimension ($D_B$) is estimated from the relation

$$N(s) \approx cs^{-D_B}$$

by plotting log $N(s)$ against log $s$ (a Richardson plot), where $N(s)$ is the total number of boxes of side length $s$. There are two types of boxes: border boxes that overlap the boundary of the image and interior boxes that are contained within the image. Thus,

$$N(s) = N_{\text{border}}(s) + N_{\text{interior}}(s)$$

An unbiased estimator of the surface fractal dimension is obtained by regressing

$$\log(N(s) - 0.5N_{\text{border}}(s)) \text{ against } \log s$$

4. The fractal dimension indicates how well a space is filled. The larger the dimension, the more the body of the system fills the space; for completely filled planes, $D = 2$. Fractal dimensions vary between species and for the species/individual at different times depending on the size of the system, resource availability, microclimate, and the like. Systems with dispersed, narrow, and increasingly independent search fronts (*"guerrilla"* strategy *sensu* Schmid and Harper [78]) are border fractal structures, with $D_{Bsurface} = D_{Bmass}$ [29].

## Quantifying Diversity

Many diversity indices, whether attempting to quantify genetic variation in populations or the numbers of species within communities, make use, in effect, of density estimates, treating individuals as equal, particulate units. The difficulty of applying such an approach to fungal populations and communities, where the units may be very disparate in scale, is evident on examination of Figure 7-3. On the other hand, this figure also shows how a nested quadrat approach can be used to provide an estimate of diversity as a fractal dimension, using the number of units detected in quadrats of different size to estimate $m$ and the quadrat diameter as $r$ in the equation, $N(r) = kr^D$.

# INTERPRETING DISTRIBUTION PATTERNS FROM AN ORGANIZATIONAL VIEWPOINT

## Conceptual Themes—Heterogeneity, Ecologic Strategies, and Routine and Episodic Selection Processes

Once having gained an adequate impression of fungal distribution in natural populations, the problem of understanding how the observed patterns have arisen and how they may change in the future must be addressed. There are two main, complementary approaches to this problem—adaptational and organizational.

Adaptational approaches seek to identify what an organism must do to ensure its genetic survival in natural environments, given particular abiotic conditions and in the face of competition. These approaches have dominated evolutionary ecology during the 20th century, often to the extent that all of a "successful" organism's inherent attributes are assumed to be optimized for particular functional purposes and prescribed by specific genes. Adaptational approaches may help explain why particular attributes and behavioral responses persist and how they may be refined via selection. However, they do not in themselves explain how these attributes and responses could come into being—what the opportunities and constraints are, given a particular organizational pattern. It is to answer this question that organizational approaches have come into their own and indeed become of more primary concern in understanding distribution patterns.

Organizational approaches view organisms, and indeed social groupings, as physical systems that capture, conserve, distribute, and redistribute energy according to the way that they are bounded externally and subdivided internally. The emphasis is on how physical processes are harnessed to yield particular boundary configurations depending on internal and external parameters. From this perspective, which is perhaps especially appropriate with indeterminate systems, genes parameterize rather than prescribe an organism's attributes and responses.

The basic question when interpreting distribution patterns from an organizational or non-adaptational viewpoint is, Given that a system is organized in a certain way, how can it behave under a particular set of environmental conditions? According to one line of thinking [72], fungal mycelia operate as non-linear hydrodynamic systems in which patterns of uptake and throughput of resources are dictated by the permeability and deformability of hyphal exteriors and the partitioning and interconnectedness of protoplasm (Fig. 7-5). In other words, by harnessing the physical properties of a variably constrained, expanding fluid, the

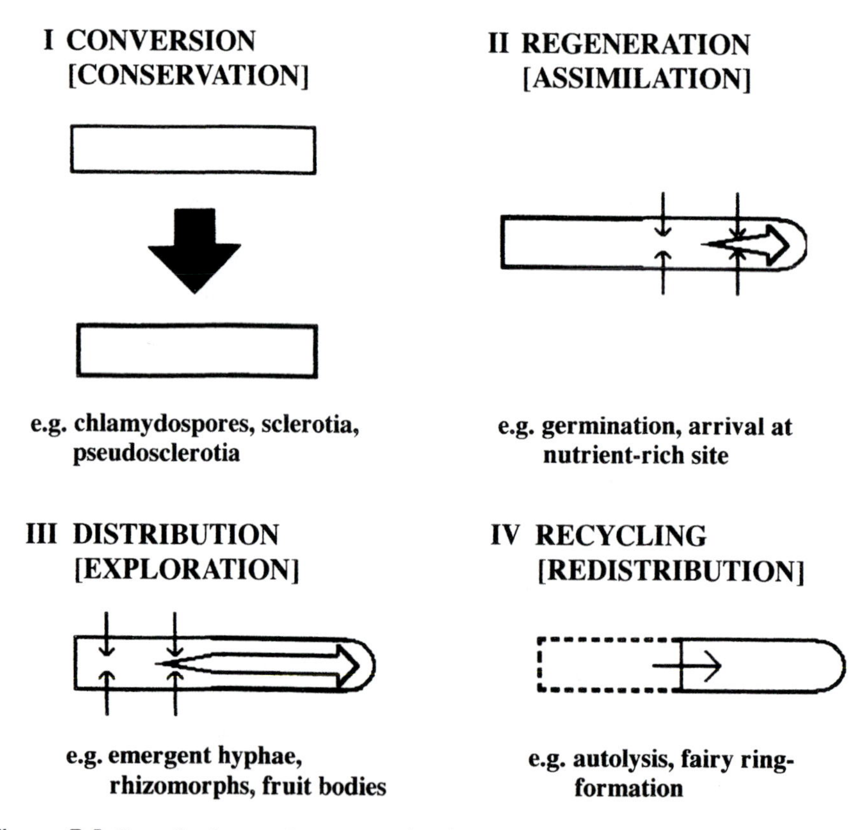


Figure 7-5 Four fundamental processes in elongated hydrodynamic systems, as determined by boundary deformability, permeability, and internal partitioning. Rigid boundaries are shown as straight lines, deformable boundaries as curves, impermeable boundaries by thicker lines, degenerating boundaries by broken lines, and protoplasmic disjunction by an internal dividing line. Simple arrows indicate input across permeable boundaries into metabolically active protoplasm; tapering arrows represent throughput due to displacement. (From Rayner [69].)

mycelium can produce a diversity of patterns by suitable adjustments of boundary-defining parameters that dictate paths of least resistance. For example, if hyphae are made more permeable and hence more capable of rapid uptake (and leakage) in high-nutrient domains, but self-seal in nutrient-depleted domains, then the respective formation of highly branched assimilative and sparsely branched distributive structures can be understood. The question that then arises concerns the mechanisms (adaptive or automatic) by which these parameters may be varied under different circumstances.

It is important to recognize the varied layers of control at which boundary parameters can be specified. There may be direct effects (e.g., on enzyme kinetics) of the environment or structure/physiologic functioning (environmental variation), or differences in content (genetic variation) or expression (epigenetic variation) of genetic information. There may also be non-genetic (i.e., non-translational, non-transcriptional) feedback between boundary components and environmental conditions. Oxidation-reduction and free radical chain reactions associated with phenol-oxidizing enzymes and reactive oxygen species could play an important role in such "hyperepigenetic" control, which provides possibilities for reiterative processes that are extremely sensitive to initial conditions. Interpretation of distribution patterns therefore depends on being able to distinguish between these control processes and their effects, either by experimentation or by deduction from observations.

From an adaptational standpoint, the basic, complementary question to the non-adaptational one posed above is, Given that a living entity is organized in a certain way, how should it respond to particular environmental conditions to maximize its fitness? To address this question, it is necessary to identify the different kinds of selection that are imposed in natural environments and the attribute sets or ecologic strategies that they most favor.

Selective conditions in environments that are either maintained in dynamic equilibrium or in which there are regular cyclic changes may be referred to as routine [20]. Following Grime [40], they can be divided into three primary categories based on the relative occurrence of disturbance, stress, and the incidence of competitors [26].

Disturbance—any environmental event or process that by enrichment of the living space or destruction of residents makes resources newly available for exploitation—selects ruderal organisms or organizational states. Such entities may be expected to be biased toward conversional and regenerative processes (Fig. 6-5) and hence to have short lives; to reproduce rapidly, without genetic diversification and using minimal resources; and to have limited organizational repertoires.

Stress—any more or less continuously imposed environmental feature other than competition—limits the productivity of the majority of organisms or states under consideration. Stress-tolerant or stress-adapted organisms or states may either emphasize conversional processes if they remain in situ or distribution and recycling if they migrate. They may diversify genetically or organizationally or both in accordance with spatially or temporally heterogeneous environments. The relative absence of disturbance and stress in an environment will select combative organisms or states that are (1) able to retain resources acquired at early states of colonization, hence emphasizing conversional processes, or are (2) able to replace former residents, hence emphasizing distribution and recycling.

Radical, non-repetitive shifts in the selective conditions that a population is exposed to constitute "episodic" selection [20]. The ability to respond to such selection will depend on the degree of genetic and epigenetic variation already present in the population or its founder and on the capacity of the latter to interconvert between recombinatorial and clonal modes of proliferation.

In view of the above considerations, the methods described in this section are aimed at uncovering the control mechanisms affecting fungal responses to environmental conditions and the degree to which particular fungi are adapted to particular kinds of selection.

## Ascertaining Genetic and Epigenetic Variability

### *Identifying Breeding Strategy—Outcrossing and Non-Outcrossing Populations*

The distribution of genetic variation in fungal populations is strongly influenced by the frequency of recombinatorial and clonal modes of proliferation. It is therefore important first to ascertain whether a fungus can reproduce asexually (by producing sporangiospores or conidia) or sexually (by producing ascospores, basidiospores, oospores, or zygospores) and second whether sexual reproduction involves outcrossing (conjugation between non-self thalli).

Procedure

1. Investigate whether the fungus in question produces an asexual stage in its life cycle. Such stages may be evident in field collections or can be detected by microscopic examination of cultures. In the latter case, it is often important to use fresh isolates since the ability to produce asexual spores frequently declines during subculturing. Exposure of cultures to near ultraviolet light often induces sporulation, as does interaction with another mycelium.
2. If an ascospore- or basidiospore-producing stage can be collected in the field or induced to form in culture, obtain single spores and, if possible, single spore isolates from a single "parent." Use somatic incompatibility/mating tests or molecular or other markers to determine whether the spores are genetically alike (hence derived by non-outcrossing) or different (hence probably derived by outcrossing). Similarly, for Mastigomycotina or Zygomycotina, use mating tests or markers to ascertain whether different mating types occur.
3. Where both recombinatorial and non-recombinatorial modes occur, it is important to know the relative condition of each to natural population structure. Deduce their contributions from the distribution of genets within the same and different locations (given a knowledge of unit restriction or non-restriction) or from the results of viable spore trapping and the like.

### *Generation of Varied Epigenetic States from a Single Genetic Source*

As has been implied elsewhere, fungal mycelia are not homogeneous and commonly exhibit an ability to produce a variety of organizational patterns either spontaneously or in response to changes in environmental circumstances. Sometimes this pattern can clearly be allocated to organizational states or "alternative phenotypes" capable of playing distinctive functional roles. Examples include mycelia and yeasts, coenocytic and septate states; dense and effuse branching patterns; diffuse and aggregated hyphal systems; and continuously and discontinuously extending colony forms [72]. More subtle but considerable variation may also be apparent within any one of these states, and it has long been known that the same fungus is often capable of producing two or more reproductive states. Where such diversity occurs in material originating from the same genetic source, it may be considered to be the result of epigenetic or even hyperepigenetic variation. It can be manifested both in homogeneous and heterogeneous culture systems.

Procedure: Homogeneous Culture Systems

1. Obtain cultures of the fungus in question from single spores of mycelial tissue (preferably single hyphal tips).
2. Subculture onto replicated series of media in suitable containers and grow in a standard environment.
3. Examine the colonies and record any variation due to sectoring, annulation, maturing, or the like.
4. Note whether morphologic changes are abrupt or continuous across the colony.
5. Subculture fragments of mycelium, hyphal tips, spores, or mycelium plus medium from distinctive zones, taking several samples from the same vicinity.
6. Determine how much variation there is among the subcultures. Can any of the variation, if present, be correlated with the sites from which they came? Do all samples from the sample vicinity behave alike?
7. Produce further generations of subcultures. Does variation continue to arise, or is a point reached where the same pattern is reiterated or where abrupt transitions are wholly superseded by continuous changes?
8. Analyze the varied states as appropriate to detect changes in genetic content or expression.

### *Examining the Capacity to Generate Varied Epigenetic States from Paired Cultures*

Encounters with others that may be potential partners (as can apply if they belong to the same species, depending on somatic or mating compatibility), competitors, or hosts are an important and frequent feature in the lives of fungal mycelia. Such encounters commonly result in responses involving fundamental changes in organizational patterns; within ascomycete and basidiomycete species such changes can involve exchange of genetic information. Pairing experiments combined with subculturing procedures and morphologic and molecular analysis as described in previous sections may therefore provide a useful approach to detecting varied epigenetic states (see [3]).

### *Growing Fungi in Heterogeneous Microcosms—Artificial Systems*

Pairing experiments provide one way of introducing heterogeneity into culture systems in such a manner as to test the organizational range of a fungus. Another approach is to grow the fungus in heterogeneous arrays of local abiotic environments. An example of a useful technique, based on this approach, is the matrix plate (Fig. 7-6).

Procedure

1. Obtain a set of plastic Repli dishes (Sterilin) divided into 25 internal compartments (or similar containers).

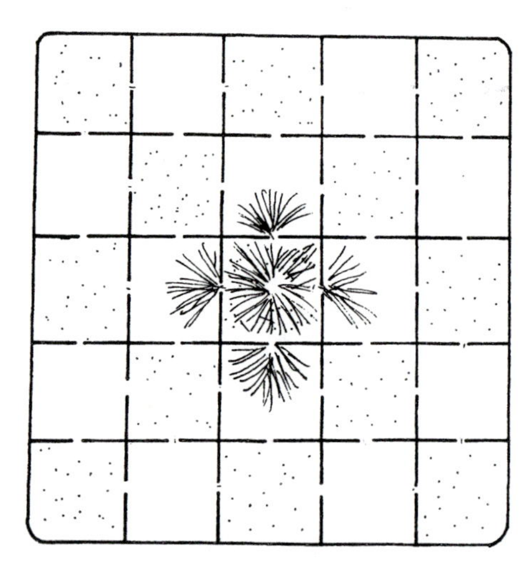

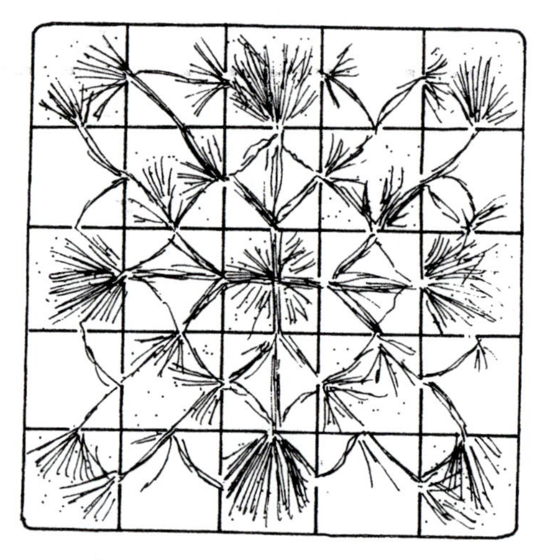

Figure 7-6 Matrix design for ordering mycelial heterogeneity (based on work by Erica Bower, Louise Owen, and Zac Watkins). In this case a 10 $cm^2$ Repli dish containing 25 chambers is filled with high- (*stippled*) and low-nutrient media. Holes have been cut in the partitions just above the level of the medium and a fungus inoculated into the central chamber. The patterns illustrated here were produced by *Coprinus picaceus* after about 10 days (*left*) and 50 days (*right*) at 20°C in darkness.

2. Using a hot needle or blade, cut holes or channels of the desired width or location (e.g., 2 mm wide through the center of each partition, as in Figure 7–6), at a depth just above the level of the medium to be added.
3. Add agar medium to each compartment in desired arrangement (e.g., alternating high- and low-nutrient concentrations).
4. Inoculate fungus in desired location (e.g., central or corner chambers), and allow to grow through the system.
5. Record the growth pattern of the mycelium as it develops over time. Often, distinctive organizational states are produced in successive chambers.
6. Use each compartment as a locale that can be analyzed separately (e.g., for metabolite production), enabling the resulting data to be correlated with particular organizational states.
7. Use seminatural substrate in the wells (e.g., cellulose or filter paper squares). Alternatively, employ such substrata (either unaltered or soaked in nutrients) in non-compartmented agar plates, though the scope for analyzing separate compartments is then lost.

### *Growing Fungi in Heterogeneous Microcosms—"Foraging" Experiments in Seminatural Systems*

To approach the real world even more closely, systems employing relatively homogeneous agar can be discarded in favor of *non-sterile* systems using entire natural components (e.g., containers of soil with arrays of wood, leaf litter, cones, and the like).

Procedure

1. It is difficult to follow patterns non-destructively in three dimensions (3-D) of growth and substratum colonization by mycelia grown in 3-D because the constituents of natural environments are opaque. Thus for examining

mycelia foraging for resources or interactions among mycelia, use two-dimensional (2-D) systems, either horizontally or vertically. Produce horizontal systems by making a flat layer (1–2 cm thick) in bioassay trays, seed trays, or the like. Compressing the soil tends to constrain mycelial growth to the surface, though the extent of growth actually within the soil depends among other things on the species and on whether or not mycelia are aggregated to form linear organs. Examine growth in a vertical plane in a vertical box/tray made of two sheets of glass or perspex, 1–2 cm apart, joined at the sides and base, but open at the top for filling. A removable front panel is ideal for dismantling the system for analysis at the end of an experiment. As an alternative, use a glass cylinder (about 4-cm diameter) with a second cylinder (about 2-cm diameter) within, with the space between the two filled with soil; this set-up is not, however, suited to dismantling the system for analysis of components.
2. Inoculate the soil system in the desired location (e.g., middle, corner, surface) with a resource precolonized with an appropriate fungus—either a naturally colonized resource or one that has been placed for several weeks on an agar culture of the fungus.
3. Examine responses to heterogenously distributed nutrient resources by adding uncolonized new resources or resources colonized by other fungi (Fig. 7-7). Use combinations of different types and sizes and varied spatial arrangements [13].
4. Record patterns of mycelial development over time, preferably photographically so that extension rates, fractal dimension, and biomass can be determined by image analysis.

## Ascertaining Responses to Stress

Responses to stress include death; changes in extension rate, biomass production, colony morphology, and enzyme production; production of survival spores and pseudosclerotial plates; and other effects. As usual these effects are visualized and quantified most easily in artificial culture, as is control of the physico-chemical environment. Again, however, considerable care must be taken when extrapolating from such artificial conditions, and wherever possible one should approximate the

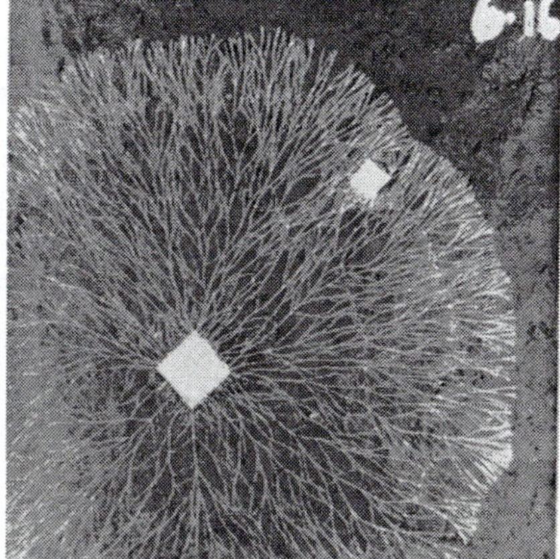

Figure 7-7 Soil trays with an uncolonized resource (1 × 1 × 0.5 cm wood block) inoculated with an *Agrocybe* species in a wood block (2 × 2 × 1 cm). *Left*, 28 days after inoculation; *Right*, 35 days after inoculation.

natural situation as closely as possible, though controlling the environment or knowing what conditions are actually being experienced by the fungi is often very difficult.

Procedure: Deduction from Natural Colonization Patterns

1. Locate stressful locations, such as desiccating twigs and branches; wood, leaves, and the like exposed to direct, intense isolation; water-saturated soil; litter subject to intense invertebrate grazing; sites polluted with toxic chemicals; and bonfire sites.
2. Describe quantitatively the stressful abiotic conditions.
3. Record the location of fungus; note the presence of special structures, such as chlamydospores, pseudosclerotial plates, and mycelial cords. Record the growth form if mycelia are extending out of natural resources.
4. Correlate stressful conditions with the fungi present and special structures or growth forms.

### *Growing Fungi at Different Water Potentials*

Water availability is a major determinant of fungal growth and activity, but is notoriously difficult to describe, measure, and control [12, 70]. At low water contents, *water potential* (MPa) is the most direct measure of water availability. It can be thought of as the negative pressure required to remove water from the substratum. Overall, water potential ($\Psi$) has several components:

$$\Psi = \Psi_m + \Psi_\pi + \Psi_p + \Psi_g$$

where $\Psi_m$ is the matric potential (which results from a solid matrix due to surface tension effects), $\Psi_\pi$ is the solute or osmotic potential resulting from the presence of dissolved solutes, $\Psi_p$ is the pressure potential, and $\Psi_g$ is the gravitational potential. $\Psi_p$ and $\Psi_g$ are usually negligible, and $\Psi_m$ or $\Psi_\pi$ often predominate. For example, in the surface layers of the sea, $\Psi_\pi$ predominates; in wood and soil, $\Psi_m$ predominates, although locally in water-filled voids and in water films where there are dissolved decomposition products, $\Psi_\pi$ can also be significant.

Theoretically there should probably be little difference in the effects of water potential altered by matric or solute potential, and the latter is more commonly employed, although not without difficulty. Solute potential (MPa) is given by

$$\Psi_\pi = -1.065T \log_{10} p/p_0$$
$$= -RT\rho m v \phi 10^{-10} p_0$$

where $R$ is the gas constant; $T$, absolute temperature; $m$, molality; $v$, number of ions per molecule; $\phi$, the osmotic coefficient at $m$ and $T$; $\rho$, density of water at $T$; $p$, vapor pressure of water under conditions obtaining; and $p_0$, vapor pressure of water at $T$.

**Solute Potential** Mycelial extension rate, survival, and the like are usually examined on agar media, the water potential of which has been altered by the addition of solutes. To ensure that the effect of adding a solute is purely an osmotic one and not a stimulation (e.g., due to a nutrient) or an inhibition (due to toxic ions) effect, repeat experiments with several different solutes. KCl and NaCl are often used, though they do cause inhibition. Sucrose is not satisfactory as it is metabolized. Glycerol has no adverse effects on growth, but can be used as an energy source by some basidiomycetes; polyethylene glycol is neither stimulatory nor inhibitory provided that an appropriate molecular weight is used (e.g., 1000).

Tables of osmotic coefficients and molality that will yield a given solute potential are available [27, 74, 79]. Take care to prevent water loss during the preparation of plates and the experiment; cover solutions while autoclaving, pour plates with cooled agar, and seal Petri dishes in plastic bags. The constituents of the agar will add to the solute potential, and the actual solute potential of prepared media should be checked (e.g., with a vapor-pressure osmometer). It is essential that temperature does not fluctuate.

**Matric Potential** The matric potential of soil can be measured and maintained using a pressure-plate apparatus. Wood's matric potential can be measured similarly, though not so easily. The matric potential of substrata will equilibrate with atmospheres of known humidity (maintained by the presence of appropriate saturated salt solutions) over long periods. However, this method is not very satisfactory since considerable care must be taken to ensure equilibration. Equilibration can be achieved more rapidly if the substratum is dried to a value close to that required before the start of the experiment [39]. Also, it is essential that temperature does not fluctuate.

Perhaps the simplest way to control the water potential of soil is to construct a calibration curve of percentage water content against matric potential; to achieve and maintain matric potential, water is sprayed on at intervals to keep the weight of soil plus water at an appropriate constant amount.

Procedure

1. Wash Whatman #42 filter papers in 0.005% (w/v) mercuric chloride (or preferably a safer antibiotic), and then dry them.
2. Place three of these washed filter papers as a stack within soil and allow to equilibrate.
3. After equilibration, remove the stack of filter papers, and weigh the central filter paper to determine the percentage water content.
4. Determine the matric potential from the calibration curve of percentage water content versus matric potential constructed by Fawcett and Collis-George [33].

The matric potential of wood or other litter components can be maintained by equilibration in soil or perlite of known matric potential as just described.

### *Growing Fungi in Different Gaseous Regimes*

Fungi growing on artificial media, in natural organic substrata, or in columns or trays of soil/litter can be exposed to different gaseous regimes in commercially available gas jars or in air-tight containers constructed to an appropriate size. Containers should have inlets and outlets in positions such that gases flow through or over all replicates in the container. Ideally gas mixtures should flow through the system more or less continuously so that the concentration is not altered appreciably by the respiring fungi or other organisms. Gas mixtures can be purchased in cylinders, which is expensive if a range of combinations are required, but devices are available for simultaneous production of a wide range of gas concentrations [95]. A continuous flow of gas has a desiccating effect, and it should therefore pass through a humidifier before entry into the container. Note that compact soil, dense organic substrata, and the like may take a long time to equilibrate with the gaseous regime. Thus, the fungi may not actually be experiencing the gaseous regime being applied. This gaseous regime could be quantified by, for example, inserting into the substratum a fine probe attached to a mass spectrometer [14].

### *Growing Fungi at Different Temperatures*

It is probably easier to assess the effects of temperature than the effects of any other abiotic variable, as growth systems can be simply incubated in controlled temperature incubators. Constant temperatures are often employed, but in nature such situations rarely occur. Incubators with the ability to cycle temperatures are available and provide conditions that are one step closer to the real world.

### *Growing Fungi at Different pH*

The pH of agar media can be altered easily by adding appropriate buffer solutions [5]. There are, however, several problems to be aware of: Extremes of pH prevent agar from setting properly, some buffers can be *stimulatory* and others *inhibitory* to fungal growth, and fungi have a considerable buffering capacity and commonly alter pH to suit themselves. Consequently, it is essential to measure the pH of plates both before and after experiments. The pH of soil can be altered in a similar way, though for litter components doing so is not really practicable.

### *Growing Fungi on Varied Nutrient Sources*

Energy sources and mineral nutrient solutions can be added easily to agar and liquid culture in known concentrations. Since the agar usually used in artificial media itself contains carbon and mineral nutrients, for precise studies a pure agar, such as Noble agar, should be used. Again there are caveats. Different compounds containing the same mineral nutrient may affect fungal growth and activity differently; for example, fungi have different abilities with regard to the uptake of nitrate- and ammonium-nitrogen. The ions associated with the mineral nutrient of primary interest may be inhibitory or stimulatory, so the concentrations of associated ions should be kept constant with the effect of attaining a particular nutrient environment [76]. Thus, a range of compounds should be employed. Presenting fungi with homogeneously supplied nutrients mimics few natural situations, and they require greater overall concentrations of homogeneously supplied nutrients than nutrients supplied as "discrete packages" [30]. Fungi often (usually?) encounter nutrients not in a soluble form but as organic complexes, and so compounds similar to those encountered in nature should be used whenever possible; for example, cellulose strips would often be a more sensible carbon source than glucose for cellulolytic fungi.

Nutrient concentrations in artificial media are commonly considerably in excess of those encountered naturally, and interesting insights can be gained by using low-nutrient concentrations. Silica gel can be used in preference to the relatively high-carbon-containing agars that are normally employed. If growth in the absence of carbon is to be investigated, considerable precautions have to be taken to exclude carbon from glassware and water used in medium preparation and to eliminate carbon compounds in the air [63].

### *Growing Fungi with Several Microclimatic Factors Varying Simultaneously*

In the real world not just one but several microclimatic factors vary simultaneously and have interactive effects [11]. Thus, it is important to examine the joint effect of variables. This can be achieved by combining the approaches described in the previous six protocols. However, such studies are intrinsically laborious and costly since they involve the use of individual plates for each combination of factors. Work can be reduced considerably in some situations by employing culture systems that incorporate solute gradients. The use of 2-D gradient plates is eminently suitable for examining spore germination and determining whether growth can occur under certain conditions, but is of limited use for determining mycelial extension rates [15]. Essentially, such 2-D plates consist of four layers of agar that are poured as wedges; each wedge contains different solutes, the actual solutes and their concentrations depending on the two physico-chemical factors being examined and the range of values required. The technique is illustrated for 2-D salt/pH plates (Fig. 7-8).

Procedure

1. Prepare appropriate quantities of 2% malt extract agar (MEA).
2. Into a wettable, square (10 × 10 cm) Petri dish, add 15 ml MEA, to which 0.3 ml $H_2SO_4$ has been added. Allow medium to solidify, with one edge of the plate raised on a 3-mm diameter glass rod on a level surface.
3. When set, place the plate on a level surface, and add 15 ml MEA containing 0.6 ml 1 M NaOH. Allow to solidify.
4. When set, rotate the plate through 90°, and repeat for layers 3 and 4; layer 3 containing, for example 1.2, 2.4, or 3.6 g KCl and layer 4 containing no additions.
5. Leave plates at room temperature for 24 h to allow layers to equilibrate vertically.
6. Determine the pH gradient at nine points that are each 1 cm apart across a sample of plates, using a flat-tipped surface pH electrode at room temperature.
7. Determine the KCl gradient using a conductivity meter. Calibrate the conductivity meter by preparing five plates that are the same thickness as the

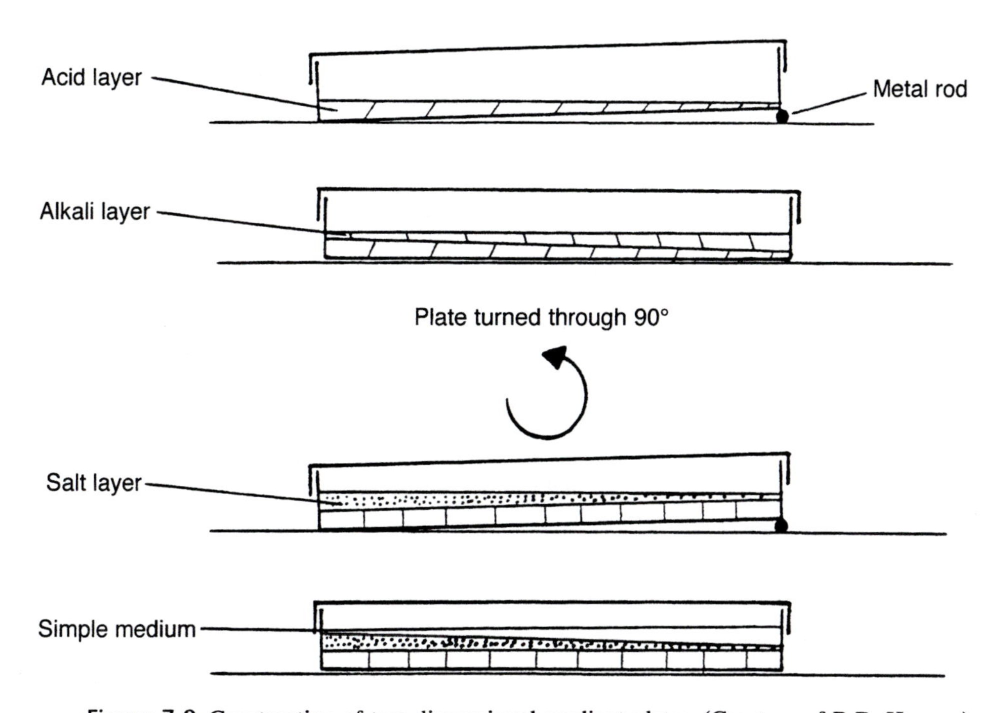


Figure 7-8 Construction of two-dimensional gradient plates. (Courtesy of R.D. Harvey.)

test plates at different KCl concentrations ranging between 0–160 g $l^{-1}$. Remove agar plugs (6 mm diameter) using a cork borer, and dissolve in 10 ml boiling distilled water. After cooling, measure conductivity and construct a calibration curve.

8. Prepare spore suspensions by adding 10 ml sterile distilled water and glass beads to sporing cultures growing in McCartney bottles and shaking for 5 min. Filter supernatant through sterile muslin into a sterile McCartney bottle, and dilute to approx. $4.0–5.0 \times 10^4$ spores $ml^{-1}$.
9. Add 0.5-ml aliquot of spore suspension to 4.5 ml half-strength, cool, molten MEA in a sterile McCartney bottle. Shake gently to distribute spores, and then pour evenly over the surface of the gradient plate.
10. Insert into each plate a polytetrafluoroethylene grid, divided into 81 $1.1 \times 1.1$ cm by 1.0 cm deep squares, to prevent decay of gradients by lateral diffusion.
11. To determine the joint effect of the third and fourth variables, incubate plates under different temperature and gaseous regimes.

## Ascertaining Responses to Competition

When considering how fungi compete with one another, it is vital to take account of the fact that the indeterminacy of mycelia prevents these systems (and their "offspring") from being treated as discrete (particulate) entities. This point continues to be neglected, causing confusion and misapplication of approaches based on animal (and even plant) ecology. An indeterminate system occupies a potentially indefinite spatiotemporal territory, and its "fitness" (if indeed this term has a useful meaning for such systems) increases as it expands its territory. It does not have finite needs. Consequently, it is important to recognize several different aspects of competition involved in the establishment of territory, which can broadly be encompassed under two headings: primary resource capture and combat [26].

*Primary resource capture* involves those processes that lead to the acquisition of freshly available resources, notably the arrival of mycelium or propagules, germination, and establishment of a branching system. With respect to the branching system, trade-offs are possible between radial and tangential proliferation, which increase extent and content, respectively, and between somatic and reproductive development. Where the balance is struck between these alternatives may be altered in different organizational states and relates to different ecologic strategies. A bias toward radial development, for example, may lead to rapid expansion of regional boundaries, but also increases susceptibility to incursion by other systems.

*Combat* involves overtly antagonistic mechanisms that operate at the interfaces between regional boundaries of mycelial systems and either results in (1) a stalemate or deadlock across a persistent demarcation zone or (2) ingress of one system into the other and hence secondary resource capture. Secondary resource capture can involve either the development of an invasion front, consisting of massed ranks of hyphae, or infiltration by sparsely branched hyphae or hyphal aggregates.

There are several approaches to deducing and experimenting on the effects of these various aspects of competition in fungal populations and communities.

### *Deduction from Natural Colonization Patterns*

Remembering that the fundamental niche—the potential to develop in an environment in the absence of other organisms—is very wide in many fungi, the influence of competition on community structure can be inferred from many different kinds of field observations. These observations can be made both in undisturbed sites and in experimental plots.

Procedure

1. Using methods that have already been described in this chapter, plot the distribution of fungi at a particular field site or in/on a particular resource.
2. Check whether there is evidence that mycelia of different individuals (of the same or different species) are in direct physical contact with one another, but are separated (e.g., by demarcation zones) into mutually exclusive domains. If so, the individuals are competing, but the underlying mechanisms cannot be deduced for certain without further investigation.
3. Check whether resilient structures, such as pseudosclerotial plates, are evident at the interfaces between individuals; such structures can provide protection against incursion.
4. Check whether there is any evidence of actual antagonism at interfaces; for example, reduced decay or demarcation zones produced between mycelia as a result of direct incubation or in paired cultures.
5. Ascertain whether there is evidence for replacement of one organism by another. There are several ways in which this can be done:

   If the same location has been observed previously to be occupied by an organism that is no longer present; in some cases, the organism may have been experimentally introduced into the field site;

   If "relics" (e.g., old sporophores, pseudosclerotial plates of an organism no longer present) can be found (see Fig. 7-2);

   However, to determine the mechanism of replacement, whether indirect or by antagonism, requires further investigation.

6. Check whether direct replacement could have occurred (e.g., using experimental pairings, observing mycelial interactions on or in incubated samples, or by noting the persistence of the replaced organism in locations where it is not exposed to the invader). Similarly, note whether any changes in environmental circumstances have occurred that could lead indirectly to replacement, such as alteration in moisture content or microfaunal infestation.

### *Pairing Experiments in Artificial Culture Systems*

It is sometimes possible to gain insights into the occurrence and mechanisms of replacement and deadlock in the field from pairing experiments in culture. However, the difficulties of extrapolating from laboratory to field settings should always be borne in mind. Equally important, the fact that mycelia are complex systems means that they do not always interact consistently, even in precisely replicated conditions, and different kinds of events may occur at various locations along an interaction interface. The utility of pairing experiments in assaying the organizational variability of mycelial systems and the basic methodology entailed (in connection with somatic incompatibility) have been described earlier.

### *Pairing Experiments and Using Natural Substrata*

The outcomes of interaction experiments in natural substrata ostensibly provide a better indication of events in the field than do artificial culture experiments. Even so, it is very difficult to reproduce the conditions under which mycelia grow and interact in nature, and the failure of establishment of inoculum is often a serious problem. The principles, however, are simple: Inoculate varied combinations of strains into natural substrata, and use the techniques already described to study their eventual distribution patterns. Foraging experiments using combinations of strains rather than a single strain often provide interesting data [31].

## SNAPSHOTS OF METABOLIC ACTIVITY

Understanding what fungi are actually doing in the natural environment is the ultimate goal of any ecologic studies. Yet, this goal is often not achieved or even attempted, probably because of the many difficulties that beset the investigator and the inadequacies of currently available techniques. Methods fall broadly into two categories: (1) those that attempt to measure activity in the field and (2) those that assess potential capabilities in the laboratory. The latter are obviously easier to attempt, a major problem with the former being the difficulty of separating fungal activity from that of other organisms. The main aspects of fungal activity that are assessed are (1) germination and growth—agar methods, growth in model soil systems, and use of image analysis have been described above; (2) uptake, translocation, and release of carbon compounds, mineral nutrients, and water; (3) use of organic substrata and production of enzymes; and (4) mycelial interactions, for which methods of study have already been described.

### Germination in Soil in the Field

Procedure

1. Obtain non-nutrient substrata, such as membrane filters, Fiberglas tape, or glass slides [50]. Organic substrata, such as wood veneers and leaves, could be used where appropriate.
2. Obtain spore suspensions of appropriate fungi (see the section, Growing Fungi with Several Microclimatic Factors Varying Simultaneously).
3. Add spore suspensions to substrata. On drying, estimate the number of spores by staining and microscopy. For opaque substrata it may be necessary to remove spores from the surface on, for example, sticky tape.
4. Bury substrata at appropriate locations.
5. Retrieve substrata after various time intervals (ranging from days to months.)
6. Assess the number of spores remaining and the percentage of germination (as in Step 3).

Note that the introduction of these foreign materials inevitably results in the failure to re-create the exact microenvironment that spores would usually encounter.

### Extension in Soil and Litter

It is possible to measure the extension rate of macroscopic-aggregated mycelia in the field, such as cord- and rhizomorph-forming fungi. The extent of systems of fairy ring formers can be deduced by mapping the positions of the fruit bodies, and in some cases their mycelium is aggregated or profuse enough to be mapped directly.

Procedure

1. Locate the system to be mapped or inoculate the soil/litter layer with the fungus to be studied. This approach is possible with saprotrophic cord-

forming fungi [32], though it is very much more difficult with other ecologic groups.
2. Remove the leaf litter carefully without disturbing the system.
3. Map the system by covering with an appropriately sized grid and drawing the location of the cords, or photograph.
4. Re-cover the system carefully with the litter that was removed.
5. Employ an image analysis system or more conventional means, such as a map measurer, to determine the extent of the system.
6. Repeat Steps 2 to 5 at intervals (between 3 months and 1 year) to determine extension rates and the like.

## Uptake, Translocation, and Release of Carbon Compounds, Mineral Nutrients, and Water in the Laboratory

Isotopic labeling provides a ready means for following the rate of uptake, translocation, and subsequent release of carbon compounds, mineral nutrients, and water. With some isotopes, monitoring can be achieved non-destructively, as well as destructively. Non-destructive monitoring has considerable advantages since the same system can be followed over time.

### Procedure

1. Set up experimental systems in such a way that the labeled compound can be added to a specific part of the system without "leaking" to other parts. For example, in studies of foraging mycelia inoculum or bait resources to which the tracer (labeled compound) is to be added at a later date, place the bait in a small dish of soil within the larger soil trays (Fig. 7-9).
2. Add tracer when the experimental systems have developed to the required extent. Ideally, the radiolabeled compound should be similar to that used in nature. For example, if examining the movement of carbon from plants into mycorrhizal fungi, supply the carbon as $^{14}CO_2$, or label wood and leaf litter by incorporating a radioisotope into the nutrient sources of growing plants. In practice, radiolabeled compounds are often supplied as compounds that are readily taken up by fungi and in luxury amounts. Determine the amount of radiolabeled compound to be added by preliminary experimentation; it should be sufficient to be detectable at appropriate sampling positions after appropriate incubation times. For example, 370 k Bq of $^{32}P$ in 2 $cm^{-3}$ of 10 mM $K_2HPO_4$ was suitable for 4 $cm^3$ wood block inocula, from which mycelial cords emanated and colonized new resources 5 or 6 cm away and which were monitored for about 60 days.
3. Where possible, as with $^{32}P$, monitor tracers non-destructively during the experiment, since a single sampling at the end can miss many changes. If this is not possible, then sample replicates at different times, although the usual problems of variation between systems then come into play. Monitor $^{32}P$ non-destructively using an anthracene scintillation probe connected to a scaler ratemeter. It is essential that readings are taken from exactly the same place at each time; this can be achieved by permanently attaching the probe to a dish lid that can be aligned by marks on the lid and base. Correct counts for background and interference from the source by subtracting the count from similarly arranged controls without the fungus.
4. At the end of the experiment harvest all components of the system, separately where appropriate. Dry material and remove any adhering soil and the like. Weigh, grind (coffee grinders are inexpensive and suitable), and

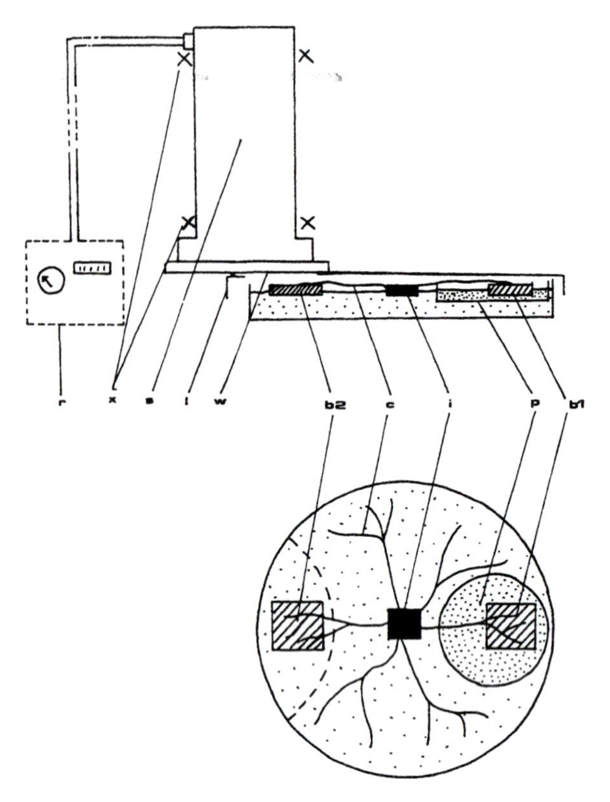


Figure 7-9 Model soil systems (containing non-sterile soil) inoculated with a wood block precolonized by *Phanerochaete velutina* (i) and two wood baits (b1, b2). Bait b1 is positioned within an inner Petri plate (p) in which radiotracer is added. Mycelial cords (c) have grown out and colonized the bait. The scintillation probe (s), attached to ratemater (r) was attached permanently to the lid of a large (14-cm diameter) Petri dish and supported on clamps (x), so that the window of the probe was always in the same position (w). (From Hughes and Boddy [49].)

then determine the amount of tracer. For radiotracers, weigh ground samples into scintillation vials containing distilled water (about 10 $cm^3$) and allow 2 days for thorough wetting. Add $H_2O_2$ (0.5 $cm^3$ of 30%) to bleach samples. Drive off excess $O_2$ after 24 h by heating to 80°C. Obtain counts in a liquid scintillation counter. Correct counts for physical and chemical quenching, background radiation (by constructing calibration curves using a known amount of tracer added to vials with the appropriate amount of fungal or litter material), and radioactive decay. Radioactive decay is calculated as

$$A = A_0 e^{-lt}$$

where A is activity at time t, $A_0$ is initial activity, and l is 0.693/half-life of the radionuclide.

## Uptake, Translocation, and Release of Carbon Compounds, Mineral Nutrients, and Water in the Field

In the field, radiolabels with short half-lives, such as $^{32}P$, or non-radioactive labels, such as AIB (aminoisobutyric acid), should be used. For further details see Read et al. [73] and Wells and Boddy [92].

Procedure

1. Obtain all of the necessary legal permissions.
2. Locate mycelial systems or inoculate mycelial systems.
3. Mark out the site with all of the necessary warnings.
4. Add the labeled compound to the part of the system being studied (e.g., wood block inocula, colonized litter) ensuring that the compound does not escape and that all of it is taken up. It is sensible to create a well into which the labeled compound is added (e.g., drill a hole in wood).
5. After an appropriate time (e.g., several days), determine the extent of movement of the tracer. $^{32}P$ is useful in that it can be detected non-destructively with a Geiger-Muller probe. It is useful to monitor it in conjunction with mapping the system.
6. Harvest *all* components containing the labeled compound. Collecting all radiolabeled material is essential not only scientifically but for safety.
7. In the laboratory, assay components to determine the quantity of labeled material; for example, use liquid scintillation counting for $^{32}P$-labeled compounds.
8. Construct maps and calculate fluxes of labeled compounds.

## Decomposition of Organic Substrata: Weight Loss and Chemical Change

The rate of decomposition is commonly estimated by determining weight loss or strength loss of a resource (e.g., wood or leaf litter) or by determining the loss or change of individual chemical components of the resource over time. Similar methods can be employed in the laboratory and field, although in the latter arena, considerable care must be taken to ensure that the resources can be recovered in their entirety (i.e., no fragments are lost) at sampling.

Procedure

1. Determine the initial state of decay. An estimate of oven-dry (105°C) weight forms a good baseline, but since oven drying can cause considerable structural and chemical changes to litter (as well as killing the fungi present!), determine a conversion factor for air-dry (or fresh) weight to oven-dry weight at the start of the experiment by drying representative samples of twigs, leaves, and the like. For bulky, slowly decaying substrata, such as wood, estimate the density—dry mass per unit volume of undried wood—for a sample of experimental units.
2. Determine the fresh weight (or weight and volume for food) of each individual resource being used in the experiment. Then estimate the oven-dry weight of each (but not of wood) by employing the air-dry to oven-dry weight conversion factor.
3. Place resources in the field or laboratory experimental systems either directly, tethered, or in litter bags. Litter bags with different mesh sizes have often been used to ascertain the effects of excluding different-sized components of the soil microfauna; however, different mesh sizes result in different moisture regimes, which can confound experiments considerably. In field experiments, mark carefully the location of the experimental resource units; otherwise it is often extremely difficult to locate them again.
4. Retrieve replicate samples after different times and determine their oven-dry weight or density.

5. Determine weight loss by comparing the oven-dry weight of each individual resource at the sample time with the estimated oven-dry weight at the start of the experiment. For wood, determine the density of each sample and estimate the percentage weight loss as

$$\frac{\text{(density of wood at start—density of decayed wood)}}{\text{density of wood at start}} \times 100$$

Note that the decay rate obtained by either method underestimates actual decay since no account is taken of the mycelium present within the source.

6. As an alternative to a change in weight or density, the change in tensile strength is sometimes suitable [36, 43]. Also, attach dyes to particular substrates (e.g., cellulose strips), and measure their release or residual retention to estimate decay rate [57]. Quantify gross chemical changes using standard procedures for ecologic materials [4]. Specific sites of chemical change can be revealed by microscopy and staining [36].

## Decomposition of Organic Substrata: Gaseous Exchange

The methods described in the previous section are all destructive, but it is often extremely valuable to follow the decomposition of individual resources over time by measuring $CO_2$ uptake or release.

### Procedure

1. Allow decomposition to proceed in model laboratory systems or the field.
2. When the decay rate is to be estimated, enclose the resource in a sealed chamber for a period of time (e.g., 2 h) long enough to allow a measurable amount of $CO_2$ to accumulate or of $O_2$ to be removed.
3. Measure $CO_2$ inexpensively by absorbing the gas in concentrated alkali (e.g., KOH) and then back-titrating with acid [50]. Multi-channel manometric and electrolytic respirometers are also available [89, 90], but most are suited to small-volume samples. Gas chromatography, using a katharometer detector, a glass column of Poropak Q beads (150–180 mm) with helium as a carrier gas, is a very useful, rapid, and sensitive technique for $CO_2$ measurement [10]; employing an additional column of Poropak Q held at −78°C [9] or a molecular sieve at room temperature allows simultaneous measurement of $O_2$ [86]. Infrared gas analysis (IRGA) is also a sensitive technique for $CO_2$, and mass spectrometry is useful for atmospheric and dissolved gases [14].
4. Take care when interpreting these "instantaneous" measurements of decay rate, since they may reflect the metabolism of storage compounds, in addition to substratum decomposition at the time of measurement.

## Enzyme Activity

There are many tests for extracellular enzyme activity [36, 70, 85]. Most of these involve the addition of substrates, so laboratory tests evaluate potential rather than actual activity. Enzyme activity can be measured in the field, though it is rarely possible to separate that of fungi from other organisms [85].

## REFERENCES


1. Adams, T.J.H., Williams, E.N.D., Todd, N.K., and Rayner, A.D.M. 1984. A species-specific method of analyzing populations of basidiospores. Trans. Br. Mycol. Soc. 82: 359–361.

2. Ainsworth, A.M. and Rayner, A.D.M. 1990. Aerial mycelial transfer by *Hymenochaete corrugata* between stems of hazel and other trees. Mycol. Res. 94:263–266.
3. Ainsworth, A.M., Beeching, J.R., Broxholme, S.J., Hunt, B.A., Rayner, A.D.M., and Scard, P.T. 1992. Complex outcome of reciprocal exchange of nuclear DNA between two members of the basidiomycete genus *Stereum*. J. Gen. Microbiol. 138: 1147–1157.
4. Allen, S.E., Ed. 1989. Chemical Analysis of Ecological Materials, 2nd ed. Blackwell Scientific Publications, Oxford.
5. Altman, P.L. and Dittmer, D.S. 1964. Biological Handbook: Biology Data Book. Federation of the American Society of Experimental Biology, New York.
6. Anderson, J.M. 1978. The preparation of gelation-embedded soil and litter sections and their application to some soil ecological studies. J. Biol. Educ. 12:82–88.
7. Baker, J.H. 1988. Epiphytic bacteria. In Methods in Aquatic Bacteriology (Austin, B., Ed.), pp. 171–191. John Wiley, Chichester.
8. Barrett, D.K. 1978. An improved selective medium for isolation of *Phaeolus schweinitzii*. Trans. Br. Mycol. Soc. 71:507–508.
9. Blackmer, A.M., Baker, J.H., and Weeks, M.E. 1974. A simple gas chromatographic method for separation of gases in soil atmospheres. Soil Sci. Soc. Am. J. 38:689–691.
10. Boddy, L. 1983. Carbon dioxide release from decomposing wood: effect of water content and temperature. Soil Biol. Biochem. 15:501–510.
11. Boddy, L. 1984. The micro-environment of basidiomycete mycelia in temperate deciduous woodlands. In Ecology and Physiology of the Fungal Mycelium (Jennings, D.H. and Rayner, A.D.M., Eds.), pp. 261–289. Cambridge University Press, Cambridge.
12. Boddy, L. 1986. Water and decomposition processes in terrestial ecosystems. In Water, Plants and Fungi (Ayers, P.G. and Boddy, L., Eds.), pp. 375–398. Cambridge University Press, Cambridge.
13. Boddy, L. 1993. Cord-forming fungi: warfare strategies and other ecological aspects. Mycol. Res. 97:641–655.
14. Boddy, L. and Lloyd, D. 1989. Portable mass spectrometry: a potentially useful ecological tool for simultaneous, continuous measurement of gases in situ in soils and sediments. In Nutrient Cycling in Terrestrial Ecosystems: Field Methods, Applications and Interpretation (Harrison, A.F., Ineson, P., and Heal, O.W., Eds.). Elsevier Applied Science, New York.
15. Boddy, L., Wimpenny, J.W.T., and Harvey, R.D. 1989. Use of gradient plates to study germination with several microclimatic factors varying simultaneously. Mycol. Res. 93:106–109.
16. Bolton, R.G. and Boddy, L. 1993. Characterisation of the spatial aspects of foraging mycelial cord systems using fractal geometry. Mycol. Res. 97:762–768.
17. Bolton, R.G., Morris, C.W., and Boddy, L. 1991. Non-destructive quantification of growth and regression of mycelial cords using image analysis. Binary 3:127–132.
18. Booth, C. 1971. Introduction to general methods. In Methods in Microbiology, Vol. 4 (Booth, C., Ed.), pp. 1–47. Academic Press, London.
19. Booth, C. 1971. Fungal culture media. In Methods in Microbiology, Vol. 4 (Booth, C., Ed.), pp. 49–94. Academic Press, London.
20. Brasier, C.M. 1987. The dynamics of speciation. In Evolutionary Biology of the Fungi (Rayner, A.D.M., Brasier, C.M., and Moore, D., Eds.), pp. 231–260. Cambridge University Press, Cambridge.
21. Brasier, C.M. and Rayner, A.D.M. 1987. Whither terminology below the species level in the fungi? In Evolutionary Biology of the Fungi (Rayner, A.D.M., Brasier, C.M., and Moore, D., Eds.), pp. 379–388. Cambridge University Press, Cambridge.
22. Bridge, P.D., Boddy, L., and Morris, C.W. 1994. Information resources for pest identification—an overview of computer aided approaches. In Identification and Characterization of Pest Organisms (Hawksworth, D.L., Ed.), pp. 153–167. CAB International, Wallingford, United Kingdom.
23. Brundrett, M.C., Piche, Y., and Peterson, R.L. 1984. A new method for observing the morphology of vesicular-arbuscular mycorrhizae. Can. J. Bot. 62:2128–2134.
24. Clark, G., Ed. 1981. Staining Procedures, 4th ed. Williams & Wilkins, Baltimore.

25. Coates, D. and Rayner, A.D.M. 1985. Fungal population and community development in cut beech logs. I. Establishment via the aerial cut surface. New Phytol. 110:153–171.
26. Cooke, R.C. and Rayner, A.D.M. 1984. Ecology of Saprotrophic Fungi. Longman, London.
27. Dallyn, H. and Fox, A. 1980. Spoilage of materials of reduced water activity by xerophilic fungi. In Microbial Growth and Survival in Extremes of Environment (Gould, G.W. and Corry, J.E.L., Eds.), pp. 129–139. Academic Press, New York.
28. Dieffenbach, C., Ed. 1995. PCR Primer: A Laboratory Manual. Cold Spring Harbor Laboratory Press, Plainview, NY.
29. Donnelly, D.P., Wilkins, M.F., and Boddy, L. 1995. An integrated image analysis approach for determining biomass, radial extent and box-count fractional dimension of macroscopic mycelial systems. Binary 7:19–28.
30. Dowding, P. 1981. Nutrient uptake and allocation, during substrate exploitation by fungi. In The Fungal Community, Its Organization and Role in the Ecosystem (Wicklow, D.T. and Carroll, G.C., Eds.), pp. 621–635. Marcel Dekker, New York.
31. Dowson, C.G., Rayner, A.D.M., and Boddy, L. 1988. The form and outcome of mycelial interactions involving cord-forming decomposer basidiomycetes in homogenous and heterogenous environments. New Phytol 109:423–432.
32. Dowson, C.G., Rayner, A.D.M., and Boddy, L. 1988. Inoculation of mycelial cord-forming basidiomycetes into woodland soil and litter. I. Initial establishment. New Phytol. 109:335–341.
33. Fawcett, R.G. and Collis-George, N. 1967. A filter paper method for determining the moisture characteristics of soil. Aust. J. Exp. Agric. Animal Husbandry 7:162–167.
34. Frankland, J.C. 1974. Importance of phase-contrast microscopy for estimation of total fungal biomass by the agar film technique. Soil Biol. Biochem. 6:409–410.
35. Frankland, J.C. 1975. Estimation of live fungal biomass. Soil Biol. Biochem. 7:339–340.
36. Frankland, J.C., Dighton, J., and Boddy, L. 1990. Methods for studying fungi in soil and forest litter. In Methods in Microbiology, Vol. 22 (Grigorova, R. and Norris, J.R., Eds.), pp. 343–404. Academic Press, London.
37. Franklin, J.C., Latter, P.M., and Poskett, J. 1995. Merlewood Research & Development Paper No.
38. Gray, T.R.G. 1990. Methods for studying the microbial ecology of soil. In Methods in Microbiology, Vol. 22 (Grigorova, R. and Norris, J.R., Eds.), pp. 309–342. Academic Press, London.
39. Griffith, G.S. and Boddy, L. 1991. Fungal decomposition of attached angiosperm twigs. III. Effect of microclimate on fungal growth, survival and decay of wood. New Phytol. 117:259–269.
40. Grime, J.P. 1979. Plant Strategies and Vegetation Processes. John Wiley, New York.
41. Gurr, E. 1965. The Rational Use of Dyes in Biology. Leonard Hill, London.
42. Halk, E.L. and De Boer, S.H. 1985. Monoclonal antibodies in plant disease research. Annu. Rev. Phytopathol. 23:321–350.
43. Harrison, A.F., Latter, P.M., and Walton, D.W., Eds. 1988. Cotton Strip Assay: An Index of Decomposition in Soils. Institute of Terrestrial Ecology, Grange-over-Sands, United Kingdom.
44. Hawksworth, D.L. 1974. Mycologists Handbook. Commonwealth Mycological Institute, Kew, United Kingdom.
45. Hawksworth, D.L., Sutton, B.C., and Ainsworth, G.C. 1983. Ainsworth and Bisby's Dictionary of the Fungi, 7th ed. Commonwealth Mycological Institute, Kew, United Kingdom.
46. Hedger, J.N. 1990. Fungi in the tropical forest canopy. Mycologist 4:200–202.
47. Hering, T.F. and Nicholson, P.B. 1964. A clearing technique for the examination of fungi in plant tissue. Nature (Lond) 201:942–943.
48. Holden, M.M. 1982. Guide to the literature for the identification of British fungi. Bull. Br. Mycol. Soc. 16:36–55, 92–112.
49. Hughes, C.M. and Boddy, L. 1994. Translocation of $^{32}P$ between wood resources recently colonized by mycelial cord systems of *Phanerochaete velutina*. FEMS Microbiol. Ecol. 14:201–212.

50. Johnson, L.F. and Curl, E.A. 1972. Methods for Research on the Ecology of Soil-Borne Pathogens. Burgess Publishing, Minneapolis.
51. Kormanik, P.P. and McGraw, A.C. 1982. Quantification of vesicular-arbuscular mycorrhizae from plant roots. In Methods and Principles of Mycorrhizal Research (Schenck, N.C., Ed.), pp. 37–45. American Phytopathology Society, St. Paul, MN.
52. Kubiena, W.L. 1932. Uber Fruchtkorperbildung und engere Standortwahl von Pilzen in Bodehohlraumen. Arch. Mikrobiol. 3:507–542.
53. Kubiena, WL. 1938. Micropedology. Collegiate Press, Ames, IA.
54. Leslie, J.F. 1993. Fungal vegetative compatibility. Annu. Rev. Phytopathol. 31:127–150.
55. Mandelbrot, B.B. 1982. The Fractal Geometry of Nature. Freeman, San Francisco.
56. Mitchell, L.A. 1985. Monoclonal Antibodies and Immunochemical Techniques: Applications in Forestry Research. Canadian Forestry Service, Victoria, British Columbia.
57. Moore, R.L., Bassett, B.B., and Swift, M.J. 1979. Developments in the Remazol Brilliant Blue dye-assay for studying the ecology of cellulose decomposition. Soil Biol. Biochem. 111:311–312.
58. Morrell, J.J., Gibson, D.G., and Krahmer, R.L. 1985. Effect of fluorescent-labelled lectins on visualization of decay fungi in wood sections. Phytopathology 75:329–332.
59. Morris, C.W. and Boddy, L. 1995. Artificial neural networks in identification and systematics of eukaryotic microorganisms. Binary 7:70–76.
60. Nicholas, D.P., Parkinson, D., and Burges, N.A. 1965. Studies of fungi in a Podzol. II. Application of the soil-sectioning technique to the study of amounts of fungal mycelium in soil. J. Soil Sci. 16:258–269.
61. Nobles, M.K. 1965. Identification of cultures of wood-inhabiting hymenomycetes. Can. J. Bot. 43:1097–1139.
62. Parkinson, D., Gray, T.R.G., and Williams, S.T. 1971. Methods for Studying the Ecology of Soil Microorganisms. IBP Handbook No. 19. Blackwell Scientific Publications, Oxford.
63. Parkinson, S.M., Wainwright, M., and Killham, K. 1989. Observations on oligotrophic growth of fungi on silica gel. Mycol. Res. 93:529–534.
64. Pearce, R.B. 1984. Staining fungal hyphae in wood. Trans. Br. Mycol. Soc. 82:564–567.
65. Pearce, R.B. and Rutherford, J. 1981. A wound-associated suberized barrier to the spread of decay in the sapwood of oak (*Quercus robur L.*). Physiol. Plant. Pathol. 19: 359–369.
66. Phillips, J.M. and Hayman, D.S. 1970. Improved procedures for clearing roots and staining parasitic and vesicular-arbuscular mycorrhizal fungi for rapid assessment of infection. Trans. Br. Mycol. Soc. 55:158–161.
67. Poppe, J. and Welvaert, W. 1983. Identification of *Hymenomycetes* in pure-culture by characterization of their mycelia and trials for artificial fructification. Med. Fac. Landbouww-Rijksuniv. Gent. 48:901–912.
68. Rayner, A.D.M. 1991. The challenge of the individualistic mycelium. Mycologia 83: 48–71.
69. Rayner, A.D.M. 1996. Has chaos theory a place in environmental mycology? In Fungi and Environmental Change (Frankland, J.C., Magan, N., and Gadd, G.M., Eds.), pp. 317–341. Cambridge University Press, Cambridge.
70. Rayner, A.D.M. and Boddy, L. 1988. Fungal Decomposition of Wood: its Biology and Ecology. John Wiley International, Chichester.
71. Rayner, A.D.M. and Todd, N.K. 1979. Population and community structure and dynamics of fungi in decaying wood. Adv. Bot. Res. 7:333–420.
72. Rayner, A.D.M., Griffith, G.S., and Wildman, H.G. 1994. Differential insulation and the generation of mycelial patterns. In Shape and Form in Plants and Fungi (Ingram, D.S., Ed.), pp. 293–312. Academic Press, London.
73. Read, D.J., Francis, R., and Finlay, R.D. 1985. Mycorrhizal mycelia and nutrient cycling in plant communities. In Ecological Interactions in Soil. Plants, Microbes and Animals (Fitter, A.H., Atkinson, D., Read, D.J., and Usher, M.B., Eds.), pp. 193–217. Blackwell Scientific Publications, Oxford.
74. Robinson, R.A. and Stokes, R.H. 1959. Electrolyte Solutions, 2nd ed. Academic Press, New York.

75. Rose, D.R. 1993. ROTTERS—An expert key for identification of wood-rotting fungi in culture. Binary 5:9–12.
76. Rorison, I.H. 1968. The response to phosphorus of some ecologically distinct plant species. I. Growth rates and phosphorus absorption. New Phytol. 67:913–923.
77. Roser, D.J., Keabe, P.J., and Pittaway, P.A. 1982. Fluorescent staining of fungi from soil and plant tissue with ethidium bromide. Trans. Br. Mycol. Soc. 79:321–329.
78. Schmid, B. and Harper, J.L. 1985. Clonal growth in grassland perennials. I. Density and pattern-dependent competition between plants with different growth forms. J. Ecol. 73:793–808.
79. Scott, W.J. 1953. Water relations of *Staphylococcus aureus* at 30°C. Aus. J. Biol. Sci. 6:549–564.
80. Sims, R.W., Freeman, P., and Hawksworth, D.L. 1988. Key to Works to the Fauna and Flora of the British Isles and North Western Europe, 5th ed. Clarendon Press, Oxford.
81. Smith, M.L., Bruhn, J.N., and Anderson, J.B. 1992. The fungus *Armillaria bulbosa* is among the largest and oldest living organisms. Nature 356:428–431.
82. Soderstrom, B.E. 1977. Vital staining of fungi in pure cultures and in soil with fluorescein diacetate., Soil Biol. Biochem. 9:59–63.
83. Soderstrom, B.E. 1979. Some problems in assessing fluorescein diacetate-active fungal biomass in soil. Soil Biol. Biochem. 11:147–148.
84. Stalpers, J.A. 1978. Identification of wood-inhabiting Aphyllophorales in pure culture. Stud. Mycol. (Baarn) 16:1–248.
85. Tabatabai, M.A. 1982. Soil enzymes. In Methods of Soil Analysis. Part 2, Chemical and Microbiological Properties, 2nd ed. (Pace, A.C., Miller, R.H., and Keeney, D.R., Eds.), pp. 903–947. American Society of Agronomy, Madison, WI.
86. Tadmor, U., Applebaum, S.W., and Kafir, R. 1971. A gas chromatographic micromethod for respiration studies on insects. J. Exp. Biol. 54:437–440.
87. Taylor, J.B. 1974. Biochemical tests for identification of mycelial cultures of basidiomycetes. Ann. Appl. Biol. 78:123.
88. Tiffin, A.I. 1987. Monoclonal antibodies and their use in microbiology. J. App. Bact. Symp. Supp. 127S–139S.
89. Tribe, H.T. and Maynard, P. 1989. A new automatic electrolytic respirometer. Mycologist 3:24–27.
90. Umbreit, W.W., Burris, R.H., and Stauffer, J.F. 1964. Manometric Techniques, 4th ed. Burgess Publishing, Minneapolis, MN.
91. Visser, S. and Parkinson, D. 1975. Fungal succession on aspen-poplar leaf litter. Can. J. Bot. 53:1640–1651.
92. Wells, J.M. and Boddy, L. 1995. Phosphorus translocation by saprotrophic basidiomycete mycelial cord systems on the floor of a mixed deciduous woodland. Mycol. Res. 99:977–980.
93. Wilcox, W.W. 1964. Preparation of decayed wood for microscopial examination. U.S. Department of Agriculture Forestry Service Research Note, FPL-056.
94. Wilcox, W.W. 1968. Changes in wood microstructure through progressive stages of decay. U.S. Department of Agriculture Forestry Service Research Paper, FPL70.
95. Wimpenny, J.W.T., Boddy, L., Harrison, J., and Williams, T.N. 1990. A novel gradient mixing device for simultaneous production of a wide range of gas concentrations. J. Microbiol. Meth. 115:115–120.

# 8

G.F. ESTEBAN
K.J. CLARKE
B.J. FINLAY

# Rapid Techniques for the Identification of Free-Living Protozoa

The general availability of light microscopes at the turn of the century fueled the widespread passion for natural history and enabled its extension toward the smallest "animals"—the protozoa [5, 21]. Later, the growth of ecology in the 1930s, and in particular the development of new ideas concerning coherent ecosystem function, also embraced protozoan communities in the notable attempts by Picken [19] and others. The intensification of interest in microbial community structure and function in the 1980s produced some new and radical ideas, particularly with respect to the function of microbial processes in the open waters of lakes and oceans. The microbial loop, and later the microbial food web, circumscribed a pivotal role for the protozoa as the principal consumers of bacterioplankton.

This interest in planktonic food webs also stimulated renewed interest in the biology of the organisms themselves; a notable example is the comprehensive understanding obtained of the physiologic ecology of filter feeding by unicellular phagotrophs [6]. Studying the biology of the organisms themselves has repeatedly been shown to confer its own benefits to a comprehensive understanding of the role of protozoa in aquatic ecosystems. For example, an expanded role for the planktonic ciliates in the open ocean was revealed by the realization that a large proportion had sequestered chloroplasts [22]; and the ability of some protozoa to complete their life cycles in anoxic habitats has been shown to be due to novel biochemical adaptations in their mitochondria and to syntrophic associations with endosymbiotic methanogenic bacteria that act as $H_2$ sinks [10].

If we consider all free-living protozoa, a large proportion have other organisms living inside them or on their external surfaces, and the ecologic function of the protozoon is best understood by considering the complete symbiotic association. In addition, often closely related species are involved in symbiotic associations with different types of symbiotic partners [3].

These considerations point clearly in one direction—if we are to understand the ecologic function of such microorganisms as protozoa, we have to have some clear idea of the identities of the organisms concerned.

Although many of the methods used to study the ecology of free-living protozoa, such as grazing rates [6], total abundance [11, 12], and respiration rates [7], are now fairly well described, a large question mark still hangs over such fundamental issues as the identities of the protozoan species in a natural community, the phenotypic plasticity of defined species, and, perhaps most fundamental of all, the relationship between phenotypic and genotypic "species."

Protozoa are unusual among microorganisms in the richness of the diversity of their morphologic phenotypes, and a suite of methods are available to tap into this diversity for purposes of identification. There are some dangers in this approach, such as convergent evolution toward certain phenotypes (completely unrelated ciliates may be morphologically indistinguishable from each other), and these problems may in most cases be resolvable only with modern molecular methods.

The purpose of this chapter is to introduce some quick and relatively simple methods that can be used to identify protozoa. The reader may also be interested in consulting other texts [9, 13, 17, 18] for a wide range of other methods useful in microscopic techniques and studying the ecology of free-living protozoa.

## IDENTIFYING CILIATED PROTOZOA WITH THE SILVER CARBONATE IMPREGNATION TECHNIQUE

Ciliated protozoa are, within the group of protozoa, some of the most challenging organisms in the study of biodiversity. They present an enormous variety of shapes and sizes, and their phenotypic characterization is usually based on the study of the oral and somatic infraciliature. All of the classical methods used to identify ciliates use one of the silver salts to impregnate and reveal the species-specific characteristics of this infraciliature. The most rapid methods employ ammoniacal silver carbonate for this purpose [8, 15]. One advantage of this technique is that it does not require a large amount of glassware and special equipment; simply a water bath (any size) and the reagents indicated in this section are needed. Hence, the silver carbonate technique can be carried out immediately after taking the sample, which is especially useful when we need to know the ciliated protozoa species living at the moment of sampling at the sampling place [4]. Two methods are described in this chapter: (1) for activated sludge—and any other biologic wastewater treatment that supports extraordinarily high ciliate abundances and (2) for natural water samples taken from fresh water and seawater (Figs. 8-1 to 8-4). Each procedure takes about 15 minutes.

### Activated Sludge

Keep the living sample in a refrigerator (~ 4°C) if it is impossible to deal with it within 1 h of sampling. Do not fix the sample with formalin as it may stick the ciliates to the sludge flocs, making subsequent identification difficult.

In handling specified volumes of reagents, sufficient accuracy is obtained by counting the drops dispensed from wide-bore glass Pasteur pipettes. One drop is roughly equivalent to 50 μl.

#### Procedure

1. Pour 10 ml distilled water into a 50-ml beaker.
2. Add three drops undiluted formalin.
3. Add 1–2 ml raw sludge sample.
4. Leave for 2 min at room temperature.

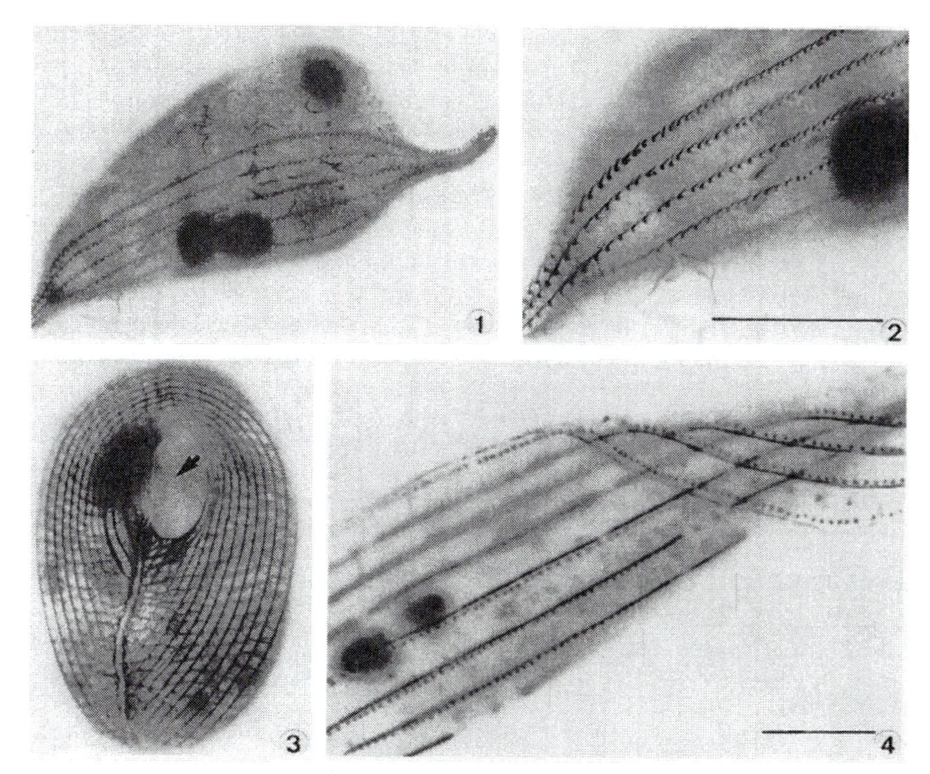


**Figures 8-1 to 8-4** Ammoniacal silver carbonate impregnation of some ciliated protozoa.
8-1. Morphology of the infraciliature in *Litonotus* sp. Whole cell 60 μm.
8-2. Posterior part of *Litonotus* sp. from Figure 8-1. Scale bar = 10 μm.
8-3. *Frontonia minuta.* Whole cell 65 μm. Arrow to the oral aperture.
8-4. Detail of the infraciliature of the marine ciliate *Kentrophoros* sp. Scale bar = 16 μm.

5. Add 20 to 25 drops 5% bactopeptone (see below for preparation of solutions).
6. Add 10 drops undiluted pyridine (this should be done out of doors or in a fume cabinet).
7. Add 2 ml ammoniacal silver carbonate solution (see below for preparation of solutions).
8. Add 20 ml distilled water. Stir constantly between steps 5 and 8 inclusive.
9. Place the beaker in a water bath at 58–62°C. While the liquid is heating in the water bath, take an evaporating dish (of any diameter ≥10 cm) and pour 50 ml distilled or deionized (but not tap) water into it. This water will be used to stop the reaction that is taking place in the beaker.
10. Wait until the color of the liquid in the beaker changes to that of dark brandy or even black. This reaction takes about 5 to 10 min, but some samples require a longer time.
11. As the liquid becomes progressively darker, pour out small quantities into the evaporating dish to give a series of grades of impregnation of the ciliates.
12. Allow the ciliates to settle in the evaporating dish, which usually takes about 10 min.
13. Use a wide-bore glass Pasteur pipette, or any other type, to collect the sedimented ciliates and transfer to a 10-ml screw-top tube (or go straight to step 15).

14. Add distilled water c.s if the volume is smaller than 9 ml. Add 1 ml of formalin. Preparations can be kept in this way for several months [14].
15. Ciliates stain a distinctive yellow to brown. For observation under the light microscope, pick the ciliates out of the tube—or directly from the evaporating dish in case they were not transferred into a tube—with a Pasteur pipette, and place them into a dish or any container (Sedewage-Rafter chambers would do fine). Then, pick out the cells individually, using drawn-out Pasteur pipettes and a low-power stereo microscope (e.g., magnification 20 to 40x).
16. Place the ciliates onto a microscope slide. Flatten them between the microscope slide and cover slip, observe, and photograph. The permanency of so-called permanent preparations is unreliable.

Further details of this method and some of its variants can be found in Fernández-Galiano [8] and Foissner [15]—the latter also includes other valuable staining and microscopic techniques for ciliated protozoa.

### *Preparation of Solutions*

Solution 1: Dissolve 50 g $Na_2CO_3 \cdot 10H_2O$ in 1 liter distilled water.

Solution 2: Dissolve 5 g $AgNO_3$ in 50 ml distilled water.

Solution 3

1. Mix the entire volume of Solution 2, proceeding slowly, with 150 ml of Solution 1. A milky suspension will form.
2. Use a 10-ml pipette to add $NH_4OH$ drop by drop, and shake regularly until the white suspension disappears and the solution becomes transparent. Take care not to add an excess of ammonium hydroxide.
3. Transfer to a 1-liter measuring cylinder.
4. Add Solution 1 to the 800-ml mark. This is the ammoniacal silver carbonate solution.
5. Keep it in a brown bottle at room temperature. It is stable for up to 3 years.

Solution 4

1. Dissolve 5 g bactopeptone (DIFCO) in 100 ml distilled water.
2. Add 1 ml formalin as a preservative.
3. Keep in a brown flask at 4°C and make a fresh solution every 2 months.

## Silver Carbonate Staining of Natural Water Samples

### *Freshwater Samples*

For any freshwater sample other than activated sludge, satisfactory results are obtained using this procedure.

Procedure

1. Fix a 5 ml sample with three drops of formalin in a beaker.
2. Leave for 2 min at room temperature.

3. Add 17 to 19 drops of 5% bactopeptone. It is advisable to increase the amount of 5% bactopeptone to up to 20 to 25 drops for organically polluted freshwater samples.
4. Now follow Steps 6–16 of the protocol for the silver staining of activated sludge.

### *Marine Samples*

The problem with staining marine samples is that the silver carbonate with the seawater salts forms a white precipitate and no reaction is produced during heating in the water bath. To avoid this problem, it is necessary to first dilute the salts present in the natural sample, as follows:

Procedure

1. Fix 10 ml of sample with 0.5 ml of formalin in a glass centrifuge tube.
2. Centrifuge at 1500–1600 rpm for 2 to 5 min at room temperature.
3. Wash the pellet two to three times with distilled water, centrifuging after each wash.
4. Transfer the last pellet to a beaker with 5 ml distilled water and three drops of formalin.
5. Complete Steps 5–16 described of the protocol for the silver staining of activated sludge.

Alternatively, it is possible to fix either micro-pipetted ciliates or 0.5 ml of marine sample in a beaker containing three drops of formalin in 10 ml distilled water, but the procedure described above has rendered best results (Fig. 8-4).

## RAPID SPECIMEN PREPARATION METHODS FOR ELECTRON MICROSCOPY

Some important morphologic characteristics of protozoa are recognizable only with the high magnifications and resolving power of electron microscopy (EM). The chrysomonad genus *Paraphysomonas*, for instance, contains different species that appear identical when using light microscopy, but are remarkably different under the electron microscope (see Figs. 8-6, 8-7, 8-10, and 8-11). Many specimen preparation techniques for electron microscopy are directed at the preservation of ultrastructure. These methods are generally very time-consuming with such standard procedures as freeze-substitution taking weeks to complete. However, when the rapid identification of microorganisms becomes necessary, as in ecologic surveys of whole microbial communities, some fundamental information can usually be obtained relatively quickly and easily.

Several simple EM techniques are described here (Fig. 8-18). They will aid in the quick and accurate identification of a surprisingly large number of the microorganisms that lend themselves to this type of processing (e.g. testate and scale-bearing amoebae, heterotrophic (especially scale-bearing) flagellates, some prasinomonads and prymnesiomonads, dinoflagellates, and diatoms; Figs. 8-5 to 8-17).

### Safety

Use appropriate precautions when handling and disposing of the chemicals routinely used in electron microscopy, especially osmium tetroxide, uranyl acetate,

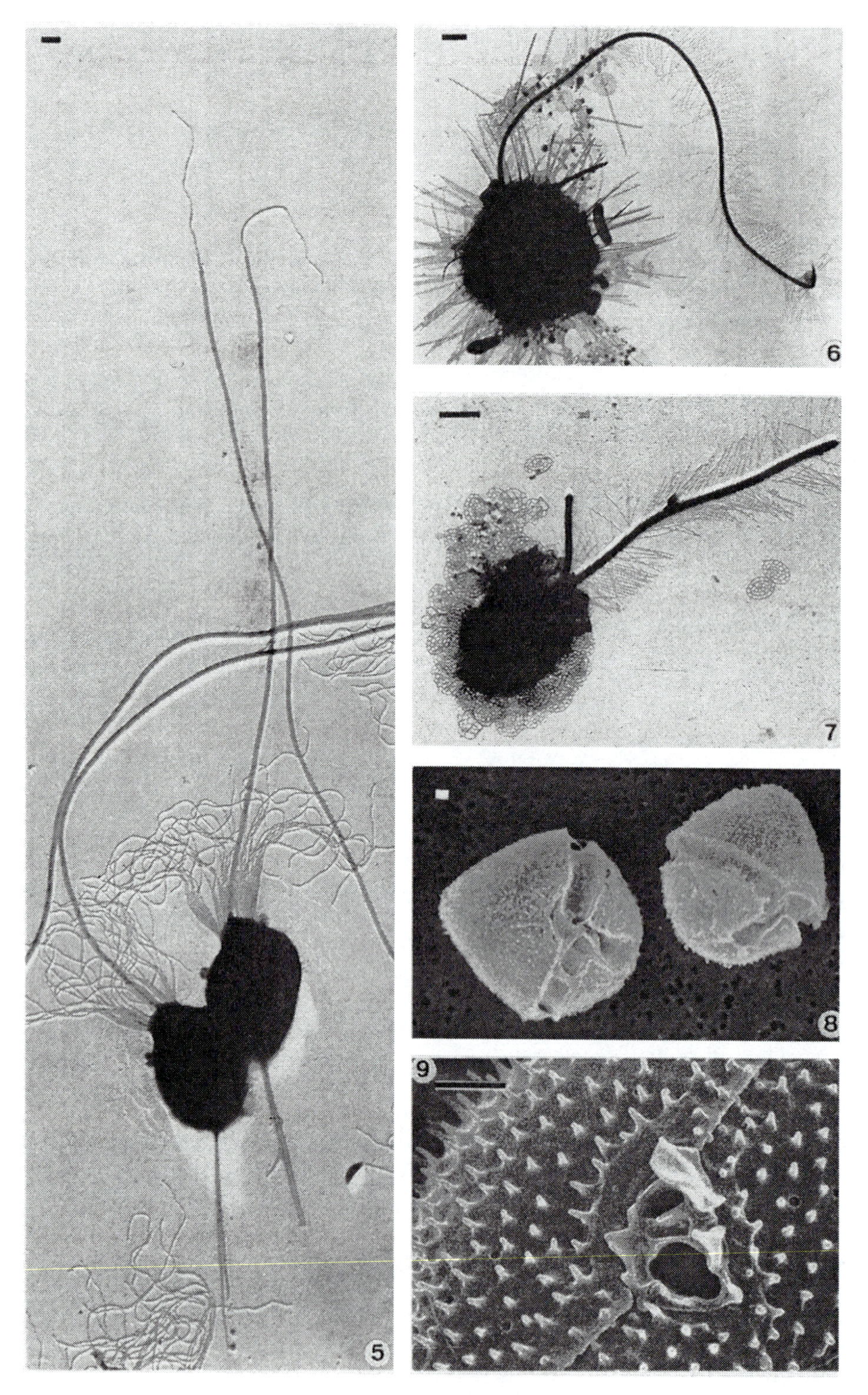

Figures 8-5 to 8-9 Morphology of whole cells. Specimens prepared by rapid EM techniques. Bar represents 1 μm.

8-5. *Sphaeroeca volvox.* Shadow-cast with chromium; TEM.

8-6. *Paraphysomonas vestita.* Shadow-cast with gold; TEM.

8-7. *Paraphysomonas butcheri.* Shadow-cast with gold; TEM.

8-8. *Peridinium sp.* Freeze-dried from water; SEM.

8-9. *Prorocentrum balticum.* Critical-point dried from ethanol and carbon dioxide; SEM.

and sodium cacodylate. Protective laboratory coat and gloves should be worn and all chemicals handled only in a fume cabinet.

## Rapid Protocols for Transmission Electron Microscopy (TEM)

### *Basic Protocol*

These methods give detailed ultrastructural information on cell appendages. Shadow-casting is best used for whole-cell configuration, body-scales, spines, bristles, and related structures. Negative staining is best for high-resolution TEM of flagella and their associated structures for mastigonemes, trichocysts, and the like.

#### Procedure

1. Concentrate sample in a 15-ml tube by gentle centrifugation for 3–5 min. Avoid speeds in excess of 2500 rpm (or over 1000 g) as they can remove cell appendages. Use standard glass centrifuge tubes (some cells stick to plastics).
2. Using a disposable pipette, decant and discard about 60% of the supernatant. Resuspend the concentrated sample in the remaining liquid.
3. Attach Formvar-coated TEM grids by their rims to the extreme edge of a small strip of double-sided sticky tape secured in a Petri dish. The grids should lie flat. Carefully place a drop of the resuspended sample onto each grid.
4. Add three drops of 2% osmium tetroxide (in distilled water or buffer) to the inside of the lid of the Petri dish. Place the lid over the lower part of the dish and its contents. Leave in place for 1 min. The vapor from the osmium tetroxide will have killed and partly fixed the microorganisms present. Remove the lid and replace with a clean lid. Allow the fixed cells to settle for a further 15–20 min.
5. With a fine drawn glass pipette, carefully add a drop of distilled water to the sample on each grid. Do not overflow the grid area. Allow drops to stand for 5 min, and then remove carefully with a cut point of fine filter paper without disturbing the settled specimen material.
6. Add a further drop of distilled water, and repeat this washing procedure three to six times.
7. Allow the sample material on the TEM grid to air dry.

### *Shadow-Casting Option*

Shadow-casting the sample with evaporated metal gives contrast modeling effects similar to cross-illumination [2].

#### Procedure

1. When the grids are dry, transfer the lower half of the Petri dish with the grids to the stage of a vacuum evaporator for shadow-casting.
2. Evaporate the shadowing metal, and continue to pump evaporator until suspended vaporized metal is pumped safely away.
3. Detach the grids from the sticky tape and examine in the TEM.

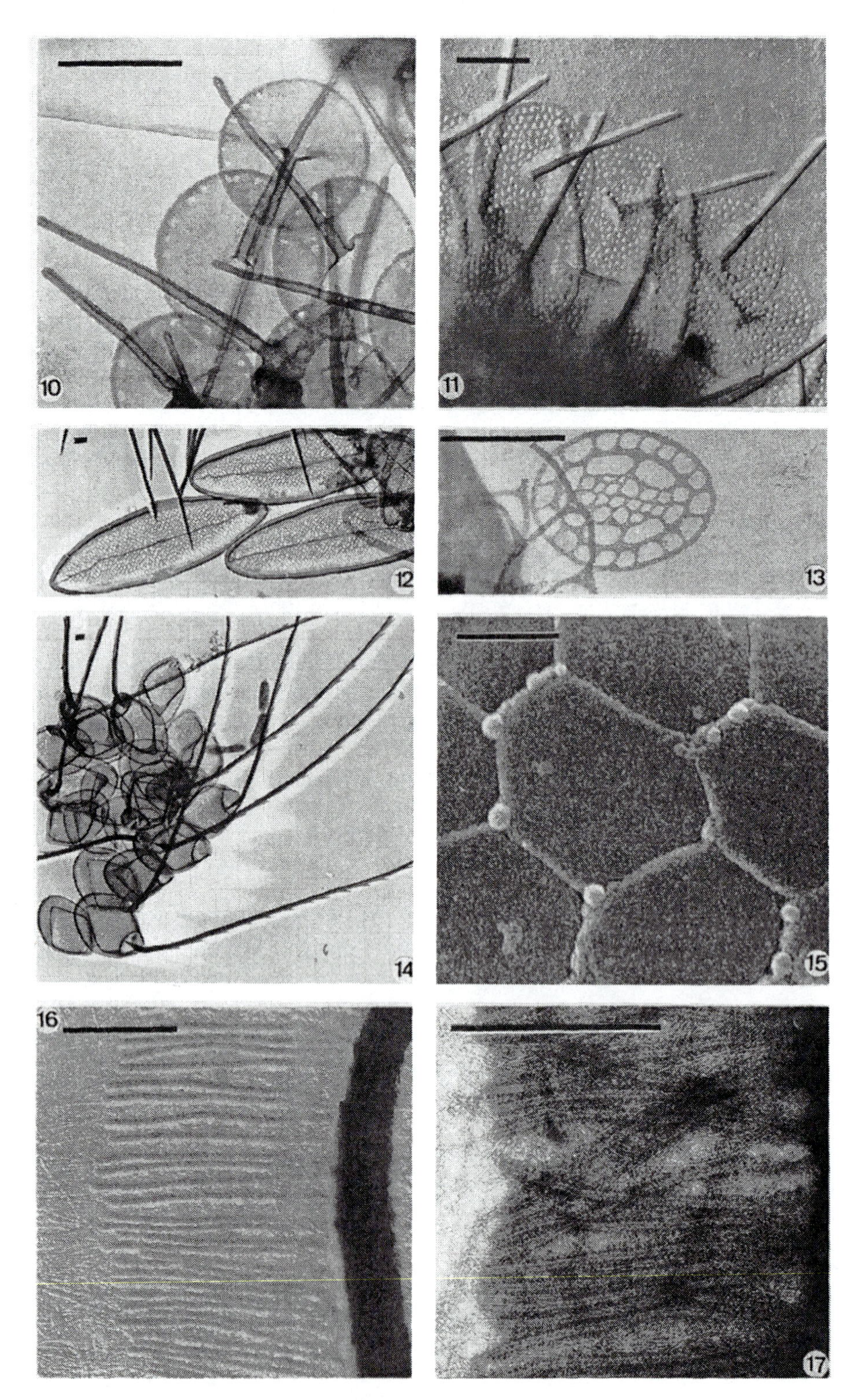


**Figures 8-10 to 8-17** Species-specific structures of cells. Specimens prepared by rapid EM techniques. Bar represents 0.5 μm.

8-10. Scales of *Paraphysomonas circumforaminfera.* Lightly shadow-cast with chromium; TEM.

8-11. The similar scales of *Paraphysomonas foraminifera.* Shadow-cast with gold; TEM.

8-12. Scales of the heliozoon *Raphidiophrys symmetrica.* Shadow-cast with chromium; TEM. *Legend continued on opposite page.*

Notes

Using chromium as a shadow source material yields average to good results. It is tolerant of wide variation in the amount of deposition, and fine control of the metal evaporation is unnecessary to obtain an acceptable result.

If gold, gold/palladium alloy, or platinum are used, sharper, higher-contrast shadows can be obtained, but care must be taken not to under- or over-shadow-cast the specimen.

With any shadowing material, use a high-shadow angle (30–35°) for bulky specimen cells and a low angle (15–20°) for fine scales, flagella, and the like. Shadow at 20–30° to the horizontal (the grid surface) for routine samples of mixed type.

### *Negative Staining Option*

Negative staining the sample dramatically increases the resolution of the exterior details of the microorganisms present [2].

Procedure

1. Use all electron-dense stains in a similar manner. Using a syringe with an attached 0.2-μm Nucleopore polycarbonate syringe-filter, add a drop of stain to the surface of the grid. Leave the drop in place for a few seconds and then remove using a cut point of filter paper. Leave the surface damp with stain.
2. Allow grids with residual stain to air-dry, and examine by TEM.

Notes

Use either (1) 1% to 2% uranyl acetate in distilled water for 20–30 s; (2) 1% to 2% potassium phosphotungstate in distilled water for 30 s; (3) 1% to 2% sodium dodeca tungstosilicate in distilled water for 30 s; or (4) 2% methylamine tungstate for 30–120 s.

The choice of stain will depend on the size, type, and concentration of microorganisms in the sample and the degree of detail to be resolved. It is best chosen by experimentation. Uranyl acetate can give the most contrast and methylamine tungstate the least, but the latter stain has the potential to resolve the finest detail. The pH of all of these stains can be altered carefully to obtain degrees of effect between positive and negative staining, and results which best suit the material under examination should again be sought by experimentation.

All stains must be filtered through 0.2-μm pore-size polycarbonate filters before use. Uranyl acetate is photolabile, so staining with this material should be carried

←

8-13. Fine scales of *Paraphysomonas butcheri.* Lightly stained with methylamine tungstate; TEM.
8-14. Spined scales of *Mallomonas aereolata*. Shadow-cast with chromium; TEM.
8-15. The surface of *Rhodomonas lacustris* showing trichocyst ports in the corners of each plate. Freeze-dried from ethanol; SEM.
8-16. Mastigonemes on the long flagellum of *Lepidochrysis glomerifera.* Shadow-cast with chromium; TEM.
8-17. Mastigonemes on the long flagellum of *Ochromonas villosa.* Negative stained with uranyl acetate; TEM.

out in subdued lighting conditions. Addition of 50 μg/ml of the polypeptide antibiotic bacitracin to any of these stains reduces its surface tension, enabling the stain to contact the sample surface and enhancing the potential resolution of the technique [16].

Heating of dense areas of stain by the electron beam can disrupt the grid Formvar coating when viewed by TEM. To avoid this, approach each new area of a grid at a low illumination level, gradually increasing the condenser lens brightness until the desired level is reached.

In most cases, whole microorganisms examined by these methods are too thick or dense to allow observation of their internal structure. However, flagella, spines, bristles, and scales are fine enough for characteristic detail to be resolved. The surface structure of a microorganism can be examined in the area around the cell's periphery.

## Rapid Protocols for Scanning Electron Microscopy (SEM)

### *Basic Protocol*

These methods enable examination of the whole cell shapes of microorganisms with their characteristic configuration of surface scales and other structures. However, the SEM methods do not have the same potential for high-resolution as negative staining does for TEM. The basic protocol should also be used for the examination of hydrated material in the new variable vacuum and environmental SEMs, and the basic protocol together with the concentration option used for TEM examination of marine microorganisms.

#### Procedure

1. Concentrate sample by gentle centrifugation in a 15-ml glass tube (see the TEM protocols). Use a pipette to decant most supernatant, and gently resuspend the sample in the remaining 0.5–1.0 ml of liquid.
2. Using a pipette, rapidly inject 1–2 ml 2% osmium tetroxide in 0.1 M sodium cacodylate buffer (pH 6.8) onto the sample.
3. Leave in a fume cabinet for 30 min.
4. Cover the top of the tube with cling-film, and centrifuge to gently concentrate material (3–5 min; less than 1000 g). Remove osmium tetroxide supernatant with a disposable pipette.
5. Resuspend cells in 10 ml distilled water, and leave in place for 5 min before recentrifuging and removing supernatant. Replace with distilled water and resuspend sample material. Repeat this washing procedure three to six times.

### *Available Options*

#### Procedure: Concentration

Add drops of concentrated, resuspended sample to coated TEM grids, and treat as in the TEM protocol.

#### Procedure: Freeze-Drying from Water

1. Place a small ceramic crucible in a shallow Dewar bowl and fill both with liquid nitrogen to a level where the crucible's rim is just above the liquid. Wait for the liquid nitrogen to stabilize.

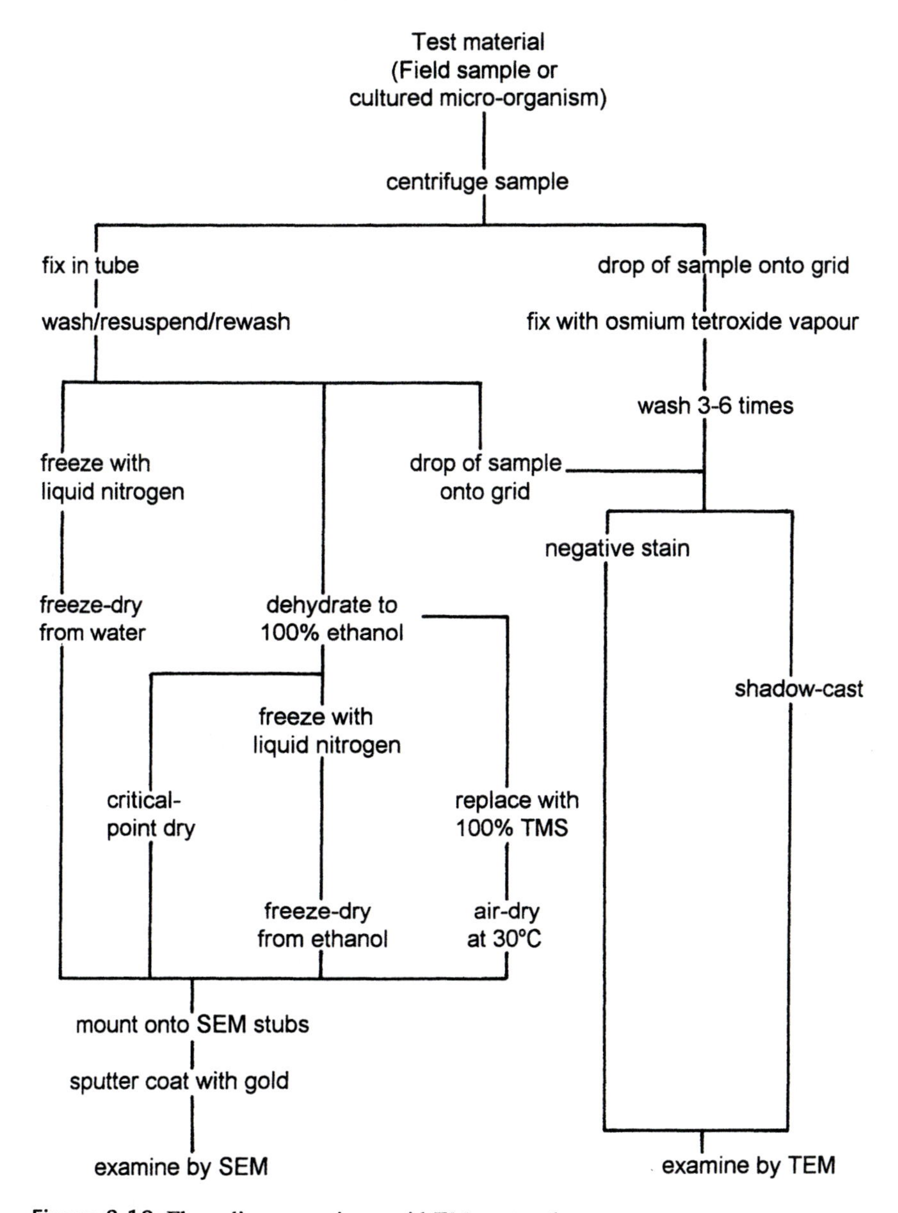


Figure 8-18 Flow diagram using rapid EM protocols.

2. Using a drawn glass pipette, allow small drops of sample suspension to fall into the liquid nitrogen within the crucible. These drops will briefly 'fizz' on the nitrogen surface before sinking.
3. Take a 1.0-ml plastic Eppendorf-style tube with a screw-cap; push a number of holes through the cap, and remove it.
4. Take a 45-mm diameter 0.2-μm pore-size Nucleopore polycarbonate filter, fold it in half, turn in each end toward the center, and staple each to hold in place. This now forms a pocket.
5. Place this pocket under liquid nitrogen in the Dewar bowl, and open it using tweezers. Still using tweezers, rapidly transfer the frozen droplets into the pocket and fold down the top.
6. Freeze the open Eppendorf tube in the liquid nitrogen, and using tweezers, fold and push the closed pocket of frozen sample drops into it. Screw on

the tube cap with its perforations, and transfer to the chilled stage of a freeze-dryer.

7. Freeze-dry the sample.
8. When dry (10–12 h), remove the sample. Using a stereo-microscope, scalpel, and tweezers, cut up and mount pieces of the pocket, with the attached sample exposed, onto SEM specimen stubs using a carbon-loaded adhesive, such as Leit-C. Alternatively, tip sample material directly from the pocket onto semiset carbon adhesive coated on SEM specimen stubs or onto Tempfix adhesive precoated onto a specimen stub surface (see detailed instructions enclosed with each pack of this adhesive). Tempfix is an ideal mountant for holding dried microorganisms for SEM.
9. Allow mounts to dry, sputter-coat with gold, and examine by SEM.

Procedure: Freeze-Drying from Ethanol

1. Dehydrate the sample material through an increasing concentration series of ethanol in water until 100% ethanol is reached. Dehydration should be carried out using centrifugation, liquid replacement, and resuspension as in the washing procedures in the basic protocol for SEM, Step 5. Begin the ethanol concentration series with a 30% solution, then 50%, 70%, 95%, and 100%. The sample time in each should be more than 10 min, and two final changes in 100% ethanol should be given.
2. Treat as in the procedure of Freeze-Drying from Water. Reduce the freeze-drying time to approximately 2 h.

Procedure: Critical-Point Drying from Ethanol

1. Dehydrate as in the procedure, Freeze-Drying from Ethanol.
2. Transfer the sample material to a Nucleopore filter pocket as in the procedure, Freeze-Drying from Water. During all transfers the sample material MUST be kept wet with ethanol to prevent damage to cells.
3. Transfer the pocket containing the sample to the specimen boat of a critical-point dryer. Critical-point dry the sample.
4. After drying is complete, mount the sample as in the procedure, Freeze-Drying from Water.

Procedure: Air-Drying from Tetramethylsilane (TMS)

1. Note that TMS is an inert substance with a low viscosity and boiling point (it is also highly flammable). Its addition to a sample reduces air-drying stresses in the material to a minimum [1].
2. Dehydrate as in the procedure, Freeze-Drying from Ethanol.
3. Add 1–2 ml TMS to the sample. Leave in place for 10 min.
4. Gently centrifuge the sample (2–3 min; less than 1000 g), remove the TMS with a pipette, and resuspend the sample in another 2 ml of TMS.
5. Recentrifuge the sample, and using a pipette decant most of the TMS. Transfer the sample to a Nucleopore filter pocket (as in the procedure, Freeze-Drying from Water), and dry at 30°C.
6. When dry, mount and sputter-coat the sample as in the procedure, Freeze-Drying from Water.
7. Examine by SEM.

Procedure Notes

Air-drying of delicate biologic specimens from water causes distortion and disruption as the air-water interface moves through the drying material. These surface tension effects can be dramatically reduced or eliminated by the methods described in this section.

Microorganisms examined by SEM should retain their living shape with all appendages in place. Processed material should not appear distorted; however, freeze-drying can reduce cell volume by up to 20% and critical-point drying by up to 30%.

Many critical-point drying manuals maintain that "flushing" with liquid carbon dioxide is an essential procedural stage in the drying process. We consider that flushing is not necessary when drying from ethanol and that it can be an important cause of sample damage or loss when processing microorganisms. A phase-separation boundary between ethanol and liquid carbon dioxide can be observed through the window of the critical-point dryer, and careful elimination of this boundary by draining can ensure complete replacement of ethanol by carbon dioxide without the need to flush the sample.

Use extreme care when mounting dried samples for SEM examination. Static-charge effects can cause the loss of sample material, especially when handling dried cells on polycarbonate filters. Use a low-power (10–40×), long working distance stereo-microscope when transferring specimens to SEM specimen stubs; an eyelash glued to the end of a cocktail stick can be extremely useful for moving large cells, cell aggregates, or pieces of sample to the stub for mounting.

## Cryo-Scanning Electron Microscopy

The simple preparation protocols described in this section can be applied when using any transmission or scanning electron microscope. However, the SEM with an attached cryo facility is now in common use. It enables microorganisms to be prepared and examined quickly in a fully or partially hydrated but frozen state, minimizing cell disruption and loss of cell volume [20].

Different cryo-SEMs have different procedures for the manipulation of frozen samples, so for this reason detailed preparation protocols are not described here. The basic methods involve either (1) chemically fixing the sample, washing and resuspending it in distilled water, and rapidly freezing a droplet of it placed on a specimen stub, or (2) rapidly freezing a concentrated droplet of the living sample (physical fixation) on a specimen stub. The coolant used is frequently a slurry of solid and liquid nitrogen at −210°C, produced from liquid nitrogen in apparatus ancillary to the cryo-SEM.

Once the frozen sample is transferred to the microscope, microorganisms can be revealed by slowly raising the temperature of the stage and sample to between −130°C and −100°C to gently sublime away surrounding ice. Rapidly lowering the temperature again will halt sublimation and hold the exposed cells at the desired degree of hydration. Cryo-sputtering, in which the still-frozen sample is transferred under vacuum from the SEM to an adjoining chamber for metal sputtering before returning to the SEM, enhances the resolution of the image of the newly exposed cells.

A sample that has been fixed chemically has the advantage that it can be washed free of detritus and dissolved chemicals before its freezing and introduction to the SEM. The main disadvantage of this procedure is the several hours of time required to process the material during chemical fixation. However, living material that is physically fixed by freezing has the advantage of very rapid processing.

Exact cryo procedures differ for each instrument, but a general rule is that the faster the freeze, the better the sample preservation. The coolant used, the bulk of the specimen, and its support all affect the cooling rate. The smaller the sample, the more rapid the freeze, but also a slurry of solid and liquid nitrogen at −210°C will freeze a sample faster than liquid nitrogen at −196°C. Excellent sample preservation can be obtained using liquid ethane or propane as a coolant, but use of these liquid gases can be extremely hazardous and pose the risk of explosion and fire. Their use is unnecessary where the SEM examination of a specimen is required only for its identification.

## SUPPLIERS

Most reagents and consumable items used in these electron microscopy procedures can be obtained from any good supplier. Two such suppliers are Agar Scientific Ltd., 66a Cambridge Road, Stansted, Essex, CM24 8DA, UK, and Electron Microscopy Services, 321 Morris Road, Box 251, Fort Washington, PA 19034, USA.

## REFERENCES

1. Dey, S., Basu Baul, T.S., Roy, B., and Dey, D. 1989. A new rapid method of air-drying for scanning electron microscopy using tetramethylsilane. J. Microsc. 156: 259–261.
2. Dykstra, M.J. 1993. A Manual of Applied Techniques for Biological Electron Microscopy. Plenum Press, New York.
3. Embley, T.M. and Finlay, B.J. 1994. The use of small subunit rRNA sequences to unravel the relationships between anaerobic ciliates and their methanogen endosymbionts. Microbiology 140: 225–235.
4. Esteban, G.F. and Finlay, B.J. 1996. Morphology and ecology of the cosmopolitan ciliate *Prorodon viridis*. Arch. Protistenkd. 147: 181–188.
5. Fauré-Fremiet, E. 1924. Contribution a la connaissance des infusoires planktoniques. Bull. Biol. Fr. Belg., Suppl. 6: 1–171.
6. Fenchel, T. 1986. Protozoan filter feeding. Prog. Protistol. 1: 65–113.
7. Fenchel, T. and Finlay, B.J. 1983. Respiration rates in heterotrophic, free-living protozoa. Microb. Ecol. 9: 99–122.
8. Fernández-Galiano, D. 1994. The ammoniacal silver carbonate method as a general procedure in the study of protozoa from sewage (and other) waters. Wat. Res. 28: 495–496.
9. Finlay, B.J. 1993. Behavior and bioenergetics of anaerobic and microaerobic protists. In Handbook of Methods in Aquatic Microbial Ecology (Kemp, P.F., et al., Eds.), pp. 59–65. Lewis Publishers, Boca Raton, FL.
10. Finlay, B.J. and Fenchel, T. 1992. Methanogens and other bacteria as symbionts of free-living anaerobic ciliates. Symbiosis 14: 375–390.
11. Finlay, B.J. and Guhl, B.E. 1992. Plankton sampling—freshwater. In Protocols in Protozoology (Lee, J.J., and Soldo, A.T., Eds.), pp. B-1.1. Society of Protozoologists, Lawrence, KS.
12. Finlay, B.J. and Guhl, B.E. 1992. Benthic sampling—freshwater. In Protocols in Protozoology (Lee, J.J. and Soldo, A.T., Eds.), pp. B-2.1. Society of Protozoologists, Lawrence, KS.
13. Finlay, B.J., et al. 1988. On the abundance and distribution of protozoa and their food in a productive freshwater pond. Eur. J. Protistol. 23: 205–217.
14. Finlay, B.J., Tellez, C., and Esteban, G. 1993. Diversity of free-living ciliates in the sandy sediment of a Spanish stream in winter. J. Gen. Microbiol. 139: 2855–2863.
15. Foissner, W. 1992. The silver carbonate methods. In Protocols in Protozoology (Lee, J.J. and Soldo, A.T., Eds.), pp. C-7.1. Society of Protozoologists, Lawrence, KS.

16. Gregory, D.W. and Pirie, B.J.S. 1973. Wetting agents for biological electron microscopy 1. General considerations and negative staining. J. Microsc. 99: 251–265.
17. Kemp, P.F., Sherr, B.F., Sherr, E.B., and Cole, J.J. (Eds.) 1993. Handbook of Methods in Aquatic Microbial Ecology, Lewis Publishers, Boca Raton, FL.
18. Lee, J.J. and Soldo, A.T. 1992. Protocols in Protozoology. Society of Protozoologists, Lawrence, KS.
19. Picken, L.E.R. 1937. The structure of some protozoan communities. J. Ecol. 25: 368–384.
20. Robards, A.W., and Sletyr, U.B. 1985. Low temperature methods in biological electron microscopy. Vol. 10: Practical Methods in Electron Microscopy (Glauert, A.M., Ed.), Elsevier, Amsterdam.
21. Saville Kent, W. 1881–2. A Manual of the Infusoria. I and II. David Bogue, London.
22. Stoecker, D.K., Michaels, A.E. and Davis, L.H. 1987. Large proportion of marine planktonic ciliates found to contain functional chloroplasts. Nature (Lond) 326: 790–792.

9

ROBERT V. MILLER

# Methods for Enumeration and Characterization of Bacteriophages from Environmental Samples

The natural ecology of bacteria-bacteriophage interactions is not well known. For many years, microbiologists dismissed bacterial viruses as unimportant factors in determining the characteristics of an ecosystem and its microbial components. This view stemmed mainly from the fact that the estimates of virions present in the environment suggested that interactions between viruses and their hosts were rare events [17]. However, recent reports have changed this attitude. Several studies in the past 5 years have revealed that bacteriophage concentrations in many aquatic environments are quite high, often being one to two orders of magnitude higher than bacterioplankton concentrations [3–5, 21]. These observations have brought into question many of our conclusions about the ecology of aquatic microbial populations. They suggest that bacteriophage-bacterial interaction may in fact be a common process that affects bacterial abundance in many natural ecosystems. Certainly, many of our conclusions concerning phage-host interactions based solely on laboratory studies are naive at best. As we learn more about environmental interactions between bacteria and their viruses, many questions arise concerning the influence of these intracellular parasites on the makeup of bacterial communities, environmental horizontal gene transfer, and the impact of various environment factors, such as solar ultraviolet (UV) radiation. These and other questions point out the importance of the study of the environmental interactions among bacteria and their viral pathogens and the influences these interactions have on the development, maintenance, and decline of various ecosystems.

In this chapter, various techniques for the study of bacteriophages are discussed. These techniques include those necessary for the cultivation and characterization of bacteriophages in the laboratory and for exploring phage-host interaction in situ. Many of these latter techniques are rapidly evolving as more is learned about the natural ecology of bacteriophages. Various reviews and monographs have been written on the subject of phage ecology [6, 7, 11, 12, 14, 15, 17, 19, 23, 26].

## LABORATORY TECHNIQUES

### Enumeration of Bacteriophages (Titering)

Recognition of the presence of bacteriophages (and in truth of all viruses) in a particular sample depends on the cytopathic effects of the virus on its host organism. Remember that viruses are obligate intracellular parasites that cannot grow outside the host. For bacteriophages, the most easily identifiable symptom of infection is the lysis of the host cell, which is identified most easily by the formation of a plaque on a lawn of bacteria growing on solid medium. Each plaque arises from a single viral particle in the original sample. By counting the number of plaques produced, the number of infective virus particles in the original sample can be estimated. Since the particle concentration may be great, enumeration often requires the preparation of a 10-fold dilution series of the environmental sample for analysis. This method provides an estimation of the number of virus particles capable of producing a plaque (i.e., plaque-forming unit; PFU) in a particular sample and so may underestimate the total number of virus particles in the sample.

Usually, medium that will support growth of the host organism is sufficient for use in assessing phage-host interactions. However, the addition of divalent cations may be necessary for efficient adsorption of virions onto the host-cell receptor [28]. Temperature of incubation is also an important factor. Maximal infection is often obtained at the temperature of the environment from which the virus was isolated and not necessarily at the maximal growth temperature of the host [24]. It should also be remembered that not all strains of a particular species of bacterium are sensitive to infection by the same viruses. If the contents of the sample are unknown (i.e., the specific viruses have not been characterized), it may be advantageous to use several strains of the same species to enumerate virus particles capable of causing infection.

For accurate results, dilutions should be enumerated in triplicate. Only one dilution should be used to determine the titer of PFUs. This dilution should produce between 20 and 200 plaques on the enumeration plate. This is a statistically significant number that is unlikely to be affected by multiple infection of the same host cell.

#### Procedure

1. Make a serial dilution of the sample suspected of containing viruses in a rich medium, such as nutrient broth. Rich medium is used because virions are stabilized by the presence of proteinaceous material and other macromolecules. They often lose infectivity rapidly in water or minimal salts solutions [14].
2. Prepare a culture of the host organism by growing the bacteria in a liquid culture of the same medium (usually a rich medium) onto which the final enumeration mixture will be plated. Grow this culture to the late exponential phase to ensure that a thick lawn of growth will be established on the enumeration plates. The temperature of growth should be appropriate for maximal infection by the phage.
3. Samples of this culture should be dispensed into small sterile test tubes (13 mm). Usually, 0.1–0.2 ml samples are sufficient.
4. Add 0.1 ml of the various dilutions of the sample to be analyzed to the tubes containing host organisms.
5. Allow time for adsorption by incubating the tubes at an appropriate temperature. Often the temperature at which maximum absorption takes place is that of the environment from which the phage was isolated. It may be

necessary to test several temperatures to determine the one that is most effective. The time may vary with the specific phage, but 20 min is usually sufficient.

6. Add 2.5 ml of top agar (0.75% agar prepared in the culture medium) that has been melted and cooled to 50°C.
7. Mix gently by rolling slowly between the palms of your hands), and pour immediately onto a fresh, thick agar plate. Rotate the plate to spread the mixture evenly over its entire surface.
8. Cool for 30 min or until the top agar is set, invert, and incubate at an appropriate temperature overnight.
9. Count the plaques and calculate the titer from the amount plated and dilution using this formula:

$$n = y/(v)(x)$$

where n is the titer (PFU/ml) of the original sample, y is the average number of plaques counted, v is the volume plated, and x is the dilution plated.

This procedure allows the determination of the number of PFUs capable of initiating productive infections of their host bacterium in the original sample. For statistically significant measurements, only one dilution that gives between 20–200 plaques should be used, and each dilution should be plated in at least triplicate.

## Preparation of Phage Stocks (Phage Lysates)

The optimal method for the preparation of phage stocks or lysates varies significantly for different phages. A general method is presented here, which may be modified as needed. The propagation of most phages is best achieved by the soft-agar overlay method [1]. With some phages, however, better results may be achieved by a liquid method (see Alternate Method).

### *Soft-Agar Overlay Plate Method*

In the soft-agar overlay method, bacteriophage and host are mixed and plated in soft agar onto Petri dishes in a fashion similar to that outlined for the procedure, Enumeration of Bacteriophages. However, the goal of this procedure is total infection and lysis of the potential bacterial lawn and not the production of well-separated, countable plaques.

#### Procedure

1. Grow the host strain to the mid-exponential phase in an appropriately rich medium, such as L-broth [16] or nutrient broth.
2. Mix a 0.2-ml sample of the bacteria with the phage-containing sample at a multiplicity of infection (MOI, phage-to-bacterium ratio) of approximately 0.1.
3. Allow an appropriate adsorption period (see Enumeration of Bacteriophages).
4. Add melted and cooled 2 ml top agar.
5. Mix the mixture gently, and pour onto a thick 1.5% agar plate containing an appropriate growth medium.
6. Allow the top agar to solidify, and incubate at an appropriate temperature overnight (see Enumeration of Bacteriophages).

7. The next day, scrape off the top agar layer with a sterilized spatula and transfer to a screw-capped, sterilized centrifuge tube containing 0.5 ml of chloroform; wash the plate with 2–3 ml of medium that is added to the centrifuge tube.
8. Cap the centrifuge tube, shake vigorously by hand, and spin in a benchtop centrifuge for approximately 15 min to remove agar and unlysed cells.
9. Transfer the supernatant fluid to another tube containing 0.5 ml $CHCl_3$. *Note*: some phages are sensitive to chloroform and rapidly lose infectivity if stored over $CHCl_3$. Lysates of these phages should be filter-sterilized by passing through a 0.45-μm pore-size nitrocellulose filter.
10. Titer this lysate as described above and store at 4°C. Lysates may have titers as high as $10^{10}$–$10^{14}$ PFU/ml. *Note*: Many phages do not freeze well, and in general, freezing of lysates should be avoided.

### *Alternate Method (Liquid Method)*

This method allows phage-host interaction and phage propagation to take place in a liquid environment. It is particularly suited for propagation following plaque purification of a phage.

1. The method is essentially the same as the Soft-Agar Overlay Plate Method, except that, after the adsorption period, add 3–6 ml of rich medium in place of the top agar. Incubate this liquid culture overnight (with shaking when appropriate).
2. After incubation, add $CHCl_3$ (0.5 ml), and centrifuge the culture at 10,000 × g for 15 min to remove unlysed host cells.
3. Collect the supernatant fluid into a fresh, sterile tube. *Note*: Cultures may appear not to have grown because of significant lysis by phage infection.
4. Process this supernatant fluid in the same manner as for the Soft-Agar Overlay Plate Method.

## Plaque Purification of Phages

Often it is desirable to isolate a pure sample of the progeny of a single viral particle. In the plaque purification method, which is analogous to the preparation of a pure culture of bacteria, a single plaque is picked and its progeny propagated.

### Procedure

1. Prepare an overnight culture of the bacterial host used to enumerate the viruses in a sample.
2. Inoculate a 2-ml culture with 0.01 ml of this overnight culture.
3. Using the point of a sterile toothpick or inoculation needle, poke the center of the desired plaque and inoculate the 2-ml culture.
4. Incubate overnight.
5. Prepare lysate as described above.

### Notes

Unlike titering, plaques should be picked from a plate containing a minimum number of plaques. Plaques separated by as much as the entire diameter of a Petri dish have been shown to be capable of contaminating each other. For best results, repeat plaque purification at least twice to reduce the potential for contamination to a minimum!

## Direct Enumeration of Bacteriophage Particles (Electron Microscopy)

Direct observation of viral particles is best done using electron microscopy. These methods allow determination of the total number of particles in a sample and are not dependent on identifying a susceptible host bacterium. There are numerous methods for the preparation of samples for electron microscopy [3–5, 7, 18, 21, 29], and the development of specific methods is best done in consultation with a professional electron microscopist. However, two standard methods are presented that are adaptable to many situations. For an excellent in-depth presentation of staining and visualization methods for viruses, see Hayat and Miller [8].

### Procedure: Method I

1. Prepare a purified preparation of phages from a lysate (see Identification of Lysogens) in 10 mM ammonium acetate containing 1 mM $MgCl_2$, pH 7.0.
2. Mix this solution with an equal volume of 2% sodium phosphotungstate (pH 7.0), and spray with a nebulizer on a carbon-coated collodion film rendered hydrophilic by a 5- to 10-s, 5000-V glow discharge at 50 μm pressure.
3. Use this sample for electron microscopic examination under conditions of minimal beam exposure [29].

### Procedure: Method II

1. Deposit a purified phage preparation from a phage lysate (see Identification of Lysogens) as a small drop on the carbon-coated electron microscopic grid prepared as in Method I.
2. Draw off excess fluid by touching a piece of filter paper to the edge of the grid.
3. Then treat the grid's surface by adding a drop of freshly prepared 2% uranyl formate, and remove the excess liquid as before with a piece of filter paper.
4. Examine the grids in the electron microscope under conditions of minimal beam exposure [29].

## Identification of Sensitive Hosts (Cross-Streak Test)

The ability of a virus to infect a host cell is dependent on the presence of a unique receptor on that host [14, 17]. Therefore, some viruses have a broad host range if the receptor is present in many species or genera of bacteria, whereas others are limited to only some strains of a specific species if the receptor is not essential for growth and can be lost through mutation. In addition, expression of receptors may be regulated and depend on the growth conditions of the host. For instance the receptor for *Escherichia coli* phage λ is the inducible maltose receptor, which is only expressed if the host bacterium is grown in the presence of maltose. Therefore, it is often wise to apply a simple test for sensitivity of a bacterium to a specific virus before it is used in extensive studies. The easiest of these tests is the cross-streak test.

### Procedure

1. Grow isolated colonies of the bacteria to be tested on agar medium and under the culture conditions to be employed in future experiments.
2. Freshly prepare thick 7.5% agar plates of the culture medium.

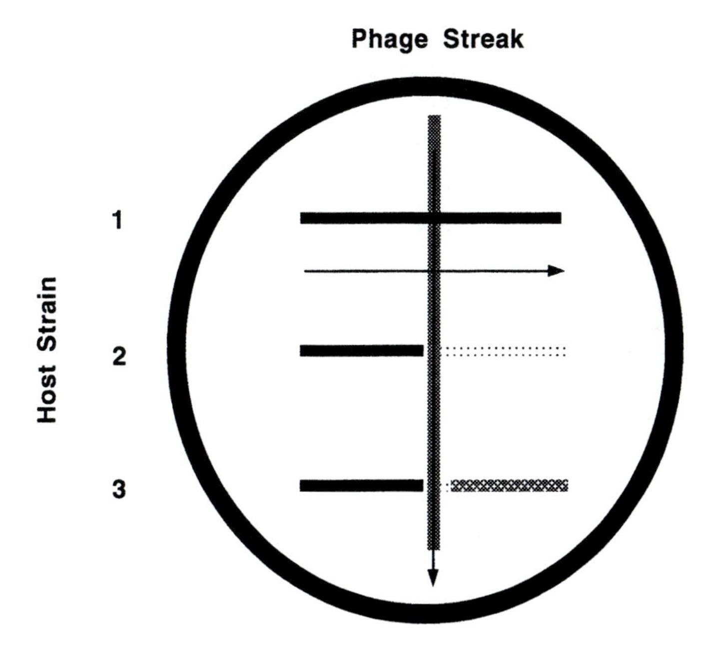


Figure 9-1 Cross-streak plate showing response of sensitive and resistant bacteria to infection by a bacteriophage. *Streak 1*, Resistant or lysogenic strain. *Streak 2*, Sensitive strain infected by a virulent phage. *Streak 3*, Sensitive strain infected by a temperate phage. Colonies arising to the right of the phage streak are lysogens.

3. Extract 50 μl of a high-titer lysate of the phage to be investigated using an automatic micro-pipetter.
4. Using the pipetter, deliver the 50 μl of phage solution to the agar plate as a streak running from the top to the bottom of the Petri dish (see Fig. 9-1). Allow the streak to dry thoroughly before use.
5. Using a sterilized toothpick, select a colony of the bacterial strain to be tested from the plate grown in Step 1, and cross-streak it across the phage streak from left to right at right angles to the streak of phage (see Fig. 9-1). Cross the phage streak **only once**. As the bacteria pass over the phage streak, the bacteria on the toothpick will become inoculated with the phage, and reduced growth of the streak will be observed to the right of the phage streak.

Notes

Resistant strains, which lack the viral receptor, will show no reduction of growth (streak 1). The amount of reduction of growth in sensitive strains will depend on the titer of virus particles in the phage streak and on whether the virus is virulent (streak 2) or temperate (streak 3). Virulent phages will show little or no survivors (colonies) to the right of the infection streak. Temperate phages will allow significant growth (see Fig. 9-1) due to the growth of lysogens. Lysogens will look like resistant bacteria when cross-streaked against the lysogenizing phage.

## Identification of Lysogens

Many bacteriophages are temperate and can establish a latent infection of their host termed *lysogeny* [11, 14, 17, 19]. In this state, the viral genome is known as a prophage, and the host is said to be lysogenized or a lysogen. The viral genome is carried in a quiescent state in the host either integrated into the bacterial chromosome or autonomously replicating as a plasmid. In either case, the viral genes are, in general, not expressed, and the bacterial host grows apparently normally. Many environmental isolates have been shown to be lysogenic for one or more bacteriophages [15, 17, 19]. Preliminary identification of lysogens can be made by a modification of the cross-streak test.

When cross-streaked against a phage for which they are lysogenic, bacteria appear as if they are resistant to infection (Fig. 9-1). This is because the molecular mechanisms that keep the prophage from being expressed inhibit expression of the superinfecting phage genomes. This condition is called immunity since the receptor is still present and adsorption of superinfecting virions can still take place. Therefore, any strain that appears resistant to a temperate phage may, in fact, be a lysogen and should be tested for lysogeny.

Populations (cultures) of a lysogen constantly release low levels of virions due to the spontaneous induction of the prophages harbored in a small fraction of individual cells. Therefore, lysogeny can be tested by testing for the presence of these virions.

Procedure

1. Streak a colony of a strain known to be *sensitive* to the phage from the top to bottom of a Petri dish containing the appropriate agar medium using a toothpick or inoculating loop. Best results are usually obtained by using a loop-full of an exponentially growing liquid culture of the host bacterium.
2. Allow the streak to air-dry completely before use.
3. Pick colonies of the suspected lysogens with a toothpick as described in the Cross-Streak Test, and cross-streak perpendicularly to the streak of the sensitive strain.

Notes

Virions released from the lysogen will infect the sensitive strain, and a zone of killing will be observed at the intersection of the lysogen and the sensitive strain. Because this is often a subjective observation, it is imperative to streak known sensitive, resistant, and lysogenic control strains on each plate whenever possible.

## Induction of Prophages Contained in Lysogenic Bacteria

Many natural and artificial forms of DNA-damaging stress activate prophages to express their genetic information and produce progeny virions with accompanying cytopathic effects to the host bacterium [14]. Induction of prophages can be accomplished by treating lysogens with various agents. The most commonly used agents in the laboratory are ultraviolet (UV) light and various antimicrobials, such as mitomycin C or nalidixic acid, that target nucleic acid-related processes. Methods for the induction of lysogens with each of these classes of stressor are presented here.

Procedure: UV Irradiation

1. Incubate cultures (approximately 15 ml) of the lysogen to be induced in a rich medium, such as Luria broth [16] or nutrient broth, at an appropriate temperature until they reach the early exponential phase. It is important that

the culture not be too dense or the large number of cells will act to shade each other from UV irradiation and reduce its effectiveness.

2. Centrifuge the culture to pellet the cells, and wash them twice with saline or other minimal salts solution. Washing removes the rich medium, as aromatic compounds present in this medium will act to absorb the UV light and protect the cells from inducing damage.
3. Perform all operations under reduced light conditions or in the dark to protect from photoreactivation and the reversal of activation [11, 14]. A GBX-2 safelight can be used to provide illumination to the work area.
4. Place the cells, suspended in the minimal salts solution, in a sterile Petri dish, and expose them to 256 nm UV radiation. As the dose ($J/m^2$) required for maximal induction and phage production varies significantly with the prophage and species of bacterium, use various doses initially. (Maximal induction will most likely take place in the range of between 1 and 100 $J/m^2$).
5. After each dose, withdraw a 1-ml sample from the dish and add to 4 ml of rich medium to allow growth of the induced bacteriophages. Since some spontaneous induction of prophages will occur under almost all conditions, withdraw an unexposed control sample at the beginning of the experiment and analyze it to determine the uninduced levels of virions in the lysate.
6. Incubate samples with shaking (if appropriate) in the dark for 2–3 h. This is done to allow expression of viral genes and the production of virions.
7. After this expression period, centrifuge the samples to remove unlysed cells, and collect the supernatant fluid (containing the virions) in a fresh, sterile tube. This lysate should be handled, titered, and stored in the same way as described for other lysates (i.e., stored over $CHCl_3$ or filter-sterilized).

Procedure: Induction Using DNA-Damaging Antimicrobial Agents

1. Inoculate a culture of appropriate growth medium with the lysogenic strain to be induced.
2. Grow to the early exponential phase.
3. Divide the culture into 1.0-ml aliquots in foil-covered tubes (to protect from photoreactivation of drug-induced DNA damage).
4. Add mitomycin C or nalidixic acid at various concentrations to the series of culture aliquots. Concentrations of antimicrobials vary depending on the sensitivity of the host to the drug and should range between 0.001 and 10 μg/ml for mitomycin C and between 1 and 1000 μg/ml for nalidixic acid. Include a no-drug control to allow estimation of the non-induced concentrations of virions in the culture. (*Note*: Extreme caution should be used when employing mitomycin C. It has been shown to be mutagenic and a possible carcinogen.)
5. Incubate cultures for 2–3 h in the dark.
6. After incubation, centrifuge the lysates to remove unlysed cells and collect the supernatant fluid (containing the virions). These lysates should be handled, titered, and stored in the same way as described for other lysates (stored over $CHCl_3$ or filter-sterilized).

## GENETIC METHODS: TRANSDUCTIONAL ANALYSIS

Many viruses mistakenly package host DNA elements (either chromosomal or plasmid DNA) in a few capsids during the maturation process of virions to form transducing particles (see [11] or [14] for a complete review). This is a rare event that usually only occurs approximately once in the production of every $10^6$ virions. The DNA packaged is approximately the same molecular size as the viral genome. Hence, for an average bacterium, between 1% and 3% of the genome can be packaged in one transducing particle. Anywhere from 30–200 kilobasepairs of DNA may be packaged depending on the specific phage that is being tested.

Most bacteriophages that carry out this form of generalized transduction do not package viral DNA in transducing particles. However, the attachment and absorption process of viruses are properties of the protein capsid, and transducing particles can inject the DNA they contain into new bacterial cells sensitive to the parental virus. Thus, transducing particles do not, in general, cause pathologic events to occur in the recipient bacterium, but do allow the introduction of new DNA with potentially new genetic determinants into the bacterial cell.

Transduction is a very useful technique for the fine structure mapping of bacterial chromosomes and the transfer of conjugation-deficient plasmids [23]. Because only 1% to 3% of the bacterial genome is packaged into a transducing particle, only closely linked loci can be inherited together. The process has also become important in evaluating the mechanism of genetic diversity and evolution in natural populations of bacteria [23, 26].

## Preparation and Titering of Transducing Lysates

A transducing lysate is simply a phage stock that has been grown on the appropriate genetic donor from which the investigator wishes to obtain the genetic locus or loci to be transferred into the recipient strain.

### Procedure

1. Prepare lysates using either the soft-agar overlay plate or liquid method. Use the genetic donor strain as the host strain for the production of this transducing lysate.
2. Titer the lysate as described in the procedure, Enumeration of Bacteriophages (p. 219).
3. If the source of the phages used to prepare this stock was a lysate grown on a genetically different host strain, repeat the process using the lysate prepared in Step 1 as the source of the phage particles. Repetition ensures that the final lysate contains transducing particles carrying genetic markers derived from the desired donor host and not from earlier hosts.
4. After titering, use this lysate for transduction experiments.

## Generalized Transduction

### Procedure

1. Grow a 10-ml culture of the recipient bacterium in rich medium to the mid-exponential phase. (*Note*: Preliminary experiments should be done to determine the approximate viable count of colony-forming units [CFU]/ml obtained at this phase of growth.) Spin down the culture, and resuspend it in 1.0 ml of the same medium or an appropriate adsorption medium for the transducing phage.
2. Mix together 0.1 ml of bacteria and 0.1 ml of an appropriate dilution of the transducing lysate (see above) to obtain an MOI of 0.1. This is a reasonable MOI to ensure maximal infection with minimal co-infection of transducing

particles or virions. However, some viruses may give higher transduction frequencies at other MOIs. Therefore, do preliminary experiments using a variety of MOIs to determine the best ratio for the specific system under study.
3. Incubate for 10–20 min to allow adsorption of the transducing particles.
4. Pellet the cells in a table-top centrifuge for 5–10 min at 10,000 × g, and wash them twice with minimal salts solution to remove unabsorbed phages and transducing particles.
5. Suspend the cells in 0.1–0.2 ml of minimal salts, and plate on various selective agar media plates developed to be selective for the genetic determinants under study. (*Note*: Plates should be poured thick and used fresh to increase the recovery of transductants.)
6. Incubate plates for at least 2 days before counting the transduced colonies. Slower-growing species may require extended incubation.
7. Also plate as controls the recipient cells that have not been exposed to the transducing lysates, as well as samples of the phage lysate. This is done to estimate the spontaneous reversion rates of the genetic loci and to establish the bacterial sterility of the transducing lysate, respectively. No growth should be observed on these plates. If it is, the frequency of these colonies should be subtracted from the frequency of colonies on the transduction plates to obtain the true frequency of transduction.

Transduction frequency is usually reported as transductants/PFU although transductants/recipient bacterium is sometimes used (especially in environmental studies).

## ENVIRONMENTAL METHODS

### Screening of Environmental Samples for the Presence of Bacteriophages (Most Probable Number Assay)

Often it is impossible or impractical to carry out a plaque assay on an environmental sample to determine the presence of bacteriophages infecting the host of interest. For instance, many aquatic and marine organisms may not be culturable on solid medium. In these cases most probable number (MPN) assays can be used to determine the concentration of lytic phages [27]. The assay is simple and sensitive with a detection limit of as little as 0.01 PFU/ml. It calls for the serial dilution of the environmental sample to be enumerated and exposure of the dilutions to the sensitive bacterium. The greatest dilution in which no growth (or growth followed by lysis) of the host organism can be observed is considered to contain at least one virus. By comparing the results to an MPN table [10], the number of infective units in the original sample can be estimated.

#### Procedure

1. Prepare 500 ml of host culture by transferring 50 ml of an exponentially growing culture into 450 ml of fresh medium.
2. As soon as exponential growth is observed, transfer 4 ml of culture into each of 55, 7-ml sterile, capped culture tubes.
3. While waiting for the culture to achieve exponential growth, filter 25 ml of the environmental sample to be analyzed through a 0.45-μm pore-size polycarbonate filter into a filter flask that has been prerinsed with filtrate

to remove bacteria and other non-viral contaminants. (*Note*: In some habitats, some bacteria may be small enough to pass through this filter. Be sure to incubate several tubes of the filtered sample to determine if growth is observable.)

4. Place 18 ml of culture medium into four 50-ml culture tubes or flasks.
5. To the first tube, add 2 ml of the filtered sample. Mix well.
6. Take 2 ml from the first tube ($10^{-1}$ dilution), and add it to the second tube ($10^{-2}$ dilution). Using a fresh pipette for each transfer, continue in this manner until the entire dilution series (up to a $10^{-4}$ dilution) has been completed.
7. Add 1 ml each of the undiluted sample to 10 of the culture tubes prepared in step 2.
8. Repeat Step 7 for each of the four dilutions.
9. Do not inoculate the final five tubes. They will serve as controls.
10. Observe daily for growth. This observation may include introducing the culture tubes into a spectrophotometer to measure turbidity or, in the case of phytoplankton, into a fluorimeter to measure chlorophyll fluorescence [27].
11. In most cases, cultures that have *not* grown (or lysed after growth) in 7 days are assumed to contain viruses. Record the number of such tubes in each dilution.
12. Express the data as the number of tubes from each dilution in which no growth or lysis occurred; for example, 10, 10, 5, 1, 0. Calculate the MPN of viruses in the original sample by using a published table, such as the one published by Koch [10], or with the assistance of computer programs designed for the purpose [9].

## Enrichment of Environmental Samples for Bacteriophages

Often the goal of analysis of an environmental sample is to isolate a bacteriophage capable of infecting a specific host bacterium. In these cases, enrichment for such viruses is appropriate. The enrichment of environmental samples for the presence of bacteriophages infecting a specific host can be carried out quite simply by inoculating environmental samples with the host organism and waiting for lysis to occur. These samples can then be treated as a "phage lysate" for further testing.

### Procedure

1. Filter the environmental sample to be enriched through a 0.45-μm pore-size polycarbonate filter into a sterile filter flask.
2. Place 5 ml of the sample in a sterilized culture tube.
3. Add 5 ml of an appropriate medium to support growth of the host organism.
4. Inoculate this mixture with 0.1 ml of an exponentially growing culture of the host organism.
5. Incubate under appropriate conditions. Incubation may have to be extended because of slow generation times of hosts or long latent periods of viruses, and the culture should be sampled daily. Growth followed by lysis may be observed.
6. Analyze daily samples for the presence of phages. This analysis can be done by plate assay to determine the titer of PFUs, electron microscopy to obtain the total number of virus-like particles, MPN analysis for bacterial hosts that are difficult to culture on solid medium, or other methods.

## Ultrafiltration of Viruses to Increase the Probability of Host-Virus Interaction

Although environmental samples, especially from aquatic environments, contain high levels of bacteriophages, viruses capable of infecting a particular host may be low in number. The natural virus community can be concentrated by ultracentrifugation [27]. This process increases the concentration of any minority population and thus the probability of specific host-viral interaction. It can be used in conjunction with the enrichment method used to isolate host-specific viruses from nature. However, ultrafiltration is expensive and time-consuming.

### Procedure

1. Dispense 20–100 liters of water sample into a pressure vessel. (*Note*: These vessels can often be purchased or rented from soft-drink wholesalers.)
2. Pressure-filter the water at <130 mm Hg through a 142-mm diameter glass fiber (1.2-μm nominal pore size) and a "Durapore" membrane (either 0.45-μm or 0.22-μm pore size, Millipore) filter. (*Note*: 0.45-μm filter may let many environmental bacteria through and will make ultrafiltration much slower. However, the 0.22-μm pore size will retard *many* of the larger complex bacteriophages and will certainly eliminate many of the algal viruses from the sample. Filter selection should be made with these characteristics and limitations in mind and will depend on the primary goal of the study.)
3. Concentrate the filtrate 100- to 1000-fold using an ultrafiltration cartridge (30,000 MW; Amicon). (*Note*: At a back-pressure of 1000 mm Hg, flow rates ≈850 ml/min.)
4. Use the concentrated sample in various analysis or enrichment procedures described elsewhere in this chapter.

## Ultracentrifugational Concentration Method for Observation of Naturally Occurring Bacteriophages in Environmental Samples by Electron Microscopy (EM)

Environmental samples from aquatic habitats often contain high concentrations of bacteriophages. However, they may not be of high enough concentration for critical EM examination. Several concentration methods have been developed to increase the number of phage particles for grids for EM. These methods have been developed by several laboratories, and several excellent references on the subject exist [2–4, 27].

One disadvantage to this type of analysis is that it is strictly physical (i.e., biologic activity is not measured) so that the infectivity and even the identity of the particles must be inferred. Therefore, it has become a standard practice to refer to these entities as "virus-like particles" (VLPs). The method described here has been adapted from Suttle [27] and relies on filtration and ultracentrifugation for the concentration of samples for EM examination.

### Procedure

1. Collect and prepare water samples as quickly as possible. Remember that biologic processes including phage-host interactions will continue in unprocessed samples and may influence the concentration of VLPs. Treat samples with 1–2% glutaraldehyde (EM grade), and store at 4°C in the dark. Aggregates of viruses can often be disrupted by the addition of a

surfactant, such as Tween 80, at low concentrations (≈10 μg/ml). This treatment may also be helpful in detaching viruses from particulate matter.

2. Prepare grid platforms specifically adapted for the EM grids and centrifuge tubes to support the grids during centrifugation. These platforms need to be sturdy, but not so dense as to compromise the safety of the rotor. Suttle [27] suggests making them from epoxy resin or machining them from a hard plastic, such as Plexiglas. Include a threaded hole in the platform so that it can be removed from the centrifuge tube. A small rod can then be inserted into the hole so that it can be pulled from the tube after the experiment. The surface of the platform should have a grid-size depression to hold the grid. A small notch in the surface surrounding the grid will facilitate its removal from the support. The platform should sit snugly into the bottom of the tube.
3. Prepare or purchase carbon-coated Formvar copper EM grids, and place them, carbon side up, on the grid platforms in an appropriate swinging-bucket ultracentrifuge tube. Treat the surface of the grid films by exposure to UV radiation before loading. Doing so will equalize the hydrophilicity of the grids and facilitate more uniform distribution of VLPs on them.
4. Run samples for a period of time and at a speed that will allow 100% efficiency of sedimentation of a particle size of 80S. This will ensure that small virus particles are sedimented. The time of centrifugation for a specific rotor and speed run at 20°C can be calculated by this formula [27]:

$$T = \frac{\left(\frac{1}{s}\right)\left(\ln\left[\frac{r_{max}}{r_{min}}\right]\right)}{60\omega^2}$$

where T is the time of centrifugation in minutes, s is the sedimentation coefficient in seconds (an 80S particle has a sedimentation coefficient of $80 \times 10^{-13}$ s), $r_{min}$ is the distance in centimeters from the center of the rotor to the top of the sample in the centrifuge tube, $r_{max}$ is the distance to the EM grid, and $\omega$ is the angular velocity in radians and is equal to $0.10472 \times$ rpm. (*Note*: Sedimentation time will vary as a function of salinity and temperature. Suttle [27] recommends increasing sedimentation time by 12% for solutions containing 3.5% salt and 25% for each 5°C below 20°C.)
5. After centrifugation, remove the grids with forceps from the platforms, and wick off any excess water with a piece of filter paper.
6. Take special care to remove NaCl and other salts from the grid before viewing. Any remaining salts can be removed by floating the grids, sample-side down, on several drops of distilled water. Repeat this step several times. Wick off excess water after each round. (*Note*: The grid should *never* be allowed to dry.)
7. Stain samples using one of the methods described in this chapter. For enumeration of VLPs, most investigators feel that better results are obtained with a positive stain, such as uranyl acetate, than with a negatively stained preparation since the negative stain is often deposited unevenly, making it difficult to count all of the VLPs present [27].
8. For enumeration, scan grids to confirm that VLPs are distributed randomly. In addition, preliminary inspection should ensure that a minimum number of large, non-VLP particles should be present or they may interfere with the enumeration. A large enough number of VLPs should be present to ensure statistical accuracy. Usually this requires counting a minimum of 20 fields containing at least 200 VLPs [27].

9. To determine the number of particles in the original sample, first calculate a taper correction factor ($S_f$) [27]. This correction factor is necessary because particles do not sediment in a straight line, but tend to be thrown against the sides of the tube as they sediment. Calculate the taper correction factor from this formula [13]:

$$S_f = [0.5\ r_{max} - (0.5\ r_{min}^2/r_{max})]/(r_{max} - r_{min})$$

10. After calculation of $S_f$, use the following formula to determine the concentration of VLPs in the original sample [27]:

$$\text{VLP's/ml} = \frac{P_f}{(A_f)(H)(S_f)}$$

where $P_f$ is the average number of particles counted in a microscopic field, $A_f$ is the area of the grid represented by the field, and H is the height of the sample in centimeters (i.e., $r_{max} - r_{min}$).

## DNA Probing for Analysis of Bacteriophage Distribution Patterns

DNA hybridization techniques have been applied to the detection of viral DNA sequences in natural environmental samples [20]. They have been used to analyze free virus particles in the sample and to estimate the degree of lysogenization of naturally occurring bacteria by bacteriophages. These techniques have the disadvantage of requiring that the virus or virus community be isolated in the laboratory so that a suitable DNA probe can be prepared. Hence, only known viruses or mixtures of viruses can be analyzed. However, the techniques are powerful ones that can be used to determine the distribution of a specific viral species or population in a natural habitat. The presence of as few as $10^3$ total virus particles or $10^6$ lysogenic cells in the entire sample can be detected by this method [20].

### *Preparation of the DNA Probe*

This method requires that a specific DNA probe be prepared for the genome(s) of the virus(s) to be studied. Both probes to specific viruses and to mixtures of viruses isolated from a specific habitat can be produced and are useful for analysis of environmental samples for the distribution of bacteriophages.

#### Procedure

1. Use lysates of an isolated phage to produce probes specific for one species of bacteriophage, or use the viral community present in an environmental sample to produce a community probe. It is often advisable to concentrate the viruses in aquatic samples by the ultrafiltration method before preparing DNA. The host-species enrichment methods can also be used to enrich the sample for host-specific viruses and hence enrich the community DNA probe for sequences specific to viruses infecting a unique bacterial host.
2. Extract and purify DNA from the virions using standard methods for the extraction of viral DNAs. These methods have been summarized elsewhere [25].
3. After extraction, purify phage DNA by CsCl density-equilibrium ultracentrifugation [22].
4. Dialyze the CsCl-purified DNA against Tris-EDTA-sodium acetate buffer, pH 7.5.

5. To generate the radiolabeled phage DNA probe, nick-translate between 0.5 and 1 μg of DNA restricted with an appropriate restriction enzyme, such as *Sal*I, using $^{32}$P-dCTP as a source of radioactive isotopes [22]. Determine the specific activity of the radiolabeled DNA by subjecting a small fraction of the probe to analysis in a scintillation counter. The specific activity should be greater or equal to $10^8$ dpm/μg DNA.
6. Store the probe at −70°C for one to two half-lives. (*Note*: New methods and commercial kits for the production of non-radiolabeled probes have recently come on the market. Particularly useful are those that produce a chemiluminescent probe. These probes have the advantage of being safer to use and of having a significantly longer shelf-life than radiolabeled probes.)

### *Use of Phage Genomic Probes to Detect Viral Particles and Lysogenic Bacteria in Environmental Samples*

This method allows the direct analysis of samples from various habitats. By employing differential centrifugation and filtration methods described elsewhere in this chapter, separate analysis of cell- and virion-free viral DNA, intact virions, and lysogenized bacterial host can be carried out.

#### Procedure

1. After preparation of the environmental sample by concentration or differential separation by sedimentation or filtration, immobilize the free DNA, virions, or lysogenized bacteria on nylon membrane filters (e.g., "Biotrans" available from ICN Biomedicals) using a dot-blot filtration manifold. Then denature, neutralize, and permanently fix the DNA from the immobilized samples according to the method of Silhavy et al. [25].
2. Use these membranes in DNA hybridization experiments employing the radiolabeled probes prepared in the previous procedure. Hybridization methods [22] vary according to the G+C content of specific probes and the membrane used. If the G+C content of the phage is unknown, it is usually best to select conditions appropriate to the G+C content of the host bacterium, because viral DNA *usually* reflects the composition of the host's DNA in this respect [18]. Specific methods are usually available from the membrane manufacturer that maximize the sensitivity and efficiency of the membrane.

## Determination of the Frequency of Prophage-Containing Bacterial Isolates (i.e., Lysogens) by Colony Hybridization

An alternative method for evaluating and enumerating culturable lysogens in an environmental sample is colony hybridization analysis.

#### Procedure

1. Inoculate selective agar plates by spreading dilutions of the environmental sample. Incubate these plates to allow formation of bacterial colonies.
2. Replica-plate these plates onto nylon membrane discs, and hybridize to the phage-DNA probe prepared above.
3. Detect signals from colonies containing hybridizable DNA (presumptive lysogens) by autoradiography on X-ray films. (*Note*: Similar analysis is possible using chemiluminescent-labeled probes.)

4. Determine the frequency of presumed lysogenization by the probed viral prophage by determining the frequency of positive colonies among the total number of colonies on the plate. Again for statistical purposes, analyze at least 20 plates containing between 100–200 colonies each.
5. Further analyze those colonies testing positively for viral-specific DNA for release of virus particles and superinfection immunity. Pick inocula for the preparation of cultures for this analysis from the master plate used to replica-plate the hybridization membrane.

## Use of Plaque Hybridization to Determine the Frequency of Host-Specific Viruses

The frequency of a specific bacteriophage among viruses infecting a specific host species can be determined by plaque hybridization studies.

### Procedure

1. Prepare and analyze water samples for the titer of host-specific bacteriophages by the methods outlined above.
2. Once the titer of species-specific phages has been determined, prepare 20 plates of the appropriate dilution to give between 100–200 plaques/plate each.
3. Once the plaques develop, replica-plate the plates onto nylon membranes, and probe using the appropriate radiolabeled phage DNA.
4. Analyze signals from positive plaques by autoradiography on X-ray film.
5. Determine the frequency of the specific phage from which the DNA probe was generated among bacterial-species-specific phages by counting the fraction of plaques that test positively.
6. Isolate viruses from positive (homologous) or negative (heterologous) plaques by poking a sterilized toothpick into the middle of the plaque on the master replica plate and inoculating a liquid culture of the host bacterium.
7. After the appropriate growth period, prepare lysates of the phage as described above.

*Acknowledgments* This work was supported in part by cooperative agreement No. CR820060 from the Gulf Breeze Environmental Research Laboratory of the U.S. Environmental Protection Agency. The contents of this chapter do not necessarily reflect the views of the Environmental Protection Agency nor does mention of trade names or commercial products constitute endorsement or recommendation for use.

## REFERENCES

1. Arber, W., Enquist, L., Hohn, B., Murray, N.E., and Murray, K. 1983. Experimental methods for use with lambda. In Lambda II (Hendrix, R.W., Roberts, J.W., Stahl, R.W., and Weisberg, R.A., Eds.), pp. 433–458. Cold Spring Harbor Laboratory Press, Cold Spring Harbor, NY.
2. Berg, G. 1987. Methods for Recovering Viruses from the Environment. CRC Press, Boca Raton, FL.
3. Bergh, O., Borsheim, K.Y., Bratbak, G., and Heldal, M. 1989. High abundance of viruses found in aquatic environments. Nature 340: 467–468.

4. Borsheim, K.Y., Bratbak, G., and Heldal, M. 1990. Enumeration and biomass estimation of planktonic bacteria and viruses by transmission electron microscopy. Appl. Environ. Microbiol. 56: 352–356.
5. Bratbak, B., Heldal, M., Norland, S., and Thingstad, T.F. 1990. Viruses as partners in spring bloom microbial trophodynamics. Appl. Environ. Microbiol. 56: 1400–1405.
6. Goyal, S.M. 1987. Methods in phage ecology. In Phage Ecology (Goyal, S.M., Gerba, C.P., and Bitton, G., Eds.), pp. 267–287. John Wiley & Sons, New York.
7. Goyal, S.M., Gerba, C.P., and Bitton, G. 1987. Phage Ecology. John Wiley & Sons, New York.
8. Hayat, M.A. and Miller, S.E. 1990. Negative Staining. McGraw-Hill, New York.
9. Hurley, M.A. and Roscoe, M.E. 1983. Automated statistical analysis of microbial enumeration by dilution series. J. Appl. Bacteriol. 55: 159–164.
10. Koch, A.L. 1981. Growth measurements. In Manual of Methods for General Bacteriology (Gerhardt, P., Murray, R.G.E., Costilow, R.N., Hester, E.W., Wood, W.A., Drieg, N.R., and Phillips, G.B., Eds.), pp. 179–207. American Society for Microbiology, Washington, DC.
11. Kokjohn, T.A. 1989. Transduction: mechanism and potential for gene transfer in the environment. In Gene Transfer in the Environment (Levy, S.B. and Miller, R.V., Eds.), pp. 73–98. McGraw-Hill, New York.
12. Kokjohn, T.A. and Miller, R.V. 1992. Gene transfer in the environment: transduction. In Release of Genetically Engineered and Other Micro-Organisms (Fry, J.C. and Day, M.J., Eds.), pp. 54–81. Cambridge University Press, Cambridge.
13. Mathews, J. and Buthala, D.A. 1970. Centrifugal sedimentation of virus particles from electron microscopic counting. J. Virol. 5: 598–603.
14. Miller, R.V. 1992. Methods for evaluation transduction: an overview with environmental considerations. In Microbial Ecology: Principles, Methods, and Applications (Levin, M.A., Seidler, R.I., and Rogul, M., Eds.), pp. 229–251. McGraw-Hill, New York.
15. Miller, R.V. 1993. Genetic stability of genetically engineered microorganisms in the aquatic environment. In Aquatic Microbiology: An Ecological Approach (Ford, T.E., Ed.), pp. 483–511. Blackwell Scientific Publishers, Boston.
16. Miller, R.V. and Ku, C.-M.C. 1978. Characterization of *Pseudomonas aeruginosa* mutants deficient in the establishment of lysogeny. J. Bacteriol. 134: 875–883.
17. Miller, R.V. and Sayler, G.S. 1992. Bacteriophage-host interactions in aquatic systems. In Genetic Interactions among Microorganisms in the Natural Environment (Wellington, W.M.H. and van Elsas, J.D., Eds.), pp. 176–193. Pergamon Press, Oxford, United Kingdom.
18. Miller, R.V., Pemberton, J.M., and Richards, K.E. 1974. F116, D3, and G101: temperate bacteriophage of *Pseudomonas aeruginosa* PAO. Virology 59: 566–569.
19. Miller, R.V., Ripp, S., Replicon, J., Ogunseitan, O.A., and Kokjohn, T.A. 1992. Virus-mediated gene transfer in freshwater environments. In Gene Transfers and Environment (Gauthier, M.J., Ed.), pp. 51–62. Springer-Verlag, Berlin.
20. Ogunseitan, O.A., Sayler, G.S., and Miller, R.V. 1992. Application of DNA probes to analysis of bacteriophage distribution patterns in the environment. Appl. Environ. Microbiol. 58: 2046–2052.
21. Proctor, L.M. and Fuhrman, J.A. 1990. Viral mortality of marine bacteria and cyanobacteria. Nature 343: 60–62.
22. Sambrook, J., Fritsch, E.F., and Maniatis, T. 1989. Molecular Cloning: A Laboratory Manual, 2nd ed. Cold Spring Harbor Laboratory Press, Cold Spring Harbor, NY.
23. Saye, D.J. and Miller, R.V. 1989. Gene transfer in aquatic environments. In Gene Transfer in the Environment (Levy, S.B. and Miller, R.V., Eds.), pp. 223–254. McGraw-Hill, New York.
24. Seeley, N.D. and Primrose, S.B. 1980. The effect of temperature on the ecology of aquatic bacteriophages. J. Gen. Virol. 46: 87–95.
25. Silhavy, T.J., Berman, M.L., and Enquist, L.W. 1984. Experiments with Gene Fusions. Cold Spring Harbor Laboratory Press, Cold Spring Harbor, NY.
26. Stotzky, G. 1989. Gene transfer among bacteria in soil. In Gene Transfer in the Environment (Levy, S.B. and Miller, R.V., Eds.), pp. 165–222. McGraw-Hill, New York.

27. Suttle, C.A. 1993. Enumeration and isolation of viruses. In Aquatic Microbial Ecology (Kemp, P.F., Sherr, B.R., Sherr, E.B., and Cole, J.J., Eds.), pp. 121–134. Lewis Publishers, Boca Raton, FL.
28. Tolmach, L.J. 1957. Attachment and penetration of cells by viruses. Adv. Virus Res. 4: 63–110.
29. Williams, R.C. and Fisher, H.W. 1970. Electron microscopy of tobacco mosaic virus under conditions of minimal beam exposure. J. Mol. Biol. 52: 121–123.

# PART II

# ENVIRONMENTAL TECHNIQUES

# 10

J.K. FREDRICKSON
D.L. BALKWILL

# Sampling and Enumeration Techniques

Sampling of environmental media is one of the most important aspects of investigating the relationships between microorganisms and their environment, yet is often given little attention. Obtaining samples that are representative of a particular environment in a quantity that is sufficient to define the variability is a crucial component of microbial ecology studies. However, there are no set rules on the size and number of samples necessary to quantify microorganisms and their processes in environmental samples. Such decisions are dependent on numerous factors, including the specific environment being sampled, the types of microorganisms being quantified, the hypothesis being tested, the inherent heterogeneity of the environment being sampled, the resources available to the investigator for conducting the analyses, and the ease with which the environment can be sampled. Contrasting examples of the last factor are surface soils that are readily sampled using a hand auger versus subsurface sediments that require more sophisticated drilling and coring techniques. Or, sampling soils to determine lateral spatial variability using geostatistical approaches is relatively easy, whereas conducting similar analyses with deep aquifer or marine sediments is inherently more difficult.

Therefore, the methods used for sampling, not surprisingly, vary considerably depending on the environment to be sampled. There are, however, certain common-sense rules that should be taken into account regardless of the environment to be sampled: ensuring that tools used for sampling and containers for holding samples are clean and disinfected, sampling in a manner that ensures that representative samples are taken, maintaining samples to prevent or limit microbial and geochemical changes before analysis, preventing the contamination of samples with exogenous microorganisms, and processing samples in a timely manner to limit artifacts due to post-sampling changes. One example of how these general rules can be employed involves the sampling of anaerobic sediments. On exposure to the atmosphere, $O_2$ will begin to diffuse into anaerobic sediments, oxidizing reduced geochemical constituents, such as Fe(II) and $H_2S$, and inhibiting anaerobic microbial metabolism. In addition, exposure of some anaerobic bacteria to $O_2$, depending on their physiologic state, can be lethal. Therefore, approaches for sampling such environments should prevent or limit exposure to air. The use of anaerobic glovebags with $O_2$-free atmospheres and other types of anaerobic vessels can aid in maintaining the anaerobicity of environmental samples during sample processing and transport. It should be emphasized that practically any sampling endeavor can disturb the sample, causing changes in physical and chemical gradients that can affect its microbiologic properties. Therefore, it is highly desirable to employ sampling procedures that minimize such disturbances.

There has been increasing interest in sampling microorganisms from novel or extreme environments [9, 22, 26, 30] in order to define the ecology of microorganisms in such systems and to obtain novel microorganisms that have potential biotechnologic applications. Extremes in these environmental conditions are often of interest: pH, temperature, salinity, and pressure. Sampling of extremes in pH and salinity generally does not require special precautions, although such samples can be corrosive, requiring careful selection of sample containers. In contrast, sampling of high-pressure environments for the study of barophilic microorganisms can be a complex process requiring specialized sampling approaches and tools to maintain samples under high pressure. Some barophilic and thermophilic bacteria [22] can survive exposure to ambient surface atmospheric pressures and temperatures, but some obligate extremophiles may not survive even short exposures to ambient conditions at the earth's surface. In addition, if one is interested in assessing the in situ activities of such microorganisms, it is desirable to maintain samples as close to actual environmental conditions as possible.

Little consideration is often given to field sampling design. Several factors can constrain the number and variety of samples to be collected, including budget, staff resources, and inaccessibility of the environment to be sampled. Regardless, attention should be paid to environmental characteristics and anticipated variation in the properties of interest so that the proper number of samples can be collected to maximize the interpretation of results. For some applications it may be desirable to employ geostatistical approaches to evaluate the spatial variation in microbiologic properties. Traditionally, collection of samples for microbiologic analyses has assumed that variability in microbiologic properties is purely random. However, in many environments microbial properties are spatially correlated. Geostatistical approaches take advantage of this fact by quantifying spatial and intervariable correlations. Based on such analyses, properties at unsampled locations can be estimated.

Because the types of environments that microbiologists study vary tremendously, it is beyond the scope of this chapter to outline methods for sampling for all such environments. Therefore, this chapter focuses on generic guidelines for microbiologic sampling and on sampling approaches for select environments with which the authors are familiar.

## METHODS FOR SAMPLING ENVIRONMENTAL MEDIA

### Soils

Soils are typically heterogeneous in both the vertical and horizontal directions, but the changes in properties are more often gradational than abrupt. Unsurprisingly, variations in the physical and chemical properties of soil have a major impact on its microbiologic properties. Sampling soils for microbiologic analysis requires some prior knowledge or estimate of the expected variation in these properties so that the numbers of samples needed to obtain a given precision can be determined. Different approaches for systematic sampling of soil have been described elsewhere [35]. Soil sample size is dependent upon the specific analyses undertaken, but is typically 100 g or less.

Procedure

1. Identify the area and depth of soil at the site from which samples are to be obtained, and estimate the variability in microbiologic measurements to be made. Using this information, select a sampling plan and identify the numbers of samples needed to obtain the desired precision.
2. Select a hand soil auger or other soil-coring device based on the ample size required and the physical nature of the soil being sampled. Coring devices are generally preferable to a shovel-and-pail approach for soil sampling because they enable more precise sampling with less compaction. Before coring and in between the collection of different cores, thoroughly clean all surfaces of the sampling device that come into contact with the sample. (*Note*: It is generally not necessary to sterilize auger surfaces before sampling because of the relatively high numbers of microorganisms in soil; augers can be used with or without core liners.)
3. Core the soil to the desired depth, and place the sample in a sterile container or Whirl-pak® bag. It may be desirable to collect composite samples; in this case multiple cores can be placed in the same container.
4. Process samples as expeditiously as possible because changes in microbiologic properties can occur rapidly. Time is generally more critical for activity measurements than for enumerations, but populations shifts can occur rapidly after the disturbance caused by sampling. It is desirable to limit the exposure of samples to high temperatures. However, there is debate as to whether keeping samples on ice before processing is beneficial or not.

## Subsurface Sediments and Rocks

The sampling of subsurface sediments and rocks presents unique challenges compared with sampling surface soils, waters, and sediments. Due to its inaccessibility, sampling of the subsurface requires specialized technologies, including drilling and coring. Multiple approaches can be used for coring, depending on the depths and lithologies to be sampled. Hollow-stem auger drilling is the most common method used for shallow (i.e., less than 50 m) subsurface coring, whereas mud or air rotary drilling with wireline coring are the most common methods for sampling deeper strata. Cable-tool drilling has also been used successfully for sampling subsurface sediments less than 300 m in depth. The reader should refer to additional references for additional details regarding these methods and for the advantages and disadvantages of each [8, 10, 11, 12a, 28, 29, 34].

One of the principal concerns with subsurface sampling is the contamination of samples either by overlying soils or sediments or by the introduction of organisms and solutes associated with circulating drilling fluids. For many operations it is desirable to include microbial and chemical tracers in the drilling fluids in order to quantify the extent of contamination. Fluorescent latex microspheres added to drill muds or placed in the shoe of a core barrel (Fig. 10-1) [24] and the coliform bacteria [4] fortuitously present in them have been employed as microbial tracers. Ideally, it is best to use a microorganism with a distinguishable phenotype not found in the subsurface as a tracer. However, local regulations should be carefully checked to ensure that the use of live microorganisms as tracers is permitted. Careful paring to remove outer core material that has the greatest potential for being contaminated should be standard practice, especially

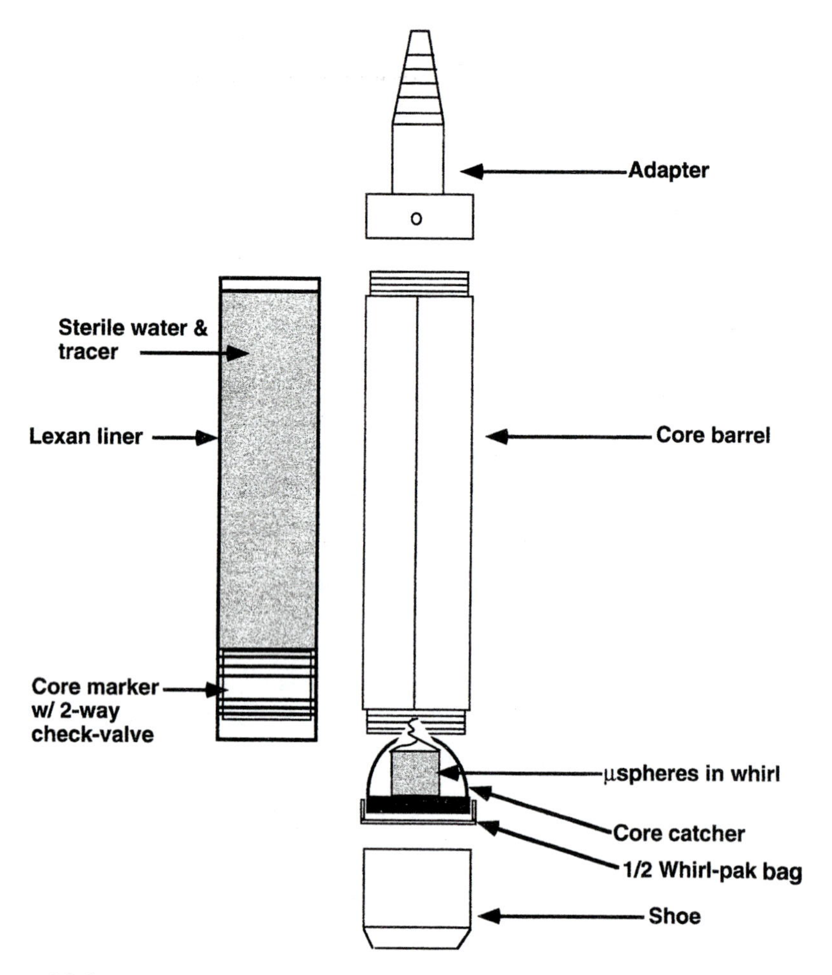


Figure 10-1 A split-spoon core barrel used for sampling subsurface sediment. Note that concentrated solute tracers such as LiBr can be placed in the inner liner in conjunction with a leak-proof core marker. Particulate tracers such as fluorescent latex microspheres can be placed in the shoe, the liner, or both.

for low-biomass subsurface environments, in which case it is desirable to limit even low levels of contamination during sampling.

When sampling chemically reduced subsurface sediments, it may be necessary to take precautions against geochemical oxidation of sediments and to prevent the inactivation of strict anaerobes. The use of Lexan or polyvinylchloride (PVC) core liners helps prevent the exposure of sediments to atmospheric $O_2$. As cores in intact core liners are removed from the core barrel, plastic caps can be placed over the ends of the cores to further limit exposure to the atmosphere. At this point the cores can be transferred into an anaerobic chamber containing an inert atmosphere, such as Ar or $N_2$, where they can be processed with little or no exposure to $O_2$. In general, it is not desirable to use $H_2$ as a component in the glovebag atmosphere as it is an excellent electron donor for microorganisms and hence could stimulate growth and metabolic activity in samples. However, if it is necessary to remove all $O_2$ from the glovebag, some type of catalyst must be employed, as even with extensive flushing of airlocks during sample exchange some $O_2$ will enter the chamber.

An additional concern regarding the sampling of subsurface materials for microbiologic analyses is that such samples seem to be particularly susceptible to perturbation by the sampling process itself. Typically, periods ranging from hours to days may elapse between the time that samples are collected and the time that analyses are initiated. Several studies have shown that significant changes in microbial population sizes and activities can occur in subsurface samples over relatively short time spans [6, 13, 17, 20]. Thus, it is important to limit the amount of time between coring and initiation of microbiologic analyses and experiments.

Wireline coring with circulating drill mud has been selected as the example method; however, many of the individual steps also apply to other types of subsurface sampling.

Procedure

1. Ensure that all surfaces that come into contact with subsurface sediments are clean and, where practical, disinfected. Steam clean all core barrels, drill rod, auger flights, shoes and the like before they are placed down the borehole. Autoclave core liners and end caps, or disinfect with a bleach solution.
2. If tracers are used, introduce them to the borehole immediately before coring. When using circulating fluids, inject directly into the fluids or mix in with the muds as they are made up to maintain a desired concentration. Bromide, as LiBr, and perfluorocarbons can be used as solute tracers. Microspheres can be used as bacterial tracers, but because of their prohibitive cost they are not usually mixed directly with large volumes of muds. However, microspheres can be introduced by placing them into a Whirl-pak bag that can then be placed into the shoe of the core barrel (see Fig. 10-1). As the core barrel contacts the formation, the bag is ruptured when the core is pushed through the prongs of the core catcher, and the microspheres are released.
3. Carefully assemble the core barrel with the liner as shown in Figure 10-1 and fill with sterile water. Add solute or microsphere tracers to the water at this point, if desired. Then send the core barrel to the bottom of the borehole where it engages with the drill rod and bit. The drill rod and bit are rotated to advance the borehole, and the cored rock or sediment fills the liner. After the desired depth has been reached, the core barrel is retrieved by a wireline.
4. After the core barrel is removed from the borehole, open the barrel and remove the inner core liner containing the core. Mark the depth of the core on the liner, and cap the ends to limit exposure of the core to air. Then place the capped core into a glovebag containing an inert atmosphere, such as Ar or $N_2$. This can be done at the site, or the enclosed core can be returned to the investigator's laboratory. (**Note:** It is generally not desirable to use a $H_2$-palladium system to scrub $O_2$ as $H_2$ can serve as an electron donor for many anaerobic as well as aerobic bacteria. However, if there is concern that strict anaerobes, such as methanogens, may be inactivated by trace quantities of $O_2$ it may be necessary to use an $O_2$ scrubbing system. Cores greater than 35 cm in length may require specially constructed airlocks. Cragg et al. [12] describe a simple method for maintaining samples for an extended period of time using nylon-aluminum-polyethylene laminated bags that can be heat-sealed and used to maintain cores anaerobically.)
5. Once the core is inside the glovebag, open the liner using a utility knife or a small battery-powered, hand-held saw. After breaching the liner, carefully pare the outer portion of the core with sterile tools or an extruding device to remove potentially contaminated outer core material. Then use the inner core material for microbiologic and geochemical analyses. Sample and analyze parings and inner core material for the various tracers.

## Groundwater

The need to obtain subsurface solids in order to accurately characterize subsurface microbial populations is generally accepted. The population density and community composition are generally different in groundwater samples than in subsurface sediments obtained from the same formation [18, 20], probably due to the selective attachment of some cells to the surface of solids. In addition, the process of drilling and developing a well results in the introduction of exogenous microorganisms and overall changes in the biogeochemical environment in and near the well [8, 15]. There are, however, situations where sampling groundwater for microbiologic analysis is desirable or where sampling solids is impractical. Such situations include analyzing for water quality indicators or for pathogens or monitoring changes in microbial populations and activities during bioremediation.

### Procedure

1. Disinfect or sterilize a submersible pump and tubing and place downhole; pump a minimum of three well volumes. Alternatively, if sampling from an artesian well, open the valve and purge for at least three well volumes. In addition to extensive purging for at least three well volumes, monitor Eh, pH, and the concentration of dissolved $O_2$ in the groundwater. These geochemical parameters are often quite different in the well bore than they are in the formation, and monitoring them until they stabilize helps ensure that one is sampling groundwater that is representative of the formation.
2. Collect water in a sterile container by placing the tube at the bottom of the container to minimize the mixing of water with air. Fill the container completely, leaving no headspace to limit exposure of the groundwater to $O_2$. This step prevents inactivation of $O_2$-sensitive anaerobic microorganisms and the oxidation of reduced chemical species. Seal the container tightly until analyses are initiated.

In uncontaminated, oligotrophic groundwater the population density of microorganisms is often quite low, requiring some type of concentration before analysis. Depending on the volume of water that is to be concentrated, various types of filtration systems can be used, ranging from simple membrane filtration systems that require a vacuum or pressure system for small volumes (i.e., milliliters to a few liters) to ultrafiltration systems that can concentrate cells from large volumes (hundreds to a few thousand liters) of water [14, 16]. Stevens et al. [31] used flow-through sand columns to study the microbiology of a deep, anaerobic basalt aquifer.

## Aquatic Environments

Microbiologic sampling of freshwater, marine, or estuarine habitats is beyond the scope of this chapter. However, an excellent review of methods for microbiologic sampling of the air-water interface, water column, and sediments has been conducted by Herbert [19].

# TOTAL COUNTING AND BIOMASS METHODS

A measurement that is fundamental to studying the microbial ecology of essentially any environment is the quantitative enumeration of microorganisms. Enumeration of general and specific groups of microorganisms and of total microbial biomass is important in defining how microorganisms interact with their surrounding environment. In addition, such enumerations often provide the necessary basis for interpreting microbial activity measurements. For bioremediation, enumeration of microorganisms is one attribute necessary for assessing the effectiveness of in situ stimulation and, ultimately, the success of bioremediation.

Although the enumeration of microorganisms in environmental samples seems to be relatively simple and straightforward, there are many difficulties, as evidenced by the effort that microbiologists have expended on developing and refining methods over the years. One problem inherent in many enumeration methods is how to release microorganisms that are attached to particles. Microorganisms in soils and sediments are often firmly attached to particle surfaces. This feature complicates their enumeration by such techniques as direct microscopy or viable plate counts. A variety of approaches have been used to promote the release of organisms from surfaces, including physical disruption by shaking or blending and the addition of surfactants or dispersing agents, such as sodium pyrophosphate. In addition to these difficulties, viable counts also suffer from inherent selectivity. Although a variety of media have been developed to optimize the number of heterotrophic microorganisms cultured from environmental samples, it has been estimated that fewer than 1% of the total population of microorganisms in soil is typically recovered by such methods. Although all methods for enumeration of microorganisms have inherent limitations, they remain useful tools in microbial ecology as long as the user is aware of their limitations and how they may affect the interpretation of results.

For the purposes of this chapter, methods for enumeration of microorganisms can be broadly categorized as those that directly enumerate total numbers of microorganisms, methods that estimate total microbial biomass, and viable counting methods.

## Dispersion-Separation-Filtration Method for Epifluorescence Microscopic Enumeration of Total Bacteria in Soils

This procedure was developed and is described in detail by P. J. Bottomley [5]. It is based on three major functional steps: (1) separation of bacteria from the soil structure by physical dispersion, (2) separation of bacteria from large mineral particles and organic debris by selective filtration, and (3) concentration of bacteria on membrane filters for staining with fluorescent dyes. Although designed for the direct enumeration of total bacteria in soils, the method should be applicable (perhaps with minor alterations) to aquatic sediments, subsurface materials, and the like.

### Procedure

1. Suspend quadruplicate 3-g samples of soil in separate 160-ml acid-washed screw-capped dilution bottles containing a monolayer of acid-washed 3-mm glass beads and 27 ml of filter-sterilized 0.15 M NaCl.
2. Shake the bottles vigorously by hand until the soil structure has been destroyed completely (typically 10–15 min).
3. Allow the soil suspension to settle for exactly 5 min.
4. Using a pipette, carefully recover 10 ml of the supernatant fluid above the sediment layer in each bottle, and transfer to 490 ml filter-sterilized deionized water (thereby creating a 1:500 final dilution of soil).
5. Filter a 50-ml portion of each dispersed and diluted soil sample through a separate 47-mm diameter, 8.0-μm pore-size polycarbonate membrane filter under −0.05 MPa of suction. *Note*: The membrane filter is placed in a

Buchner funnel connected to a vacuum pump via a side-arm flask, vacuum gauge, and moisture trap for this step.

6. Filter each 8.0-μm filtrate from Step 5 through a separate 47-mm diameter, 3.0-μm pore-size polycarbonate filter under −0.05 MPa of suction.
7. Add 0.1 ml of formalin to a 1.9-ml portion of each 3.0-μm filtrate from Step 6 (final concentration, 2% wt/vol).
8. Add 0.2 ml of filter-sterilized (0.2-μm pore size) acridine orange solution (0.1% wt/vol in 0.1 M citrate buffer, pH 6.6) to each formalin-fixed sample from Step 7.
9. Incubate in the dark at room temperature for 10 min.
10. Add 0.25 ml of each acridine-orange-treated sample (from Step 9) to 8.0 ml of filter-sterilized 0.15 M NaCl. (This is done to promote the uniform dispersion of bacterial cells onto the filter in the next step of the procedure.)
11. Filter each 8.25-ml sample from Step 10 through a blackened 25-mm diameter, 0.45-μm pore-size, polycarbonate filter under −0.05 MPa of suction. **Notes:** A filtration unit that makes use of a glass chimney-filter support is used for this step. Blackened filters, which must be used to reduce background fluorescence, can be purchased commercially.)
12. While the filters are still on the filter supports and under a slight vacuum, destain them by sequentially passing 20 ml portions of the following solutions through each filter.
    - prefiltered 0.1 M sodium citrate buffer, pH 6.6
    - prefiltered 0.1 M sodium citrate buffer, pH 5.5
    - prefiltered 0.1 M sodium citrate buffer, pH 4.0
    - deionized water
13. Prepare each filter for viewing as follows:
    - Place a drop of Cargille immersion oil (Type A or B) on a glass microscope slide.
    - Place the filter on top of the drop of immersion oil.
    - Apply a second drop of immersion oil to the upper side of the filter.
    - Apply a cover slip.
14. Examine the slides with an epifluorescence microscope equipped with a fluorescein filter set, and count the number of bacterial cells in 25 to 30 representative fields. **Notes:** A 100x, 1.25 NA, planachromat oil-immersion objective lens is recommended for prevention of eye fatigue caused by field distortions and edge effects. The procedure described above will provide 50 to 100 bacteria per field of view if the soil initially contains about $10^9$ cells/g. With soils containing a lower density of cells, the number of cells per field can be increased by filtering a larger volume of acridine-orange-stained suspension or reducing the extent of soil dilution before passage through the 8.0-μm filters (Step 5).
15. Calculate the number of bacteria per gram of oven-dry soil according to the extended equation (1), which can be reduced to equation (2):

$$N = \frac{n \times B \times 2.2 \times 500 \times (27 + X + Sv)}{a \times 0.25 \times 1.9 \times 10 \times Sw} \quad (1)$$

$$N = \frac{nB1100(27 + X + Sv)}{4.75a \times Sw} \quad (2)$$

where N = the number of bacteria per gram of oven-dry soil, n = the number of bacteria per field of view, B = the effective filtering area of the membrane filter ($\mu m^2$), a = the area of the microscopic field of view ($\mu m^2$), 0.25 = the volume (ml) of the acridine-orange-stained sample that was filtered for enumeration, 2.2 = the final volume (ml) of the 1.9-ml portion of the 500-fold dilution that was mixed with formalin (0.1 ml) and stained with acridine orange (0.2 ml), 500 = the total volume (ml) of the 500-fold dilution, 1.9 = the initial volume (ml) of the 500-fold dilution that was fixed for staining, 10 = the volume (ml) of the initial soil suspension that was used to make the 500-fold dilution, X = the volume of water (ml) in the 3-g soil sample, Sv = the volume (ml) of soil particles (assume a particle density of 2.7 g $cm^{-3}$), and Sw = the weight (g) of oven-dry soil.

## Substrate-Induced Respiration

Total microbial biomass in environmental samples can be determined by a variety of approaches in addition to direct microscopic counting. One approach pioneered by Anderson and Domsch [2] for determining the total microbial biomass in soil is based on measuring the initial respiratory response after the addition of a readily metabolizable substrate. This approach has been used mainly for soils and has been calibrated against other methods, such as direct microscopy, chloroform fumigation [23], and adenosine triphosphate (ATP) concentrations. Use of select microbial [1] inhibitors can be used to estimate relative contributions of bacteria and fungi to the total biomass. A major advantage of this method is that it is relatively simple and repeatable. Some of the disadvantages of this method are that it has been used mainly for soil, and considerable background work and cross-referencing with other methods are necessary before applying the method to sediment and water samples. However, Kieft and Rosacker [25] did use the glucose-induced respiration method for estimating total microbial biomass in subsurface sediments. Another confounding factor is that there is no "universal" substrate, and thus no one substrate, including glucose, will probably be used by all heterotrophic microorganisms within a given sample. The method described in this section is derived from the approach of Horwath and Paul [21]. The reader is encouraged to read this reference for additional information on glucose-induced respiration and other methods for measuring total microbial biomass in soil.

Procedure: Glucose-Induced Respiration in Soil

1. Determine the minimum glucose concentration required to give the maximum respiratory response for the soil(s) of interest. This value is typically between 5 and 400 $\mu$mol $g^{-1}$.
2. Weigh soil into clean, gas-tight containers. Serum bottles work well for this purpose. The size of the container depends upon the mass of soil used, which ranges from 10–100 g depending on the organic matter concentration. The higher the organic matter content, the less soil required.
3. Add sterile glucose solution to achieve the desired glucose concentration and water content. A water content of 50% of maximum water-holding capacity is a good range for most soils, but sometimes more water is added to ensure an even distribution of substrate throughout the soil. Between three and five replicates are usually adequate.
4. Crimp-seal serum bottles containing soils with butyl-rubber septa and incubate at 22°C. Measure $CO_2$ evolved as a function of time by gas chromatography, infrared gas analyzer, on other methods for quantifying $CO_2$. Typically, hourly measurements are made of $CO_2$ headspace concentrations.
5. Estimate microbial biomass C concentrations based on the work of Anderson and Domsch [2], correlating glucose-induced respiration to total biomass as determined by the chloroform fumigation method. Use the expression

$$x = 40.04y + 0.37 \quad (r^2 = 0.96)$$

where x = total microbial biomass C and y = maximum initial rate of $CO_2$ respiration (ml of $CO_2$ $g^{-1}$ dry weight of soil).

## Total Adenosine Triphosphate (ATP)

ATP, being a universal energy carrier in biologic systems, has been used as an estimate of total microbial biomass in soil and sediment [27]. Because the relationship between ATP content and cell biomass varies considerably depending on a variety of factors, including nutrient availability and environmental conditions, it has been difficult to consistently relate ATP content to total biomass [21]. Although wide ranges in biomass C/ATP ratios have been reported for soils, Brookes et al. [7] reported that biomass-ATP concentrations of approximately 10 μmol ATP $g^{-1}$ biomass C varied little, even on prolonged incubation of a grassland soil without substrate addition. Hence, the use of ATP content for measuring microbial biomass should be done with caution and only in conjunction with one or more alternative methods of biomass measurement. Advantages of using ATP as a measure of biomass include rapidity, direct measurement, and usefulness for comparison of samples from similar environments or of samples from the same environment upon which different treatments have been imposed.

### Procedure: Extraction and Measurement of ATP

1. Weigh soil or sediment samples (1–2 g wet weight) into triplicate sterile centrifuge tubes, and add 10 ml of extractant containing 670 mM $H_3PO_4$, 20 mM EDTA, 2 M urea, and 240 mM dimethylsulfoxide (DMSO). **Note:** Some researchers advocate adding a detergent, such as polyoxyethylene 10 lauryl ether at a final concentration of 50 g $l^{-1}$, to aid in removing cells from surfaces and promoting their lysis. Adenosine can also be added at a final concentration of 200 mg $l^{-1}$ to block ATP binding sites in the soil and improve extraction efficiencies.
2. Sonicate the samples on ice for 5 min, centrifuge at 12,000 × g for 10 min at 5°C, and collect the supernatant. At this point the extract can be analyzed immediately or frozen at −80°C and analyzed at a later time.
3. Quantify ATP concentration in the extracts by measuring the amount of light emitted from the luciferin-luciferase light reaction using a photometer or luminometer. In a reaction vessel suitable for the specific photometer to be used, add 50 μl of sample, 100 μl of firefly luciferase (available from Analytical Luminescence Laboratory or Sigma), and 50 μl of pH 7.8 Tricine buffer (containing 25 mM Tricine, 5 mM $MgSO_4$, 1 mM EDTA, and 1 mM dithiothreitol). Concentrations are expressed in terms of ρmol of ATP $g^{-1}$.
4. Determine the efficiency of extraction of ATP by spiking soil samples with 0.5 μmol ATP (available from Sigma), and then treating these spiked samples in a manner identical to that described in the procedure.

## Total Lipid Phosphate

Lipid phosphate is a major component of the plasma membrane of microbial cells and hence can be used as a measure of microbial biomass [32, 33]. However, there are several inherent problems associated with converting phospholipid concentration to cell numbers. The phospholipid content varies with organism type, growth conditions, and physiologic status of cells. Most conversion factors that have been developed are based on the phospholipid content of pure cultures. For some environments, conversion factors derived in this manner can provide a reasonably accurate measure of total cell numbers [3]. For additional discussion of phospholipid fatty acids (PLFA) and detailed methods for analyses, see Chapter 14.

Procedure: Microbial Lipid Phosphate from Sediment and Soil [33]

1. Weigh between 5–15 g of sediment or soil into a 250-ml separatory funnel containing a solvent mixture of 30 ml of 5 mM phosphate buffer (pH 7.4), 75 ml of anhydrous methanol, and 37.5 ml of chloroform.
2. Shake the suspension vigorously on a wrist-action shaker for 2 h. Add an additional 37.5 ml each of chloroform and sterile deionized water, and mix briefly. Allow the mixture to separate for 24 h.
3. Carefully remove the upper (aqueous/methanol) phase, and decant the chloroform layer, filtering through Whatman 2E filter paper. Transfer a portion of the chloroform phase to an acid-washed, 18 × 150 mm test tube, and remove the solvent using a stream of $N_2$ and heating the tube to 40°C.
4. Digest the dried lipid with 1.5 ml of 35% perchloric acid heated to 180–200°C for 2 h. Allow the mixture to reflux, with the vapor condensing approximately 5 cm above the heated base. Do not allow the mixture to go to dryness. Cool to room temperature.
5. Add 2.4 ml of molybdate reagent (4.4 g of ammonium molybdate, 14 ml of concentrated $H_2SO_4$ per liter of water) and mix. Then add 2.4 ml of ANSA (30 g of sodium bisulfite, 2 g sodium sulfite, 0.5 g of 1-amino-2 naphthol-4-sulfonic acid dissolved in 200 ml of water, stored in the dark at 4°C, and diluted 1:2 with water before use). Heat the mixture in a boiling water bath for 20 min, and then allow to cool to room temperature.
6. Read the absorbance of the mixture at 830 nm and compare to a standard phosphate curve prepared from phosphate—perchloric acid solutions treated in a manner identical to that in Step 5. *Note*: Additional portions of the solvent can be added to the test tubes and removed to increase the sensitivity. Samples should be analyzed in triplicate.

## Viable Count Methods

### *Blending Method for Dispensing Soil and Subsurface Sediment Samples before Enumeration of Viable Microorganisms by Plate Counting or MPN Determination*

This procedure is designed to produce a well-dispersed suspension of soil or sediment that can be diluted readily for inoculation of plates or most probable number (MPN) tubes. It has been found to work effectively on a wide variety of soil and sediment types. The method may also serve to release microbial cells from particle surfaces, but the extent to which it does so depends on the physical and chemical nature of the sample being processed.

Procedure

1. Aseptically transfer 10 g of sample material to a sterile blender head. **Note:** Metal "semimicro" (25–250 ml) blender heads (Eberbach 8580) are recommended because they blend 100-ml samples more effectively than do the larger heads that come with most blenders.
2. Add 95 ml of sterile 0.1% sodium pyrophosphate ($Na_2P_4O_7 \cdot 10H_2O$; pH 7.0) to the sample in the blender. This produces a 1:10 suspension, assuming approximately 50% pore space in the sample.

3. Blend the sample for 1 min (two 30-s bursts separated by a 30-s rest interval).
4. *Rapidly* transfer the blended material to a sterile container that can be stoppered tightly and shaken (e.g., a flask, bottle, or Fleaker™). *Note*: It may be necessary to pour the liquid back and forth rapidly between the blender head and sterile container in order to transfer as much of the solid material to the container as possible.

### *Procedure for Plate-Count Enumeration of Culturable Microorganisms in Soil and Subsurface Sediment Samples*

This is a generic procedure for spread-plating a well-dispersed suspension of soil or sediment. Various methods, including the blending approach described in this chapter, may be used to produce a well-dispersed suspension of soil or sediment. No specific nutrient medium is named in this procedure because the medium must be chosen according to the nature of the sample being examined.

#### Procedure

1. Prepare serial 10-fold dilutions of the blended sample in sterile phosphate-buffered saline (PBS) as follows:
   a. Thoroughly shake the dispersed (preferably 1:10) suspension of soil or sediment, and immediately (to minimize settling of solids) pipette 10 ml of the shaken suspension into a dilution bottle containing 90 ml sterile PBS, thereby producing a $10^{-2}$ dilution. *Note*: The use of 90-ml dilution blanks is preferred because they permit the use of a 10-ml inoculum at each dilution, thereby minimizing errors due to settling of heavy solids in the pipettes. Also, the use of relatively large (10-ml pipettes helps reduce clogging of the pipette tips with sand or silt.
   b. Thoroughly shake the dilution made in Step 1a, and immediately pipette 10 ml of the shaken dilution to a dilution bottle containing 90 ml sterile PBS, thereby producing a $10^{-3}$ dilution.
   c. Repeat Step 1b to produce the desired number of 10-fold dilutions. Dilutions from $10^{-1}$ (the original well-dispersed suspension) to $10^{-5}$ are sufficient for most subsurface sediment samples. However, soil samples may require $10^{-1}$ to $10^{-7}$ dilutions. Unnecessary dilutions can be eliminated by trial and error when similar samples are plated repeatedly.
2. Spread-plate each dilution on the desired nutrient medium in triplicate as follows:
   a. Thoroughly shake the dilution, and *immediately* pipette 0.1 ml of the shaken dilution (as an inoculum) to each of three nutrient plates. *Notes*: This should be done as quickly as possible in order to minimize errors caused by the settling of sands and other heavy solids in the pipette. The use of a 0.1-ml inoculum increases the effective dilution on the plates by a factor of 10. For example, a plate that is inoculated with 0.1 ml from a $10^{-2}$ dilution is actually counted as a $10^{-3}$ dilution plate.
   b. Using a sterile glass spreading rod, spread the inoculum over the surface of each plate as evenly as possible.
   c. Repeat Steps 2a and 2b until all of the dilutions have been plated.
3. Incubate the plates (inverted) aerobically, preferably at the ambient temperature of the environment from which the soil or sediment sample was taken.
4. Check the plates periodically for the development of microbial colonies; continue incubation until new colonies cease to appear. *Notes*: Most colonies

develop on concentrated or nutrient-rich media within 3 to 7 days; full colony development on dilute or nutrient-poor media may require 2 to 8 weeks. Extended incubation times may require that the plates be wrapped in parafilm or stored in plastic Petri dish sleeves to prevent excess drying of the agar.

5. Count the number of colonies on the plates.
6. Calculate the number of culturable microorganisms in the original sample by multiplying the average number or colonies on the plates from the "countable" dilution by the dilution factor. The countable dilution is the dilution that produces an average of 30 to 300 colonies per plate. For example,

| *Final dilution on plates* | *No. of colonies on triplicate spread plates* | *Avg. no. of colonies* |
|---|---|---|
| $10^{-2}$ | TNTC, TNTC, TNTC | TNTC |
| $10^{-3}$ | TNTC, TNTC, TNTC | TNTC |
| $10^{-4}$ | 418, 283, 322 | 341 |
| $10^{-5}$ | 37, 42, 29 | 36 |
| $10^{-6}$ | 4, 3, 0 | 2.3 |

(TNTC = too numerous to count)

Correct reported count = $36 \times 10^5 = 3.6 \times 10^6$ colony-forming units (CFU)/gram (wet wt).

**Note:** Results should be reported as CFU per g of sample, rather than the number of cells per gram, because one cannot be certain that each colony on the plate arose from a single cell.

## EXAMPLES OF MEDIA FOR PLATING SOIL AND SUBSURFACE SEDIMENT SAMPLES

### Soil Extract Agar

#### *Soil Extract Solution*

Procedure

1. Suspend 100 g soil (or subsurface sediment) in 200 ml tap water.
2. Autoclave for 1 h at 121°C.
3. Cool the autoclaved suspension to room temperature, and allow the solids to settle.
4. Pour off the fluid, and retain it (discard settled solids).
5. Centrifuge the fluid at approximately 3500 g for 10 min.
6. Pour off the supernatant fluid and retain it (discard pellet).
7. Freeze the supernatant fluid (to remove colloids).
8. Thaw the frozen fluid and filter through Whatman # 4 filter paper.
9. Add tap water to the filtered fluid to a final volume of 200 ml.
10. Store at 4°C until use (can also be autoclaved for long-term storage).

#### *Soil Extract Agar*

| | |
|---|---|
| Soil extract solution (above) | 50 ml |
| Tap water (or groundwater) | 950 ml |
| Agar | 15 g |

## PTYG Agar—Concentrated and Dilute Formulas

| | *PTYG agar* | *5% PTYG agar* | *1% PTYG agar* |
|---|---|---|---|
| Difco peptone | 5.0 g | 0.25 g | 0.05 g |
| Difco tryptone | 5.0 g | 0.25 g | 0.05 g |
| Difco yeast extract | 10.0 g | 0.5 g | 0.1 g |
| Glucose | 10.0 g | 0.5 g | 0.1 g |
| $MgSO_4 \cdot 7H_2O$ | 0.6 g | 0.6 g | 0.6 g |
| $CaCl_2 \cdot 2H_2O$ | 0.07 g | 0.07 g | 0.07 g |
| Agar | 15.0 g | 15.0 g | 15.0 g |
| Distilled water | 1.0 liter | 1.0 liter | 1.0 liter |

## REFERENCES


1. Anderson, J.P.E. and Domsch, K.H. 1973. Quantification of bacterial and fungal contributions to soil respiration. Arch. Mikrobiol. 93: 113–127.
2. Anderson, J.P.E. and Domsch, K.H. 1978. A physiological method for the quantitative measurement of microbial biomass in soils. Soil Biol. Biochem. 10: 215–221.
3. Balkwill, D.L., Leach, F.R., Wilson, J.T., McNabb, J.F., and White, D.C. 1988. Equivalence of microbial biomass measures based on membrane lipid and cell wall components, adenosine triphosphate, and direct counts in subsurface aquifer sediments. Microb. Ecol. 16: 73–84.
4. Beeman, R.E. and Suflita, J.M. 1989. Evaluation of deep subsurface sampling procedures using serendipitous microbial contaminants as tracer organisms. Geomicrobiol. J. 7: 223–233.
5. Bottomley, P.J. 1994. Light microscopic methods for studying soil microorganisms. In Methods of Soil Analysis. Part 2: Microbiological and Biochemical Properties. (Weaver, R.W., Angle, J.S., and Bottomly, P.S., Eds.), pp. 81–105. Soil Science Society of America, Madison, WI.
6. Brockman, F.J., Kieft, T.L., Fredrickson, J.K., Bjornstad, B.N., Li, S.W., Spangenburg, W., and Long, P.F. 1992. Microbiology of vadose zone paleosols in south-central Washington state. Microb. Ecol. 23: 279–301.
7. Brookes, P.C., Newcombe, A.D., and Jenkinson, D.S. 1987. Adenylate energy charge measurements in soil. Soil Biol. Biochem. 19: 211–217.
8. Chapelle, F.H. 1993. Ground-Water Microbiology & Geochemistry. John Wiley & Sons, New York.
9. Chastain, R.A. and Yayanos, A.A. 1991. Ultrastructural changes in an obligately barophilic marine bacterium after decompression. Appl. Environ. Microbiol. 57: 1489–1497.
10. Clark, R.R. 1988. A new continuous sampling wireline system for acquisition of uncontaminated minimally disturbed soil samples. Ground Water Monitor. Rev. 8: 66–72.
11. Colwell, F.S., Stormberg, G.J., Phelps, T.J., Birnbaum, S.A., Mckinley, J., Rawson, S.A., Veverka, C., Goodwin, S., Long, P.E., Russell, B.F., Garland, T., Thompson, D., Skinner, P., and Grover, S. 1992. Innovative techniques for collection of saturated and unsaturated subsurface basalts and sediments for microbiological characterization. J. Microbiol. Meth. 15: 279–292.
12. Cragg, B.A., Bale, S.J., and Parkes, R.J. 1992. A novel method for the transport and long-term storage of cultures and samples in an anaerobic atmosphere. Lett. Appl. Microbiol. 15: 125–128.

12a. Fredrickson, J.K. and Phelps, T.J. 1996. Subsurface drilling and sampling. In Manual of Environmental Microbiology (Hurst, C.J., Knudsen, G.R., McInerney, M.J., Stetzenbach, L.D., and Walters, M.V., Eds.), pp. 526–540. American Society for Microbiology, Washington, D.C.

13. Fredrickson, J.K., Li, S.W., Brockman, F.J., Haldeman, D.L., Amy, P.S., and Balkwill, D.L. 1995. Time-dependent changes in viable numbers and activities of aerobic heterotrophic bacteria in subsurface samples. J. Microbiol. Meth. 21: 253–265.
14. Fry, N.K., Fredrickson, J.K., Fishbain, S., Wagner, M., and Stahl, D.A. 1997. Population structure of microbial communities associated with two deep, anerobic, alkaline aquifers. Appl. Environ. Microbiol. 63: 1498–1504.
15. Ghiorse, W.C. and Wilson, J.T. 1988. Microbial ecology of the terrestrial subsurface. Adv. Appl. Microbiol. 33: 107–173.
16. Giovannoni, S.J., DeLong, E.F., Schmidt, T.M., and Pace, N.R. 1990. Tangential flow filtration and preliminary phylogenetic analysis of marine picoplankton. Appl. Environ. Microbiol. 56: 2572–2572.
17. Haldeman, D.L., Amy, P.S., White, D.C., and Ringelberg, D.B. 1994. Changes in bacteria recoverable from subsurface volcanic rock samples during storage at 4°C. Appl. Environ. Microbiol. 60: 2697–2703.
18. Hazen, T.C., Jimenez, L., de Victoria, G.L., and Fliermans, C.B. 1991. Comparison of bacteria from deep subsurface sediment and adjacent groundwater. Microb. Ecol. 22: 293–304.
19. Herbert, R.A. 1988. Sampling methods. In Methods in Aquatic Bacteriology (Austin, B., Ed.), pp. 3–25. John Wiley & Sons, New York, NY.
20. Hirsch, P. and Rades-Rohkohl, E. 1988. Some special problems in the determination of viable counts of groundwater microorganisms. Microb. Ecol. 16: 99–113.
21. Horwath, W.R. and Paul E.A. 1994. Microbial biomass. In Methods of Soil Analysis Part 2: Microbiological and Biochemical Properties (Weaver, R.W., Angle, J.S., and Bottomly, P.S., Eds.), pp. 753–773. Soil Science Society of America, Madison, WI.
22. Jannasch, H.W., Wirsen, C.O., Molyneaux, S.J., and Langworthy, T.A. 1992. Comparative physiological-studies on hyperthermophilic Archaea isolated from deep-sea hot vents with emphasis on *Pyrococcus* Strain Gb-D. Appl. Environ. Microbiol. 58: 3472–3481.
23. Jenkinson, D.S. and Powlson, D.S. 1976. The effect of biocidal treatments on metabolism in soil-I. Fumigation with chloroform. Soil Biol. Biochem. 8: 167–177.
24. Kieft, T.L., Fredrickson, J.K., McKinley, J.P., Bjornstad, B.N., Rawson, S.A., Phelps, T.J., Brockman, F.J., and Pfiffner, S.M. 1995. Microbiological comparisons within and across contiguous lacustrine, paleosol, and fluvial subsurface sediments. Appl. Environ. Microbiol. 61: 749–757.
25. Kieft, T.L. and Rosacker, L.L. 1991. Application of respiration- and adenylate-based soil microbiological assays to deep subsurface terrestrial sediments. Soil Biol. Biochem. 23: 563–568.
26. Norton, C.F., Mcgenity, T.J., and Grant, W.D. 1993. Archaeal halophiles (Halobacteria) from two British salt mines. J. Gen. Microbiol. 139: 1077–1081.
27. Oades, J.M. and Jenkinson, D.S. 1979. The adenosine triphosphate content of soil microbial biomass. Soil Biol. Biochem. 11: 210–204.
28. Phelps, T.J., Fliermans, C.B., Garland, T.R., Pfiffner, S.M., and White, D.C. 1989. Recovery of deep subsurface sediments for microbiological studies. J. Microbiol. Meth. 9: 267–280.
29. Russell, B.F., Phelps, T.J., Griffin, W.T., and Sargent, K.A. 1992. Procedures for sampling deep subsurface microbial communities in unconsolidated sediments. Ground Water Monitor. Rev. 12: 96–104.
30. Stetter, K.O., Huber, R., Blochl, E., Kurr, M., Eden, R.D., Fielder, M., Cash, H., and Vance, I. 1993. Hyperthermophilic Archaea are thriving in deep North Sea and Alaskan oil reservoirs. Nature 365: 743–745.
31. Stevens, T.O., McKinley, J.P., and Fredrickson, J.K. 1993. Bacteria associated with deep, alkaline, anaerobic groundwaters in southeast Washington. Microb. Ecol. 25: 35–50.
32. Tunlid, A. and White, D.C. 1992. Biochemical analysis of biomass, community structure, nutritional status, and metabolic activity of microbial communities in soil. In Soil Biochemistry (Statzky, G. and Bollag, J.-M., Eds.), pp. 229–262. Marcel Dekker, New York.

33. White, D.C., Davis, W.M., Nickels, J.S., King, J.D., and Bobbie, R.J. 1979. Determination of the sedimentary microbial biomass by extractable lipid phosphate. Oecologia 40: 51–62.
34. Wilson, J.T., McNabb, J.F., Balkwill, D.L., and Ghiorse, W.C. 1983. Enumeration and characterization of bacteria indigenous to a shallow water-table aquifer. Groundwater 21: 134–142.
35. Wollum, A.G. 1994. Soil sampling for microbiological analysis. In Methods of Soil Analysis. Part 2: Microbiological and Chemical Properties (Weaver, R.W., Angle, J.S., and Bottomly, P.S., Eds.), p. 81–105. Soil Science Society of America, Madison, WI.

# 11

DAVID C. WHITE
DAVID B. RINGELBERG

# Signature Lipid Biomarker Analysis

The classical microbiologic approach that was so successful in public health for the isolation and culture of pathogenic species is clearly not satisfactory for many environmental samples. Microbes may still be infectious and antigenic, even though they are not culturable. It has been repeatedly documented in the literature that viable or direct counts of bacteria from various environmental samples may represent only 0.1% to 10% of the extant community [12–15, 17]. Finally, classical microbial tests are time-consuming and provide little indication of the nutritional status or evidence of toxicity that can affect pathogenicity, and these tests do not give accurate estimates of microbial fragments or other components that can act as antigens or immune potentiators.

The signature lipid biomarker (SLB) assay described in this chapter does not depend on growth or morphology. Instead, the microbial biomass is determined in terms of universally distributed biomarkers that are characteristic of all cells. The extraction process provides both a purification and concentration of these lipid biomarkers.

The SLB method is based on the extraction of "signature" lipid biomarkers from the cell membranes and walls of microorganisms [4]. Lipids, which are recoverable by extraction in organic solvents, are an essential component of the membrane of all cells and play a role as storage materials. They can be extracted directly from a wide range of materials, such as ceiling tiles, soils, and filters. Even polycarbonate filters can be extracted using a modified one-phase hexane/isopropanol procedure [5].

## LIPID EXTRACTION

SLB analysis can be useful to the microbiologic researcher by quantitatively providing (1) an estimate of viable microbial biomass by measuring the amount of cellular membrane, (2) an outline of community structure by identifying signature biomarkers indicative of prokaryotic and eukaryotic taxa, and (3) an indication of microbial physiologic status by analyzing for known stress indicators. This procedure uses the single-phase chloroform:methanol:water extraction system of Bligh and Dyer [1] as modified by White et al. [16] to quantitatively extract the lipid soluble components from viable cells.

Samples to be analyzed must be stored at −20°C or fixed in buffered formalin until extraction. Ideally, samples are lyophilized and weighed before extraction. If this is not possible, wet weights are recorded. If the samples have been grown in culture, the medium should be centrifuged and the resulting cell pellet rinsed twice with 0.05 M phosphate buffer (pH 7.5) before lyophilization.

The apparatus needed for the extraction consists of glass separatory funnels with Teflon stopcocks and ground-glass apertures (see Note 1). The standard size is 250 ml, but smaller volumes require smaller separatory funnels. A glass test tube is taped under the stopcock drain to catch drips. Round-bottom flasks (of the size appropriate to the volume being reduced, usually 250 ml) with glass stoppers and cork ring seats are used to recover the lower organic phase.

The lipids are filtered and dried by passing the organic phase through Whatman #2$^{v}$ folded paper filters (12.5 or 18.5 cm) placed in glass funnels to fit round-bottom flasks. The lipid is recovered in a rotating glass solvent evaporator with a temperature-controlled water bath, which removes the solvent. It is transferred into Pyrex test tubes with Teflon-lined screw caps. Solvent is removed in a stream of nitrogen in a gas blow-down apparatus in test tubes with a temperature-controlled water bath. The lipids are stored under nitrogen in the dark at −20°C until processed further.

The reagents used in the extraction are organic solvents, which include chloroform and methanol. All solvents should be of the purest grade possible (Burdick and Jackson GC$^{2}$ or equivalent). An aliquot of each new lot of chromatography solvents should be concentrated by a suitable factor (e.g., 1000) and analyzed by capillary gas chromatography to be sure it meets the manufacturer's specifications for organic residue. The water is nanopure-filtered organic-free deionized distilled water. All distilled water should be chloroform-extracted using approximately 200 ml chloroform per 4 liters of distilled water. Any aqueous solution used in these analyses should be stored over chloroform in this same ratio. In the extraction 50 mM phosphate buffer, pH 7.4: phosphate buffer is made to a working concentration by dissolving 8.7 g $K_2HPO_4$ (dibasic) in 1 liter nanopure distilled water and adjusting to pH 7.4 with approximately 3.5 ml 6 N HCl. Phosphate buffer should be chloroform-extracted using 50 ml chloroform per 1 liter buffer.

## The Modified Bligh/Dyer (mB/D) First-Phase Solvent

Reagent

A combination of chloroform, methanol, and 50 mM phosphate buffer mixed to a volume ratio of methanol:chloroform:phosphate buffer, 2:1:0.8.

Procedure: Cells

1. Weigh lyophilized cells and add directly to a separatory funnel where they are extracted.
2. Add extraction solvents in this order: buffer, chloroform, and methanol. Observe the ratio of 1 mg lyophilized cells to 1 ml chloroform in the first-phase mB/D extraction (i.e., 37 mg cells in 75 ml methanol, 37.5 ml chloroform, and 30 ml phosphate buffer).
3. Allow first-phase extractions to proceed for a minimum of 2 h (up to 18 h) at room temperature.

Procedure: Sediments and Soils

1. Before extraction, using a mortar and pestle, lyophilize and thoroughly homogenize sediment samples.
2. Weigh and transfer samples to glass centrifuge bottles where they are extracted. Observe the ratio of 1 g lyophilized sediment to 1 ml chloroform in the first-phase mB/D extraction (i.e., 37 g sediment in 75 ml methanol, 37.5 ml chloroform, and 30 ml phosphate buffer). If lyophilization is not possible, subtract the amount of water in the sample from the amount of buffer added.
3. Sonicate sediments for no more than 2 min.
4. Once the extraction is complete, centrifuge the bottles (30 min at 2000 rpm) to separate sediment from solvent, and decant the mB/D first phase into a separatory funnel.

## The Modified Bligh/Dyer (mB/D)

Procedure

1. Once the first-phase extraction is complete (2–18 h), add chloroform and water to split the phase and to provide a final solvent volume ratio of 1:1: 0.9 for chloroform:methanol:water/buffer. Portions of water and chloroform are added to equal the amount of chloroform added in the first phase [4, 6].
2. Shake the separatory funnel vigorously, vent it, and allow to separate overnight (approximately 18 h) or until the aqueous (upper) phase is no longer cloudy.
3. Remove the organic phase (lower) via the stopcock through a Whatman #2[v] filter supported by a glass funnel into a glass round-bottom flask.
4. Drain the organic phase until the interface between water and solvent just meets the stopcock, making certain none of the aqueous phase drains through. If analysis of lipopolysaccharide hydroxy fatty acids (LPS OH-FA) is to be performed, retain the aqueous phase.
5. Remove the solvents from the organic fraction in the round-bottom flask under vacuum (<37°C) with a rotary evaporator. Take care not to exceed 37°C, since heat breaks down the unsaturated fatty acids. In addition, never expose lipids to air, as oxygen will react with the double bonds, further breaking down the unsaturation. Christie [2] and Kates [7] recommend using nitrogen gas to evacuate the rotary evaporator. Avoid exposure to light, especially from fluorescent lights, as much as possible.
6. Transfer the dried total lipid in the round-bottom flask to test tubes with 3 × 2 ml chloroform washes.
7. Sample cloudiness upon the addition of chloroform indicates the presence of water. Add methanol (approximately 0.5 ml) until the cloudiness disappears, and redry the sample on the rotovap. This procedure may need to be repeated.
8. Once transferred, remove the solvent from the test tube under constant nitrogen flow using the dry-down with a water bath temperature of less than 37°C.
9. Store the total lipid under nitrogen at −20°C until lipid class separation is achieved.

Notes

1. All glassware used for lipid analysis must be ion- and organic-free. Scrupulous cleaning of all glassware is necessary for contaminant-free analysis.

If glassware is made dirty, immediately immerse it fully into a washtub full of hot water and phosphate detergent (S/P Micro or equivalent). Scrub the glassware with a brush and rinse four times each with cold tap water and deionized water. Allow glassware to dry completely before wrapping in aluminum foil and heating in a clean muffle furnace for a minimum of 4 h at 450°C. Disposable glassware, such as pipettes and silicic acid columns, need not be washed, but should be fired similarly. Take care not to recontaminate fired glassware. Rinse items that will not tolerate heating to 450°C with methanol and then chloroform, and allow to dry.

2. Organic-free technique is different from sterile technique. Meticulous technique must be practiced to ensure contaminate-free analyses. This is the cardinal rule; No materials other than fired glass and solvent-rinsed Teflon may come into contact with lipid solvents. Finger lipids, hair, stopcock grease, oils, and hydrocarbons are all potential contaminants.

## SILICIC ACID COLUMN CHROMATOGRAPHY

Silicic acid column chromatography is used to separate total lipid extracts into general lipid classes (neutral lipid, glycolipid, and polar lipid). This procedure uses three solvents of increasing polarity (chloroform < acetone < methanol) to selectively elute the lipid classes from the silicic acid stationary phase. Reagents for this procedure include chloroform, acetone, methanol, and silicic acid (powder). Safe handling of these materials is described in the material safety data sheet (MSDS) literature. Samples—the total lipid extract obtained from the lipid extraction—should be stored in a test tube at −20°C until use. The test tubes must warm slowly to room temperature before the caps are opened.

### Equipment

Glassware includes test tubes with Teflon-lined screw caps, 10-ml beakers, Pasteur pipettes. All glassware is fired in a muffle furnace (450°C for at least 4 h).

Chromatography columns are constructed from large-volume dispo-pipets (Fisher #13-678-8) packed with glass wool plugs inserted into the bottom and fired as above. Prepacked silicic acid columns (Burdick and Jackson inert solid-phase extraction system #7054G) are a commercial alternative.

Suitable racks to hold the assembled columns.

Nitrogen gas blow-down for solvent removal from test tubes with a temperature-controlled water bath.

### Reagents

1. Organic solvents include chloroform, methanol, and acetone. All solvents should be of the purest grade possible (Burdick and Jackson $GC^2$ or equivalent)
2. Silicic acid, 100–200 mesh powder—Unisil (Clarkson Chemical Co., Williamsport, PA) or equivalent—activated at 100°C for a minimum of 1 h in a fired test tube or flask (see Note 1)

### Procedure

1. If commercial columns are not used, construct a suitable column from dispo-pipets by wetting the glass wool in the bottom of the dispo-pipet with chlo-

roform and transferring a silicic acid slurry (0.5 g silicic acid suspended in 5 ml chloroform in a 10-ml beaker) by Pasteur pipette. There should be no sign of air pockets within the bed. If there are, add additional chloroform, and agitate the bed with a Pasteur pipette until the air bubbles rise to the surface. Whichever column is used, do not allow the packing to dry or disturb the surface of the bed once the procedure has begun.

2. Resuspend the total lipid in a minimal volume of chloroform (100–200 ml), and load onto the top of the silicic acid bed with a Pasteur pipette. Repeat three times for a quantitative transfer. Take care not to disturb the surface of the bed once total lipid has been applied (see Note 2).
3. Once the column is loaded, use a series of three solvents of increasing polarity to separate the lipid classes: neutral lipids, 5 ml chloroform; glycolipids, 5 ml acetone; and phospholipids, 5 ml methanol [4]. Collect lipid classes in separate test tubes set up below the column.
4. Once each fraction is collected, remove the solvent with the nitrogen gas blow-down, and store the lipid under nitrogen at −20°C for further analysis.
5. If necessary, perform sterol analysis on the neutral lipid fraction and PHA analysis on the glycolipid fraction. Phospholipid fatty acid methyl esters are prepared from the phospholipid fraction.

Notes

1. Silicic acid is slightly acidic precipitated silica. Silanols (active sites on the silicic acid granules) contain -OH groups directly bound to the silicon atom. The silanols interact with the polar groups of the lipid classes, whereas the non-polar end of the lipid molecule contributes little to separation. As the polarity of the solvents increases, the lipid classes are selectively eluted from the silanols, thereby effecting separation. Silicic acid is easily hydrated and must be dehydrated at 100°C for at least 1 h before use.
2. Do not overload the silicic acid columns. By saturating the active sites with lipid, quantitative recovery of lipid fractions is diminished, and a lower biomass estimate will result. It may be necessary to increase the amount of silicic acid if the samples are of exceptionally high biomass or are highly pigmented. If the mass of silicic acid is increased, volumes of eluting solvents must also increase accordingly to retain a 1:10 ratio (g silicic acid: ml eluting solvent).

## PREPARATION OF FATTY ACID METHYL ESTERS FROM ESTERIFIED LIPIDS

Mild alkaline methanolysis is used to cleave the fatty acids from the phospholipid glycerol backbone and replace the glycerol bonds with methyl groups, creating fatty acid methyl esters (FAMEs). The purpose of this procedure is to prepare FAMEs from the esterified lipids found in either total lipid extracts or the individual lipid classes. Reagents for this procedure include chloroform, hexane, methanol, toluene, acids, and bases. Safe handling of these materials should be followed as described in the MSDS literature. Samples are either the total lipid extract or the individual lipid classes. Most commonly, the polar lipid fraction is used.

Equipment

a 37°C heating block for incubation
table-top centrifuge

vortx mixer
nitrogen gas blow-down
litmus paper

Reagents

1. Methanolic potassium hydroxide: 0.2M KOH in methanol, made fresh before each use; 0.28 g KOH per 25 ml methanol (or direct proportion thereof); guard against aqueous contamination (see Note 1)
2. 1 N glacial acetic acid: 5.72 ml concentrated (17.5 N) glacial acetic acid per 100 ml nanopure distilled water
3. Organic solvents: toluene:methanol (1:1, vol/vol; i.e., 125 ml toluene to 125 ml methanol) and hexane:chloroform (4:1, vol/vol; i.e., 200 ml hexane to 50 ml chloroform)
4. Nanopure distilled water

Procedure

1. Prepare fatty acid methyl esters from esterified lipids in the lipid fraction by mild alkaline methanolic transesterification as reported by Guckert et al. [4].
2. Redissolve the dried lipid in 1 ml toluene:methanol (1:1, vol/vol) and 1 ml methanolic KOH.
3. Vortex the mixture briefly, and incubate the samples for at least 15 min at no greater than 37°C.
4. After the samples have cooled to room temperature, add 2 ml hexane:chloroform (4:1, vol/vol), and mix the sample.
5. Then neutralize the sample (pH 6-7) with approximately 200 ml 1 N acetic acid, and analyze the pH with litmus paper (see Note 2).
6. Add 2 ml nanopure distilled water to break, phase, and mix the sample for at least 30 s on a vortex mixer. The phases (upper: organic containing the fatty acid methyl ester (FAME), lower: aqueous) are separated by centrifugation (5 min at approximately 2000 rpm).
7. Transfer the upper phase to a clean test tube (see Note 3).
8. Re-extract the lower phase with 2 ml hexane:chloroform (4:1, vol/vol), centrifuge, and transfer as above, twice more.
9. Remove the solvent with the nitrogen gas blow-down, and store the FAME under nitrogen at −20°C until separation and quantification.

Notes

1. Any water in the reaction will act as a reagent by attacking double bonds in the long-chain fatty acids. Water will also compete with the methanol for the fatty acids, yielding free fatty acids rather than methyl esters. Because potassium hydroxide is hygroscopic and will absorb water out of the air, it must be stored in a sealed container, and a quick transfer from the balance to the methanol is required.
2. The sample is neutralized because (1) methanolysis is incomplete at a higher pH and (2) FAMEs have a higher affinity for water at a higher pH. Usually, 200 ml 1 N acetic acid is sufficient to neutralize the sample. One sample from the set is pH-analyzed by drawing a small amount of the lower phase into a Pasteur pipette and spotting the sample onto litmus paper. The litmus paper should indicate that pH = 7.

3. For this operation, it is best to hold both test tubes in one hand while pipetting with the other hand. Take care to avoid transferring any water with the organic phase. It is not necessary to retrieve all of the organic phase each time, since the aqueous phase is rinsed three times.

## SEPARATION, QUANTIFICATION, AND IDENTIFICATION OF ORGANIC COMPOUNDS

The purpose of this procedure is to separate, quantify, and identify the organic compounds isolated through the various procedures by gas chromatography (GC) and to confirm identification of these compounds by gas chromatography/mass spectrometry (GC/MS). Reagents for this procedure are iso-octane, cholestane, and nonadecanoic acid. Safe handling of these materials should be followed as described in the MSDS literature. Samples are either (1) fatty acid methyl esters (FAMEs), (2) poly-β-hydroxyalkanoates (PHAs) from (3) trimethyl-silyl (TMSi) derivatives of 3β-ol sterols, (4) TMSi derivatives of lipopolysaccharide hydroxy fatty acids (LPS OH-FA), or (5) dimethyl disulfide (DMDS) derivatives of mono-unsaturated FAMEs.

### Equipment

Separation: nitrogen gas blow-down and capillary gas chromatograph (GC) with optional autosampler and controller; the mobile phase for the GC is hydrogen gas, obtained from the purest possible source (99.999% pure or above).

Quantification: signal output from the GC retained and initially processed by a data system (P.E. Nelson or equivalent).

Identification: preliminary identification of compounds based on comparison to retention times of standards; mass spectrometry used for verification of compound structure.

### Reagents

1. Iso-octane: This solvent should be of the purest grade possible (Burdick and Jackson $GC^2$ or equivalent). An aliquot of each new lot should be concentrated by a suitable factor (e.g., 1000) and analyzed by capillary GC for any organic contaminants
2. 50 pmol/ml C19:0 internal standard: 15.6 mg C19:0 (nonadecanoic acid methyl ester, M.W. 312) in 1.000 liter iso-octane
3. 50 pmol/ml cholestane internal standard: 18.7 mg cholestane (M.W. 287.86) in 1.000 liter iso-octane

### Procedure Separation

1. Separate the compounds to be analyzed for quantification using capillary GC with flame ionization detection. Use a 60-meter non-polar cross-linked methyl silicone column (i.e., Restek RTX-1) with a suitable temperature program (see Note 1).
2. Before dilution with internal standard solution, shoot a rangefinder using iso-octane as the solvent to determine the correct dilution of the sample (see Note 2).

3. Before injection, remove any remaining solvent with the nitrogen gas blow-down, and dilute the sample in the appropriate internal standard solution. Generally, 1 ml is injected.

Procedure: Quantification

1. Base quantification on a comparison to an internal injection standard (FAME, C19:0, LPS OH-FA, C19:0; sterols, cholestane; PHAs, malic acid). Equimolar responses are generally assumed within the range of microbial FAMEs (12:0–24:0) and sterols (22C–30C); however, tables of molecular weight correction factors are available [2].
2. Results obtained from the GC will be quantified areas under each sample peak, including the internal injection standard. For each peak, do the following calculation to obtain molar or weight amounts per sample. To normalize these amounts to a per gram dry weight basis, use appropriate dilution factors and mass measurements. The calculations done for each compound are

$$C_X = (A_X/A_{ISTD}) * C_{ISTD} * D$$

where $C_X$ is the calculated concentration of compound X (moles or weight per unit volume), $A_X$ is the GC area of compound X (unitless), $A_{ISTD}$ is the GC area of the internal injection standard as determined by the GC data system (unitless), $C_{ISTD}$ is the concentration of the internal injection standard as given above, and D is the appropriate dilution factor.
3. Assuming an average phospholipid content of $10^{-4}$ moles of PLFA per $5.9 \times 10^{12}$ bacterial cells (based on *E. coli*) and $10^{-4}$ moles PLFA per $1.2 \times 10^{10}$ algal cells (based on *Chlorella*), obtain an estimate of bacterial and algal cells by multiplying calculated picomolar concentrations of PLFAME by the appropriate factor (1 to $5.9 \times 10^4$ cells/pmol for bacteria, $1.2 \times 10^2$ cells/pmol for algae), yielding cells per gram.

Procedure: Identification

1. FAMEs. The use of a linear temperature program for the separation of FAMEs permits the use of equivalent chain length (ECL) analysis for FAME identification. This technique, detailed by Christie [2], is based on the linear relationship between the retention times of a homologous series of straight-chain saturated FAMEs against the number of carbons in the FAME chain. Because ECLs are a constant property of a specific FAME as long as the temperature program is linear, published ECLs in a library of FAMEs can be used to help identify specific FAME.

   This identification is preliminary, however, and selected samples should be further analyzed by (1) GC/MS as detailed in Guckert et al. [4] and [2] DMDS derivatization of monounsaturated double bonds [10].

   Fatty acid nomenclature is of the form A:BwC where A designates the total number of carbon atoms, B the number of double bonds, and C the distance of the closest unsaturation from the aliphatic end of the molecule. The suffixes **c** for *cis* and **t** for *trans* refer to geometric isomers. The prefixes **i** and **a** refer to iso and anteiso methyl-branching, respectively [7].
2. PHAs. Comparison of unknown peaks to a prepared standard (usually poly-β-hydroxybutyrate) allows preliminary identification of PHAs. However, structural identification requires GC/MS analysis as detailed by Findlay and White [3].

3. Sterols. Due to variations in chromatographic variables, identification of sterols requires the calculation of relative retention times (RRT) based on cholesterol and sitosterol [8]. The RTT for each peak is calculated by the following formula:

$$RRT_X = 1 + [0.63 * (RT_X - RT_c)]/(RT_S - RT_C)$$

where $RRT_X$ is the relative retention time of the unknown peak, $RT_X$ is the retention time of the unknown peak, $RT_C$ is the retention time of cholesterol, and $RT_S$ is the retention time of sitosterol.

   By comparing the calculated $RRT_X$ of an unknown sterol to a library of RRT for known sterols under the given chromatographic conditions, preliminary identification of individual compounds is possible. This identification is preliminary, however, and selected samples should be analyzed further by GC/MS as detailed in Nichols et al. [9].
4. LPS OH-FA. A bacterial fatty acids standard mixture containing a- and β-hydroxy fatty acids may be obtained from Matreya (cat# 1114) and preliminary identification of hydroxy fatty acids achieved by comparison to the standard mixture. Identification should be considered tentative, however, and GC/MS analyses should be performed according to Parker et al. [11].

Notes

1. FAME temperature program: 100°C for 0 min, 10°/min to 150°C for 1 min, 3°/min to 282°C for 5 min. Injector temperature = 270°C, detector temperature = 290°C. Total run time = 55 min.
2. Sterol temperature program: 200°C for 0 min, 10°/min to 280°C for 0 min, 2°/min to 310°C for 5 min. Injector temperature = 290°C, detector temperature = 290°C. Total run time = 28 min.
3. PHA temperature program: 60°C for 10 min, 10°/min to 280°C for 0 min. Injector temperature = 220°C, detector temperature = 290°C. Total run time = 32 min.
4. LPS OH-FA temperature program: Same as FAME temperature program.
5. Rangefinders are shot to ensure that the internal standard is within a factor of the sample peaks. Generally, for 37 mg lyophilized bacterial isolate or 37 g dry sediment, a rangefinder is shot at 1:1000 μl iso-octane (no internal standard). Adjustments are made and the sample diluted in internal standard only when the proper dilution is determined.

## PREPARATION OF POLY-β-HYDROXYALKANOATES FROM GLYCOLIPIDS

Poly-β-hydroxyalkanoates (PHAs) are bacterially synthesized endogenous storage polymers. During their preparation, an acid ethanolysis is used to cleave the polymer and form ethyl esters of the constituent monomers found in the glycolipid fraction and to prepare these compounds for quantification and identification. Reagents for this procedure include absolute ethanol, diethyl ether, malic acid, chloroform, and strong acids. Safe handling of these materials should be followed as described in the MSDS literature. Samples are dried glycolipid collected during silicic acid chromatography.

Equipment

a 100° heating block
table-top centrifuge
vortex mixer
nitrogen gas blow-down

Reagents

1. 2 mM malic acid internal standard: 0.1341 g malic acid (MW 134.1) in 500.00 ml nanopure distilled water
2. Organic solvents: diethyl ether, absolute ethanol, chloroform
3. Concentrated hydrochloric acid (12M); no dilution from the stock bottle required
4. Nanopure distilled water

Procedure

1. Prepare ethyl esters of PHAs from the glycolipid fraction by a strong acid ethanolysis as reported by Findlay and White [3].
2. To the dried glycolipid, add 100 ml diethyl ether and 100 ml (accurate) 2 mM malic acid internal standard, and completely dry the sample under a stream of nitrogen.
3. Once dry, dissolve the sample in 500 ml chloroform, cap tightly, and heat at 100°C for 10 min to dissolve the PHAs (see Note 2).
4. After removal from the heating block, gradually release pressure from the test tube by slowly unscrewing the cap.
5. To the hot chloroform mixture add 1.7 ml absolute ethanol and 200 ml concentrated HCl, and replace the cap tightly.
6. Vortex the test tube briefly, and return to the heating block for at least 4 h at 100°C.
7. After the samples have cooled to room temperature, add 2 ml chloroform and 2 ml nanopure distilled water to split phase.
8. Vortex the sample for at least 30 s, and separate the phases (upper: aqueous, lower: organic, containing the PHAs) by centrifugation (5 min at approximately 2000 rpm).
9. Remove and discard the upper phase (see Note 3).
10. Add an additional 1 ml nanopure distilled water, vortex, and centrifuge the sample as above.
11. Remove the lower organic phase to a clean test tube (see Note 4).
12. Re-extract the remaining upper phase with 1 ml chloroform, centrifuge, and transfer as above, twice more.
13. Remove the solvent from the combined organic phases under nitrogen until approximately 100 ml sample is left (see Note 5).
14. Store the PHA under nitrogen at −20°C until separation and quantification.

Notes

1. If the sample is of exceptionally high biomass or is highly pigmented, the dried glycolipid must be rinsed with 2 × 2 ml absolute ethanol washes followed by 2 × 2 ml diethyl ether washes. Solvent is gently applied to the top of the test tube and allowed to wash down the side of the tube. It is removed from the test tube with a Pasteur pipette, taking precaution not to remove any of the PHA. If any white flocculent material is present in the

pipette, the sample is redried on the blow-down and the rinsing procedure resumed.

2. PHAs dissolve completely in hot chloroform. Using Teflon-lined screw caps with complete circular indentations from the test tube in the Teflon lining should prevent any sample from escaping during the heating process. If a test tube does leak, replacing the cap with another one usually curbs this loss; however, transferring the sample to a different test tube may be necessary.
3. Since no lipid soluble organic compounds are present in the upper aqueous phase, this phase may be discarded with no fear of losing valuable sample. This step washes the organic phase, removing most water-soluble contaminants.
4. This is accomplished by bubbling air through a Pasteur pipette as it passes through the upper phase, drawing off the lower phase, and dripping solvent through the upper phase as the pipette is removed from the liquid.
5. Loss of volatile ethyl esters will occur if the sample is allowed to dry completely.

## PREPARATION OF STEROLS FROM ESTERIFIED LIPIDS

Sterols are a stable class of compounds used as signature biomarkers for microeukaryotes, such as fungus and algae. During this procedure, an alkaline saponification is used to derivatize sterols found in either the neutral lipid or total lipid fraction and to prepare these compounds for quantification and identification. Reagents for this procedure include methanol, chloroform, hexane, bases, and BSTFA. Safe handling of these materials should be followed as described in the MSDS literature. Samples are either neutral lipids or total lipid extract.

### Equipment

a temperature-controlled 60°C heating block for derivatization
table-top centrifuge
vortex mixer
nitrogen gas blow-down

### Reagents

1. Organic solvents: hexane:chloroform (4:1, vol/vol; i.e., 200 ml hexane to 50 ml chloroform)
2. Methanolic potassium hydroxide: 5% (w/v) KOH in methanol:water (80:20); 10 g KOH in 160 ml methanol and 40 ml nanopure distilled water
3. Nanopure distilled water
4. N,O-*bis*(Trimethylsilyl)trifluoroacetamide (BSTFA); Pierce Chemical Company, Rockford, IL)

### Procedure

1. Form trimethyl-silyl (TMSi) derivatives of 3β-ol sterols from either the neutral lipid or total lipid fraction by alkaline saponification as described by Nichols et al. [9].
2. To the dried lipid, add 3 ml 5% KOH in methanol:water (80:20, vol/vol), and heat the samples for 2 h at 60°C.

3. After the samples have cooled to room temperature, add 1 ml nanopure distilled water and 2 ml hexane:chloroform (4:1, vol/vol), and vortex the sample for a minimum of 30 s.
4. Separate the phases (upper: organic containing the sterols, lower: aqueous) by centrifugation (5 min at approximately 2000 rpm).
5. Transfer the upper phase to a clean test tube.
6. Re-extract the lower phase with 1 ml hexane:chloroform (4:1, vol/vol), centrifuge, and transfer as above, twice more.
7. Remove the solvent with the nitrogen gas blow-down, and store the sterols under nitrogen at −20°C.
8. Within 24 hours before GC analysis, add 100 ml BSTFA (see Note 1), and heat the sterols for 30 min at 60°C.
9. Remove the sample from the heating block, dry under nitrogen, and store at −20°C until separation and quantification.

Notes

1. BSTFA replaces -OH groups with trimethyl-silyl groups on the sterol molecule, improving separation and identification during gas chromatography. This bond is unstable, however, and samples must be analyzed before the reverse reaction occurs.

## PREPARATION OF LIPOPOLYSACCHARIDE HYDROXY FATTY ACIDS

Lipopolysaccharide hydroxy fatty acids (LPS OH-FA) are signature biomarkers found in gram-negative bacteria that are useful in community structure investigations. During this procedure, the lipopolysaccharide present in the aqueous phase of the separated modified Bligh/Dyer (mB/D) extraction is hydrolyzed by mild acid hydrolysis, and the conjugal fatty acids are methylated by acid methanolysis before purification by thin-layer chromatography (TLC). The hydroxy fatty acids are silylated just before separation and quantification. Reagents for this procedure include chloroform, methanol, hexane, diethyl ether, fatty acid standards, BSTFA, rhodamine 6G, hydroxy fatty acid standards, and acids. Safe handling of these materials should be followed as described in the MSDS literature. Samples are the aqueous phase collected from the separated Bligh/Dyer lipid extract.

### Equipment

Extraction: nitrogen gas blow-down, round-bottom flasks and separatory funnels, roto-vap solvent evaporator.

Hydrolysis: reflux apparatus, round-bottom flasks and separatory funnels, nitrogen gas blow-down, rotary solvent evaporator, and Teflon-lined screw-cap test tubes.

Acid methanolysis: 100°C heating block for incubation, Teflon-lined screw-cap test tubes, vortex mixer, and table-top centrifuge.

Thin-layer chromatography: TLC plates (Whatman LK$^6$ silica gel, 250 mm thick with preabsorbent zone), chromatography paper (Whatman 4 mm), solvent-rinsed TLC developing tanks, drying oven, Manostat, 100-ml capillary pipettes, rhodamine spraying apparatus, large-volume dispo-pipets (Scientific Products #P5240-1 or equivalent) packed with glass wool plugs inserted into the bottom of the pipette and fired, vacuum pump, ultraviolet lamp, 60°C heating block.

Reagents

1. Magic methanol: MeOH:$CHCl_3$:conc. HCl (10:1:1, vol/vol/vol); 10 ml each chloroform and concentrated HCl in 100 ml methanol
2. 1 N hydrochloric acid: 8.3 ml concentrated (12 N) HCl per 91.6 ml nanopure distilled water
3. Organic solvents: iso-octane, hexane:diethyl ether (1:1, vol/vol; i.e., 25 ml hexane to 25 ml diethyl ether), hexane:chloroform (4:1, vol/vol; i.e., 200 ml hexane to 50 ml $CHCl_3$), and chloroform:methanol (1:1, vol/vol; i.e., 125 ml $CHCl_3$ to 125 ml MeOH)
4. N,O-*bis* (trimethyl-silyl) trifluoroacetamide (BSTFA; Pierce Chemical Company, Rockford, IL)
5. 14C a- and β-hydroxy fatty acid plating standard: 10 mg each of methyl 3-hydroxytetradecanoate (3-OH14:0) and methyl 2-hydroxytetradecanoate (2-OH14:0) in 10 ml chloroform for a final concentration of 1 mg/ml
6. Rhodamine 6G spray reagent: 0.01% (wt/vol) rhodamine 6G in water; 25 mg rhodamine 6G (chloride salt, M.W. 479) in 250 ml nanopure distilled water

Procedure

1. Prepare fatty acid methyl esters from lipopolysaccharide lipid A present in the residue of the aqueous portion of the separated Bligh/Dyer lipid extraction by acid hydrolysis and acid methanolysis as reported by Parker et al. [11].
2. Extract samples using the Bligh/Dyer lipid extraction, and separate according to procedures outlined earlier.
3. Remove the lower chloroform phase and collect for analysis.
4. Drain the upper aqueous phase containing the lipid A into a 250 ml round-bottom flask and evaporate to dryness on a rotary evaporator (see Note 1).
5. The hydrolysis is done as follows: To the dried residue, add 30 ml 1 N hydrochloric acid, and heat the sample at reflux at 100°C for a minimum of 2 h. After the samples have cooled to room temperature, transfer the acid mixture to a 250-ml separatory funnel. Rinse the reflux apparatus with two 10-ml portions methanol then add two 25-ml portions chloroform, adding each rinse to the separatory funnel. Allow the phases to separate overnight, and collect the organic phase (lower) into a round-bottom flask. Remove the solvents under vacuum with a rotary evaporator. Resuspend the hydrolyzed LPS in 3 × 2 ml chloroform washes and transfer to Teflon-lined screw-cap test tubes. Remove the solvent under nitrogen, and store the hydrolyzed LPS at −20°C until acid methanolysis.
6. The acid methanolysis is done using the following procedure: To the dried sample, add 2 ml magic methanol (MeOH:$CHCl_3$:conc. HCl 10:1:1, vol/vol/vol), and heat at 100°C for 1 h. Once the samples cool to room temperature, add 2 ml hexane:chloroform (4:1, vol/vol), and vortex the sample for a minimum of 30 s. Separate the phases (upper: organic containing the LPS FA, lower: aqueous) by centrifugation (5 min at approximately 2000 rpm), and transfer the upper phase to a clean test tube. Re-extract the lower phase with 2 ml hexane:chloroform (4:1, vol/vol), twice more. Remove the solvent with the nitrogen gas blow-down, and store the methylated LPS

FAME under nitrogen at −20°C until purification by thin-layer chromatography.

7. Thin-layer chromatography (TLC): Rinse lipopolysaccharide FAMEs from the test tube and apply to a prepared and cleaned (see Note 2) TLC plate with three 75-ml washes of chloroform. Apply the sample in a straight line to the preabsorbent zone using a manostat fitted with a 100-ml capillary pipette. The OH-FA plating standard (100 ml) is similarly applied to narrow lanes on both sides of the TLC plate. TLC plates are developed in 50 ml hexane:diethyl ether (1:1, vol/vol). When the solvent front is within 2 cm of the top of the plate (approximately 20 minutes), note the farthest extent of the front, and remove the plate from the tank and allow to dry.
8. Spray the narrow lanes containing plating standard with rhodamine G6, and scrape and recover the hydroxy fatty acid bands (0.25 cm above the 2-OH FA and 0.5 cm below the 3-OH FA standards) in an inverted Pasteur pipette plugged with glass wool by suction.
9. Elute hydroxy fatty acids from the silica gel scrapings with 5 × 1 ml portions of chloroform:methanol (1:1, vol/vol).
10. Remove the solvent with the nitrogen gas blow-down, and store the LPS OH-FAME under nitrogen at −20°C.
11. Within 24 hours of gas chromatographic analysis, add 100 ml BSTFA, and heat the sample for 30 min at 60°C.
12. Remove the sample from the heating block, dry under nitrogen, and store at −20°C until separation and quantification.

Notes

1. There are two acceptable procedures for accomplishing this task: (1) Methanol present in the sample is removed by rotary evaporation with the water bath set at 45°C. The water bath temperature is then increased to 55°C and the remaining water removed to dryness. Do not exceed 55°C in the water bath. (2) Methanol is removed by rotary evaporation at 40°C, and the remaining liquid is frozen and lyophilized.
2. TLC developing tanks are solvent-rinsed with hexane:diethyl ether (1:1, vol/vol) before use. Chromatography paper is cut to fit the tank (20 × 55 cm) and placed inside with 50 ml hexane: diethyl ether (1:1, vol/vol) to saturate the paper with solvent. Before precleaning, a line is scraped free of silica gel down the middle of the plate to allow for two samples to be plated at the same time. TLC plates are precleaned by developing the plate in solvent before spotting with sample. Once dry, plates are activated in a drying oven at 100°C for 15 min.

## PREPARATION OF DIMETHYL DISULFIDE ADDUCTS OF MONOUNSATURATED FAME

Verification of monounsaturated fatty acid double bond position is made possible by dimethyl disulfide (DMDS) derivatization and subsequent GC/MS analysis. During this procedure, DMDS derivatives are prepared for GC/MS analysis by an iodine-catalyzed addition of DMDS to monounsaturated FAMEs. Reagents for this procedure include hexane, iodine, sodium thiosulfate, chloroform, DMDS, and iso-octane. Safe handling of these materials should be followed as described in the MSDS literature. Samples are fatty acid methyl esters.

Equipment

a table-top centrifuge
vortex mixer
nitrogen gas blow-down
2-ml autosampler vials with Teflon-lined screw-cap lids

Reagents

1. Organic solvents: hexane, hexane:chloroform (4:1, vol/vol; i.e., 200 ml hexane to 50 ml chloroform)
2. Sodium thiosulfate solution: 5% (wt/vol) sodium thiosulfate in water; 0.5 g $Na_2S_2O_4$ in 10 ml nanopure distilled water
3. Iodine solution: 6% (wt/vol) iodine in diethyl ether; 0.6 g elemental iodine in 10 ml diethyl ether
4. Dimethyl disulfide (DMDS; Gold label, Aldrich Chemical, Milwaukee)

Procedure

1. Form DMDS adducts of monounsaturated FAMEs according to procedures outlined by Nichols et al. [9].
2. Quantitatively transfer samples with 3 × 0.5 ml washes of hexane:chloroform 4:1 (vol/vol) to standard 2-ml autosampler vials fitted with Teflon-lined screw-cap lids and dried under nitrogen.
3. Dilute dried FAME in 50 ml hexane, and add 100 ml DMDS and one to two drops of 6% iodine solution (wt/vol).
4. Heat the sample in a GC oven at 50°C for a minimum of 48 h.
5. Once cool, dilute the sample in 500 ml hexane.
6. Remove the iodine by adding 500 ml sodium thiosulfate solution (5%, wt/vol) and vortexing until none of the original iodine color is present.
7. Separate the phases (upper: organic containing the FAME, lower: aqueous) by centrifugation (5 min at approximately 2000 rpm), and remove the upper organic layer to a clean 2-ml autosampler vial.
8. Re-extract the lower phase with 500 ml hexane:chloroform (4:1, vol/vol), centrifuge, and transfer as above, twice more.
9. Remove the solvent under nitrogen, and dilute the DMDS derivatives in the original volume of iso-octane before GC/MS analysis.

## PREPARATION OF ARCHAEBACTERIAL OR *THERMODESULFOBACTERIUM COMMUNE* ETHER LIPIDS FOR SUPERCRITICAL FLUID CHROMATOGRAPHY

Ether membrane lipids are not cleaved from their phosphate head groups by the mild alkaline methanolysis used to prepare fatty acid methyl esters (FAMEs) for GC. In this method, a modification of a mild alkaline methanolysis is used to prepare FAMEs for GC analysis, the FAMEs are separated from phospho-ethers by a second silicic acid column, and a strong acid methanolysis is used to cleave the ether lipids from their phosphate groups for supercritical fluid chromatography (SFC) analysis. Reagents for this procedure include chloroform, hexane, methanol, toluene, silicic acid, acids, and bases. Safe handling of these materials should be followed as described in the MSDS literature. Most commonly, the polar lipid fraction from silicic acid column chromatography are used as samples.

Equipment

a table-top centrifuge
vortex mixer
nitrogen gas blow-down
2-ml autosampler vials with Teflon-lined screw-cap lids
100°C heating block

Reagents

1. Methanolic potassium hydroxide: 0.2M KOH in methanol, made fresh before each use; 0.28 g KOH per 25 ml methanol (or direct proportion thereof); guard against aqueous contamination (see Note 1)
2. 1 N glacial acetic acid: 5.72 ml concentrated (17.5 N) glacial acetic acid made to 100 ml with nanopure distilled water
3. Organic solvents: toluene:methanol (1:1, vol/vol; i.e., 125 ml toluene to 125 ml methanol) and hexane:chloroform (4:1, vol/vol; i.e., 200 ml hexane to 50 ml chloroform)
4. Nanopure distilled water
5. Magic methanol: methanol:chloroform:concentrated hydrochloric acid (10:1:1, vol/vol/vol); 10 ml each chloroform and concentrated hydrochloric acid in 100 ml methanol, made fresh daily

Procedure

1. Utilize FAMEs prepared from esterified lipids in the lipid fraction by mild alkaline methanolic transesterification as reported by Guckert et al. [4] and described above. Only the differences are explained here.
2. After the addition of methanolic KOH and incubating for 15 min at 40°C, remove the samples from the heating block, and allow to cool to room temperature.
3. To the sample, add 2 ml *chloroform* (not hexane:chloroform), and vortex the sample.
4. Then neutralize the sample.
5. Separate the phases (lower: organic chloroform containing the FAMEs: and the unaffected phospho-ethers, upper: aqueous) by centrifugation (5 min at approximately 2000 rpm).
6. Transfer the lower phase to a clean test tube.
7. Re-extract the aqueous phase with 2 ml chloroform, centrifuge, and transfer as above, twice more.
8. Remove the solvent with the nitrogen gas blow-down and the FAMEs, and store phospho-ethers under nitrogen at −20°C until fractionation by silicic acid column chromatography.
9. Separate the FAMEs from the unchanged phospho-ethers by silicic acid column chromatography, except that the acetone elution is omitted.
10. Recover FAMEs in the neutral lipid (chloroform) fraction and phospho-ethers in the polar lipid (methanol) fraction. Separate and quantify FAMEs as described above.
11. Use strong acid methanolysis to cleave the ether lipids from the methanol eluate from their phosphate head groups.
12. Add 1 ml magic methanol to each sample in a screw-top test tube with a Teflon-lined cap (see Note 2).
13. Screw the cap on tightly, and heat the tube on the 100°C heating block for 1 h.

14. After the tubes cool, add 2 ml each water and hexane:chloroform (4:1).
15. Transfer the upper organic phase to a clean test tube, and extract the lower aqueous phase twice more with hexane:chloroform (4:1). Blow-down the combined organic extracts, and store at −20°C until analysis by SFC.

Notes

1. Any water in the reaction will act as a reagent by attacking double bonds in the long-chain fatty acids. Water will also compete with the methanol for the fatty acid, yielding free fatty acids rather than methyl esters. Because potassium hydroxide is hygroscopic and will absorb water out of the air, it must be stored in a sealed container, and a quick transfer from the balance to the methanol is required.
2. The condition of the test tube and cap is very important in the strong acid methanolysis step. A very slight leak will allow the solvent and the ether lipids to escape. Each test tube should be examined for chips, cracks, or deformities in the lip or threads. The Teflon cap liner must be free of cuts or gouges. The test tube and cap must screw together freely and snug down tight. If a high-pitched crack is heard while tightening the cap, it should be removed to look for cracks in the tube's lip.

## REFERENCES


1. Bligh, E.G. and Dyer, W.J. 1959. A rapid method of total lipid extraction and purification. Can. J. Biochem. Physiol. 37: 911–917.
2. Christie, W.W. 1989. Gas chromatography and lipids. The Oily Press, Ayr, Scotland.
3. Findlay, R.H. and White, D.C. 1983. Polymeric beta-hydroxyalkanoates from environmental samples and *Bacillus megaterium.* Appl. Environ. Microbiol. 45: 71–78.
4. Guckert, J.B., Antworth, C.P., Nichols, P.D., and White, D.C. 1985. Phospholipid, ester-linked fatty acid profiles as reproducible assays for changes in prokaryotic community structure of estuarine sediments. FEMS Microbiol. Ecol. 31: 147–158.
5. Guckert, J.B. and White, D.C. 1988. Evaluation of hexane/isopropanol lipid solvent system for analysis of bacterial phospholipids and application to chloroform-soluble nucleopore (polycarbonate) membranes with retained bacteria. J. Microbiol. Meth. 9: 131–137.
6. Guckert, J.B. and White, D.C. 1988. Phospholipid, ester-linked fatty acid analyses in microbial ecology. Proceedings of the Fourth International Symposium of Microbial Ecology, pp. 455–459. Slovene Society for Microbiology, Ljubljvana, Yugoslavia.
7. Kates, M. 1986. Techniques in Lipidology: Isolation, Analysis, and Identification of Lipids. 2nd ed. Elsevier, Amsterdam.
8. Nes, W.D. and Parish, Edward J. 1989. Analysis of Sterols and Other Biologically Significant Steroids. Academic Press, San Diego.
9. Nichols, P.D., Volkman, J.K., and Johns, R.B. 1983. Sterols and fatty acids of the marine unicellar alga, FCRG 51. Phytochemistry. 22: 1447–1452.
10. Nichols, P.D., Guckert, J.B., and White, D.C. 1986. Determination of monounsaturated fatty acid double-bond position and geometry for microbial monocultures and complex microbial consortia by capillary GC/MS of their dimethyl disulfide adducts. J. Microbiol. Meth. 5: 49–55.
11. Parker, J.H., Smith, G.A., Fredrickson, H.L., Vestal, J.R., and White, D.C. 1982. Sensitive assay, based on hydroxy fatty acids from lipopolysaccharide lipid A, for gram negative bacteria in sediments. Appl. Environ. Microbiol. 44: 1170–1177.
12. Tunlid, A. and White, D.C. 1991. Biochemical analysis of biomass, community structure, nutritional status, and metabolic activity of the microbial communities in soil. In Soil Biochemistry (Bollag, J.-M. and Stotzky, G., Eds.), pp. 229–262. Marcel Dekker, New York.

13. White, D.C. 1983. Analysis of microorganisms in terms of quantity and activity in natural environments. In Microbes in Their Natural Environments (Slater, J.H., Whittenbury, R., and Wimpenny, J.W.T., Eds.), pp. 37–66. Society for General Microbiology Symposium 34, Cambridge University Press, New York.
14. White, D.C. 1986. Environmental effects testing with quantitative microbial analysis: chemical signatures correlated with *in situ* biofilm analysis by FT/IR. Toxicity Assess. 1: 315–338.
15. White, D.C. 1988. Validation of quantitative analysis for microbial biomass, community structure, and metabolic activity. Adv. Limnol. 31: 1–18.
16. White, D.C., Davis, W.M., Nichols, J.S., King, J.D., and Bobbie, R.J. 1979. Determination of sedimentary microbial biomass by extractable lipid phosphate. Oecologia 40: 51–62.
17. White, D.C., Ringelberg, D.B., Hedrick, D.B., and Nivens, D.E. 1993. Rapid identification of microbes from clinical and environmental matrices by mass spectrometry. In Identification of Microorganisms by Mass Spectrometry (Fenselau, C., Ed.), pp. 8–17. American Chemical Society Symposium Series 541, Washington, DC.

# 12

ANDREW OGRAM

# Isolation of Nucleic Acids from Environmental Samples

Some of the methodological limitations involved in analysis of the structure and activity of microbial communities are being overcome through the application of molecular genetics to microbial ecology (see Chapter 13). Before many of these techniques may be applied, however, nucleic acids of sufficient quality and quantity must be recovered from the sample of interest. Several methods for isolating nucleic acids from environmental samples have been reported over the last 15 years, with modifications being incorporated into updated protocols to increase their simplicity, speed of purification, and efficiency of recovery.

Two general strategies for the isolation of nucleic acids are currently in use, with either being the appropriate choice depending on the application. The first approach was pioneered by Torsvik [23] and colleagues and involves the separation of cells from the environmental matrix by successive washings and differential centrifugations (cell fractionation), followed by cell lysis and purification of DNA by standard methods. The second approach is based on the lysis of cells in the presence of particulates or soil particles (direct lysis), followed by extraction of the released nucleic acid from the particulates and separation of the nucleic acid from organic carbon present in the sample [12]. Both of these approaches have advantages and disadvantages that are discussed in a later section of this chapter and should be considered when choosing a given application.

Regardless of whether direct lysis or cell fractionation is used, isolation of nucleic acids from many environmental matrices is not a trivial matter, and several factors should be considered before proceeding. The most important factors affecting the composition and quality of the recovered DNA or RNA are the lysis efficiency of the target microorganisms, the efficient separation of target microorganisms or nucleic acid (depending on whether cell fractionation or direct lysis is used) from particulates, and separation of the nucleic acid from organic contaminants. These factors are affected by the nature of the environmental matrix, and it is unlikely that any single procedure that is currently available will work equally well with all applications and all environmental samples. For this reason, users should become familiar with the different strategies used to overcome potential problems and be able to modify existing procedures for use with their samples. Much of the following discussion is related to the isolation of nucleic acids from soils and sediments, but general principles apply to the isolation of nucleic acids from all environments.

## FACTORS AFFECTING THE RECOVERY OF NUCLEIC ACIDS

### Biomass

Before any strategies for nucleic acid isolation are considered, a general assessment of the available biomass should be made so that a sample size is used that is sufficient for recovery of nucleic acids for the desired application. Sufficient biomass for recovery of mg quantities of nucleic acids is generally not a problem with surface soils (which may harbor between $10^7$–$10^{10}$ cells per gram), but may be problematic with low-biomass environments, such as many aquatic samples and subsurface sediments [2, 15]. Isolation of nucleic acids from low-biomass aquatic samples may require processing of hundreds of liters, and planktonic biomass may be concentrated before nucleic acid extraction by filtration through either 0.2-mm filters or tangential flow membranes [2]. Either of these approaches may be subject to slow filtration rates due to clogging of filters by suspended particulates that are present in many rivers and lakes. Because concentration of biomass by filtration before extraction of nucleic acids is not an option with solid matrices, such as low-biomass subsurface sediments, isolation of nucleic acids from such samples should maximize the efficiency of extraction.

The efficiency of nucleic acid extraction from soils and sediments is generally higher with smaller sample sizes. In general, no more than 2–4 g soil or sediment should be processed per extraction. Two grams of surface soil generally yields between 10 and 50 mg DNA, which is a sufficient amount for most applications [12, 24]. If more DNA or RNA is needed for a given application or if the matrix contains very low amounts of biomass (such as is the case with deep subsurface sediments), 50 g or more may be processed as individual 3–4 g aliquots, and nucleic acids from the individual aliquots can be pooled at the end of the procedure [15].

### Environmental and Cellular Nucleases

Enzymes responsible for the degradation of RNA and DNA are thought to be ubiquitous in most environments, as well as in microbial cells, and may be responsible for the significant loss of nucleic acids during purification. Standard precautions, such as the use of autoclaved reagents, are usually sufficient for purification of DNA, although it is likely that treatment of the sample during lysis with a detergent, such as sodium dodecyl sulfate (SDS) and heat (65°C), aids in inactivating some nucleases. RNA is much more susceptible to attack by nuclease than DNA, and enzymatic degradation of RNA is probably the greatest single source of RNA loss encountered during any isolation procedure. Protein denaturants, such as diethylpyrocarbonate (DEPC), guanidinium isothiocyanate, phenol, and detergents, such as SDS, are commonly used to inactivate RNases [19]. Most RNA isolation procedures incorporate at least some protein denaturants during lysis to protect the released nucleic acid from environmental nucleases, and most of the denaturants listed above also aid in lysis. All glassware and reagents used in the isolation of RNA should be treated with 0.1% DEPC followed by autoclaving. DEPC has been shown to modify purines in nucleic acids and should be removed from solutions by autoclaving before exposure to RNA.

## Cell Lysis

The efficiency of lysis of microorganisms varies considerably among species and may vary among cell ages within the same species. In general, gram-negative species are more easily lysed than gram-positive ones, older cells may be more easily lysed than younger cells, and larger cells may be lysed more easily than small cells. Some fungal species and gram-positive endospores may be extremely difficult to lyse compared with most bacterial cells. Complete lysis of all microorganisms in a target sample is required if the recovered nucleic acids are to be truly representative of the sample, although this may be an unattainable ideal for many applications or may not be desirable for other applications. If the organisms of interest are lysed easily, the researcher may wish to use a gentle lysis procedure that will enrich the target sequences in the recovered nucleic acid. If all of the nucleic acids present in a sample are of interest, however, a very harsh lysis procedure should be used that may also damage DNA from easily lysed cells. Take care to select the appropriate lysis procedure for the desired application and to understand the limitations of the chosen procedure.

Most lysis procedures may be divided into two very general categories: chemical lysis and mechanical disruption. Chemical lysis involves treatment with a detergent and boiling, whereas mechanical procedures include bead-mill homogenization (homogenization with zirconium/silica beads [11, 12], freeze-thaw cycles [24], French pressure cells, heating with microwave ovens [3, 17], and sonication [17]. Chemical lysis procedures are generally thought to be less efficient than mechanical procedures, but have the advantage that recovered DNA has a higher molecular weight distribution. Procedures that fragment DNA to smaller sizes may not be desirable for cloning and may also increase the formation of chimeric amplification products during polymerase chain reaction (PCR) that would confound analysis of rDNA sequences. Most lysis procedures combine elements of both chemical and mechanical procedures.

Many, but certainly not all, bacteria may be lysed by a combination of enzymatic digestion of peptidoglycan by lysozyme, solubilization of cell wall constituents by a detergent such as SDS at 65°C, and physical disruption of cell walls by multiple rounds of freezing and thawing. This procedure may be used in direct lysis, and several studies have indicated that it decreases acridine orange direct counts (AODC) by more than 90% [24]. This procedure is considered to be relatively gentle, and the DNA recovered is typically of high molecular weight (greater than 30 kbp [5]). This method does not lyse endospores, however, and may not be efficient at lysing fungi and some difficult-to-lyse bacteria.

Another relatively gentle lysis procedure is boiling in SDS with optional rounds of freeze-thaw cycles [6, 15]. This procedure has been used to lyse cells isolated from both aquatic and terrestrial samples; however, the efficiency of lysis has not been evaluated at this time. An advantage of this procedure is that less degradation and higher efficiencies of recovery of RNA have been observed relative to lysis in SDS at 65°C and mechanical disruption. These results may be due both to higher efficiencies of lysis and inactivation of RNases present in the sample.

Aside from freeze-thaw cycles, the most commonly used mechanical disruption procedure is bead-mill homogenization. In this process, the environmental sample or cells is mixed with either glass or zirconium/silica beads in a homogenizer, such as the Bead-Beater (BioSpec Products, Bartlesville, OK). Shear force generated in the homogenizer disrupts cell walls and lyses most cells and endospores. In a comprehensive study comparing bead-mill homogenization with an SDS/freeze-thaw method, Moré et al. [11] found that bead-mill homogenization was much more efficient at lysing *Bacillus* endospores and somewhat more efficient at total lysis than the SDS/freeze-thaw method. Homogenization in a bead-mill is the most rigorous lysis procedure yet described and is likely to yield the least-biased recovery of nucleic acids from a sample, but it has certain disadvantages. DNA isolated by most bead-mill methods may be sheared to fragments of less than 10 kbp [8, 12], possibly increasing the potential for artifact formation during the PCR [7]. DNA recovered by this approach is suitable for hybridization with nucleic acid probes, however, as hybridization is not generally adversely affected by fragment size if the DNA is not degraded completely. Bead-mill homogenization may not be the most appropriate method for isolation of RNA, as increased amounts of RNA degradation relative to a boiling-lysis procedure have been observed [6]. This degradation is likely to be due to active RNases rather than to mechanical shearing because the relatively small size and compact conformation of RNA due to its secondary structure make it less available to mechanical shearing than DNA.

Very rigorous lysis methods should be used with caution in some cases with direct lysis procedures since they may yield nucleic acids from a variety of organisms, including non-microbial species. Surface soils frequently contain roots and root hairs that cannot be removed easily by sieving, and bead-mill homogenization might liberate nucleic acids from plants as well as microorganisms. The presence of plant DNA may confound some analyses, such as community cross-hybridization and random amplified polymorphic DNA (RAPD) fingerprinting. If one is interested only in microbial nucleic acids to the exclusion of non-microbial species, the possible contamination of microbial DNA with plant DNA should be checked. One approach to screening recovered DNA for the presence of plant DNA is by amplifying sequences related to the ribulose bisphosphate carboxylase gene, a gene found in all known plants [10]. However, this gene is highly conserved among most photosynthetic organisms, including bacteria, and this approach may not be suitable if high numbers of prokaryotic photosynthesizers are present in the matrix under study.

## Separation of Nucleic Acids from Particulates

If a direct lysis procedure is used, nucleic acids released by cell lysis may be absorbed to particulates in the sample, which may result in inefficient extraction of DNA and RNA. The extent of adsorption of nucleic acids to particulates is dependent on several factors, including the mineralogy, ionic composition, and pH of the matrix, and may be very significant in some matrices [13, 14]. Some soils have been shown to adsorb a maximum of over 500 mg DNA/g soil (A. Ogram, unpublished observations), and some pure clays may adsorb several times their own weight of DNA [4]. An understanding of the factors controlling adsorption may aid in designing appropriate extraction schemes.

DNA is a polyanion at most environmental pHs because of the presence of the phosphate groups in the sugar-phosphate backbone (pKa approximately 1). Of the components of DNA, the anionic phosphate groups probably have the greatest influence on the adsorption of DNA to particulates and may participate in several sorptive mechanisms. They may interact with divalent cations, such as calcium and magnesium, to form cation bridges to the predominantly anionic surface of particulates [9]. Chelating agents, such as EDTA, have been shown to release DNA adsorbed to sand and should be incorporated into extractants for this reason.

Iron and aluminum oxides are the most important sorbents of phosphate in soils and sediments [20] and are probably the most significant factors affecting the adsorption of nucleic acids to particulates. It is likely that a significant mechanism of adsorption is through interaction of phosphate groups in the sugar-phosphate backbone of nucleic acids with iron and aluminum groups on the surface of particulates. Preliminary studies have shown that increasing the amount of phosphate in extractants decreases the adsorption of DNA, presumably because of competition between phosphate in solution and phosphate in DNA for specific adsorption sites (A. Ogram et al., unpublished observations). Extractants containing phosphate should therefore be used to saturate as many phosphate-binding sites as possible to prevent the adsorption of nucleic acids. It should be noted that extraction of DNA from soils containing very high amounts of iron (greater than 10% iron) is extremely difficult, and alternative strategies would likely have to be developed for these soils.

The purine and pyrimidine bases of DNA and RNA may be protonated at pHs below 4, giving DNA both a cationic and anionic character. Cation exchange sites are much more prevalent than anion exchange sites on particulates, and cationic purines and pyrimidines would allow the nucleic acid to participate in cation exchange reactions with the soil surface, thereby increasing adsorption greatly. Increases in DNA adsorption at a low pH are particularly significant in matrices containing smectite clays (e.g., montmorillonite). DNA has been shown to enter the interlayer spaces of these clays, thereby expanding them and resulting in adsorption of very large amounts of DNA (16 mg DNA may be adsorbed by 1 mg of pure clay at a low pH [4]). To counteract this mechanism, the pH of extractants should be buffered so that the pH of the extractant/particulate mixture is more than 6. Extractants buffered to pH 8 are typically used for DNA, which should not adversely affect RNA. A disadvantage of using alkaline extractants is that higher amounts of organic carbon are also extracted from the matrix, thereby increasing the difficulty of purifying the nucleic acids.

For these reasons, most extractants are composed of alkaline phosphate buffers containing EDTA. We have found that sequential extractions with 0.2 M sodium phosphate buffer (pH 8.0) with 0.1 M EDTA will quantitatively extract DNA from many soils and sediments. Three or four extractions using two parts extractant to one part soil are generally sufficient to remove more than 90% of the free DNA from many samples.

## Purification

In contrast to the isolation of nucleic acids from pure cultures, the purification of nucleic acids from environmental samples is primarily concerned with the removal of humic substances, rather than proteins and phospholipids. Humic compounds are structurally different from proteins and phospholipids, and standard strategies for the purification of nucleic acids from pure cultures are typically ineffective in many environmental applications. These compounds are very large molecules of indeterminate structure and are operationally defined as the organic carbon component of soils, sediments, and natural waters that is alkaline-extractable. Humic compounds may be divided further into humic acids (alkaline-extractable and acid-precipitable) and fulvic acids (alkaline-extractable and acid-soluble [16]). Both humic and fulvic acids are thought to be formed from condensation reactions involving lignin and cellulose degradation products. They are large, water-soluble anionic polymers like DNA and RNA which co-purify with DNA and RNA in many purification schemes, and have been shown to be important inhibitors of the enzymatic manipulation of environmental nucleic acids, such as required for restriction digestion, radiolabeling, reverse transcription, and the PCR [22]. A substantial amount of data have also indirectly suggested that humic acids may form complexes with nucleic acids, and in some cases these complexes must be disrupted by chaotropic salts before complete purification is achieved. The exact nature of the postulated humic-DNA complex is unknown, but may be at least partially disrupted by treatment with 8 M urea, indicating that dipole-dipole interactions may be involved [23].

The degree of purification that is required is dependent on the application. Hybridization of specific sequences with labeled probes will generally tolerate some degree of humic contamination, whereas enzymatic manipulation, such as is required for PCR, restriction enzyme digestion, or reverse transcription of RNA, may be highly sensitive to the presence of humic contaminants [22]. The purification scheme may therefore be optimized based on the degree of contamination that the desired application can tolerate.

Several different strategies have been suggested to separate humic compounds from nucleic acids, but at this writing no single technique is available that will purify all samples of all humic compounds. The exact nature of humic compounds is likely to vary from environment to environment, and modifications of existing procedures are frequently required to obtain humic-free nucleic acids. A variety of approaches have been taken, including chromatographic separations [13, 23], electrophoretic separations [26], selective precipitations [25], extractions with organic solvents, and density-gradient centrifugation [5]. The particular method or combination of methods of choice is dependent on the nature of the sample, and in many cases the researcher should use his or her judgment as to the best approach. The principles and uses of some of the more commonly used techniques are described below.

### *Chromatographic Separations*

A variety of chromatographic methods have been used with varying degrees of success to separate humic substances from nucleic acids. Most of these have been open column chromatography applications, although more efficient and as yet untested high-performance liquid chromatography (HPLC) methods may meet with more success. In the first paper describing isolation of DNA from soils, Torsvik [23] used an ion exchange-sized exclusion resin (DEAE-Sepharose CL-6B; Pharmacia) intended to take advantage of differences in size and charge density between DNA and humic compounds. This approach achieved some degree of purification, but some humics still co-purified with DNA. Many humic compounds were separated from the DNA by elution along a potassium chloride gradient, but a significant fraction of the DNA co-eluted with humic compounds. Since humic and fulvic acids are composed of a wide range of sizes of individual molecules, they were distributed throughout the column.

Hydroxyapatite chromatography has also been used with moderate success in the purification of DNA from soils [12, 23]. DNA is bound to hydroxyapatite [$Ca_{10}(PO_4)_6OH_2$] through its phosphate groups, and humic compounds are eluted from the column by low molarities of phosphate buffer in 8 M urea. Hydroxyapatite has effectively separated humics from DNA in some cases, but is rarely used because hydroxyapatite open column chromatography is subject to clogging and very slow flow rates.

Several readily prepared, disposable chromatography columns are commercially available and have been used with environmental samples with mixed success [11, 17]. Most of these systems were developed for the rapid purification of high copy number plasmids from pure cultures of *E. coli* and may not be very efficient at purifying environmental DNA. Many of these columns may also select for small fragments of DNA, with the concomitant loss of some larger fragments. The use of these columns to purify environmental RNA has not been reported.

### *Selective Precipitations/Complexations*

A variety of compounds have been incorporated into mixtures of DNA and humic compounds in an effort to disrupt potential associations and to selectively precipitate one group of compounds while leaving the other in solution. Sodium acetate was originally incorporated into procedures for the isolation of DNA from pure culture to disassociate protein-nucleic acid interactions and may have a similar effect on humic-nucleic acid interactions. Ammonium acetate at a final concentration of between 1.5–2.5 M has been shown to precipitate significant amounts of humic compounds while leaving some DNA and RNA in solution (C. Guo and A. Ogram, in preparation). DNA may be trapped in the precipitate, however, resulting in the significant loss of nucleic acids. Another salt, cetyltrimethylammonium bromide (CTAB), forms complexes with proteins and polysaccharides in solutions of sodium chloride above 0.5 M and also aids in the removal of humic compounds from soil extracts. Protein/polysaccharide:CTAB and humic:CTAB complexes are then removed by extraction with $CHCl_3$ [1].

Polyvinylpolypyrrolidone (PVPP), a water-insoluble polymer, is commonly incorporated into protocols for the isolation of nucleic acids from plants because of its usefulness in removing phenolic compounds that co-purify with nucleic acids. PVPP has also been useful in the removal of some humic compounds, particularly in cell-fractionation procedures. PVPP is added either to a lysate or mixture of whole cells and humic compounds, and the resulting PVPP:humic complex is removed by centrifugation [5].

Selective precipitation of nucleic acids has been less successful in general than precipitation of humic compounds. Precipitation of nucleic acids with ethanol, isopropyl alcohol, and polyethylene glycol has been incorporated into many procedures as a means of concentrating nucleic acids, but does not remove many humics from the precipitate.

### *Extraction with Organic Solvents*

Extractions with phenol and chloroform are commonly used as denaturants of protein and for the solubilization of lipopolysacharides. They may be useful in removing some relatively non-polar humics, but in general are not very useful. Phenol:chloroform extractions may be most valuable in the purification of samples that have a relatively high biomass and relatively low humic content, such as in the isolation of nucleic acids from cells isolated from surface soils by the cell-fractionation method.

### *Agarose Gel Electrophoresis*

Agarose gel electrophoresis separates molecules on the basis of conformation, charge, and size and has been useful in separating DNA from humic compounds. Gels consisting of 0.8% agarose with 2% polyvinylpyrrolidone (PVP) may be particularly efficient at removing contaminants [26]. PVP, a water-soluble polymer, complexes with phenolic compounds (including the loading dye bromophenol blue), thereby retarding their movement through the gel. Humics electrophorese much more slowly than DNA in the presence of PVP for this reason, and purified DNA may be recovered from the gel. If DNA is to be recovered from gel slices, the DNA should move as a single band during electrophoresis to aid in its recovery. Gel purification may not be appropriate if a harsh lysis procedure is used that results in the shearing of DNA to a range of fragment sizes.

### *CsCl-Ethidium Bromide Density-Gradient Ultracentrifugation*

CsCl-ethidium bromide ultracentrifugation is commonly used in the separation of chromosomal, or linear, fragments of DNA from supercoiled plasmid DNA. The separation of the two forms of DNA is dependent on the intercalation of ethidium bromide to a greater degree in relaxed DNA than in supercoiled DNA. The greater the intercalation of ethidium bromide, the lower the density of the complex. A density gradient is established by centrifugation of CsCl at very high speeds, and each complex seeks its own density within the gradient [19]. Humic compounds represent a range of densities, and density-gradient centrifugation can separate some humic compounds from DNA [5]. As with most of the approaches discussed, this process must also be used frequently in combination with other approaches to effect complete purification of nucleic acids if the sample contains large amounts of humic compounds [12].

## CELL FRACTIONATION VERSUS DIRECT LYSIS

Two general strategies have been proposed for the isolation of nucleic acids from environmental samples: cell fractionation and direct lysis. Cell fractionation is based on the separation of microbial cells from the environmental matrix by differential centrifugation before lysis, whereas direct lysis relies on the lysis of cells in the presence of the environmental matrix and the subsequent extraction of the nucleic acids from the matrix. Each of these approaches has advantages and disadvantages, and the user should employ the approach suitable for his or her application.

In cell fractionation, bacterial cells are separated from fungal cells and soil particulates by first gently homogenizing the sample in a buffer designed to separate cells from particulates. A low-speed centrifugation of the homogenized mixture removes fungi and soil particles, leaving the bacterial fraction in solution. Bacteria may then be recovered from the supernatant by high-speed centrifugation. Multiple rounds of homogenization and differential centrifugation are generally used to increase the recovery of cells from a sample. This procedure results in the extraction of low amounts of humic compounds, some of which may be removed by treatment with the insoluble polymer PVPP. The purified bacterial fraction is lysed and the nucleic acids isolated by standard methods. Some humic compounds typically co-purify with nucleic acids isolated by this approach and must be removed as discussed in the previous section. The primary advantages of cell fractionation methods are that it may extract much fewer humic compounds than are typically extracted in direct lysis procedures, thereby facilitating purification; cells recovered by this method may be enriched in new, growing cells; very high molecular weight DNA (50 kbp) may be recovered; no extracellular DNA is recovered; and prokaryotes may be separated efficiently from eukaryotes, such as plant material and fungi, before lysis. Disadvantages of this approach are that it is time-consuming and is biased against the recovery of cells that are tightly adhered to soil particles [5].

Direct lysis methods were originally developed on the presumption that cell-fractionation methods may be biased against cells that are tightly bound to particulates [12]. It was thought that the lysis of cells before separation from particulates would yield nucleic acids that were more representative of the entire community. Most direct lysis approaches involve rigorous lysis procedures followed by extraction and subsequent purification of the extracted nucleic acids. Advantages of direct lysis methods over cell-fractionation methods are that more nucleic acids are recovered (five to six times more DNA is recovered), the recovered nucleic acids are less biased with regard to content (DNA from fungi and bacterial endospores may be recovered by direct lysis), and it is less time-consuming. The primary disadvantage of direct lysis is that large amounts of humic compounds may be extracted with the alkaline buffers used to extract nucleic acids, thereby increasing the difficulty of purification. Other disadvantages are that extracellular DNA must be pre-extracted from soil and sediment samples, and the potential exists for all of the nucleic acids present in a system to be extracted, including those present in bacterial endospores, fungi, dead cells, and plant material. The inclusive nature of direct lysis methods may be controlled to a certain extent by the choice of lysis procedures as discussed in the section, Cell Lysis.

## METHODS

Selected methods for the isolation of nucleic acids from environmental matrices are presented in this section. Most of these methods were developed for the isolation of nucleic acids from soils and sediments, but may be modified for use in many different matrices. The user is encouraged to modify any existing protocol to accommodate his or her individual samples and applications. Isolation of RNA from any source requires special handling due to the presence of RNases, and before attempting the isolation of RNA, the user should become familiar with precautions required when handling RNA [1, 19].

### Direct Lysis Methods

#### *Isolation of DNA*

The following procedure was optimized for use in surface soils containing moderate amounts of organic carbon (1% to 5%). This procedure has been used successfully to isolate DNA from a range of environmental matrices, including surface soils, compost [10], and ancient desiccated human feces [21]. The efficiency of recovery of purified DNA from surface soils is approximately 10%. It may not be appropriate for low-biomass samples or for the extraction of RNA.

Materials and Reagents

1. Sterile 50-ml polycarbonate Oak Ridge tubes
2. Washing buffer (120 mM sodium phosphate, pH 8): Mix 0.88 g $NaH_2PO_4 \cdot H_2O$ with 16.13 g $Na_2HPO_4$ and bring to 1 liter with $H_2O$; the pH of the resulting buffer should be pH 8.
3. Dry ice/ethanol bath or access to −70°C freezer
4. For SDS/freeze-thaw lysis:
   Lysis Solution I: 150 mM NaCl, 100 mM EDTA (pH 8), 10 mg/ml lysozyme
   Lysis Solution II: 100 mM NaCl, 500 mM Tris HCl (pH 8), 10% SDS
5. For bead-mill/SDS lysis: bead-mill homogenizer, 0.1-mm zirconium/silica beads, and Lysis Solution II
6. Extraction buffer: 0.12 M sodium phosphate buffer (pH 8), 10 mM EDTA
7. 5 M NaCl
8. 10% CTAB (hexadecyltrimethyl ammonium bromide) in 0.7 M NaCl
9. 13% polyethylene glycol (8000 MW) in 1.6 M NaCl
10. $CHCl_3$:isoamyl alcohol (24:1)
11. 10 M ammonium acetate
12. 70% ethanol in water
13. 95% ethanol
14. TE buffer: 10 mM Tris HCl, 1 mM EDTA (pH 8)

15. Sterile 1.5-ml microfuge tubes
16. Microcentrifuge

Note that all glassware and reagents except ethanol and chloroform should be sterilized by autoclaving.

Procedure: Removal of Extracellular DNA and Dispersal of Soil

1. Add 10 ml washing buffer to a 50-ml Oak Ridge centrifuge tube containing 1 to 5 g soil.
2. Vortex 1 min, and then let stand for 10 min with occasional mixing.
3. Centrifuge at 8000 rpm (7500 × g) for 10 min and discard supernatant.
4. Repeat once.

### *Lysis of DNA*

Two alternative lysis procedures are presented here. The first (SDS/freeze thaw) is gentler, yields the highest molecular weight DNA, and reduces AODC by greater than 90% [24]. The second (bead-mill/SDS) is much more rigorous, yields DNA of a range of fragment sizes, and results in a 94% decrease in AODC [11]. The bead-mill/SDS procedure may also be much more efficient at lysing eukaryotes [8].

Procedure: SDS/Freeze-Thaw Lysis

1. Add 8 ml of Lysis Solution I and mix.
2. Incubate at 37°C with occasional mixing for 1–2 h.
3. Add 8 ml of Lysis Solution II, and perform three cycles of freezing at −70°C or in dry ice/ethanol bath for 20 min and thawing at 65°C for 20 min.
4. Centrifuge at 8000 rpm (7500 × g) for 10 min and discard pellet.
5. There may be very small amounts of particulates present in the supernatant that may carry through the rest of the procedure and inhibit PCR. Remove these particulates by filtering through a pre-wet Kimwipe into a fresh Oak Ridge tube. Extraction of the soil pellet twice with 5 ml of extraction buffer improves recovery efficiency, but is generally not required for high-biomass surface soils.

Procedure: Bead-mill/SDS Lysis [11]

1. Add 0.5 g soil and 0.5 ml Lysis Solution II to 2-ml screw-cap polypropylene microcentrifuge containing 2.5 g 0.1-mm diameter zirconium/silica beads. Add 0.25 ml lysis buffer to this mixture.
2. Shake each tube at high speed for 10 min in a bead-mill homogenizer, such as the BioSpec Mini-Bead Beater (BioSpec Products, Bartlesville, OK).
3. Remove tubes from the bead-mill, centrifuge for 3 min at 12,000 × g, and collect the supernatant. Extraction of the soil pellet twice with 5 ml of extraction buffer will improve recovery efficiency but is generally not required for high-biomass surface soils.

Procedure: Purification

1. To supernatant, add 2.7 ml 5 M NaCl, 2.1 ml 10% CTAB in 0.7 M NaCl (final concentrations should be approximately 0.7 M NaCl and 1% CTAB). Mix and incubate at 65°C for 10 min. Add an equal volume of $CHCl_3$:

isoamyl alcohol (24:1), and mix by inversion. Centrifuge at 5000 rpm (3000 × g) for 5 min, and transfer upper phase to a clean tube. Discard lower phase.
2. Add an equal volume of 13% polyethylene glycol (PEG) in 1.6 M NaCl to the upper phase and mix. Keep on ice for 10 min. Centrifuge at 10,000 rpm (12,000 × g) for 15 min, and discard the supernatant. Wash pellet with 70% ethanol. Dry pellet, dissolve nucleic acids in 750 μl TE, and transfer to 1.5 ml microfuge tube.
3. Add 110 μl (190 ml for very dark suspensions) 10 M ammonium acetate (final concentration of between 1.5 M to 2.5 M ammonium acetate); keep on ice for 10 min. Centrifuge for 15 min, and transfer the supernatant to a clean microfuge tube. Discard pellet. *Note*: A significant amount of DNA and RNA may be lost in this step due to co-precipitation with humic compounds. In our experience, ammonium acetate precipitation is required in this procedure for samples containing organic contents greater than 1%, and efficiency of recovery is sacrificed for a higher degree of purification.
4. Add 2 vol ethanol to the upper phase. Centrifuge for 15 min, discard the supernatant, and wash the pellet with 0.5 ml 70% ethanol. Allow the pellet to dry, and dissolve in 200 μl TE. DNA recovered at this stage is generally amplifiable by PCR.
5. If DNA is still not amplifiable, electrophorese in 0.8% low-melting temperature agarose containing 2% PVP [26]. Cut DNA band from gel, add an equal volume of water, and melt at 65°C. PCR may be performed in the melted gel slice. Alternatively, bypass the agarose step and repeat the first step in purification followed by ethanol precipitation to obtain amplifiable DNA (E. Jutras, personal communication).

### *RNA Isolation from Sediments by Direct Lysis*

The following procedure was modified from existing protocols for the isolation of RNA from marine samples for use with surface soils containing organic carbon contents of between 1% to 5% [6]. The efficiency of recovery of mRNA by this method ranges from over 25% for samples containing negligible amounts of organic carbon (less than 0.1%) to less than 10% for samples containing over 1% organic carbon [15]. The presence of nucleases is of prime concern during the isolation of RNA by direct lysis methods, and much of this procedure is devoted to their inactivation. The lysis procedure described here is thought to be more efficient at inactivating nucleases than either the SDS/freeze-thaw method or the bead-mill/SDS homogenization procedure, and is therefore recommended for RNA isolations. Its efficiency of lysis has not yet been evaluated.

#### Materials and Reagents

*Note*: Autoclave all glassware and plasticware, and treat all solutions with 0.1% DEPC from a 10% DEPC stock, followed by autoclaving to inactivate nucleases. Wear gloves during the entire procedure.

1. Autoclaved 50-ml polycarbonate Oak Ridge tubes
2. Extraction buffer: 0.2 M sodium phosphate buffer (pH 8), 0.1 M EDTA: Prepare 0.2 M sodium phosphate buffer (pH 8) by mixing 1.46 g $NaH_2PO_4$ and 26.88 g $Na_2HPO_4$ with water for a final volume of 1 liter
3. 10% SDS
4. Guanidinium isothiocyanate-phenol-sarkosyl solution (GIPS [18]): 4 M guanidinium isothiocyanate, 0.5% sarkosyl, 25 mM sodium citrate (pH 7)

5. Buffered phenol (molecular biology grade)
6. $CHCl_3$:isoamyl alcohol (24:1)
7. 5 M NaCl
8. 10% CTAB (hexadecyltrimethyl ammonium bromide) in 0.7 M NaCl
9. 13% polyethylene glycol (8000 MW) in 1.6 M NaCl
10. 10 M ammonium acetate
11. 70% ethanol in water
12. 95% ethanol
13. TE buffer: 10 mM Tris HCl, 1 mM EDTA (pH 8)
14. DNase-free RNase (10 mg/ml)
15. Optional: RNase-free DNase (10 mg/ml)
16. Boiling water bath
17. Sterile 1.5-ml microfuge tubes
18. Microcentrifuge

Procedure: Lysis

1. Add 1 g of soil to 50-ml Oak Ridge tubes containing 8 ml of extraction buffer and 1.5 ml 10% SDS, and place the mixture in a boiling water bath for 5 min. Transfer the boiled samples to either a dry ice/ethanol bath or a −80°C freezer for 30 min followed by thawing in a 65°C water bath for 15 min.
2. Extract nucleic acids from the sediment by centrifuging the lysate at 12,000 × g at 10°C for 15 min.
3. Transfer the supernatant to a fresh tube, and resuspend the pellet in 5 ml of lysis buffer for a second round of boiling as described above.
4. Centrifuge the boiled mixture as described above, and pool the supernatants.
5. Discard the pellet.

Procedure: Inactivation of Nucleases

1. Add 11.2 ml GIPS, 11.2 ml phenol, and 6.4 ml $CHCl_3$:isoamyl alcohol (24:1) to the pooled supernatants.
2. Centrifuge the mixture at 6000 × g at 4°C for 10 min.
3. Transfer the aqueous upper phase to a clean Oak Ridge tube.

Procedure: Purification

1. Remove humic compounds from the extracts as described in Step 3 of the Purification procedure.
2. Note that all reagents must be treated with 0.1% DEPC before use with RNA.

Procedure: Digestion of DNA and Recovery of RNA

1. After the recovery of all nucleic acids by precipitation with ethanol as described in Step 3 of the Purification procedure, remove DNA by digestion with RNase-free DNase I.
2. Add DNase I to a final concentration of 10 mg/ml, and incubate for 1 h at 37°C.
3. Remove DNase I by extraction with 0.5 vol phenol:chloroform:isoamyl alcohol (25:24:1).
4. Mix the phenol:$CHCl_3$ gently with the aqueous phase by inverting the tube several times, and centrifuge for 30 s at high speed in microcentrifuge.

5. Transfer the upper (aqueous) phase to a clean microcentrifuge tube, and discard the lower (organic) phase.
6. Recover RNA by precipitation of the RNA with ethanol and resuspension in TE as described above.

## Cell-Fractionation Method

The following procedure is based on a previously published protocol for isolation of bacterial nucleic acids from a starting sample size of 50 g soil [5].

### Materials and Reagents

1. Six Waring blenders with 40 fl oz (1.2 liter) glass jars
2. Ice/water bath for six blender jars
3. Twelve 250-ml centrifuge bottles
4. Superspeed centrifuge
5. Six small paint brushes
6. Twelve 50-ml polycarbonate Oak Ridge tubes
7. Vortex mixer
8. Water bath set at 37°C
9. Water bath set at 65°C
10. Refractometer to measure initial density of cesium chloride gradient
11. Ultracentrifuge
12. Twelve ultracentrifuge tubes
13. 10x Winogradsky's Salt Solution (10x WS):
    Dissolve 5.0 g $K_2HPO_4$ in 880 ml $H_2O$.
    Dissolve the following in a separate 800 ml $H_2O$:

    | | |
    |---|---|
    | $MgSO_4 \cdot 7H_2O$ | 5.0 g |
    | NaCl | 2.5 g |
    | $Fe_2(SO_4)_3 \cdot H_2O$ | 0.050 g |
    | $MnSO_4 \cdot 4H_2O$ | 0.050 g |

    Combine the above and adjust to pH 6.0 with concentrated HCl.
    Bring the final volume to 2 liters with $H_2O$. Before use, dilute 1:10 and autoclave.
14. Homogenization solution: 1x Winogradsky's salt solution, 0.2 M sodium ascorbate
15. Acid-washed PVPP:
    Slowly add 300 g PVPP to 4 liters of 3 M HCl while stirring. Cover beaker and continue stirring overnight.
    Filter the suspension through Miracloth or several layers of cheesecloth, and resuspend the PVPP in 4 liters $H_2O$. Mix for 1 hour, and filter a second time through Miracloth or cheesecloth.
    Resuspend the PVPP in 4 liters of 20 mM potassium phosphate buffer (pH 7.4), and mix for 1 to 2 h. Repeat filtration and washes with 20 mM phosphate buffer until the suspension has a pH of 7.0. After the final filtration, spread the PVPP on lab paper and let air-dry overnight.
16. TE: 33 mM Tris, pH 8.0; 1 mM EDTA, pH 8.0
17. 5 M NaCl
18. 20% sarkosyl
19. Tris/sucrose/EDTA: 50 mM Tris, pH 8.0; 0.75 M sucrose; 10 mM EDTA, pH 8.0
20. 40 mg/ml lysozyme solution in TE: Prepare 5 ml on the day of use and store on ice until ready for use.

21. Pronase E: 50 mg/ml pronase (type XXV from *Streptomyces griseus*) in TE: Preincubate for 30 min at 37°C before use to inactivate any nucleases contaminating the Pronase stock.
22. Ethidium bromide: 10 mg/ml in TE: Stir overnight by magnetic stirrer.
23. CsCl balance solution ($R_f$ = 1.3885):
    Add 250 g molecular biology grade CsCl to 250 ml $H_2O$, and mix until dissolved.
    Add 12.5 ml of the 10 mg/ml ethidium bromide stock solution, and mix. Adjust the mixture as necessary to a final $R_f$ of 1.3885. Alternatively, the density of the final solution should be 1.58.

Procedure: Cell Fractionation

1. Mix 50 g soil sample with 200 ml of homogenization solution and 15 g acid-washed PVPP in a blender jar. Homogenize for 1 min followed by cooling in an ice/water bath for 1 min. Repeat two times.
2. Pour the homogenized solution into a 250-ml centrifuge bottle and centrifuge at 640 × g for 15 min at 4°C. Carefully pour the supernatant into a clean 250-ml centrifuge bottle, and collect the bacterial fraction by centrifugation at 14,740 × g for 20 min at 4°C.
3. Add 200 ml homogenization buffer to the soil pellet, and repeat the homogenization and centrifugations steps (1 and 2) twice. Combine all bacterial pellets.
4. Wash the cell pellet by carefully resuspending the combined cell pellet from Step 3 in 200 ml TE using a small clean paint brush. Centrifuge at 14,470 × g for 20 min at 4°C, and discard the supernatant.

### *Lysis*

This lysis procedure is designed for recovering very high molecular weight DNA (>50 kbp) from a variety of bacterial species. If high molecular weight DNA is not required, substitution of the lysis procedures presented above may be more efficient at lysing a broader range of species.

Procedure

1. Resuspend the cell pellet in 20 ml TE using a small clean paint brush, and transfer the suspension to a 50-ml Oak Ridge tube. Mix with 5 ml 5 M NaCl and 125 ml 20% sarkosyl. Incubate at room temperature for 10 min.
2. Centrifuge at 11,220 × g for 20 min at 4°C. Gently resuspend the cell pellet in 3.5 ml Tris/sucrose/EDTA with a paint brush. Add 0.5 ml pronase E, and mix by vortexing. Incubate at 37°C for 30 min without shaking.
3. Place in a 65°C water bath for 10 min, and then add 250 ml 20% sarkosyl. Incubate at 65°C for 40 min. Transfer to an ice bath, and allow to stand for at least 30 min.
4. Centrifuge at 25,260 × g for 1 h at 4°C. Transfer the supernatant to a clean Oak Ridge tube, and discard the pellet.

### *Purification*

This purification procedure is based on separating humic compounds remaining in the lysate by CsCl-ethidium bromide density-gradient ultracentrifugation. It is much simpler than those purification schemes required for direct lysis procedures, and those procedures should not be substituted here unless impure nucleic acids are recovered after ultracentrifugation.

Procedure

1. Add 9 ml sterile distilled $H_2O$, 12.7 g CsCl, and 1.5 ml 10 mg/ml ethidium bromide. Mix by inversion until the CsCl is dissolved. Adjust the refractive index to between 1.3865 and 1.3885 or alternatively to a density of between 1.55 and 1.58 by the addition of CsCl or $H_2O$ as needed.
2. Transfer the mixture to ultracentrifuge tubes, seal, and centrifuge at 255,800 × g for between 9 and 16 h at 18°C. Remove the DNA band, extract ethidium bromide, and desalt by standard procedures [1, 19].

*Acknowledgments* This work was partially funded by a grant from the US Environmental Protection Agency.

## REFERENCES

1. Ausubel, F.M., Brent, R., Kingston, R.E., Moore, D.D., Seidman, J.G., Smith, J.A., and Struhl, K. 1990. Current Protocols in Molecular Biology. Greene Publishing Association and Wiley-Interscience, New York.
2. Fuhrman, J.A., Comeau, D., Hagstrom, A., and Chan, A.M. 1988. Extraction from natural planktonic microorganisms of DNA suitable for molecular biological studies. Appl. Environ. Microbiol. 54:1426–1429.
3. Goodwin, D.C. and Lee, S.B. 1993. Microwave miniprep of total genomic DNA from fungi, plants, protists and animals for PCR. Biotechniques 15:438–444.
4. Greaves, M.P. and Wilson, M.J. 1969. The adsorption of nucleic acids by montmorillonite. Soil Biol. Biochem. 1:317–323.
5. Holben, W.E. 1994. Isolation and purification of bacterial DNA from soil. In Methods of Soil Analysis, Part 2: Microbiological and Biochemical Properties (Weaver, R.W., Angle, J.S., Bottomley, P.S., Eds.). Soil Science Society of America, Madison, WI.
6. Jeffrey, W.H., Nazaret, S., and von Haven, R. 1994. Improved method for recovery of mRNA from aquatic samples and its application to detection of mer expression. Appl. Environ. Microbiol. 60:1814–1821.
7. Kopczynski, E.D., Bateson, M.M., and Ward, D.M. 1994. Recognition of chimeric small-subunit ribosomal DNAs composed of genes from uncultivated microorganisms. Appl. Environ. Microbiol. 60:746–748.
8. Leff, L.G., Dana, J.R., McArthur, J.V., and Shimkets, L.J. 1995. Comparison of methods of DNA extraction from stream sediments. Appl. Environ. Microbiol. 61:1141–1143.
9. Lorenz, M.G. and Wackernagel, W. 1994. Bacterial gene transfer by natural genetic transformation in the environment. Microbiol. Rev. 58:563–602.
10. Malik, M., Kain, J., Pettigrew, C., and Ogram, A. 1994. Purification and molecular analysis of microbial DNA from compost. J. Microbiol. Meth. 20:183–196.
11. Moré, M.I., Herrick, J.B., Silva, M.C., Ghiorse, W.C., and Madsen, E.L. 1994. Quantitative cell lysis of indigenous microorganisms and rapid extraction of microbial DNA from sediment. Appl. Environ. Microbiol. 60:1572–1580.
12. Ogram, A., Sayler, G., and Barkay, T. 1987. The extraction and purification of microbial DNA from sediments. J. Microbiol. Meth. 7:57–66.
13. Ogram, A., Sayler, G., Gustin, D., and Lewis, R. 1988. DNA adsorption to soils and sediments. Environ. Sci. Technol. 22:982–984.
14. Ogram, A.V., Mathot, M.L., Harsh, J.B., Boyle, J., and Pettigrew, C.A. 1994. Effects of DNA polymer length on its adsorption to soils. Appl. Environ. Microbiol. 60: 393–396.

15. Ogram, A., Sun, W., Brockman, F.J., and Fredrickson, J.K. 1995. Isolation and characterization of RNA from low-biomass deep-subsurface sediments. Appl. Environ. Microbiol. 61:763–768.
16. Page, A.L., Ed. 1986. Methods of Soil Analysis, 2nd ed. Vol. 2: Agronomy Monograph 9. American Society of Agronomy, Madison, WI.
17. Picard, C., Ponsonnet, C., Paget, E., Nesme, X., and Simonet, P. 1992. Detection and enumeration of bacteria in soil by direct DNA extraction and polymerase chain reaction. Appl. Environ. Microbiol. 58:2717–2722.
18. Pichard, S.L. and Paul, J.H. 1993. Gene expression per gene dose, a specific measure of gene expression in aquatic microorganisms. Appl. Environ. Microbiol. 59:451–457.
19. Sambrook, J., Fritsch, E.F., and Maniatis, T., 1989. Molecular Cloning: A Laboratory Manual, 2nd ed. Cold Spring Harbor Press, Cold Spring Harbor, NY.
20. Sanyal, S.K. and De Datta, S.D. 1991. Chemistry of phosphorus transformations in soil. Adv. Soil Sci. 16:1–120.
21. Sutton, M.Q., Malik, M., and Ogram, A. Experiments on the determination of gender from coprolites by DNA analysis. J. Archaeol. Sci., in press.
22. Tebbe, C.C. and Vahjen, W. 1993. Interference of humic acids and DNA extracted directly from soil in detection and transformation of recombinant DNA from bacteria and a yeast. Appl. Environ. Microbiol. 59:2657–2665.
23. Torsvik, V.L. 1980. Isolation of bacterial DNA from soil. Soil Biol. Biochem. 12: 15–21.
24. Tsai, Y.-L. and Olson, B.H. 1991. Rapid method for direct extraction of DNA from soils and sediments. Appl. Environ. Microbiol. 57:1070–1074.
25. Xia, X., Bollinger, J., and Ogram, A. 1995. Molecular genetic analysis of the response of three soil microbial communities to applications of 2,4-D. Mol. Ecol. 4:17–28.
26. Young, C.C., Burghoff, R.L., Keim, L.G., Minak-Bernero, V., Lute, J.R., and Hinton, S.M. 1993. Polyvinylpyrrolidone-agarose gel electrophoresis purification of polymerase chain reaction-amplifiable DNA from soils. Appl. Environ. Microbiol. 59:1972–1974.

13

ROBERT S. BURLAGE

# Molecular Techniques

Molecular biology is now an indispensable facet of microbial research. The examination of DNA and RNA can provide a wealth of information about microorganisms and microbial communities. For instance, the appropriate DNA probes can show which types of bacteria are found in certain environments, whereas reverse transcriptase-polymerase chain reaction (RT-PCR) can demonstrate which genes are functional in those samples. However, the proper use and interpretation of the data are essential and limits of resolution must always be considered.

Molecular biology spans a wide range of techniques, and an attempt has been made to put these several techniques into an understandable framework. Some techniques, such as PCR analysis or DNA hybridization, are common to all microorganisms since they are dependent on the isolation of DNA and not on the microorganism itself. Other techniques are strain-dependent. For instance, transformation or electroporation of DNA into a new isolate will not have been previously described in the literature, and therefore the researcher should start with a known technique and modify it as needed. These techniques have been shown to be efficacious for specific strains, but there is no guarantee that they will work for your strain. This is especially true for plasmid cloning vectors, in which suitable hosts are often found by trial and error. However, the techniques presented in this chapter make an excellent place to start.

Genetic manipulation of bacteria is usually carried out on Gram-negative microorganisms, which is not difficult to understand. If genetic manipulation presents a difficult task, it pays to have an organism that is easy to work with. Yet, many of the most interesting species are difficult to work with, either because they are slow growers or because they have fastidious growth conditions. The researcher should refer to the chapters in Part I to determine appropriate growth conditions. These chapters also provide some information on genetic analysis, particularly DNA probes. A comprehensive analysis of genetic techniques for each genus of bacteria would require several volumes, and therefore only the basic guidelines are presented here. However, references for specific genera are not difficult to find [8, 14, 21, 28, 32].

The objective of this chapter is to provide the microbial ecologist with the "work-horse" tools of molecular biology so that he or she can apply the most useful techniques immediately. For this reason, reference is made at every opportunity to commercial kits or simple techniques that save time. Some techniques require expensive, specialized equipment or highly skilled personnel. These techniques, such as DNA sequencing or PCR primer synthesis, are not described in great detail since they require a greater commitment to molecular research. The researcher who is contemplating such a venture would be better off attending one of the many fine workshops where the techniques are taught by experienced workers on a one-to-one basis. In addition, there are companies that provide these services, and it is probably more efficacious for the novice to utilize them for small jobs. However, when your molecular biology needs expand significantly, it may be time to learn the techniques yourself and purchase the equipment.

The techniques that are presented here, although not a comprehensive treatment, will enable the researcher to perform many of the most useful procedures. Excellent laboratory manuals are available for those who need greater detail or for those who wish to learn more about the field of molecular biology [5, 44]. In many cases, there is scant knowledge on the genetic manipulation of a particular species, and the researcher should use these general guidelines to determine which techniques will be useful.

## GENE CLONING—AN OVERVIEW

The heart of molecular biology is the cloning of a fragment of DNA. This DNA fragment may encode a gene for a desired trait, may provide a DNA probe for a specific group of microorganisms, or perhaps will create a fusion gene that makes the microorganism easier to study or trace. Cloning entails removal of the DNA from one organism and introduction into another organism and has several basic steps:

1. extraction of DNA from a microorganism (see Chapter 12)
2. restriction digestion of the DNA
3. separation of the DNA fragments by gel electrophoresis (omitted if shotgun cloning is used)
4. recovery of specific DNA fragment from gel (omitted if shotgun cloning is used)
5. extraction and restriction digestion of the cloning vector
6. ligation of the DNA fragment and cloning vector

After construction of the recombinant DNA molecule, it is introduced into a host strain by the process of transformation (see the section, DNA Transfer). This is an inefficient process, and therefore a strain with tolerable transformation efficiency is used first. This strain is usually *E. coli*, and several strains are commercially available for this step. After the successful introduction of cloned DNA into *E. coli*, the plasmid clones are extracted by a mini-prep technique (Chapter 12 or [5, 44]) and screened for the proper DNA fragment either by restriction digestion or colony hybridization. The correct clone can be moved into the final host, such as the wild-type strain, by electroporation or conjugation. Verification of the introduction of the cloned DNA can be carried out by DNA hybridization (often also using restriction digestion and separation of fragments on gels) or PCR amplification.

## DNA Extraction and Quantification

### *DNA Extraction*

In general, DNA is DNA. This is fortunate, since the same techniques can be applied to a sample of bacterial DNA as to yeast DNA or even to human DNA. Once you have isolated DNA from your microorganism or community or environmental sample (see Chapter 12), you can apply these standard techniques to analyzing the DNA. However, for several reasons manipulation of DNA from wild-type bacteria will probably be more difficult than from an attenuated strain of *E. coli*. The DNA may be protected by a DNA modification system, which might interfere with the ability of certain restriction enzymes to cut it. The bacterial strain is likely to have its own restriction enzymes, making transformation a very inefficient process. Enzyme-inhibiting substances in soil and groundwater may be present, particularly for humic acids and fulvic acids, which co-purify with nucleic acids. Recent advances in nucleic acid purification have largely eliminated this difficulty (see Chapter 12), although the researcher should avoid complacency in DNA extraction from any environmental sample; it is still partly art and partly science.

### *DNA Quantification*

Measurement of DNA concentration can be performed in several ways [5, 44]. For small concentrations of DNA, I recommend the use of Hoechst dye 33258 (Sigma Chemical, St. Louis). Unlike other methods, as little as 50 ng can be detected [36]. Hoechst 33258 is weakly fluorescent in solution, but increases in fluorescence when bound to DNA. This increased sensitivity means that less sample DNA needs to be sacrificed to obtain a concentration.

Procedure

1. Prepare Hoechst dye 33258 as a stock solution of $1.5 \times 10^{-4}$ M in distilled water. Store in a dark bottle (e.g., wrapped in aluminum foil) at 4°C. If the stock solution is older than 1 to 2 weeks, make a fresh solution.
2. Make a working solution by diluting the stock solution to $1.5 \times 10^{-6}$ M.
3. Add a fraction of the DNA solution (e.g., 1–5 μl) to 2 ml of distilled water to create the unknown sample.
4. Add 1 ml of the dye working solution to the unknown DNA solution in a fluorimeter cuvette, and mix quickly by inversion. Let stand in the dark for 5–10 min before reading.
5. Measure fluorescence in a fluorimeter (Hoefer Scientific Instruments, San Francisco).
6. Compare fluorescence measurement to a series of DNA standards (see Notes below).
7. If the DNA concentration is low (sub-microgram), use a working solution of $1.5 \times 10^{-7}$ M. Better results are obtained with this solution because the unbound dye still has a weak fluorescence.

Notes

1. Make DNA standards with a commercially prepared DNA source, such as herring sperm or salmon sperm DNA. Start with a 1 mg/ml solution, and create 10-fold dilutions. Fragmentation of the DNA is recommended for the

initial solution. Use either a brief sonication, or pass the DNA solution through a small-bore needle several times.
2. Use negative control as well. Add DNase to a sample to determine whether any fluorescence is due to the presence of contaminants.

## DNA Digestion Using Restriction Enzymes

Digestion (cutting) of DNA with restriction enzymes is the work-horse technique of molecular biology. Among other things, it is useful for isolating small regions of DNA that contain genes of interest, building new and useful vectors for genetic analysis, isolating fragments for cloning or for labeling for DNA probe hybridization, screening clones, and mapping DNA. DNA digestion is now such a routine technique that it is not unusual to find demonstrations in high-school classrooms. If the DNA is free from impurities few things can go wrong with this procedure. If something does go wrong, there are two major possibilities: (1) You forgot to add one of the reagents, or (2) the enzyme is no longer effective because it is too old or because someone left it out at room temperature. If you suspect (1), then repeat the digestion. If you suspect (2), try another digestion with a sample of DNA that you know from experience will cut with that enzyme.

### Procedure

1. In a sterile Eppendorf tube, add the following reagents to a total volume of 20 μl.

| | |
|---|---|
| Sterile distilled water | 16 μl |
| 10X restriction buffer | 2 μl |
| DNA sample | 1 μl |
| Restriction enzyme | 1 μl |

2. Incubate the tube at the appropriate temperature for 1 h. Placing the tube in a water bath is usually best.
3. Stop the reaction by heat-killing the enzyme for 5 min at 65°C or by adding 0.5 M EDTA (pH 8.0) to a final concentration of 10 mM. This step is usually unnecessary if the samples are analyzed immediately.
4. In many cases the DNA needs to be cut by two different enzymes. Generally, cut with one enzyme at a time, and start with the best cutter (e.g., *Eco*RI, *Bam*HI). If the enzyme buffers are comparable, add the second enzyme after the first enzyme is done. If the buffers are significantly different, precipitate the DNA between digests. After restriction digestion the DNA sample is ready for concentration by DNA precipitation, which is a preparation before ligation and subsequent cloning, or for separation of DNA fragments by agarose gel electrophoresis.

### Notes

1. The described conditions work well for DNA samples between 0.2 and 2.0 μg. If your sample is more dilute, adjust the volumes of DNA and distilled water accordingly. If you wish to cut more DNA, it is probably best to make duplicate tubes and combine them later before precipitation.
2. The volume used here is convenient for most DNA manipulations. To handle these small volumes, use a set of good-quality micropipettors and sterile pipette tips.
3. Most suppliers of restriction enzymes also supply the restriction buffers. These buffers are convenient to use and have passed quality control testing.

However, you can make your own buffers if the need arises; most buffers are relatively easy to make, and the formulas are provided with the enzyme. Check to be sure that the manufacturer's buffer is a 10x solution.

4. Enzymes should *always* be kept on ice or in the freezer (−20°C). They will lose potency at room temperature. Some workers use latex gloves when handling enzymes in order to avoid contaminating the enzyme with DNases from the skin.
5. Digestion for longer than an hour usually does not hurt the sample, but neither does it result in more complete digestion.
6. Restriction enzymes are stored in a glycerol solution (note that the enzymes do not require thawing before use). Glycerol at high concentrations can inhibit enzyme activity, so do not overload a tube with enzyme to get faster cutting.

Many companies sell restriction enzymes. Their catalogs contain information on the appropriate storage, handling, and assay conditions for restriction enzymes, as well as other useful enzymes (e.g., ligase, polymerases) and molecular reagents. In addition, the catalogs are frequently an excellent source of information on cloning techniques and DNA analysis. Some of these suppliers are Boehringer Mannheim (Indianapolis), Gibco-BRL (Gaithersburg, MD), New England Biolabs (Beverly, MA), and Promega (Madison, WI).

DNA digestion results in two basic types of DNA ends: blunt and sticky. Blunt ends have no single-stranded overhangs, whereas sticky ends do. Overhangs are important, because sticky ends, as the name implies, can form basepairing with other molecules and thus aid in the ligation effort. Many cloning vectors include sticky-end restriction sites because of the relative ease with which cloning can be accomplished.

To facilitate the creation of sticky ends, short fragments of double-stranded DNA, called linkers, are commercially available. These linkers typically contain one or two useful restriction sites and are added to the blunt-ended DNA by ligation (Fig. 13-1). The ligation step is followed by DNA precipitation and another restriction digestion, this time for the linker site. The presence of multiple linkers on an end is thus not a problem, since all but the end unit will be cleaved off. Follow the manufacturer's recommendations for reaction conditions, since the concentration of the linker is important to the success of the procedure. The basic ligation protocol is the model for this reaction. A clean-up step is recommended to eliminate the unused linkers since they can possibly interfere with subsequent enzymatic reactions. Use one of the techniques outlined below to clean up. In a similar method, adapter molecules can be used to change one site into another, which may allow cloning into a suitable plasmid vector (Fig. 13-2).

## Gel Electrophoresis

Gel electrophoresis is used to separate fragments of DNA based on their size. The gel provides a semisolid matrix through which the DNA migrate, and an electrical current pulls the DNA through the gel. A "power supply" provides the direct current used for electrophoresis. It is actually a transformer that converts the alternating current from the wall outlet to a direct current for the unit. Electrophoresis equipment is sold by many distributors of scientific equipment and is simple enough in design that it can be built by the investigator. Most units designed for agarose gel electrophoresis have the gel in a flat (horizontal) position, which allows its convenient forming, loading, and handling.

Either agarose or polyacrylamide gels can be used for electrophoresis. In practice, the agarose gels are preferred, since they are easy to set up and handle, the

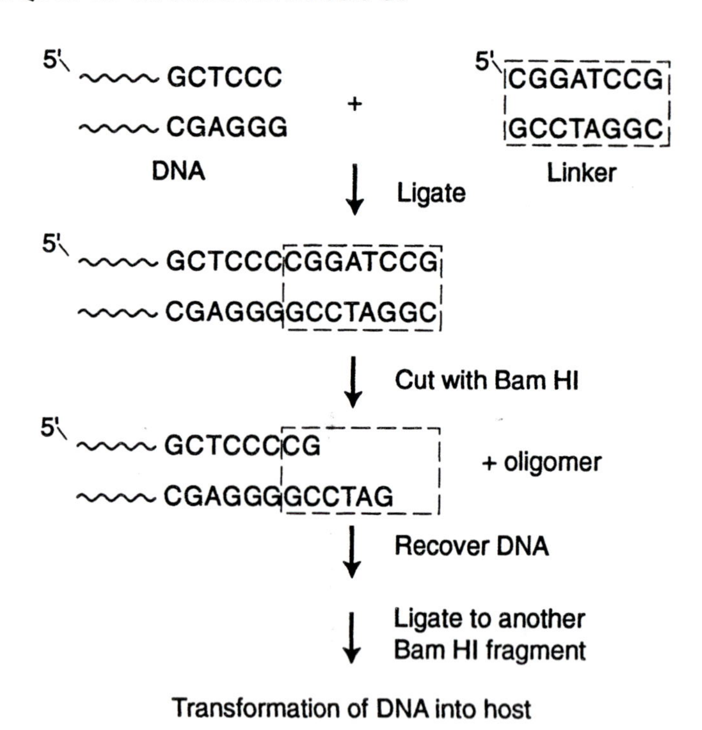


Figure 13-1 Addition of short linkers at the termini of DNA strands allows incorporation of new restriction sites that aid in cloning.

reagents are few and relatively inexpensive, and the gels are fairly durable. Polyacrylamide gels are very useful for low molecular weight DNA (i.e., less than 500 basepairs) or for resolving small differences between low molecular weight bands (e.g., sequencing gels). However, they require some practice to assemble and are not especially easy to handle, and one of the reagents (acrylamide) is rather toxic. For most experiments agarose gels work quite well, and so polyacrylamide gels are not described here. Simple agarose gels can be used to visualize DNA fragments between approximately 200 basepairs and 30 kilobasepairs.

DNA migrates through the agarose gel according to size, with the smallest fragments traveling the farthest and the larger fragments trailing behind. For DNA fragments in the range of 0.25–2.0 kb, use a 1.5% agarose gel; for DNA between 0.4–7 kb, use a 1.0% gel; for DNA between 1–12 kb, use a 0.6% gel. Resolution of the fragments increases slightly if the gel is run at lower voltages; the bands appear sharper and more defined.

DNA markers are especially useful for gel electrophoresis. These markers are DNA fragments with known sizes that are run alongside the samples. Their presence makes it easier to distinguish the approximate size of the fragments. The most popular marker is Lambda phage that has been cut with *Hin*DIII enzyme. This marker can be obtained commercially or prepared easily in the laboratory with intact Lambda phage DNA. The DNA Ladder series (Gibco-BRL) is especially helpful. At least one marker lane should be included on each gel, and it is often preferable to include several. Application of 1 $\mu$l of the stock marker ($\leq$1 $\mu$g) is sufficient. Markers are available that have been labeled, such as with digoxigenin. These markers can be used in Southern blots and developed in the same manner as the DNA probe, providing the researcher with convenient markers at the end of the procedure.

Agarose gels are used for the separation of DNA fragments on the basis of molecular weight. This separation is useful for mapping of a fragment of DNA

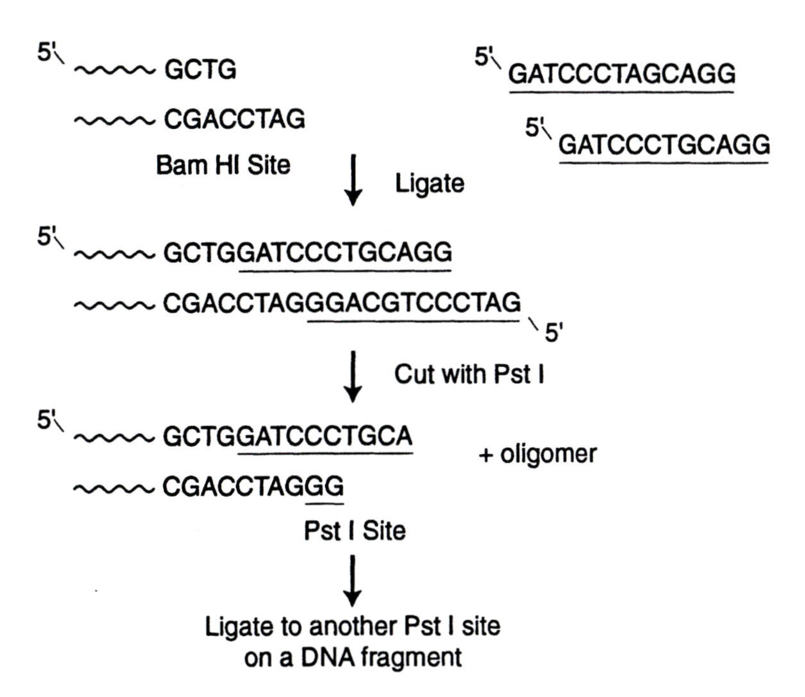


Figure 13-2 An adapter molecule allows the researcher to change one restriction site into another that may be more convenient for cloning.

(such as a plasmid or cloned gene), visualization of amplified products after PCR analysis, and isolation of specific fragments of DNA before extraction and cloning.

### *Preparing DNA Samples*

Samples that are ready for electrophoresis (DNA digests, PCR samples) should be mixed with a loading dye, commonly referred to as "Blue Stuff." The loading dye provides a convenient marker for the progress of the electrophoresis and facilitates the even loading of DNA in the wells. A 10x solution is described here; the final concentration should be 1x (e.g., add 2 μl to a typical 20-μl restriction digestion). The blue dye will migrate approximately as fast as a 300- to 400-bp fragment.

### *Power*

The electric field of the electrophoresis is the distance between the electrodes. Measure this distance and then apply an electric current equivalent to a voltage between 1–5 volts/cm. You may want to momentarily increase the voltage to observe the bubbles that rise from the cathode end (closest to the wells); there should be a "curtain" of bubbles here and only a few at the anode end (furthest from the wells). Keep the voltage in the specified range, or excess heat will be generated, which might be sufficient to melt the agarose.

### *Creating a 1% Horizontal Agarose Gel*

Procedure

1. In a 500-ml Erlenmeyer flask, mix 1 g of agarose in 100 ml of running buffer (1x TAE; see formula). Swirl to mix.

2. Heat the flask in a boiling water bath or a microwave oven until the mixture begins to boil. Do not allow the flask to boil so violently that it overflows.
3. Swirl the flask gently (WEAR GLOVES!). If the agarose has dissolved completely, you will see a clear solution. The presence of any particles or semisolid structures indicates that more heating (probably only a few seconds) is needed.
4. While the agarose is cooling, assemble the gel mold. This mold is often simply a tray with its opposite ends missing. Seal the open ends with tape (common lab tape or even masking tape will do). Place the comb at one end, about 1 cm from the end of the gel. Make sure that the teeth of the comb do not contact the gel mold.
5. Let the agarose cool until the flask is just cool enough to hold without discomfort. At this point it is possible to add ethidium bromide stain directly to the gel. Add 0.5 μg/ml of agarose from a 0.5 mg/ml stock (1000x solution). See the Hints below.
6. Pour the agarose carefully into the gel. Remove any bubbles (especially around the comb) with a pipette tip or similar device.
7. Let the gel harden for 30–45 min and then gently pry the comb out by wiggling it back and forth and lifting up at one end. Remove the tape from the ends, and set the gel and mold into the gel box.
8. Cover the gel with running buffer. Never mix different types of running buffers. The gel is now ready for loading.

After running the gel, it can be photographed to create a permanent record. Camera mounts for this purpose are commercially available, usually using Polaroid film (Type 55 or 57 for a simple black-and-white photo). A transilluminator-emitting shortwave ultraviolet (UV) light allows the DNA to fluoresce. It is not unusual for the film to detect bands that are difficult to see when viewing the gel on the transilluminator. *Exposure to shortwave UV light can cause first-degree burns and severely damage your eyes. Make use of the UV shielding that is available on the transilluminator, or use goggles that are specially designed to filter out UV light.*

Hints

1. If you forget to add the ethidium bromide or would like to stain longer to see faint bands, immerse the gel in a bath of buffer containing 0.5 mg/l ethidium bromide and examine again after 30–60 min.
2. If the gel is overstained, destain in plain buffer for 1–2 h or in a pan of cold water overnight.
3. It is not unusual for the bottom of the gel (closest to the anode) to be depleted of ethidium bromide after electrophoresis. Further staining may be required to see small bands.

### *Buffers and Reagents*

**Ethidium Bromide** DNA cannot be directly detected on gels; another compound must be present that is detectable. Ethidium bromide is a fluorescent compound that intercalates into the strands of DNA and causes it to fluoresce brightly under UV light. The stock solution is 0.5 mg/ml in water. Keep in a labeled, light-tight container.

CAUTION: Ethidium bromide is a powerful mutagen. Use care when working with this chemical. If it contacts your skin, flush the area with copious amounts of water. Always wear gloves when handling ethidium bromide, as well as solutions and gels containing ethidium bromide. The disposal of ethidium bromide and materials contaminated with ethidium bromide may be regulated in your community. Follow all applicable laws regarding waste disposal of this chemical.

### Stock (10x) Blue Stuff

0.25% bromophenol blue
50% (w/v) sucrose in water.

Store stock at 4°C. A working solution can be kept at room temperature for at least a week.

### Stock (50x) TAE Running Buffer

242 g Tris base
57.1 ml glacial acetic acid
100 ml EDTA 0.5 M (pH 8.0)

## Recovery of DNA from Agarose Gels

Recovery of DNA from agarose gels is an important step for both DNA cloning and for the preparation of DNA probes. The advantage of isolating DNA fragments is that very specific fragments can be recovered in fairly pure form, especially when plasmid DNA is separated on gels. Two relatively easy approaches that have good yields are the use of low-melt agarose and the use of agarase digestion. The latter technique is particularly valuable, since it allows the isolation of DNA fragments of 100 kb and higher. For most prokaryotic cloning, this technique is more than sufficient.

### *Low-Melt Agarose*

Gel electrophoresis is done using low-melt agarose. Suppliers of low-melt agarose are not difficult to find (e.g., Boehringer Mannheim and Gibco-BRL).

Procedure

1. Visualize the DNA band by long-wave UV light, and excise the appropriate DNA band from the agarose gel as quickly as possible. A typical gel piece can have a volume of 200 μl.
2. Add 5 vol of a solution of 20 mM Tris Cl (pH 8.0), 1 mM EDTA.
3. Melt the gel in a 68°C water bath for 5–10 min.
4. Cool to room temperature. The agarose should have been diluted in the buffer and will not resolidify.
5. Extract the DNA with an equal volume of phenol:chloroform:isoamyl alcohol (25:24:1).
6. Centrifuge briefly to separate phases, and recover the upper phase into a clean tube.
7. Extract with an equal volume of chloroform. Centrifuge briefly to separate phases, and recover the upper phase into a clean tube.
8. Perform ethanol precipitation to recover the DNA pellet.

### *Agarase Method*

Agarase is an enzyme derived from *Pseudomonas atlantica* that degrades agarose, thus releasing the DNA fragment that is contained in an isolated gel piece. High-purity agarase is commercially available (e.g., New England Biolabs) and usually comes with the appropriate 10x buffer. Use low-melt agarose to separate the DNA bands; first melt the agarose and then add the enzyme at a temperature that is low enough to allow the enzyme to function, but high enough so that the agarose will not resolidify [11].

Procedure

1. Visualize the DNA band by long-wave UV light, and excise the appropriate DNA band from the agarose gel as quickly as possible. A typical gel piece can have a volume of 200 μl.
2. Equilibrate in a solution of 5 mM EDTA and 100 mM sodium chloride. (If a buffer is supplied, use it at a final concentration of 1x.) Keep on ice for 30 min.
3. Remove buffer, and melt the agar at 68°C for 10–15 min in a water bath; then move to a 37°C water bath to cool.
4. Add agarase enzyme (1–5 units), and incubate at 37°C for 1–2 h.
5. Adjust the solution to 0.5 M NaCl. Precipitate the remaining agarose by spinning in a microfuge for 15 min.
6. Pour off supernatant into a clean tube and concentrate DNA by ethanol precipitation.

Notes

1. If you believe that the equilibration step results in a loss of your DNA by diffusion, try collecting the buffer and precipitating the DNA.
2. For large fragments of DNA (over 10 kb), do not precipitate the DNA before subsequent enzymatic reactions; amend the solution as necessary after Step 4 and add the next enzyme; for instance, for subsequent restriction enzyme digestion. This should work well with DNA fragments up to 50 kb.
3. Several commercial products do a good job in purifying DNA: GlassMax (Gibco-BRL); SpinBind extraction units (FMC, Rockland, ME) useful for a range of 0.1- to 23-kb fragments; Elutip-D columns (Schleicher and Schuell, Keene, NH); and ChromaSpin columns (Clontech, Palo Alto, CA), which are good for very small fragments.

### *Shotgun Cloning*

If you know that your gene of interest is in a particular organism, but do not know where it is, then shotgun cloning may be your answer. The name comes from the idea that a large number of fragments (like shotgun pellets), rather than a specific fragment, are placed into the ligation mix. Shotgun cloning is a very easy procedure to perform, since it does not require the separation and isolation steps just described. However, undoubtedly there are a huge number of recombinant molecules that do not have your gene of interest, and therefore a good screening procedure is necessary.

The screening procedure may be based on the gene of interest, especially if it will produce a specific color change on the selective agar. However, the cloning is dependent on cutting with restriction enzymes, and there is no guarantee that your gene will not also be cut. Or you might clone only a fragment of your gene. Therefore, it is more likely that you will screen using the colony hybridization technique, especially if a suitably homologous probe is available.

This technique is useful for creating libraries of clones (usually phage-mediated clones) that include an entire genome. Techniques for creating libraries using phage-cloning vectors and plaque hybridizations have been described [5].

Once your clone has been identified, it should be accurately mapped using restriction enzyme sites. Further proof of identity can be obtained by sequencing or Southern blotting.

## Cloning Vectors

Using plasmids is the easiest way to clone, propagate, and transfer genetic material. In some cases the cloned DNA will be used as a probe for hybridization experiments. In other cases the DNA may encode a gene that should be moved into a new host bacterium. DNA to be sequenced is easily manipulated on plasmids. Because the uses for plasmids are so varied, they are conveniently divided here into two categories: those that are most useful in *E. coli* and those that are used in other bacteria.

The cloning vector should be chosen with certain criteria in mind. First, the plasmid must be stably maintained in the final host. Many plasmids have been described that can be maintained in *E. coli*, but not in other hosts (i.e., they have a narrow host range). Plasmids are described in this section that are useful for many other genera. Second, the plasmid must be selectable; that is, it must have a marker, usually antibiotic resistance, that permits the worker to select only those clones that have taken up the incoming plasmid. This feature is especially important because transformation is always a very inefficient process—the percentage of cells that take up the plasmid may be abysmally low. Finally, the plasmid should be easy to introduce into the final host. It is often expedient to introduce the plasmid into *E. coli* initially, since reliable transformation methods have been described, and then move the plasmid into the final host by the relatively efficient process of conjugation.

Plasmids are assigned to incompatibility groups based on their ability to be stably maintained in a host without selection. Two plasmids are said to be incompatible when extended growth without selection results in the loss of one of the plasmids. In fact, incompatible plasmids have similar genetic components, and without selection for each plasmid type in the cell, it is not uncommon for cells to divide without appropriate segregation of the plasmids.

Plasmids that are recovered from wild-type bacteria can be assigned to an incompatibility group based on their ability to coexist in a bacterial strain with plasmids from other groups. Assigning plasmids is partly a process of elimination and partly a test for incompatibility. One means to test plasmids in *E. coli* has been designed by Davey et al. [12]. The test plasmids are available only from this group; refer to this paper for technique and for applicability to your system. A comprehensive treatment of plasmid vectors for Gram-negative species has been presented by Simon et al. [48], and a method to mobilize plasmids based on the incompatibility group (IncP) system has been presented by Simon [47].

Certain plasmid vectors have been developed to facilitate the cloning of DNA. Many are referred to as "shuttle vectors" because they can be used in the *E. coli* primary host and another, unrelated species. They have antibiotic resistance genes associated with them to allow selection for transformed cells. Many of these vectors enable easy screening for DNA fragment insertion (e.g., interruption of another antibiotic resistance gene or a gene with a convenient biochemical assay).

It is always important to confirm that the perspective clone actually has the appropriate fragment inserted at the right spot. Screening procedures can often give false-positive results because of point mutations. It is also important to confirm the orientation of the cloned fragment.

### *E. coli* Vectors

These plasmids have a narrow host range, sometimes confined to *Escherichia*, and are used solely to propagate DNA sequences in *E. coli*. For instance, they might be used to clone DNA fragments for sequencing, to propagate a sequence for use as a DNA probe, or for in-depth study of the cloned DNA in *E. coli*. They are typically small (less than 10 kb), present in high copy numbers, and are easy to manipulate with molecular techniques. In many instances they have been substantially genetically altered, so that favorable cloning sites and genes are present. Reliable maps of these plasmids are available, which facilitates the mapping of cloned DNA fragments. The plasmids that are presented here are typical of the many others that are available (Figs. 13-3–13-5).

Boehringer Mannheim, Gibco/BRL, New England Biolabs, and others have catalogs that contain a wealth of information, including detailed maps of many useful plasmids. They list their own constructions, as well as others in common usage.

### General Cloning Vectors for Gram-Negative Bacteria

In the second category are the plasmids that can be used in genera other than *Escherichia*. These plasmids are usually larger and have limited genetic alterations. For the relatively inefficient cloning step, the plasmid is usually introduced into *E. coli*, where clones can be screened and examined easily. The correct clone is then moved into other genera by conjugation or by electroporation with purified plasmid. *E. coli* is not always a suitable host for the study of certain genes, since the regulatory sequences for the gene of interest might not be recognized by *E. coli* enzymes. For this reason it is important to reintroduce the cloned gene into the original host.

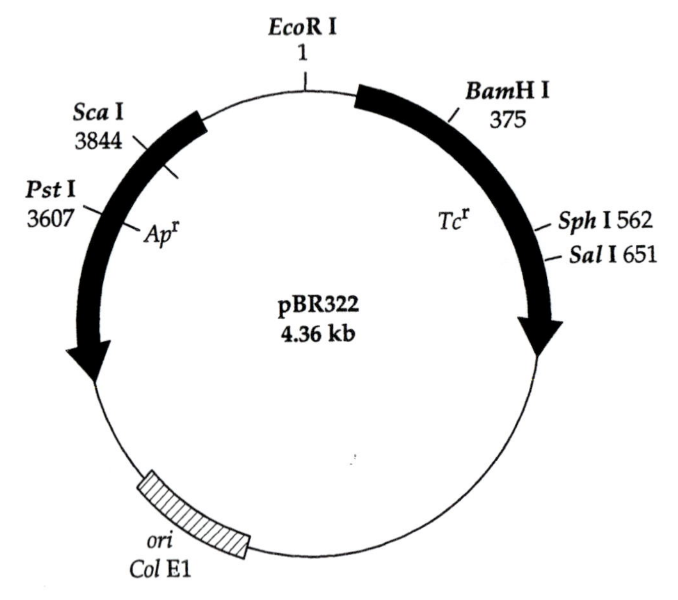


Figure 13-3 The pBR322 plasmid is one of the oldest and most useful constructed plasmids. It contains two antibiotic resistance genes, for ampicillin (Ap$^r$) and tetracycline (Tc$^r$), that can be inactivated by insertional mutagenesis. Several restriction endonuclease sites that are useful for cloning are given, along with their numerical (basepair) map positions, which aid in mapping of inserts [7].

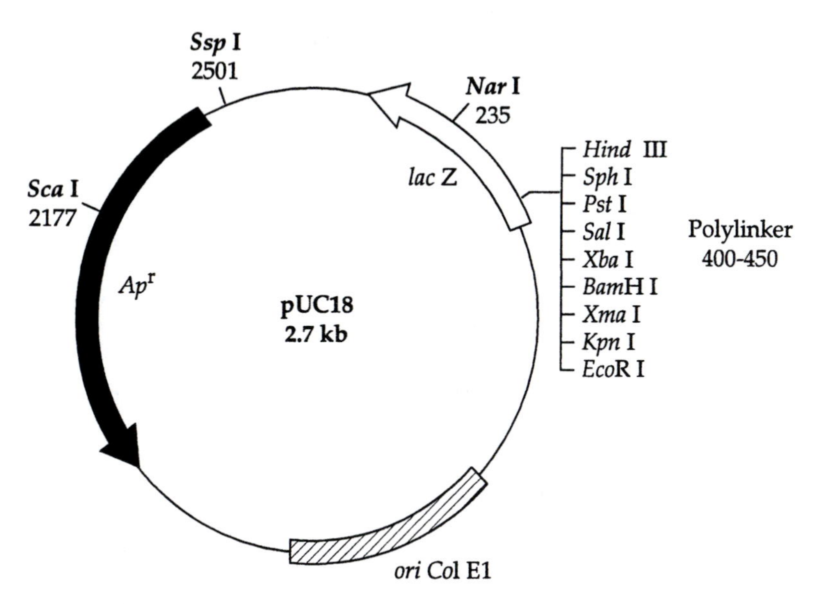


Figure 13-4 The pUC18 plasmid has a selectable marker gene for ampicillin resistance and the narrow host range ColE1 replicon. The *lacZ* gene (encoding β-galactosidase) contains a polylinker site that has several unique restriction sites. This feature permits easy insertion of DNA fragments and subsequent screening of transformants on agar plates containing X-gal or one of its cheaper variants [62].

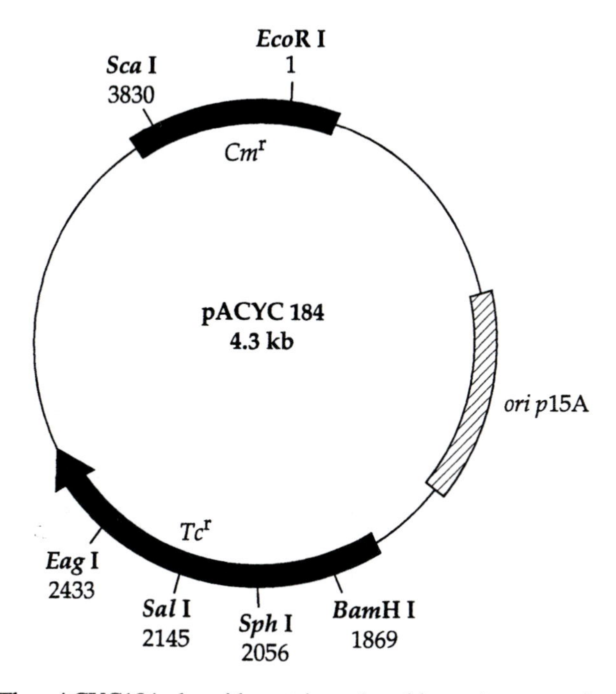


Figure 13-5 The pACYC184 plasmid contains selectable marker genes for both chloramphenicol and tetracycline. Several sites for cloning are given. Its p15A replicon allows it to be stably maintained in *E. coli* strains, even if the strain harbors a ColE1 derivative plasmid. A related plasmid, pACYC177, is another useful cloning vector. Complete sequences of both plasmids have been reported [41, 42].

Table 13–1 Genera in Which Incompatibility Groups can be Maintained Stably

| IncP[a] | | IncQ[a] | IncW[a] |
|---|---|---|---|
| *Acetobacter* | *Methylophilus* | *Acetobacter* | *Agrobacterium* |
| *Achrobacter* | *Methylosinus* | *Acinetobacter* | *Alcaligenes* |
| *Acinetobacter* | *Neiserria* | *Agrobacterium* | *Erwinia* |
| *Aeromonas* | *Paracoccus* | *Alcaligenes* | *Escherichia* |
| *Agrobacterium* | *Proteus* | *Azotobacter* | *Klebsiella* |
| *Alcaligenes* | *Pseudomonas* | *Escherichia* | *Providentia* |
| *Azospirillum* | *Rhodobacter* | *Klebsiella* | *Pseudomonas* |
| *Azotobacter* | *Rhizobium* | *Methylophilus* | *Rhizobium* |
| *Bordetella* | *Salmonella* | *Proteus* | *Serratia* |
| *Caulobacter* | *Shigella* | *Providencia* | |
| *Chromobacterium* | *Serratia* | *Pseudomonas* | |
| *Erwinia* | *Thiobacillus* | *Rhizobium* | |
| *Escherichia* | *Vibrio* | *Rhodopseudomonas* | |
| *Hyphomicrobium* | *Xanthomonas* | *Salmonella* | |
| *Klebsiella* | *Yersinia* | *Serratia* | |
| *Legionella* | *Zymomonas* | | |
| *Methylobacterium* | | | |
| *Methylococcus* | | | |

From Krishnapillai [25].
[a]IncP plasmids typically have a copy number of 4–6; IncQ a copy number of 15–20; and IncW a copy number of 4–6.

There are three incompatibility groups that have a sufficiently broad host range to be useful in many genera of Gram-negative bacteria: incompatibility groups P, Q, and W. A list of the genera in which they can be maintained stably appears in Table 13-1. This list should not be considered complete; for many species, stability must be determined empirically. If you can move the cloning vector into your species of interest, you should be able to move a cloned fragment on that vector as well.

The plasmids given here are presented for example only (Figs. 13-6–13-8). They are representative of the attributes of each incompatibility group and illustrate the requirements for use. Many more examples are available in the literature. Antibiotic resistance genes that are associated with these plasmids usually function well in these genera, although this must also be determined empirically. The inherent antibiotic resistance of each strain must be determined before introduction of the plasmid.

These plasmids can be introduced into recipient strains using transformation, conjugation, or electroporation. Protocols for each of these approaches are presented in this section. In many instances the plasmid is non-self-transmissible, but is mobilizable by a helper plasmid. A helper plasmid contains the transfer (*tra*) genes from another plasmid and is used in a triparental mating. An excellent example is pRK2013, which contains the IncP *tra* genes and which mobilizes both IncP and IncQ plasmids [16].

## Ligation

### *Precipitation of DNA*

A dilute concentration of DNA and RNA can be concentrated by precipitation of the DNA and disposal of the liquid phase, followed by resuspension in a smaller volume of liquid. There is always some loss of DNA associated with this procedure, although if it is done carefully the loss is minimal. This procedure is easily

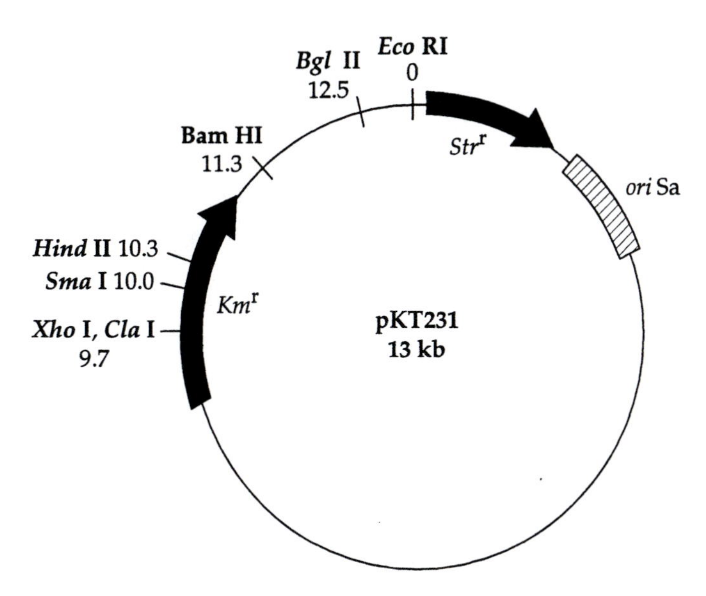


Figure 13-6 The broad host range plasmid pKT231 [6]. This is one of a family of valuable "pKT" vectors that can be used for the introduction of cloned material into many different genera. It has selectable markers for streptomycin (Str) and kanamycin (Km) resistance. Use a helper plasmid, such as pRK2013, to mobilize this plasmid into another strain. See Table 13-1 for host range.

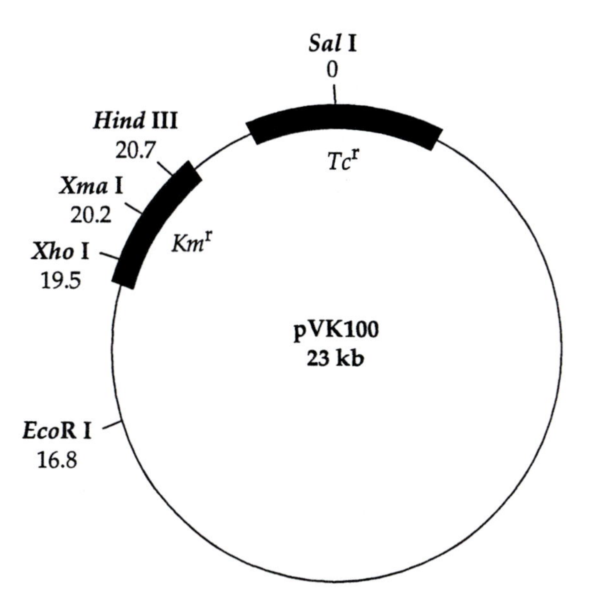


Figure 13-7 The pVK100 plasmid is a derivative of the IncP plasmid RP4. IncP plasmids have an extraordinarily large host range and a correspondingly complex replicon. This plasmid has selectable markers for tetracycline (Tc) and kanamycin (Km) resistance. Its large size makes it inconvenient for some applications. The transfer functions are not intact in this derivative, although it can be mobilized by a helper plasmid, such as pRK2013 [16]. See Table 13-1 for host range.

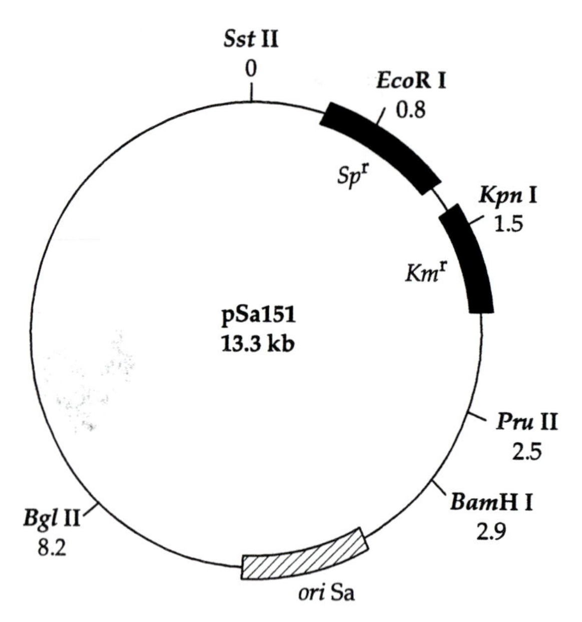


Figure 13-8 The broad host range pSa151 plasmid is a derivative of the Sa plasmid, which is in the IncW family. It encodes both streptomycin and kanamycin resistance. See Table 13-1 for host range [55].

carried out in a sterile microfuge tube. Commercial kits, such as GeneClean (Bio101, La Jolla, CA) and Qiaex (Qiagen, Chatsworth, CA), are also available.

Procedure

1. To the DNA solution, add 0.1 vol of 2.5 M sodium acetate (pH 5.2) and 2 vol of ice-cold ethanol (reagent grade).
2. Mix gently by inversion, and store at −70°C for at least 1 h or at −20°C overnight.
3. Centrifuge at 12,000 g (or top speed in a microfuge) for 15 min. (Do not allow the tube to warm before spinning.)
4. DNA may be evident as a white pellet at the bottom of the tube or may be completely invisible. Pour off liquid, and remove the residual with the tip of an absorbent wipe. The pellet should remain in the tube.
5. Wash with a small volume of 70% ethanol, and vortex briefly. It is best to reconcentrate the DNA by centrifugation as described in Step 3. Pour off ethanol, being careful not to dislodge the DNA pellet.
6. Dry the DNA thoroughly in a dessicator or a Speed-Vac apparatus.
7. Resuspend in the appropriate amount of sterile distilled water or TE buffer (10 mM Tris pH 8.0, 1 mM EDTA).

### *Dephosphorylation*

Cloning efficiency can be enhanced by dephosphorylating the vector DNA—removing the phosphate from the 5′ end of each strand of the vector. This procedure effectively stops the religation of the vector molecule, which is a common event in ligation. During the ligation reaction the two ends of the vector are always close to each other and therefore have a high probability of being involved in the ligation reaction. However, without the phosphate group the phosphate backbone of the DNA cannot be formed. Ligation of the insert fragments can still occur, as can partial ligation of the insert to the vector because the insert DNA still has phosphate groups on its 5′ ends. After transformation into a host strain, the gaps will be repaired by the host enzymes.

Use calf intestinal alkaline phosphatase (CIP) for this reaction; the enzyme is easily destroyed at the conclusion of the procedure. A suitable buffer is usually supplied with the enzyme. Typical reaction conditions are incubation of the DNA/CIP (1 unit) mixture for 1 h at 60°C. Stop the reaction by adding EDTA to a final concentration of 5 mM and then heat-inactivating it at 75°C for 10–15 min. Clean up the DNA using phenol/chloroform and ethanol precipitate.

### *Ligation*

Fragments of DNA and linearized plasmid vector must be joined together in one molecule by the process of ligation. Although there are several ligases available, T4 DNA ligase is usually used for this work. It works well for both blunt-end and sticky-end ligations. A buffer (usually 10x) is often supplied with the enzyme. Ligation buffer contains adenosine triphosphate (ATP), which is very labile at room temperature or after several freeze-thaw cycles. The absence of ATP renders the buffer useless. For this reason it is advisable to keep the buffer and enzyme on ice while these reagents are out of the freezer. You may wish to aliquot the buffer into separate tubes and store them in case one tube becomes inactive.

Procedure

1. Resuspend the dry DNA in 17 μl of sterile distilled water.
2. Add 2 μl of 10x ligation buffer.
3. Add 1 μl of T4 DNA ligase (i.e., 1–5 units of enzyme).
4. Incubate at 16°C for approximately 16 h (overnight).
5. Transform 2–5 μl into an appropriate *E. coli* strain.
6. Store the remainder at −20°C.

Notes:

1. For ligation to be successful, three molecules must come together at once: the ligase and the two ends of the DNA molecules. Molecules are more active at higher temperatures, and therefore a relatively low temperature is used for most ligation work. A low temperature is imperative for blunt-end ligation, and 4–6°C is recommended. Sticky-end ligations can be performed efficiently at room temperature in as little as 1 h.
2. Both the *amount* and the *ratio* of the individual DNA species are important to the efficiency of ligation. Total DNA in the ligation tube should be in the range of 0.2–1.0 μg. If the vector molecules predominate, you will ligate vector molecules together (but see Dephosphorylation). If the insert predominates, ligation of strings of insert will occur, which will not be maintained in the host. If the vector is dephosphorylated, a ratio of one insert to one vector molecule is best. If the vector is not dephosphorylated, a ratio of three to five inserts to one vector is recommended. In this latter case, there is a tendency to ligate inserts together, but they will not be seen after transformation. There is the possibility that two inserts will ligate to one vector, which must be determined by restriction analysis of DNA from the transformants.
3. A positive control tube is useful to test whether the ligase enzyme was functional and reaction conditions were normal. For agarose gels, many laboratories use a molecular weight marker based on Lambda phage that

has been digested with *Hin*DIII. One microliter of this DNA can be used in a control tube. After the ligation reaction, run this DNA on a 0.8% agarose gel against an unligated sample. If ligation worked, the several fragments of the Lambda DNA should have ligated into a higher molecular weight mass, which will be visible as a smear instead of the distinct bands of the unligated sample.

## DNA TRANSFER

Once a DNA molecule is created, it must be reintroduced into the appropriate bacterial strain. There are three reliable means of accomplishing this task: transformation, conjugation, and electroporation. The use of a phage to transfer large fragments of DNA, via the process of transduction, is also possible, but is not covered here because few reliable vectors exist for environmental isolates. However, transduction techniques (including packaging of DNA and transfection procedures) are very valuable for creating DNA libraries.

*Transformation* is the introduction of naked DNA into a strain, usually using a chemical modification of the cells. Transformation protocols have been described for a great many genera, although it is almost always a very inefficient process. This inefficiency means that the introduction of cloned DNA (i.e., after the ligation step) is probably doomed to failure. This fact is especially relevant when large plasmids are used, which are the norm for many Gram-negative bacteria. Generally, as the size of the plasmid increases, transformation efficiency decreases. It is possible to introduce the clone into an *E. coli* strain, for which relatively high transformation rates have been achieved, and then to purify large amounts (μg quantities) of the plasmid for a subsequent transformation into the final host. However, it is usually better to start by introducing the plasmid into *E. coli* and then using conjugation to pass the plasmid to its final host. A standard procedure for *E. coli* transformation is given in this section. Many biotech suppliers have created their own *E. coli* strains and transformation protocols that yield high transformation efficiencies. In comparing these systems, it is advisable to examine the plasmid used for transformation. Usually a very small plasmid is used, such as pUC18, and the plasmid is supercoiled to increase the transformation rate further. The strain usually has mutations that render it recombination-deficient and that remove the host restriction endonuclease systems.

*Conjugation* is bacterial mating. In this process a plasmid (or occasionally, a fragment of chromosome) is passed from one species to another. Conjugation can occur between two different genera, making it a very valuable tool for the introduction of cloned DNA from *E. coli* to another genus. In addition, conjugation is a relatively efficient process.

Conjugation is also a very efficient means of producing transposon insertion mutants. Plasmid vectors have been designed that can transfer or be mobilized from *E. coli* into another genus. The plasmid encodes a transposon and a narrow host range replicon (usually only *E. coli*) so that it cannot be maintained in the new host. Selection for the antibiotic resistance of the transposon (and against the donor strain) should result in clones in which the transposon has integrated into the host genome. These "suicide plasmids" are an effective means of generating mutants rapidly.

*Electroporation* is a means of using a high-voltage pulse of electricity to allow DNA into a cell. It is a poorly understood process, although it is thought that the electrical pulse opens up transient "holes" in the cell, allowing the DNA to migrate inside before the holes reseal. It works well with either eukaryotic or prokaryotic cells.

## Transformation

### Procedure

1. Grow the recipient *E. coli* strain in a rich medium (such as Luria-Bertani [LB] broth) overnight, using antibiotic selection if needed. Incubate at 37°C in a shaking (150 rpm) waterbath.
2. Make a 1:100 dilution of the overnight cells in fresh medium, and continue incubation as before.
3. When the cells are in the mid-exponential growth phase as determined by optical density, place 1.5-ml aliquots of cells in Eppendorf tubes.
4. Spin in a microfuge for 2–3 min to pellet the cells; discard the supernatant.
5. Resuspend in 1 ml cold 10 mM NaCl.
6. Centrifuge as above, and discard the supernatant.
7. Resuspend in 1 ml of cold 100 mM calcium chloride; leave on ice for 20–30 min.
8. Centrifuge as above, and discard the supernatant.
9. Resuspend in 0.2 ml of cold 100 mM calcium chloride, add DNA, and leave on ice for 20–30 min.
10. Heat-pulse for exactly 2 min in a 42°C water bath.
11. Add 1 ml of LB broth, and allow cells to recover at 37°C for 90 min with gentle mixing.
12. Place 10- to 100-μl aliquots on selective agar plates, spread evenly, and incubate at 37°C overnight.

### Notes

1. A positive and a negative control should always be used in this procedure. The positive control (usually 0.1 μg of purified plasmid DNA) is used to ensure that the procedure worked properly. That is, if the positive control results in few colonies on selective agar, the probability of success with your constructed plasmid is also very low. The negative control (no DNA added or an equivalent amount of sterile distilled water) establishes the background for the antibiotic selection. This background can be very important if the antibiotic is weakened by age or omitted by mistake. In the best-case scenario, the negative control will show absolutely no colonies on the selective medium.
2. A procedure by Hanahan [22] works well for *E. coli* strains DH1, DH5, and DH5α. This is a more complex procedure, however, and requires precise attention to details and the use of one of the aforementioned strains. Many commercial distributors of molecular reagents also have their own transformation strains.

## Conjugation (Bacterial Mating)

### Procedure

1. From an overnight culture, grow the donor strain (usually *E. coli*) in a rich medium (e.g., LB broth) until the mid-log phase is reached. Grow the recipient strain in rich medium until the mid-log phase. If one strain reaches this stage before the other, put it on ice until the other strain is ready.
2. Using a filtering apparatus, combine donor and recipient strains in a ratio of 1:4 (e.g. 2 ml:8 ml), and then concentrate the cells on the filter using a mild vacuum. Use a sterile 0.45-μm filter.

3. Remove the filter to a non-selective agar plate, and store inverted at a temperature suitable to both strains. Incubate for 14–16 h (overnight).
4. Place the filter in a sterile tube, and resuspend the cells in a small amount (5 ml) of buffer solution (10 mM $MgSO_4$). Vortex vigorously to resuspend cells.
5. Plate aliquots (100 μl) of cell suspension on selective agar plates, and incubate under appropriate conditions for selection of the transconjugant. Freeze the remainder of the suspension by adding sterile glycerol (15% final conc.) and storing at −70°C.

Colonies that grow on the selective agar plates are referred to as transconjugants. This is a presumptive designation, especially if any background appears on the control plates. Verification of transconjugants can be performed by examination of the DNA after extraction and restriction digestion, followed if needed by DNA hybridization, PCR analysis of clones, or colony hybridization. These procedures are outlined below. Conjugation rates can be expressed as transconjugants produced per initial donor cell at the beginning of the incubation period.

### *Transposon Mutagenesis*

Transposon mutagenesis is dependent on conjugation. The donor in this case is a bacterial strain, usually *E. coli*, that contains a plasmid with a transposon of interest. The conjugation allows the plasmid to enter the recipient strain. Plasmid vectors used for transposon delivery are usually of the narrow host range type (i.e., they can only be maintained in a narrow range of related hosts, such as *Escherichia* and *Salmonella*). When they are introduced into another species they cannot be maintained (cannot replicate), and thus the progeny cells cannot survive the antibiotic selection. The only cells to survive are those in which the transposon has copied itself onto a host genetic element (plasmid or chromosome). Naturally, there must be some selection against the donor strain (e.g., selection for antibiotic resistance on the recipient).

For Gram-negative bacteria, transposons derived from Tn5 [9] have proved very useful. Tn5 inserts at a relatively high frequency on a random basis and is usually maintained stably. In particular, the constructions of Simon et al. [49] are very useful, in that they are designed with various antibiotic resistance genes, as well as bioreporter genes and *mob* sites. Several Tn5 derivatives with various antibiotic resistance genes have been described by Sasakawa and Yoshikawa [45].

The most convenient transposon vector to work with is probably the suicide vector. The transposon is located on a plasmid that has a narrow host range. The plasmid can move into a recipient strain by conjugation, but will not be maintained in the recipient strain. Selection for the transposon means that only conjugation events that are followed by a transposition event survive on the selective medium. These plasmids are usually genetically constructed to contain the origin of replication (*ori*) from a narrow host range plasmid with the transfer genes (*oriT*, *mob*, *tra*) from a broad host range plasmid. The plasmid is either self-transmissible or mobilizable by a helper plasmid (e.g., pRK2013). The literature gives many examples of useful vectors; the vectors that are described here illustrate some important features, such as antibiotic resistance and convenient restriction sites.

If conjugation fails to introduce the plasmid, electroporation may work well enough to obtain mutants. The principles of selection are the same.

### *Other Mutagenesis*

**UV Mutagenesis** UV light induces the formation of pyrimidine dimers in nucleic acids. Repair of these dimers in some microorganisms results in a higher rate of error incorporation. However, mutation is a random process, and it cannot be guaranteed that multiple mutations were not created during the mutagenesis procedure. In addition, many species have a photoactive repair system, allowing repair to occur with few mutations if visible light is present. For this reason bright lighting should be avoided during the procedure, and plates should be incubated in the dark.

UV light sources are commercially available. Typically, these sources facilitate the application of UV doses at 10–100 $J/m^2$. The actual fluence can only be determined with a UV meter that is positioned at exactly the same distance from your source that your sample is located, and few laboratories have access to one. However, periodic calibration of equipment by a skilled technician may be a core function of your institution and is a good quality assurance practice.

Do not use UV sources that are designed for DNA cross-linking, since the UV fluence is too high.

Procedure

1. Grow strain to confluency overnight in a rich medium.
2. Use this culture to regrow the cells to the mid-exponential phase.
3. Wash the cells once by concentrating them by centrifugation, and resuspend to the same volume in phosphate-buffered saline (PBS).
4. Centrifuge the cells to a pellet again, and resuspend in the same volume of PBS.
5. Spread 5–6 ml of the suspension in a Petri dish, and swirl gently to cover the bottom.
6. Irradiate plates with UV light (254 nm) at a distance of 5 cm for various lengths of time (e.g., 10 s to 1 min). Some plates should also remain unirradiated.
7. Spread aliquots of serial dilutions on agar plates.
8. Incubate at an appropriate temperature in the dark until colonies develop.
9. Screen plates that show more than 50% lethality for mutations.

**Chemical Mutagenesis** Chemicals can cause mutations either by damaging the DNA, which allows errors to occur during repair, or by causing the misincorporation of bases during DNA replication of the damaged DNA strands. N-methyl-N′-nitro-nitrosoguanidine (MNNG or NTG) is an alkylating agent that damages DNA, causing transition mutations.

Procedure

1. Grow the cells to the mid-exponential phase.
2. Centrifuge the cells to concentrate them. Then wash them once in a sterile 30 mM sodium citrate (pH 6.0) solution.
3. Resuspend the cells in 0.5 vol of sodium citrate, and add MNNG to a final concentration of 1 mg/ml. Stock solution is 5x, made in 50 mM potassium phosphate buffer, pH 6.0.
4. Incubate for 1 h at normal growth temperature, with mild shaking.
5. Concentrate the cells by centrifugation, resuspend in 1 vol of growth medium, and incubate under normal growth conditions overnight.
6. Make appropriate dilutions, and spread on agar plates.
7. From these master plates, transfer the individual colonies to selective agars for screening.

The use of sterile toothpicks is recommended for this job. It is also useful to determine the percent survival of the mutated culture in order to determine whether the concentration of mutagen was too high or too low.

*Note*: When working with chemical mutagens, keep in mind that they are often carcinogens. Take appropriate precautions when handling them, such as wearing gloves and using a chemical fume hood when working with dry chemicals or aerosols. Dispose of the waste chemicals in an approved manner.

## Electroporation

DNA may be introduced into cells using electroporation, a relatively new procedure that acts by creating transient micropores in the cell with a strong pulse of electric current. Small molecules, such as DNA, can then enter the cells through these transient micropores. This procedure usually results in the destruction of most of the cells, although the cells that survive have a good chance to take up DNA. It requires a specialized instrument that can hold a high electrical charge in capacitance and then discharge it safely through the cells in a fraction of a second. The appropriate cuvettes and information regarding the settings for proper functioning of the instrument usually come with such instruments. For instance, the electrode gap is usually fixed on the cuvettes, eliminating one possible parameter from further experimentation.

Electroporation remains more of an art than a science, and considerable tinkering with the parameters (field strength, pulse time, washing conditions) may be needed to achieve efficiency with your microorganism. Many factors are unpredictable: Some strains electroporate better if harvested during exponential growth, whereas others are better in the stationary phase. Some strains can be frozen for future use (a potentially great timesaver), whereas others (notably *Pseudomonas*) perform poorly after freezing. The protocols that are given here are a good starting point. Reference to the literature is often helpful, especially when you are using the same strain as used in a published report.

In general, the washing procedure should remove the free ions in the solution as much as possible. It is usually done with distilled water or 10% glycerol. Failure to wash the cells effectively leads to arcing, in which the current is conducted in a narrow channel between the electrodes. Arcing is especially likely when high field strengths, such as are used with bacterial cells are employed. In addition to resulting in poor electroporation yields and causing damage to the electrodes, arcing produces a loud pop that will cause you to jump backward about 10 feet. In addition, the cell suspension should be quite dense ($10^{10}$ cells/ml) before electroporation. Unlike transformation protocols, the DNA does not need to adhere to the cell wall before the procedure.

Transformation frequency = ratio of transformed cells to total cells

Transformation efficiency = total transformed cells per μg of DNA

### Procedure: Gram-Positive Bacteria

1. Grow cells in rich medium to a high density ($OD_{600} > 1.5$).
2. Centrifuge at 3000 g for 10 min to pellet cells.
3. Pour off medium, and resuspend in an equal volume of cold sterile distilled water.
4. Centrifuge at 3000 g for 10 min.
5. Repeat Steps 3 and 4 two more times.
6. Resuspend pellet in cold sterile polyethylene glycol (PEG) 6000 (30%) to a final volume equal to 1% of the original volume.
7. Add DNA (0.2–1.0 μg) to the 100-μl aliquot. Each electroporation requires a 100-μl aliquot. Additional aliquots can be stored at −70°C until needed.

8. Transfer mix to a prechilled (on ice) cuvette.
9. Electroporate at a field strength of 12.5 kv/cm and a pulse time of 5–10 ms with exponential decay.
10. Add 2 ml of growth medium, transfer to a larger sterile tube, and incubate under normal growth conditions for at least 90 min.
11. Plate aliquots on selective medium. Typical results are in the range of $10^4$–$10^5$ transformants per μg of DNA [54].

Procedure: Gram-Negative Bacteria

1. Grow cells in rich high medium to an $OD_{600} \approx 1.0$.
2. Centrifuge at 8000 g for 10 min to pellet cells.
3. Pour off medium, and resuspend in an equal volume of cold sterile distilled water.
4. Centrifuge at 8000 g for 10 min.
5. Repeat Steps 3 and 4 two more times.
6. Resuspend pellet in cold sterile water to a final volume equal to 5% of the original volume.
7. Add DNA (0.2–1.0 μg) to the 100-μl aliquot. Each electroporation requires a 100-μl aliquot. Additional aliquots can be stored at −70°C until needed.
8. Transfer mix to a prechilled (on ice) cuvette.
9. Electroporate at a field strength of 12.5 kv/cm and a pulse time of 5 ms with exponential decay.
10. Add 2 ml of growth medium, transfer to a larger sterile tube, and incubate under normal growth conditions for at least 90 min.
11. Plate aliquots on selective medium [56].

Procedure: Yeast (*Saccharomyces*)

1. Grow cells in growth medium to a high density ($1 \times 10^7$ cells/ml).
2. Centrifuge at 1500 g for 5 min at 4°C to pellet cells.
3. Pour off medium, and resuspend in 0.1 volume of electropermeabilization (E) buffer.
4. Centrifuge at 1500 g for 5 min.
5. Repeat Steps 3 and 4.
6. Resuspend $5 \times 10^7$ cells in 48 μl of E buffer. Add 2 μl of DNA (0.01–1.0 μg).
7. Transfer mix to a prechilled (on ice) cuvette.
8. Electroporate at a field strength of 2.7 kv/cm, with a pulse time of 15 ms (square wave pulse).
9. Add 1 ml of growth medium, and incubate 1 h under normal conditions.
10. Centrifuge briefly to pellet cells, and resuspend in 100 μl of selective medium.
11. Plate aliquots on selective medium [29].

Notes

1. E (electropermeabilization) buffer: 10 mM Tris-HCl (pH 7.5), 270 mM sucrose, 1 mM $MgCl_2$
2. Incubating cells in 25 mM dithiothreitol for 10 min may significantly improve transformation efficiency.

## HYBRIDIZATION ANALYSIS

Hybridization analysis is a means of detecting the presence of specific nucleic acid sequences in extracted DNA samples and to quantify the amount there. It is a powerful technique when used correctly, especially in regard to quantification standards. Each of the procedures in this section relies on the same basic protocol. DNA is transferred and affixed to a nylon membrane, which is prehybridized with unlabeled non-homologous DNA. This non-homologous DNA serves as a "blocking" reagent so that the DNA probe does not bind non-specifically on the membrane. The membrane is hybridized in a solution containing the DNA probe, which has been labeled with a detectable marker. The membrane is washed to remove probe that is not tightly bound, and hybridized probe DNA is detected. The various protocols have been developed to allow hybridization analysis of DNA from different sources: Colony hybridization is used on whole colonies on agar plates, Southern blotting is used for DNA in agarose gels, and slot and dot blots are used for solutions of DNA.

There are many variations on the basic hybridization protocol; the literature is replete with examples. The methods and descriptions used here give you the basis for understanding when to use procedures and how those procedures work effectively. For each of the three sections the transfer of the DNA is described. Follow this procedure with the sections on prehybridization, hybridization, and detection. The section on probes may help with selection of the appropriate sequence and size of probe.

### Colony Hybridization

It is sometimes necessary to enumerate bacterial colonies in a mixed population background. Selective media are convenient to use, but are not always applicable (i.e., the bacterial strain of interest may have no antibiotic resistance, or the sample may contain many strains that are resistant to the same antibiotic). Therefore, the bacterial colonies need to be screened for the colony of interest.

Colony hybridization is a screening technique based on the hybridization of a DNA probe to the appropriate strain. Only bacteria that grow as colonies can be enumerated with this technique, but this criterion allows the use of a great many strains. This technique is also valuable for screening clones (e.g., in *E. coli* transformants) for the presence of a desired DNA fragment. For instance, if you wanted to clone gene *xyz* from *Alcaligenes* and had a gene *xyz* probe from *Pseudomonas*, you could shotgun-clone fragments into a plasmid cloning vector and screen all for the presence of that gene (or at least a fragment of it) [19].

This technique has several components: (1) the choice on an appropriate agar plate, which might be selective for the strain (using antibiotic or heavy metal resistance or such conditions as anaerobic atmosphere) or might be completely non-selective; (2) description of a DNA probe for the strain; (3) labeling of the probe, which can be radioactive or non-radioactive; and (4) hybridization with the labeled probe.

#### Procedure

1. Use nylon filters that are either cut to size or purchased precut. Precut filters are preferred since they are standardized (e.g., Biotrans membranes [ICN, Costa Mesa, Ca]).
2. Use a soft-lead pencil to label the filters so that (1) they can be matched up with the appropriate plate later and (2) the orientation of the filter with the plate is known.

3. Using sterile forceps, lay the filter on the colonies with one quick motion.
4. Leave the filter on the colonies for approximately 1 min. The moisture of the plate will soak into the filter.
5. Stretch a sheet of plastic wrap out on a flat surface. For every filter, place a 1-ml drop of denaturing solution (1.5 M NaCl, 0.5 M NaOH) in discrete areas.
6. Remove the filter with forceps, and place, colony side up, on the drop of denaturing solution. Leave for 5 min.
7. On a separate sheet of plastic wrap, place 1-ml drops of neutralizing solution (3 M Na acetate, pH 5.5). Transfer the filters to these drops, with the colony side up. Leave for 5 min.
8. Transfer filters to standard filter paper to blot-dry for about 30 min.
9. Bake filters at 80°C for 1 h to fix DNA on the filters. The filters are now ready for hybridization or may be stored.

## Southern Blotting

As described earlier, DNA fragments created by restriction enzyme digestion can be separated by agarose gel electrophoresis. To determine which fragment of the separated bands has a DNA sequence of interest, Southern blotting is performed [51]. This procedure transfers the DNA bands in the gel to a nylon membrane precisely as they appear on the gel. After hybridization to the probe DNA, only the bands that hybridize to the probe will be visible. Since the other bands, including the DNA markers, will not be visible at all, comparison will be difficult without landmarks. For this reason a labeled DNA marker may be used, and a clearly identifiable positive control lane should be run on the gel.

The procedure described here begins after the agarose gel has been run and photographed. The gel should be firm enough that it can removed from its gel box easily with a large spatula and transferred as described.

### Procedure

1. Soak the completed agarose gel in 0.25 M HCl for 30 min with gentle agitation.
2. Pour off the acid solution, and soak the gel briefly in distilled water.
3. Pour off water and, soak the gel in 0.4 M NaOH to denature the DNA. Agitate gently for 30 min.
4. Neutralize in 3 M Na acetate (pH 5.5) for 30 min.
5. Transfer to a piece of positively charged nylon membrane that has been cut to the exact size of the gel (Fig. 13-9).
6. Rinse the membrane briefly in 2x sodium chloride/sodium citrate (SSC).
7. Allow the membrane to dry thoroughly (bake 30–40 min at 80°C).
8. Fix DNA to membrane using a UV light box (e.g., 254 nm radiation for 2 min) or the Stratagene Stratalinker.
9. Use the membrane immediately, or store for at least 1 month in a plastic bag.

There are several good sources of nylon membranes (e.g. GeneScreen Plus, New England Nuclear). A Posiblot Pressure Blotter (Stratagene, LaJolla, CA) allows a faster transfer of the DNA to the membrane. Electrotransfer devices, which drive the DNA onto the membrane in the same way that gel electrophoresis moves the DNA, are also very effective.

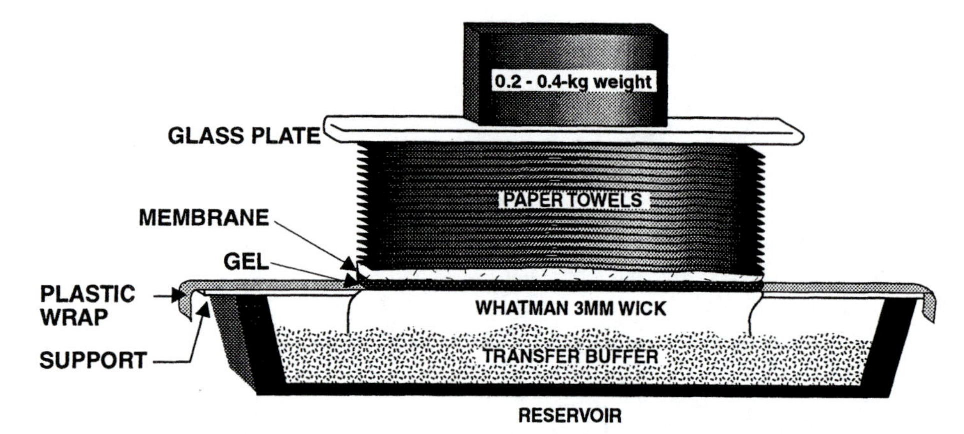


Figure 13-9 Transfer of DNA from the gel to the nylon membrane. The transfer dish is filled with 20x SSC. The gel is inverted and is placed on top of three sheets of Whatman 3MM paper saturated with 20x SSC, which is supported by a glass plate. The 3MM paper should contact the 20x SSC bath in order to keep moist. The gel is overlaid with a nylon membrane that has been cut slightly larger than the size of the gel (overhangs should be approximately 2 mm). Before application, wet the membrane thoroughly in 4x SSC. Cover the membrane with two pieces of 3MM paper that have been immersed in 4x SSC. Stack paper towels on top of the 3MM paper, and crown the pile with a glass plate and a weight. Avoid bubbles in either the gel, 3MM paper, or the membrane. Allow to sit overnight. After disassembly of the apparatus, mark the orientation of the membrane carefully. The buffer solution is drawn through the gel and into the membrane, taking the DNA with it.

## Dot and Slot Blots

Dot blots and slot blots are relatively easy to perform and can give valuable information. Dot blots are created by spotting and fixing DNA directly on a membrane and then proceeding with hybridization. They are useful for screening large numbers of DNA samples for the presence of DNA homologous to the gene probe. Slot blotting is very similar, except that a manifold is used to place the DNA solution in a defined area. This process facilitates the quantification of the DNA in the hybridized signal. In contrast, dot blots allow the DNA solution to spread in non-standard circles and thus are less effective for quantification.

DNA quantification using slot blots is performed using a solution of DNA of known identity (homology to the probe) and concentration, such as is found in a plasmid preparation. Serial dilutions of the DNA are placed in one column of wells, and the other wells are used for the unknown DNA samples. After hybridization the DNA standards show a pattern of decreasing signal strength with decreasing DNA concentration. Comparison of the unknown DNA's signal to the standards gives a good approximation of the homologous DNA concentration in the unknown sample.

DNA can be denatured on the nylon membrane, or it can be denatured before application. The latter method is useful if a dot- or slot-blot manifold is used and therefore is described here.

### Procedure

1. Denature DNA solution in 0.1 vol of 3 M NaOH, 0.1 M EDTA.
2. Incubate 60°C for 30 min.
3. Neutralize with 1 vol cold 2 M ammonium acetate.
4. Select a piece of positively charged nylon membrane of the correct size, and soak it in distilled water.

5. For slots, place it in the manifold, ensuring no leaks at the periphery (e.g., Minifold II slot blotter, Schleicher and Schuell, Keene, NH).
6. Immediately spot denatured DNA onto the membrane (dots) or into the manifold (slots), and apply a gentle vacuum.
7. Rinse each well with 0.5 ml of 0.3 M NaOH.
8. Dismantle manifold.
9. Soak membrane in 2x SSC and let air-dry.
10. Fix DNA to nylon membranes by UV (254 nm) radiation for 2 min using an UV Stratalinker (Stratagene), or bake 30–40 min at 80°C. The membrane can be used immediately for hybridization or can be stored for at least 1 month in a sealed plastic bag.

Magnagraph (Micron Separations, Westborough, MA), Zeta probe (BioRad, Hercules, CA) charged nylon membranes, and Hybond-N+ nylon membrane (Amersham, Arlington Heights, IL) make excellent hybridization membranes.

## Probe Preparation

The DNA probe must be labeled in some way that can be detected. The fastest and easiest method to label it is with a radioactive isotope, such as $^{32}P$. Since low amounts of radioactive materials are used, radioactive labels are not especially dangerous, although proper handling and disposal procedures are essential. Radioactivity is easily detected using autoradiography and produces good results with a permanent record. A drawback to this method is that $^{32}P$ has a short half-life, and so occasional users of hybridzation may waste money.

The source of DNA used in this step is critical, since all of the DNA will be labeled. If a cloned fragment of DNA is to be used as the probe and is not separated from the plasmid cloning vector, then all of the plasmid will also be labeled. In this case, the probe will detect every other gene on the plasmid, including antibiotic resistance genes. This could give false-positive responses during hybridization. It is therefore necessary to isolate the DNA of interest before labeling.

### *Nick Translation*

Nick translation of DNA is a means of incorporating radioactive nucleotides into a fragment of probe DNA. It is dependent on two enzymes: DNase I and Polymerase I. The DNase introduces breaks (nicks) in the sugar-phosphate backbone of DNA. The DNA Polymerase attempts to repair the break by excising a single strand and replacing it using the other strand as a template. Free nucleotides are incorporated. Typically, fragments of 500 bp are created during this process [40].

Another critical element in this procedure is the mix of the two enzymes. Having too much DNase in the mixture will cut the DNA to pieces. The proper amounts can be determined empirically, but it is far easier to take advantage of one of the many kits offered by suppliers that have the appropriate mixtures already made. One of these mixtures (Gibco-BRL) is shown below. The example uses labeled dATP, although any nucleotide can be labeled:

Reaction Mix

1 μg of probe DNA
5 μl of unlabelled dCTP, dGTP, and dTTP mix (each at 0.2 mM)
10 μl of alpha-labeled dATP, specific activity greater than 3000 Ci/mmole (New England Nuclear, Amersham)

5 μl of DNase/Pol I enzyme mix
sterile distilled water to bring the volume to 50 μl

Incubate at 15°C for 1 h. Stop the reaction with 5 μl of 0.3 mM EDTA (pH 8.0).

It is necessary to separate the labeled DNA from the free nucleotides and to determine the fraction of nucleotide that has been incorporated into DNA. Before separation, remove a 1-μl sample of the reaction mix to a scintillation vial. Separate the DNA and nucleotides, and then take another 1-μl sample from the DNA mix into another scintillation vial. This procedure assumes that the volume of both the reaction mix (50 μl) and purified DNA are the same. If the volume differs (e.g., if the DNA sample volume is substantially greater, causing a dilution of signal) then an adjustment needs to be made. Add 5 ml of scintillation fluid to each vial, and count on a scintillation counter. The percent incorporation is equal to the DNA counts divided by the reaction mix counts and multiplied by 100. This also gives you the value of cpm/μl for the DNA probe.

Separation of the nucleotides and the DNA can be accomplished using a Sephadex G-50 fine column. Equilibrate the column with 10 mM Tris HCl (pH 8.0), 50 mM NaCl, and 0.1 mM EDTA (pH 8.0). Other rapid methods are available commercially (e.g., NENsorb columns, New England Nuclear, Boston).

A related technique is the use of random primers (kit from Gibco-BRL). This kit uses a set of 6-mers to randomly prime the probe DNA before polymerization. The results are comparable to nick translation.

An alternative means of labeling with $^{32}P$ is to end-label the DNA probe using gamma-labeled $^{32}P$ ATP and a terminus labeling kit (Gibco-BRL). This kit uses T4 polynucleotide kinase to replace the non-radioactive phosphate of the probe with the radioactive phosphate of the ATP molecule. Follow the manufacturer's directions, especially with regard to labeling efficiency and separation of the probe and ATP.

### *Non-Radioactive Detection*

Non-radioactive detection methods are gaining favor as concerns increase about worker exposure to radiation and contaminant disposal. The Genius system (Boehringer Mannheim) has been demonstrated to be an excellent system for all types of hybridization reactions. The probe DNA is labeled with reagents that add a molecule of digoxigenin; detection is based on an antibody to the digoxigenin that then allows a chromogenic reaction to take place. Follow the manufacturer's directions for labeling of the probe, hybridization conditions, and detection.

The AttoPhos fluorescent alkaline phosphatase substrate (JBL Scientific) is another detection system that can be used. Eurogentec offers a 2-acetylamino-fluorene (AAF)-labeled *E. coli* 16S and 23S rRNA and anti-AAF monoclonal antibody. After hybridization a sheep-anti-mouse antibody labeled with alkaline phosphatase is applied.

## Probes

The number of DNA sequences that have been used as probes is so large that they could fill their own book. Many of them are mentioned in other chapters, particularly where they specify a valuable subclass of microorganisms (e.g., use of the *nif* gene for nitrogen fixation in Chapter 1). Many more are so specialized that they do not merit inclusion here, yet are easily found in the literature (e.g., *nah* genes for the detection of naphthalene-degrading microorganisms). The

Table 13-2 DNA Probes

| Probe | Sequence | Reference |
|---|---|---|
| Archaebacterial probe | TCCGGCRGGATCAACCGGAA | 17 |
| Eukaryotic probe | GGGCATCACAGACCTG | 17 |
| Eubacterial probe | ACCGCTTGT-GCGGGCCC | 17 |
| Universal probe | GWATTACCGCGGCKGCTG | 17 |
| Control probe | GTGCCAGCMGCCGCGG | 17 |
| ALF1b, alpha 16s proteo bacteria[a] | CGTTCG(C/T)TCTGAGCCAG | 27 |
| BET42a, beta 23S proteobacteria | GCCTTCCCACTTCGTTT | 27 |
| GAM42a, gamma 23S proteobacteria | GCCTTCCCACATCGTTT | 27 |
| EUB338, eubacteria 16S | GCTGCCTCCCGTAGGAGT | 2, 52 |
| Eucaryotic | ACCAGACTTGCCCTCC | 2 |

R = purine; W = A or T; K = G or T; M = A or C.
[a]Also detects some non-alpha group bacteria.

probes mentioned in Table 13-2 have more universal appeal and have demonstrated their worth in many experiments.

## Hybridization Steps

There are several steps involved in the hybridization of a DNA probe to a target. In the prehybridization step the filter is covered with a non-homologous DNA source, usually herring sperm DNA or salmon sperm DNA. This step reduces the chances of probe DNA binding to the filter non-specifically. The next step is preparation of the probe DNA, which must be denatured before hybridization. There is the hybridization step itself, which is then followed by washing. The washing step removes non-specifically bound probe and probe that is binding weakly. The temperature of the wash is a critical factor here, just as it is for the hybridization. After the washing comes detection by autoradiography or by using the reagents of the non-radioactive kits.

### *Prehybridization*

Hybridization Solution

5x Denhardt's solution
5x salt, sodium phosphase, EDTA (SSPE)
0.2% (w/v) sodium dodecyl sulfate (SDS)
500 μg/ml denatured non-homologous DNA

Procedure

1. Place the filters (or membrane) inside a sealable bag. A Seal-A-Meal bag and heater work very well for this procedure.
2. Add prehybridization solution at 10 ml for every 100 $cm^2$ of filter. This is roughly one Southern blot or two colony filters. Seal the bag.
3. Incubate in a 65°C bath for 1 h.

### *Probe Preparation*

The labeled DNA probe described above is double-stranded. It must be made single-stranded before its use in hybridization. There are two methods to accomplish that objective:

Procedure

1. Heat the DNA in a boiling water bath for 5 min, then quickly remove it, and immerse into an ice bath. The disadvantage of this method is that the boiling conditions often pop the top off the DNA tube, which may result in spillage of the radioactive contents.
2. Denature the DNA by adding 0.1 vol of 1 N NaOH and then heating for 10 min at 65°C. Then add 0.1 vol of 1 N HCl to neutralize the solution.

### *Hybridization*

Procedure

1. Keep the hybridization solution at the same temperature as the membrane to prewarm the solution.
2. Remove the prehybridization solution from the bag by snipping off one corner and squeezing the liquid out. Remove as much as possible.
3. Add 5 ml of fresh prehybridization solution for every 100 $cm^2$ of filter. Carefully pipette in the right amount of denatured probe DNA ($10^5$ dpm per ml of hybridization solution), making sure to immerse the DNA into the solution.
4. Reseal the corner, and immerse in a 65°C water bath for 12–14 h (overnight).
5. Alternatively, use a solution of 50% formamide and incubate at 42°C, followed by washing at 50°C.

**Note:** Incubation temperatures are approximate. For high-stringency hybridization, use a temperature about 5°C lower than the melting temperature (Tm), as calculated in the section, Hybridization Temperature. The definitive analysis of nucleic acid probe hybridization has been written by Stahl and Amann [52].

### *Washing*

Procedure

1. Cut the bag open, and dispose of the radioactive solution appropriately.
2. Place the membrane in a new plastic bag, and add 250 ml of wash buffer.
3. Incubate at room temperature for 30 min with vigorous agitation.
4. Pour off wash buffer, replace with the same amount of new wash buffer.
5. Repeat incubation.
6. Perform a further wash at 42°C if higher stringency is needed.
7. Alternatively, wash three times in 2x SSC at 65°C for 20 min each. Then wash once in 0.2x SSC at 55°C for 20 min. The membrane is now ready for the detection step.

### *Autoradiography*

For radioactive probes, detect the hybridized signal by autoradiography. In a darkroom, place a sheet of X-Omat film (Kodak, Rochester, NY) over the hybridization membrane and store at −70°C. The exposure time is dependent on the strength of the signal. "Hot" bands show up overnight, whereas very weak signals require 4–7 days of exposure. A new sheet of film can be used if the first exposure turns out to be too short. Kodak also offers intensifying screens that help significantly make more efficient use of the radioactive decay signal.

Procedure: Non-Homologous DNA Preparation

1. Prepare a 10 mg/ml solution of salmon or herring sperm DNA in distilled water.
2. Add 3 M NaCl to a final concentration of 0.1 M.
3. Extract once with Tris-buffered phenol and once with phenol/chloroform
4. Shear the DNA by sonication or by passing the DNA solution through a small-bore needle several times.
5. Concentrate the DNA by precipitation: add 2 vol of ethanol and keep at −20°C for 4 h.
6. Spin in a microfuge for 20 min.
7. Pour off the supernatant, and dry the DNA.
8. Resuspend the DNA in water at 10 mg/ml (as determined by $OD_{260}$).
9. Store aliquots at −20°C.
10. Before use, denature the DNA aliquot by heating in a boiling water bath for 10 min and then quickly immersing the tube in ice.

Reagents

1. 100x Denhardt's solution (per liter):
   20 g Ficoll, molecular weight (MW)
   20 g polyvinylpyrrolidone MW 360,000
   20 g bovine serum albumin, Fraction V
2. 1x SSC: 0.15 M NaCl, 0.015 sodium citrate, pH 7.0. A stock solution of 20x can be conveniently prepared.
3. SSPE, 20x stock
   3.6 M NaCl
   0.2 M sodium phosphate (pH 8.3)
   0.02 M EDTA
4. Wash buffer
   5 mM sodium phosphate (pH 7.0)
   1 mM EDTA
   0.2% SDS

## In Situ Hybridization

The technique of in situ hybridization has proven to be a powerful means of identifying bacteria in environmental samples. It uses a labeled oligonucleotide probe that hybridizes with a target inside the cell where it can be observed easily. In general, cells that are fixed are freely permeable to oligomers in the range used here. The protocol described here is very useful for in situ hybridization to the 16S rRNA [13]. In general, rRNA species are attractive targets because there are multiple copies per cell, thereby increasing the signal ($10^4$ in a rapidly growing cell). The literature is replete with examples of probes for rRNA based on phylogeny. Selection of the appropriate probe allows the identification of microorganisms according to phylogenetic breadth, from the kingdom level all the way to species. Some of the more-often utilized probes are given in Table 13-2. Another important benefit of this technique is that the signal from the probe is proportional to the amount of target (ribosome) present and therefore indicates the growth rate of the cells.

The technique uses oligomer probes that have fluorescent labels attached. Labeled probes can be produced in the laboratory, using either cloned sequences or a DNA synthesizer. However, for the occasional user, these methods are rather cumbersome and expensive in terms of time and equipment. Therefore, the oligonucleotide probes can be purchased from any of several companies that produce them to order. The oligos can then be produced quickly, with the appropriate label attached and at a known concentration.

The signal can be detected easily using the epifluorescent microscope and can be recorded using photography. Autofluorescence of the cells is a potential problem using this technique, although other fluorescent labels can be substituted that will eliminate this problem. Probes can also be prepared using a radiolabeled compound, and detection is accomplished using microautoradiography [17]. Digoxigenin and horseradish peroxidase have also been used.

The easiest way to set up these tests is on Teflon-coated glass slides that create individual wells (Cel-Line Associates, Newfield, NJ). The Teflon prevents spillage from one well to another. These slides must be pretreated before adding the fixed cells. This protocol should be applicable to most microorganisms.

Procedure: Slide Preparation

1. Soak slides in a KOH solution (10% in ethanol) for 1 h.
2. Rinse thoroughly in distilled water and air-dry.
3. Immerse the slides in a gelatin solution (0.1% gelatin, 0.01% chromium potassium sulfate at 70°C) briefly.
4. Air-dry vertically by standing slides in a rack.

Procedure: Cell Fixation

1. For liquid cultures, add 1 vol of the cell suspension to 3 vol of freshly prepared 4% paraformaldehyde in 200 mM sodium phosphate buffer (pH 7.2). If a cell pellet is used, resuspend cells in sodium phosphate buffer before adding the paraformaldehyde. Biofilm cultures can be immersed in the solution.
2. Mix by gentle vortexing, and incubate at 4°C for 3 h.
3. Centrifuge the cells to pellet them (2 min in a microfuge), and wash once with PBS to remove residual fixative.
4. Pellet the cells again, and resuspend in PBS at a concentration of $10^8$–$10^9$ cells/ml.
5. Use fixed cells immediately, or mix with 1 vol of cold ethanol and store for several weeks at −20°C.
6. Mix fixed cell suspension (or aliquot) with 0.1 vol of 1% Nonidet-40 (a non-ionic detergent from Sigma Chemical).
7. Centrifuge 1 min in a microfuge, and remove the supernatant.
8. Resuspend the cells in 1 vol 1% Nonidet-40.
9. Apply fixed cells as a drop (<5 μl) on the glass side. Allow to air-dry.
10. Dehydrate the cells by immersion of the slide in 50% ethanol for 3 min.
11. Repeat immersion in 80% ethanol for 3 min.
12. Repeat immersion in 95% ethanol for 3 min.
13. Let the slide air-dry. Slides should be very stable at room temperature.

Procedure: Hybridization

1. Prepare a moisture chamber equilibrated with 1 M NaCl (approximating isotonic conditions) by soaking a piece of filter paper in 1 M NaCl and placing it inside a screw-top test tube for 30–60 min.
2. Add 9 μl hybridization solution to each well, and incubate 30 min at the hybridization temperature.

3. Add 50 ng of labeled DNA probe (in approximately 1 μl) to the appropriate wells, and continue incubation for 2–5 h.
4. Rinse the slide gently with 5–6 ml of wash solution that was prewarmed at the hybridization temperature.
5. Immerse the slide in wash solution at the hybridization temperature for 15 min twice.
6. Rinse thoroughly with distilled water, and air-dry in the dark.
7. Store at room temperature in the dark or view immediately.

Hybridization Solution

0.9 M NaCl
50 mM sodium phosphate (pH 7.0)
5 mM EDTA
0.1% SDS
polyadenylic acid 0.5 mg/ml
10x Denhardt's solution

Wash Solution

0.9 M NaCl
50 mM sodium phosphate (pH 7.0)
0.1% SDS

### *Hybridization Temperature*

The temperature of hybridization for in situ hybridization is dependent on the oligonucleotide used as it is based on the dissociation temperature (Td) for the hybrid oligo/target DNA. Td should not be confused with the melting temperature (Tm) for the DNA, which is usually higher. The Td is dependent on the concentration of the oligo and can be determined empirically [3], although simple calculations based on the nucleotide content will give a good estimation.

$$\mathrm{Td} = 81.5 + 16.6 \log M + 0.41[\%G + C] - 820/n$$

where M = the ionic strength of the hybridization solution and n = the number of nucleotides in the oligo.

### *Probe Oligomers*

Oligomers used in in situ hybridization must be complementary to the target DNA or, as is more often the case, to the rRNA molecule. This means that the primary sequence of the rRNA must be examined for the correct complementary sequence. For examples, see the several described oligos that are used for the 16S rRNA molecule, as well as the primary structure of the *E. coli* 16S rRNA molecule [10]. The numbering system that is frequently used to describe the hybridization site of the oligomer is based on *E. coli* [10].

Typically, oligomers are used in a size range of 20–35 nucleotides. They can be synthesized in the laboratory if a DNA synthesizer and skilled technician are available. This DNA can then be labeled with a fluorescent compound (e.g., fluorescein or rhodamine) using a chemical linking process. However, these facilities may not be readily available, and it is very convenient to have one of the many DNA synthesis companies make your oligomer, with label, to order.

Fluorescence fading can be limited with a mounting solution or with Citifluor (Citifluor, London). Observe under the epifluorescent microscope. Photographs can be taken using either color or black-and-white film.

Mounting Solution

0.01 M phosphate buffer
0.15 M NaCl
0.1% (w/v) p-phenylenediamine
90% (v/v) glycerol, pH 8.0

## POLYMERASE CHAIN REACTION (PCR)

PCR has become one of the most useful techniques in molecular biology. Its objective is to amplify a specific fragment of DNA from a very high background of DNA. PCR is a relatively simple technique to perform, and several commercial kits and reagents are available to help you with the procedure. There are many variations on this technique, which demonstrates the great value that PCR has for investigators. Only the basic techniques are presented here, although an understanding of these steps will allow the worker to quickly acquire the related techniques. Atlas and Bej [4] have described several PCR applications and new applications appear in the literature on a regular basis. Quantitative PCR is a relatively new technique and has not been covered here because it is likely that the procedures will be significantly improved in the near future.

Several factors must be considered in the design of your PCR amplifications, but most people pick up the technique easily. PCR is an extraordinarily powerful technique; its power derives from the exponential increase of a target DNA fragment by replicating it in each cycle [15, 24, 31]. It can be used to detect a specific gene in single cells. PCR also requires only a minimal amount of equipment. The PCR thermal cycler is actually a "smart water bath" because it automatically changes temperatures according to a set program. A PCR machine is obligatory equipment for this procedure, since the automatic cycling is precise and labor-saving; there are many excellent machines available commercially.

Microbial ecologists find this technique useful for the detection of specific microorganisms or phylogenetic groups in environmental samples. The technique requires the extraction of DNA from the environmental sample. The extraction protocols presented in Chapter 12 yield DNA of sufficient purity for PCR. Several substances may interfere with the PCR reaction and should be avoided in the DNA preparation. These substances include heavy metals and humic acids commonly found in soil samples, as well as detergents, phenol, and EDTA (frequently used in DNA extraction). This list of substances to be avoided emphasizes the need for careful DNA preparation.

Exponential amplification of DNA cannot continue indefinitely; some component will become limiting (e.g., the nucleotides that make up the new strands), or the enzyme will deteriorate. In fact, there seems to be an upper limit to the amount of DNA that can be obtained from a typical PCR reaction [1, 53]. PCR produces enough of a specific fragment of DNA so that it can be easily detectable by agarose gel electrophoresis. The separation of DNA fragments is an essential component of PCR, since PCR may also amplify fragments of DNA other than the target.

There is one major limitation on the use of PCR: the need to know the sequence (or at least a critical part of the sequence) of the piece of DNA you wish to amplify. You need this information in order to describe the DNA primers that will flank your piece of amplified DNA. Very often this information has been published or is available through GenBank or one of the other large nucleic acid sequence databases. It is possible that you will isolate an interesting gene yourself, and then you will have to sequence it to obtain the information you need (see the section, DNA Sequencing).

PCR has several components: (1) the heat-stable DNA polymerase, usually derived from *Thermus aquaticus* (Taq); (2) two oligonucleotide primers, which flank the region to be amplified and which serve as the starting point for DNA synthesis; (3) the reaction mix, including the deoxynucleoside triphosphates (dNTPs); and (4) the appropriate series of incubation conditions, usually programmed into a PCR thermal cycler. Throughout the literature there are a range of values for each of these essential components, and subtle adjustment of them may improve PCR efficiency. Each is explained separately in this section.

The relatively new technique of in situ PCR [23] is a potentially powerful tool for the analysis of microbial communities. Performance of this technique requires some experience with PCR analysis.

## PCR Analysis of DNA

### *Reaction Mix*

Table 13-3 presents standard conditions for the reaction mix, although reference to the manufacturer's conditions is always advisable. Reaction volume varies between 25–100 μl. This section uses 50 μl for convenience.

A master mix can be prepared first, and then an aliquot is placed in each of the tubes. To use the same primers in each tube, incorporate the primers in the master mix; otherwise add the primers separately to each tube. Before starting the PCR reaction, each tube should receive one drop of sterile mineral oil. The mineral oil seals in the reaction, ensuring that the contents do not evaporate during the high incubation temperatures. To recover the amplified DNA at the end of the PCR run, plunge a pipette tip below the surface of the oil to obtain the DNA. Take as little of the oil as possible.

### *DNA Sample*

Amplified DNA fragments can vary between 100 bp and several kilobasepairs; a DNA fragment of 200–1,000 bp is most convenient to use. In all cases of PCR analysis, the objective is to determine whether the DNA sample contains your sequence of interest. It is possible that the DNA concentration may be high enough

Table 13–3 Standard Conditions for the PCR Reaction Mix

| | Add | Stock Concentration | Final Concentration | Store Stock (°C) |
|---|---|---|---|---|
| 10x buffer | 5 μl | 10x[a] (see below) | 1x | −20 |
| MgCl2 | 5 μl | 15 mM | 1.5 mM | −20 |
| dNTP | 4 μl (1 μl each) | 10 mM each | 200 uM each | −70 |
| Primers | 2 μl (1 μl each) | 25 uM | 0.5 uM | −20 |
| Taq pol | 1 μl | 5 units/μl | 1–5 units | −20 |
| DNA | 1 μl | See text | | 4 |
| Water | 32 μl | Sterile distilled | | Room temperature |
| Total | 50 μl | | | |

[a]Typical 10x buffer: 500 mM KCl, 100 mM Tris HCl, pH 8.4, 0.01% gelatin.

so that other sites can successfully compete with the target DNA for the primers. This can only be determined empirically.

### *Primers*

Primers are short fragments of DNA, usually 15–30 nucleotides in length, which are extremely important to the success of PCR. They may be as short as 10 nucleotides for some random PCR procedures. A pair of primers flanks each fragment of DNA that you wish to amplify. In the reaction mix, each primer should be present at the same concentration and should be present in excess throughout the PCR process. A range of 0.1–1.0 μM will achieve this result; a figure of 0.5 μM is typical.

The primers can be produced in-house using a DNA synthesizer. These machines are becoming more common and are fairly easy to operate. For convenience, the primers may be ordered from suppliers who specialize in the synthesis of DNA (Genosys Biotechnologies, The Woodlands, TX; Operon Technologies, Alameda, CA). These companies have good quality control, and their products arrive quickly. When designing primers, keep in mind the following general rules:

Keep the G + C content in the range of 40% to 60%. This cuts down on internal secondary structure formation.

Avoid long (over four) stretches of any one base.

Avoid complementarity between the primers, especially at the 5′ and 3′ ends. Failure to do this will result in annealing between the primers and the creation of spurious products ("primer dimers").

Avoid creating primers that are complementary to regions of secondary structure in the target, so the primers do not have to compete with the secondary structure for the site.

Keep the melting temperatures of the two primers relatively close. This makes selection of the annealing temperature of the reaction easier and more predictable.

Make the 3′ end of the primers an exact match with the DNA template.

### *Nucleotides*

The four deoxynucleoside triphosphates (dATP, dGTP, dCTP, dTTP) are the building blocks of the new strands. It is important to keep all of the nucleotides in excess during PCR and at an equal concentration. A concentration of 200 μM for each is probably the most common. In certain instances modified nucleotides can be used, especially for DNA labeling. DNA labeled with $^{32}P$ can be created by using a small amount of $^{32}P$-dATP (or other nucleotide).

Nucleotide mix is very labile. It is often a good idea to mix all four nucleotides and then apportion them into one-use tubes; keep all frozen at −70°C. The level of PCR activity in your laboratory will dictate what volume will be in the one-use tubes.

### *Reaction Buffer and Magnesium*

The reaction buffer and the $MgCl_2$ are commonly supplied along with the Taq polymerase. The buffer is usually at 10x concentration. Check the manufacturer's specifications on the $MgCl_2$, since Taq polymerase is very sensitive to Mg ion concentration. You may have to test a series of magnesium concentrations between 1 mM–4 mM to find the one that works the best.

### *DNA Polymerase*

The most commonly used polymerase for PCR is derived from Taq. However, other polymerases are available that boast greater stability or replication fidelity. The manufacturers' specifications should be followed for the amount of polymerase needed for each reaction tube, since the amount of enzyme is based on units of enzymatic activity. In general, 1 unit of enzyme per tube is sufficient.

Polymerases from non-thermophilic bacteria are not recommended, since incubation at the dissociation temperature will destroy the enzyme. Perkin Elmer (Foster, CA) has a full range of PCR reagents.

### *Incubation Conditions*

The PCR machine transfers among three temperatures:

1. *Dissociation—the temperature necessary to separate double-stranded DNA into single strands*. It is usually 94°C and is held there for 1 min.
2. *Annealing—the temperature at which the primers will anneal to the complementary single-stranded DNA*. It is determined by the melting temperature of the primer (Tm), which is dependent on the base composition of the primer. The annealing temperature should be 5°C lower than the Tm. Suppliers usually indicate the Tm of their primers. In practice, the annealing temperature is between 40–60°C. This temperature is usually held for 1 min.
3. *Extension—the temperature at which the enzyme finds the DNA: primer complex and proceeds with DNA replication*. It is usually 72°C and is held for 1–2 min. At the end of the PCR cycles, a separate 5- to 10-minute extension period at 72°C is often used.

If you amplify too many fragments of DNA (too many bands on the gel), try increasing the annealing temperature slightly (about 2°C). In some cases you can set the annealing temperature at 72°C. In this case you can program a two-step (94°C, 1 min and 72°C, 2 min) cycle, which will save some time.

The number of cycles is an important determinant of PCR success, although there is no precise figure. Between 30–40 cycles for a typical PCR run are probably sufficient for most applications, although occasionally as many as 50 cycles are used. If amplification of a very weak signal is desired, it is probably best to amplify with one set of primers for 40 cycles, then remove an aliquot (1–2 μl) to a new tube, and reamplify using a set of primers that are "nested" with respect to the first pair.

Notes

1. When handling reagents, always wear disposable gloves. Their use cuts down on the risk of contaminating your samples.
2. Use reagents (e.g., water, nucleotide mix, mineral oil) that are as sterile and as pure (DNA-free) as possible. Special "barrier-tip" pipette tips can be purchased that reduce air-borne contamination.
3. Include a positive and a negative control. The positive control tube can include a DNA preparation of the target organism or a clone of the target sequence. Use 10–100 ng of DNA. It is not unusual for one of the components of PCR to be omitted during reaction mix preparation; the positive control should detect such mistakes. The negative control tube should contain no DNA. This tube demonstrates whether contaminants are present in any of the reagents.

4. To facilitate cloning of amplified fragments, construct primers that include a convenient restriction site (e.g., *Eco*RI, *Bam*HI) on the 5′ end. After amplification these ends will be double-stranded and therefore can be digested with the appropriate enzyme. This results in sticky ends that can be inserted into many cloning vectors. Do not put the restriction site at the very end of the primer; add a few random nucleotides on the end. Some restriction enzymes cut poorly when the site is too close to the end of the fragment.
5. Take note that it may be desirable to remove the primers from the amplified DNA. Several rapid systems are available (e.g., Centricon columns, Amicon, Beverly, MA).

### *Detection*

At the conclusion of the PCR amplification the samples can be stored at 4°C. PCR products are usually analyzed by agarose gel electrophoresis.

If the amount of target in your sample is very low (less than 1000 copies), a single PCR amplification may be insufficient to produce enough product to be seen on a gel. PCR is relatively inefficient under these conditions, and the first rounds may not increase the target sequence in an exponential manner. Detection limits are dependent on the DNA probe used. There are two means of detecting low amounts of product. The agarose gel can be Southern-blotted using the appropriate DNA probe for the sequence of interest. Alternatively, a small (1–2 μl) aliquot of the amplified DNA can be reamplified using a set of nested primers. "Nested" refers to the location of the primers within the amplified fragment and results in a PCR product that is somewhat shorter than after the first amplification. The use of nested primers works better than the reuse of the original primers. Conditions for the reamplification should be the same as for the initial amplification.

## Arbitrarily Primed PCR

Arbitrarily primed PCR (AP-PCR), or random amplified polymorphic DNA (RAPD), is a means of generating a pattern of amplified DNA fragments that is unique to that species [60, 61]. The key to the success of this technique is the use of short (10-mer) primers that recognize many sites in the genome. A set of 15–20 primers is amplified with each strain, and the results are visualized on agarose gels. Only one primer is used in each PCR reaction. An amplified DNA band appears wherever the primer finds two annealing sites relatively close together (less than 5000 bp).

Sets of these primers are available (e.g., Operon Technologies, Alameda, CA). When comparing different strains, DNA should be supplied in approximately equal amounts. Annealing temperatures are more typical for 10-mer primers, usually in the 36–40°C range. Some workers prefer to run two or three cycles at a low temperature (lower than 40°C) and then the remainder of the cycles around 60°C; either protocol is acceptable. Amplification for at least 40 cycles is common. All other parameters are unchanged.

When creating a profile of various strains, it may not be critical to perform the entire set of primers for each strain. Some primers may find bands (perhaps more than one for a single primer) that are conserved throughout a series of closely related organisms, whereas other primers may find bands that are unique for a particular strain. Your needs will dictate which primers work best.

AP-PCR can be performed on several isolates to determine whether they are identical or related. Identical banding patterns of amplified DNA indicate identity. Several different primers should be used to develop a "fingerprint" of the strain (see Chapter 17 for statistical analysis). The banding pattern is also dependent on the amount of input DNA, and therefore care should be taken to equalize input DNA concentrations for all isolates tested (5–25 ng is a good range).

## Reverse Transcriptase PCR (RT-PCR)

RT-PCR was developed to allow amplification of mRNA in cells and thus the investigation of gene expression [26, 46]. The technique is essentially PCR with an added step; the isolated RNA is converted to a double-stranded form (cDNA) using the reverse transcriptase (RT) enzyme (usually derived from MMLV or AMV viruses). In the cDNA form it can be effectively amplified as described above. As with PCR, RT-PCR is dependent on a knowledge of the sequence of the mRNA for which you are searching. However, random primers can be attempted to isolate unknown genes that are being expressed. Individual reagents can be assembled for this work, although the use of a kit, such as the GeneAmp RNA PCR kit (Perkin Elmer) or the First Strand kit (Clontech) is highly recommended. Many of these kits have a poly(dT) primer that is used for the isolation of poly(A) tailed mRNA. Using this primer works wonderfully for eukaryotic mRNA, but not at all for bacterial mRNA, which is not (except in rare circumstances) polyadenylated. A different cDNA primer is required. Random hexamer mixes are sometimes effective.

If DNA is present in your RNA sample, the possibility exists that you will amplify the DNA sequence. Add RNase-free DNase to your sample to eliminate this problem. After incubation, clean your sample with phenol/chloroform, and ethanol-precipitate the RNA.

### Procedure

1. Isolate RNA as described in Chapter 12. Use an aliquot (microgram quantities are sufficient).
2. Combine the RNA and the cDNA primer (25–50 ng) in a sterile PCR tube. Use sterile distilled water to bring the total volume to 30–50 μl. The cDNA primer should be complementary to the 3′ end of the message you wish to amplify.
3. Heat the tube at 65°C for 15 min. This step breaks the basepairing that can occur in the RNA strands and then allows the primer to anneal to the RNA.
4. Place on ice, and add the RT reagents as directed by the kit.
5. Concentrate the cDNA by first cleaning with phenol/chloroform and then performing ethanol precipitation. However, simply performing PCR on 1–5 μl of the cDNA mix is usually satisfactory.
6. Amplify the cDNA with 30–40 cycles of amplification. As usual, a set of primers must be used here. The cDNA primer may be used in this step also, although a nested primer may be preferred.
7. Manipulate the amplified cDNA as described for DNA (cloned, sequenced, digested, etc.).

## rRNA Amplification

The use of rRNA as a means of phylogenetic analysis is based on the ubiquity of the 16S and 23S molecules in microorganisms [20, 34]. One of the greatest contributions of PCR technology to microbial ecology is the ability to amplify genes from a sample without the need to cultivate the bacteria, thereby identifying genetic information from non-culturable microorganisms [35, 57]. In particular, researchers have used primers for conserved regions of the 16S rRNA gene to retrieve fragments of this gene. Cloning and sequencing are then performed using standard techniques, and comparisons of the sequences are made using computer programs. Isolated rRNA genes can be sequenced and used for comparison of strains [39]. The Ribosomal Data Base project at the University of Illinois has over 1000 bacterial rRNA sequences on file and is thus a significant resource for this work.

The technique is identical to that described above. A set of primers that work well are presented in the section, Restriction Fragment Length Polymorphism. An analysis of appropriate membranes for this work has been described [38].

## RESTRICTION FRAGMENT LENGTH POLYMORPHISM (RFLP)

Restriction fragment length polymorphism (RFLP) is a convenient method for the comparison of several strains at one time [18]. It is based on a simple premise. As the name implies, the size of specific restriction fragments for several species is compared. Strains that are closely related have DNA sequences (and restriction sites) that are very similar. The farther apart phylogenetically two strains are, the greater is their differences in sequence. Ideally, DNA:DNA homology is used for the designation of species. The entire DNA sequence could be obtained and a computer program employed to test the degree of similarity between them. However, when comparing dozens of strains, subcloning and sequencing become great problems, and therefore the simpler RFLP method is employed.

Two methods are commonly used to achieve RFLP analysis. Both depend on the comparison of DNA band patterns after restriction digestion of specific fragments of DNA. (Non-specific DNA patterns can be compared through the technique of AP-PCR.) In the first method, DNA from entire genomes or from entire communities is first obtained, and the specific fragments are visualized after DNA hybridization with appropriate probes. In the second method, PCR amplification of specific fragments is the first step, which is followed by restriction digestion and visualization of the bands on gels. There is usually no need for the DNA hybridization step since the selection was performed during the PCR amplification.

### Method A

Procedure

1. Obtain DNA from the strains to be compared.
2. Cut the DNA with a set of restriction enzymes to obtain discrete fragments.
3. Separate the fragments by gel electrophoresis.
4. Transfer the fragments to a membrane, and perform Southern blot hybridization with a labeled probe.
5. Compare the patterns for similarity.

The rRNA genes have been used extensively for RFLP analysis, and therefore the example here uses gene probes based on rRNA genes (16S and 23S rRNA fraction, Boehringer Mannheim). However, it is possible to compare strains using other gene probes. For instance, comparison of thermophiles has been accomplished using RFLP analysis with a DNA polymerase probe [37] although other conserved genes might be used. The *nif* genes might be used for an analysis of nitrogen-fixing bacteria. Each step of the procedure is covered in detail below:

1. *Obtain DNA*: The technique used for this step is dependent on the DNA sequence of interest to you. If the chromosomal location of the gene of

interest is not known, then total DNA should be recovered from individual clones. If total chromosomal DNA is used, you may wish to use 6-bp restriction site enzymes, since the use of 4-bp enzymes results in a huge number of small fragments. If subclones can be obtained, they are much easier to work with.

2. *Cut the DNA*: Your fragment of DNA is probably small (less than 1.5 kb), and therefore restriction enzymes with a 6-bp recognition sequence will not cut the DNA very often (about one site for every 4000 basepairs). Use 4-bp recognition enzymes for small fragments; for example: *Alu*I, *Dde*I, *Hae*III, *Hha*I, *Msp*I, *Nla*III, *Rsa*I, and *Sau*3AI.
3. *Separate the fragments*: Since the fragments will usually also be quite small (less than 500 bp), either a high percentage (3%) agarose gel or a polyacrylamide gel should be used. Since the DNA fragments will be transferred to a nylon membrane for blotting, any DNA marker will be transferred initially, but will not be seen after hybridization. To avoid this problem, use a DNA marker that is already labeled (e.g., $^{32}P$), or use a DNA probe for the DNA marker. Use of this latter approach may result in cross-reactivity, although this is not usually a problem with the hybridization conditions that are typically used. For instance, a probe for Lambda phage DNA (used in the popular lambda/*Hin*DIII marker) does not cross-react with rRNA genes.
4. *Transfer the fragments*: Follow the procedure for Southern blot analysis.
5. *Compare the patterns*: Data interpretation may be a very simple matter if the desired end point is identity with a reference strain: Each strain will either have the same pattern as the reference or it will not. However, most RFLP users wish to compare several strains and organize them into a pictogram showing phylogenetic relatedness. Although algorithms have been published [33], a computer program is very convenient for this work; for example, GelCompar (Applied Maths, Kortrijk, Belgium) and a new software system for analyzing gels and calculating RFLP from image databases (DNA ProScan, Nashville, TN). The distance matrix that results from these comparisons can be used to construct a dendrogram using the UPGMA (unweighted pair group method with arithmetic mean) method [50]. The neighbor joining (NJTREE) method can also be used [43]. See Chapter 17 for a full analysis of these methods.

## Method B

The alternative method uses PCR amplification to obtain the fragments of interest without interference from other DNA fragments. Appropriate primer sites for PCR must be known, however, that correspond to conserved regions of the sequence. Primers that have been described for rRNA genes are shown in Table 13-4.

Table 3–4 Primers for rRNA Genes

| Primer | Sequence | Amplified Sequence | Reference |
|---|---|---|---|
| fD1 | ccgaattcgtcgacaacAGAGTTTGATCCTGGCTCAG | 16S for eubacteria, 1500 bp | 59 |
| rD1 | cccgggatccaagcttAAGGAGGTGATCCAGCC | | |
| R1n | GCTCAGATTGAACGCTGGCG | 16S, 1KB | 58 |
| U2 | ACATTTCACAACACGAGCTG | | |
| | TNANACATGCAAGTCGAICG | 16S, 1.5 kb | 30 |
| | GGYTACCTTGTTACGACTT | | |

All primers are written in a 5′ to 3′ direction. The lower case letters represent a polylinker for cloning.

Procedure

1. Obtain DNA from the strains that you wish to compare.
2. Perform PCR to amplify the fragment of interest. PCR amplification depends on the description of the appropriate primers.
3. Isolate the amplified DNA. A clean-up step is very useful because it prevents the gel from being obscured by excess primer or fragments from the initial input DNA.
4. Separate the fragments by gel electrophoresis.
5. Compare the patterns for similarity.

### Community Analysis

Several attempts have been made to analyze community heterogeneity using the RFLP technique. These approaches extract total DNA from an environmental sample and then isolate homologous genes (usually rRNA genes) from the DNA mixture. There are two methods in common use:

1. The gene of interest can be amplified from the DNA mixture using PCR primers (usually for the *rrn* genes), which is followed by cloning. The primers that include the cloning sites are especially useful. However, this technique also has drawbacks; the amplification process may skew the subsequent cloning step toward the most common genes at the expense of the rare (and often most interesting) sequences, thereby giving a false impression of the community [30, 58].

   The amplified DNA fragments are cloned into a convenient plasmid vector and transformed into an *E. coli* strain. The plasmids from each clone are isolated using standard protocols. This plasmid DNA can be treated in any of three ways: (1) The plasmid is digested with restriction enzymes; then the bands are separated on a gel and subjected to Southern blotting. (2) The insert DNA is removed from the plasmid, isolated after separation on a gel, cut with restriction enzymes, and once again separated on a gel. (3) The plasmid DNA can be subjected to another round of PCR amplification, using primers that are nested with respect to the first set. The amplified DNA is then digested with restriction enzymes and the bands are separated on a gel. Any bands that appear at this stage for (2) and (3) represent the gene of interest, so Southern blotting is not essential, although it is recommended as a good control that demonstrates the identity of the fragments.
2. The genes can be subcloned from the DNA mixture, which is followed by testing for identity using gene probes (usually oligonucleotides). This technique can be quite laborious.

## DNA SEQUENCING

Occasionally there is a need to obtain the sequence of a fragment of DNA. Sequencing is required for the description of PCR primers, for patenting of a useful gene, or for the complete study of a gene of interest. DNA sequencing requires a fully equipped molecular biology laboratory and skilled personnel. Only consider a DNA sequencing facility in your laboratory if your sequencing needs increase substantially. Otherwise, the cost of equipment, reagents, and personnel time is not justified.

In most instances it makes more sense to have the sequencing done by a professional. The companies listed in this section produce quality work in a short time and at a competitive price. All they require is a sample of the DNA that is usually cloned into a convenient plasmid vector (e.g., pUC18). Each company's requirements differ, although DNA purity equal to most commercial DNA purification kits is usually acceptable. It is more convenient for the company to sequence DNA fragments that are approximately 500 basepairs in length, and so it is preferable to subclone fragments of this size. However, most accept larger fragments and include a separate charge for designing additional sequencing primers.

LGL Ecological Genetics, Inc.
1410 Cavitt St.
Bryan, TX 77801
Tel: 409-775-2000

Commonwealth Biotechnologies, Inc.
911 E. Leigh Street
Suite G-19
Richmond, VA 23219
Tel: 1-800-735-9224

Lark Sequencing
9545 Katy Freeway
Suite 200
Houston, TX 77024
Tel: 1-713-464-7488

Retrogen
7909 Silverton Ave.
Suite 210
San Diego, CA 92126
Tel: 619-586-7918

ACGT Inc.
3605 Woodhead Dr.
Suite 110B
Northbrook, IL 60062
Tel: 708-559-8631

U.S. Biochemical
Custom Sequencing Service
P.O. Box 22400
Cleveland, Ohio 44122
Tel: 1-800-321-9322

## REFERENCES


1. Altwegg, M. 1995. General problems associated with diagnostic applications of amplification methods. J. Microbiol. Meth. 23:21–30.
2. Amann, R.I., Binder, B.J., Olson, R.J., Chisholm, S.W., Devereux, R., and Stahl, D.A. 1990. Combination of 16S rRNA-targeted oligonucleotide probes with flow cytometry for analyzing mixed microbial populations. Appl. Environ. Microbiol. 56:1919–1925.
3. Amann, R.I., Krumholz, L., and Stahl, D.A. 1990. Fluorescent-oligonucleotide probing of whole cells for determinative, phylogenetic, and environmental studies in microbiology. J. Bacteriol. 172:762–770.
4. Atlas, R.M. and Bej, A.K. Polymerase chain reaction, In Methods for General and Molecular Bacteriology (Gerhardt, P., Murray, R.G.E., Wood, W.A., and Krieg, N.R., Eds.), pp. 418–435. ASM Press, Washington, DC.
5. Ausubel, F., Brent, R., Kingston, R.E., Moore, D.D., Seidman, J.G., Smith, J.A., and Struhl, K. 1995. Current Protocols in Molecular Biology. John Wiley and Sons, New York.
6. Bagdasarian, M., Lurz, R., Ruckert, B., Franklin, F.C.H., Bagdasarian, M.M., Frey, J., and Timmis, K.N. 1981. Specific-purpose plasmid cloning vectors. II. Broad host range, high copy number, RF1010-derived vectors, and a host-vector system for gene cloning in *Pseudomonas*. Gene 16:237–247.
7. Balbas, P., Soberon, X., Merino, E., Zurita, M., Lomeli, H., Valle, F., Flores, N., and Bolivar, F. 1986. Plasmid vector pBR322 and its special purpose derivatives—a review. Gene 50:3–40.
8. Barta, T.M. and Hanson, R.S. 1993. Genetics of methane and methanol oxidation in Gram-negative methylotrophic bacteria. Antonie van Leeuwenhoek 64:109–120.

9. Berg, D.E. and Berg, C.M. 1983. The prokaryotic transposable element Tn5. Bio/Technology 1:417–435.
10. Brosius, J., Dull, T.J., Sleeter, D.D., and Noller, H.F. 1981. Gene organization and primary structure of a ribosomal RNA operon from *Escherichia coli*. J. Mol. Biol. 148:107–127.
11. Burmeister, M. and Lehrach, H. 1989. Isolation of large DNA fragments from agarose gels using agarase. Trends Genet. 5:41.
12. Davey, R.B., Bird, P.I., Nikoletti, S.M., Praszkier, J., and Pittard, J. 1984. The use of mini-gal plasmids for rapid incompatibility grouping of conjugative R plasmids. Plasmid 11:234–242.
13. Delong, E.F., Wickham, G.S., and Pace, N.R. 1989. Phylogenetic stains: ribosomal RNA-based probes for the identification of single cells. Science 243:1360–1363.
14. Elhai, J. 1994. Genetic techniques appropriate for the biotechnological exploitation of cyanobacteria. J. Appl. Phycol. 6:177–186.
15. Erlich, H.A., Ed. 1989. PCR Technology: Principles and Applications for DNA Amplification. Stockton Press, New York.
16. Figurski, D. and Helinski, D.R. 1979. Replication of an origin-containing derivative of plasmid RK2 dependent on a plasmid function provided in trans. Proc. Natl. Acad. Sci. USA 76:1648–1652.
17. Giovannoni, S.J., Delong, E.F., Olsen, G.J., and Pace, N.R. 1988. Phylogenetic group-specific oligodeoxynucleotide probes for identification of single microbial cells. J. Bacteriol. 170:720–726.
18. Grimont, F. and Grimont, P.A.D. 1986. Ribosomal ribonucleic acid gene restriction patterns as potential taxonomic tools. Ann. Inst. Pasteur/Microbiol. 137B:165–175.
19. Grunstein, M. and Hogness, D.S. 1975. Colony hybridization: a method for the isolation of cloned DNAs that contain a specific gene. Proc. Natl. Acad. Sci. USA 72: 3961.
20. Gutell, R.R., Larsen, N., and Woese, C.R. 1994. Lessons from an evolving rRNA: 16S and 23S rRNA structures from a comparative perspective. Microbiol. Rev. 58:10–26.
21. Guthrie, C. and Fink, G.R. 1991. Methods in Enzymology, Vol. 194: Guide to Yeast Genetics and Molecular Biology. Academic Press, San Diego.
22. Hanahan, D. 1983. Studies on transformation of *Escherichia coli* with plasmids. J. Mol. Biol. 166:557–580.
23. Hodson, R.E., Dustman, W.A., Garg, R.P., and Moran, M.A. 1995. In situ PCR for visualization of microscale distribution of specific genes and gene products in prokaryotic communities. Appl. Environ. Microbiol. 61:4074–4082.
24. Innis, M.A., Gelfand, D.H., Sninsky, J.J., and White, T.J., Eds. 1990. PCR Protocols: A Guide to Methods and Applications. Academic Press, San Diego.
25. Krishnapillai, V. 1988. Molecular genetic analysis of bacterial plasmid promiscuity. FEMS Microbiol. Rev. 54:223–238.
26. Mahbubani, M.H., Bej, A.K., Miller, R.D., Atlas, R.M., DiCesare, J. L., and Haff, L.A. 1991. Detection of bacterial mRNA using polymerase chain reaction. BioTechniques 10:48–49.
27. Manz, W., Amann, R., Ludwig, W., Wagner, M., and Schleifer, K. 1992. Phylogenetic oligodeoxynucleotide probes for the major subclasses of proteobacteria: problems and solutions. Sys. Appl. Microbiol. 15:593–600.
28. McBride, M.J. and Kempf, M.J. 1996. Development of techniques for the genetic manipulation of the gliding bacterium *Cytophaga johnsonae*. J. Bacteriol. 178:583–590.
29. Meilhoc, E., Masson, J., and Teissie, J. 1990. High efficiency transformation of intact yeast cells by electric field pulses. Bio/Technology 8:223–227.
30. Moyer, C.L., Dobbs, F.C., and Karl, D.M. 1994. Estimation of diversity and community structure through restriction fragment length polymorphism distribution analysis of bacterial 16S rRNA genes from a microbial mat at an active, hydrothermal vent system, Loihi Seamount, Hawaii. Appl. Environ. Microbiol. 60:871–879.
31. Mullis, K.B., Ferre, F., and Gibbs, R.A., Eds. 1994. PCR: The Polymerase Chain Reaction. Birkhauser, Boston.
32. Murrell, J.C. 1992. Genetics and molecular biology of methanotrophs. FEMS Microbiol. Rev. 88: 233–248.

33. Nei, M. and Li, W.H. 1979. Mathematical model for studying genetic variations in terms of restriction endonucleases. Proc. Natl. Acad. Sci. USA 76:5269–5273.
34. Olsen, G.J., Lane, D.J., Giovannoni, S.J., Pace, N.R., and Stahl, D.A. 1986. Microbial ecology and evolution: a ribosomal RNA approach. Annu. Rev. Microbiol. 40:337–365.
35. Pace, N.R., Stahl, D.A., Lane, D.L., and Olsen, G.J. 1986. The analysis of natural microbial populations by rRNA sequences. Adv. Microb. Ecol. 9:1–55.
36. Paul, J.H. and Myers, B. 1982. Fluorometric determination of DNA in aquatic microorganisms by use of Hoechst 33258. Appl. Environ. Microbiol. 43:1393–1399.
37. Perler, F.B., Southworth, M.W., Wilbur, D.G., and Wallace, D. 1995. Typing marine vent thermophiles by DNA polymerase restriction fragment length polymorphisms. In: Archae, A Laboratory Manual—Thermophiles (Robb, F.T., Place, A.R., Sowers, K.R., Schrier, H.J., Dassarma, S., and Fleischmann, E.M., Eds.), pp. 142–147. Cold Spring Harbor Laboratory, Cold Spring Harbor, NY.
38. Raskin, L., Capman, W.C., Kane, M.D., Rittman, B.E., and Stahl, D.A. 1996. Critical evaluation of membrane supports for use in quantitative hybridizations. Appl. Environ. Microbiol. 62:300–303.
39. Reeves, R.H., Reeves, J.Y., and Balkwill, D.L. 1995. Strategies for phylogenetic characterization of subsurface bacteria. J. Microbiol. Meth. 21:235–251.
40. Rigby, P.W., Dieckmann, M., Rhodes, C., and Berg, P. 1977. Labeling deoxyribonucleic acid to high specific activity in vitro by nick translation with DNA Polymerase I. J. Mol. Biol. 113:237–251.
41. Rose, R.E. 1988. The nucleotide sequence of pACYC184. Nucleic Acids Res. 16:355.
42. Rose, R.E. 1988. The nucleotide sequence of pACYC177. Nucleic Acids Res. 16:356.
43. Saitou, N. and Nei, M. 1987. The neighbor-joining method: a new method for reconstructing phylogenetic trees. Mol. Biol. Evol. 4:406–425.
44. Sambrook, J., Fritsch, E.F., and Maniatis, T. 1989. Molecular Cloning—A Laboratory Manual. Cold Spring Harbor Press, Cold Spring Harbor, NY.
45. Sasakawa, C. and Yoshikawa, M. 1987. A series of Tn5 variants with various drug-resistance markers and suicide vector for transposon mutagenesis. Gene 56:283–288.
46. Selvaratnam, S., Schoedel, B.A., McFarland, B.L., and Kulpa, C.F. 1995. Application of reverse transcriptase PCR for monitoring expression of the catabolic *dmpN* gene in a phenol-degrading sequencing batch reactor. Appl. Environ. Microbiol. 61:3981–3985.
47. Simon, R. 1984. High frequency mobilization of Gram-negative bacterial replicons by the in vitro constructed Tn5-Mob transposon. Mol. Gen. Genet. 196:413–420.
48. Simon, R., O'Connell, M., Labes, M., and Puhler, A. 1986. Plasmid vectors for the genetic analysis and manipulation of Rhizobia and other Gram-negative bacteria. Meth. Enzymol. 118:640–659.
49. Simon, R., Quandt, J., and Klipp, W. 1989. New derivatives of transposon Tn5 suitable for mobilization of replicons, generation of operon fusions and induction of genes in Gram-negative bacteria. Gene 80:161–169.
50. Sneath, P.H.A. and Sokal, R.R. 1973. Numerical Taxonomy. W.H. Freeman, San Francisco.
51. Southern, E.M. 1975. Detection of specific sequences among DNA fragments separated by gel electrophoresis. J. Mol. Biol. 98:503–517.
52. Stahl, D.A. and Amann, R. 1991. Development and application of nucleic acid probes. In Nucleic Acid Techniques in Bacterial Systematics (Stackebrandt, E. and Goodfellow, M., Eds.), pp. 205–248. John Wiley and Sons, NY.
53. Steffan, R.J. and Atlas, R.M. 1988. DNA amplification to enhance detection of genetically engineered bacteria in environmental samples. Appl. Environ. Microbiol. 54: 2185–2191.
54. Stephenson, M. and Jarrett, P. 1991. Transformation of *Bacillus subtilis* by electroporation. Biotechnol. Tech. 5:9–12.
55. Tait, R.C., Close, T.J., Lundquist, R.C., Hagiya, M., Rodriguez, R.L., and Kado, C.I., 1983. Construction and characterization of a versatile broad host range DNA cloning system for Gram-negative bacteria. Bio/Technology 1:269–275.
56. Trevors, J.T. and Starodub, M.E. 1990. Electroporation of pkkl silver-resistance plasmid from *Pseudomonas stutzeri* AG259 into *Pseudomonas putioa* CYM 318. Curr. Microbiol. 21:103–107.

57. Ward, D.W., Weller, R., and Bateson, M.M. 1990. 16S rRNA sequences reveal numerous uncultured microorganisms in a natural community. Nature 345:63–65.
58. Weidner, S., Arnold, W., and Puhler, A. 1996. Diversity of uncultured microorganisms associated with the seagrass Halophila stipulacea estimated by restriction fragment length polymorphism analysis of PCR-amplified 16S rRNA genes. Appl. Environ. Microbiol. 62:766–771.
59. Weisburg, W.G., Barns, S.M., Pelletier, D.A., and Lane, D.J. 1991. 16S ribosomal DNA amplification for phylogenetic study. J. Bacteriol. 173:697–703.
60. Welsh, J. and McClelland, M. 1990. Fingerprinting genomes using PCR with arbitrary primers. Nucleic Acids Res. 18:7213–7218.
61. Williams, J.G.K., Kubelik, A.R., Livak, K.J., Rafalski, J.A., and Tingey, S.V. 1990. DNA polymorphisms amplified by arbitrary primers are useful as genetic markers. Nucleic Acids Res. 18:6531–6535.
62. Yanisch-Perron, C., Vieira, J., and Messing, J. 1985. Improved M13 phage cloning vectors and host strains: nucleotide sequences of the M13mp18 and pUC19 vectors. Gene 33:103–119.

# PART III

# SPECIAL TOPICS

# 14

M.W. MITTELMAN

# Laboratory Studies of Bacterial Biofilms

Bacteria possess several adaptive mechanisms for responding to those physicochemical factors that define their environment, such as nutrient availability, pH, Eh, temperature, organic and ionic content, and the presence of antagonistic agents. Depending on the types and numbers present, bacteria can effect alterations in their physiology or physical state in response to the environment. Organic and inorganic acid production, heavy metal binding, transformation of xenobiotics, and extracellular polysaccharide production are important adaptive tools for bacteria in this regard. In most ecosystems, these activities are dependent on the ability of bacteria to attach to surfaces.

There is a large gap in understanding between the observed effects of bacterial biofilms on substrata and the nature of biofilm consortia. For the most part, studies of bacterial biofilms have involved only single organisms or undefined consortia from natural environments. Difficulties associated with the maintenance of diverse bacterial populations in continuous cultures, combined with a lack of appropriate adhesion test systems, have limited the scope of biofilm studies under conditions that approximate those of the in situ environment.

The microbial ecology of environmental niches as diverse as the deep subsurface, the gastrointestinal tract of mammals and insects, subgingival pockets in the mouth, hydrothermal vents, the rhizosphere, and domestic water systems is inextricably linked with life on surfaces. Mechanical blockage of flowing systems, corrosion activities, product contamination, and impedance of heat transfer processes are direct consequences of bacterial adhesion processes. The economic effects of these activities can be staggering. It is estimated, for example, that biological fouling of condenser tubes in power-generating operations results in annual losses exceeding $1 billion in North America alone.

This chapter presents several techniques that facilitate studies of bacterial biofilm community structure, metabolic activity, and structural characteristics. Difficulties inherent in the development of laboratory biofilms that simulate conditions of the desired in situ environment have limited more widespread studies of sessile organisms. For this reason, methods for building adhesion cells are also included in this chapter. Several excellent methodology papers and detailed reviews of biofilm analytical techniques have been published and are included in the Bibliography.

## TEST SYSTEMS

### Laminar-Flow Adhesion Cells

#### *Design and Hydrodynamic Considerations*

The following design criteria are used in the development of the flow cells:

1. Laminar flow conditions should predominate.
2. The coupons should be easily replaceable.
3. The cells should be constructed of high-quality materials that are low in extractables, highly corrosion resistant, and easy to clean and offer resistance to repeated chemical or gaseous sterilization.
4. The cells should have provisions for continuous macro- and microscopic monitoring of the coupons.
5. The cells should enable continuous monitoring of bulk-phase electrochemical parameters, such as $pO_2$, pH, and Eh.

Inlet and outlet ports to the flow channel are designed to provide a constant velocity across their respective cross sections. One way to achieve a constant velocity is to design the entry and exit sections to flare into a trumpet shape. In addition to entry-exit design, another important determinant of fluid dynamics is entry length. For flow in a channel, an entry length more than 54 times the channel depth is sufficient to allow transition from turbulent to laminar flow. Based on these calculations, and a 1 mm deep by 25 mm wide flow channel, an entry length of 55 mm between the bottom flare of the inlet and the upstream side of the coupon recess is acceptable.

Based on a channel depth of 1 mm and the assumption that the limit for laminar-flow transition to turbulent flow for a rectangular flow channel is on the order of approximately 1000 to 2000 [3], an upper limit on linear fluid velocities is given as 1500 ml/min.

#### *Construction of the Flow Cells*

High-density polyethylene (HDPE) is used for all wetted surfaces, with the exception of the viewing window and test coupon. The glass viewing window consists of a 24 × 50 mm No. 1 Corning glass coverslip (Corning, New York). The viewing window is fixed into position with a pliable, non-acetic-acid-containing silicone sealant (Mexcel Co., Paris). Test coupons with dimensions of 2.5 mm × 25.4 mm × 50 mm (depth × width × length) may be obtained from a variety of sources, including Metal Samples (Mumford, AL). As with the glass viewing windows, coupons are fixed into position with the silicone sealant. Viton or n-buna type O-rings are used to ensure a watertight seal. Thumbscrews and alignment pins are all fabricated of 316-type stainless steel (SS). Inlet and outlet connections are threaded 5 mm polypropylene compression- or slip tube-type fittings.

Fittings for electrochemical measurements (e.g., Open Current Potential) consist of 12.5 mm × 38.1 mm polypropylene screws, drilled through the center to form a 4-mm diameter lumen. A 4 mm × 4 mm Vycor glass frit (EG&G, Princeton, NJ) is used as a salt bridge at the screw tip. Schematics of the complete flow cells are shown in Figures 14-1 to 14-4. With the exception of the flow channel

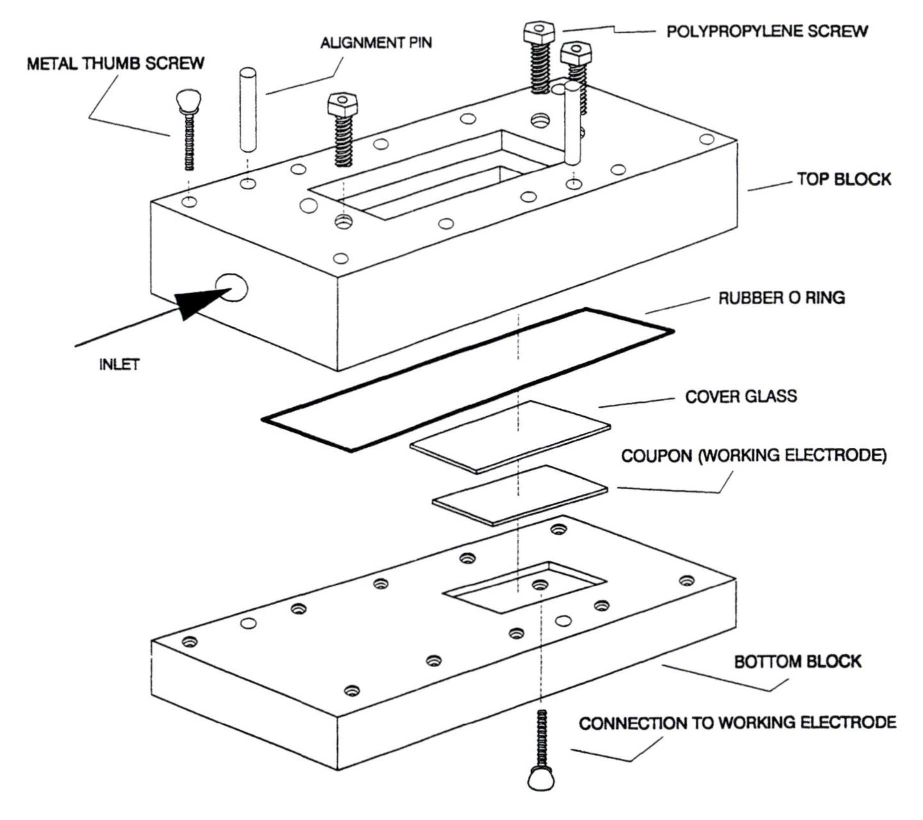


Figure 14-1 Exploded view of the laminar-flow adhesion cell illustrating its assembly.

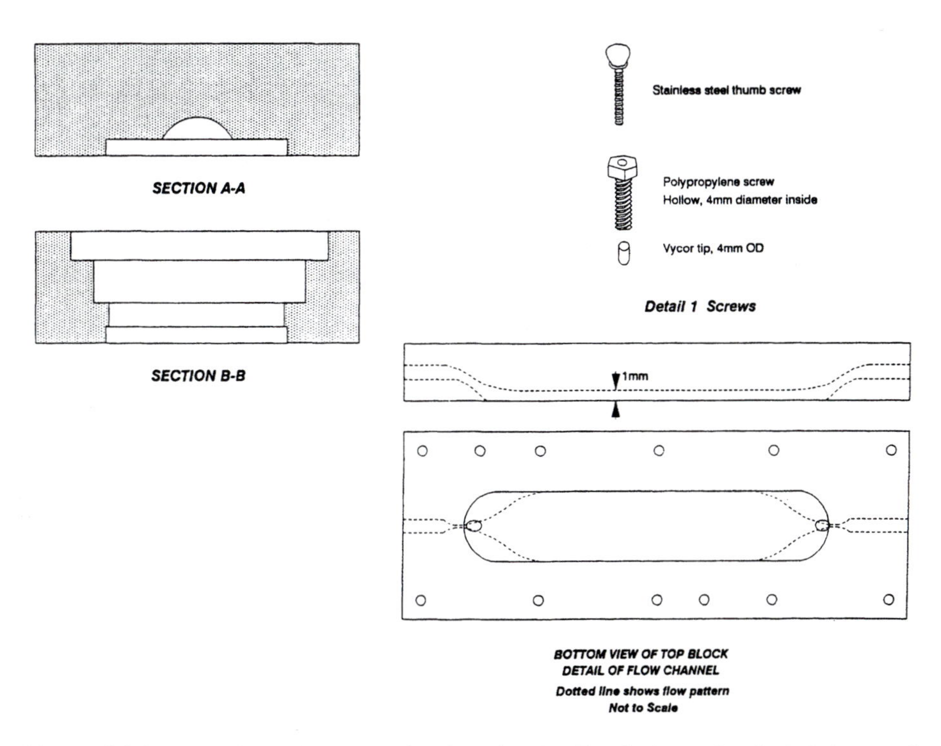


Figure 14-2 Flow channel pattern showing the details of electrochemical fittings and channel geometry.

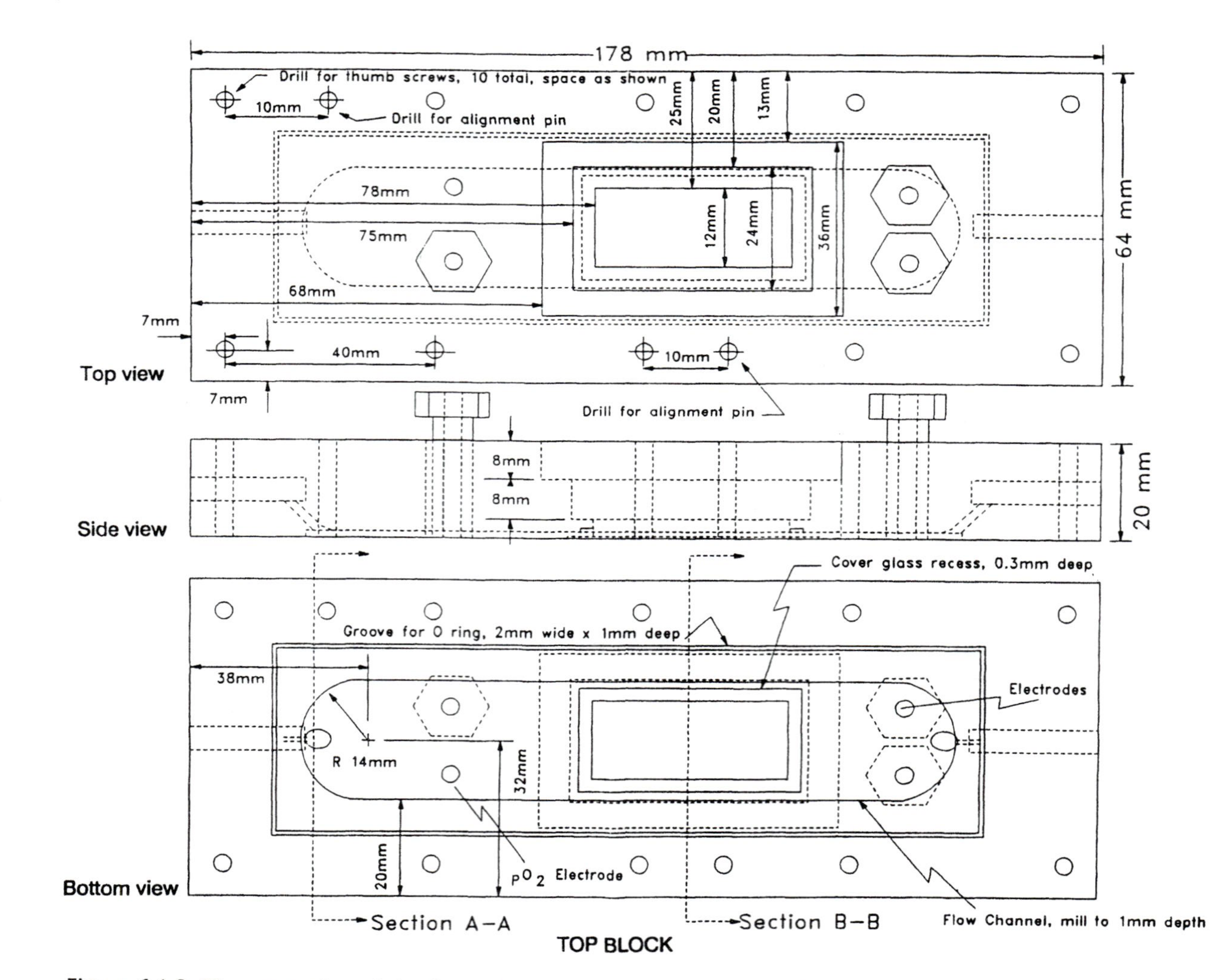


Figure 14-3 The top portion of the flow cell containing the flow channel. Dimensions are on a 1:1 scale. Top, side, and bottom views of the cell are illustrated.

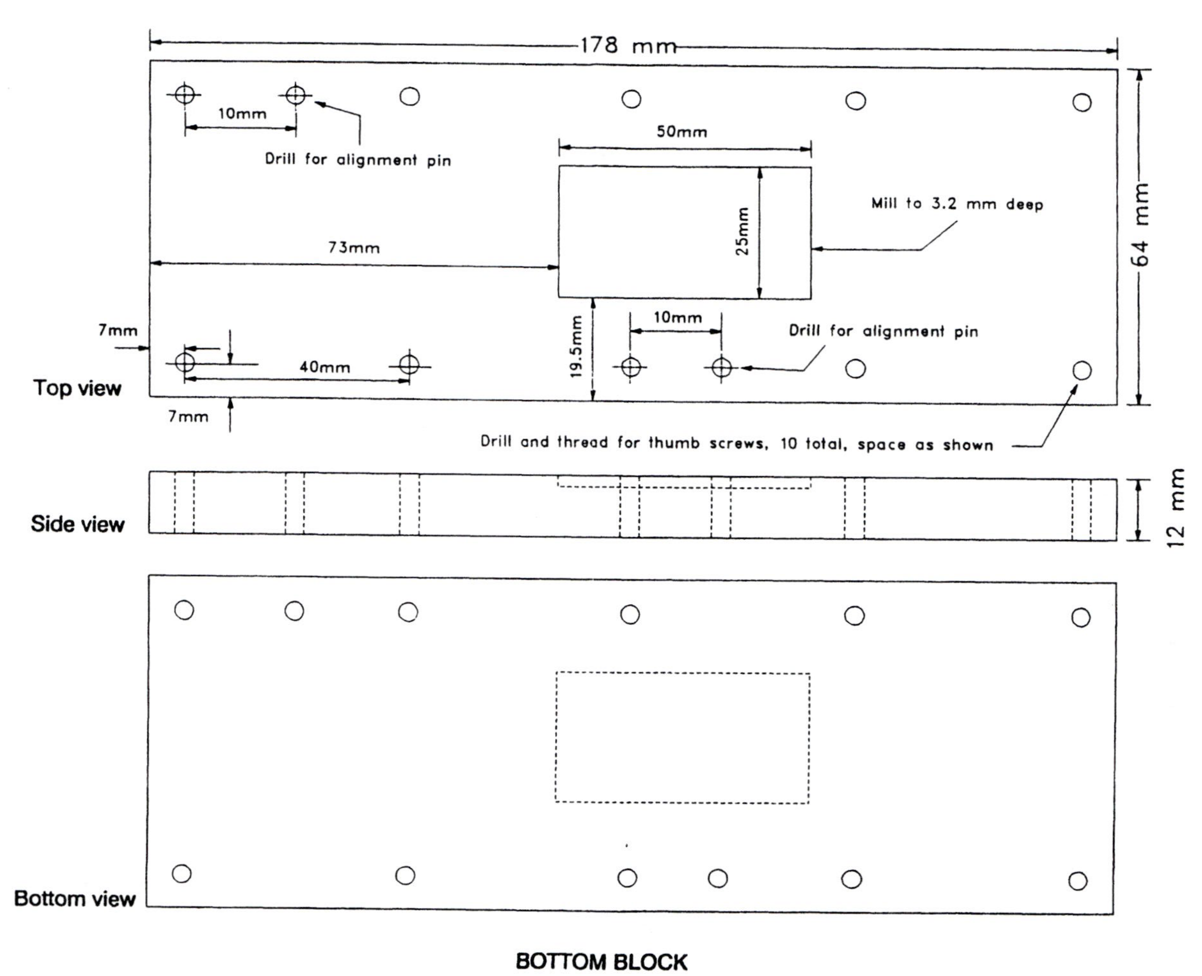


BOTTOM BLOCK

Figure 14-4 The bottom portion of the flow cell containing the coupon/working electrode. Dimensions are on a 1: 1 scale. Top, side, and bottom views of the cell are illustrated.

rendering (Fig. 14-2), all dimensions are shown at a scale of 1:1. Figure 14-1 illustrates the major components of the cell, including the three electrochemical fittings. The fittings are screwed into the top block flush with the flow channel surface. A 3-mm diameter hole is drilled adjacent to the upstream electrochemical fitting to enable insertion of a semimicro oxygen electrode. Figure 14-2 shows a cross section of the top block entry-exits and the viewing window recesses. The flow channel is also illustrated, showing the trumpet-shaped entry and exit ports. Figures 14-3 and 14-4 are detailed machining diagrams for the top and bottom blocks, respectively. A total of 10 3.2 × 25 mm thumbscrews are required to seal the cell.

The coupons and glass viewing window are inserted into 3.2 mm and 0.3 mm milled recesses, respectively. The recesses are milled slightly deeper than the actual coupon and glass thicknesses to allow for sealant addition. After insertion, both surfaces are flush with the flow channel. A thumbscrew inserted through a tapped hole in the bottom block serves both as a connection to the working electrode (the coupon) and as a means of facilitating coupon removal. A machine tolerance of ±0.075 mm should be specified for all wetting surfaces. The manufactured cost of a completed flow cell is approximately $200 when machined in lots of 8 to 10 cells.

### *Reference Electrode Preparation and Electrochemical Monitoring*

Electrodes for oxidation-reduction (redox) potential and open circuit potential measurements can be constructed in the laboratory. They are prepared by successively coating Ag wire with an AgOH slurry, followed by repeated heatings, and an anodic polarization. The half-cell used for the reference electrodes is Ag/AgCl/KCL, in which the Nernst potential is given by

$$E = E_o + 0.0591 \log a_{Cl^-} \quad (E_o = -0.224 \text{ V})$$

Procedure

1. Mix equal weight portions of $AgNO_2$ and saturated NaOH in deionized (DI) or distilled water (DW) on the surface of a Whatman No. 1 filter. Then rinse with copious quantities of DI or DW to remove excess NaOH.
2. Coat a 6-cm length of silver wire (Johnson Matthey, Ward Hill, MA) with a thin layer of AgOH. Then place in a 600°C oven for approximately 3 min to form a porous silver deposit on the surface of the 0.5-mm diameter wire.
3. Repeat the coating and heating procedure twice.
4. Anodically polarize the coated wires at approximately 1 A $cm^{-2}$ (e.g., EG&G Orteck potentiostat, Oak Ridge, TN) for 20 h in a 1 M HCl solution. Anodic polarization under these conditions results in the formation of a AgCl deposit on the surface.
5. Using the same chloride concentrations in two half-cells, calibrate the Ag/AgCl electrode potentials with a commercial $Hg/HgCl_2$ electrode (e.g., Corning, New York). Relative to the $Hg/HgCl_2$ electrode, the Ag/AgCl electrodes should yield an average potential of −0.045 V ± −0.005 V.
6. Immerse the ends of the completed electrodes into a saturated KCl solution contained within the electrochemical fittings. Seal the solution and electrode by means of a 3 mm × 4 mm rubber stopper.
7. Test each electrochemical fitting for continuity with both Ag/AgCl and saturated calomel electrodes (SCE; Corning, New York) connected to a volt-

meter (Sycopel Scientific, East Bolden, England). Discard Vycor tips observed to be leaking, plugged, or yielding unstable voltage readings.

A semimicro dissolved oxygen polarographic electrode (Microelectrodes, Londonderry, NH) monitored by a pH/ion monitor (e.g., Orion, Cambridge, MA) may be used to measure oxygen within the flow cell. An oxygen-sensitive Teflon membrane should be used for all measurements. The probe is zeroed in an atmosphere of 5% hydrogen in nitrogen; oxygen saturation may be obtained in an air-saturated atmosphere and calibrated for a 100% response at an oxygen saturation of approximately 8.5 mg/l.

### *Preparation of Flow Cells and Colonization Substrata*

Procedure

1. Before initiating any experiments, disassemble the flow cells and critically clean in an Alconox soap (Alconox, New York) solution. Place all wetted parts in a 45–50°C Alconox solution, and sonicate for 3 to 5 min in a sonicator bath.
2. Rinse the flow cells in a solvent-rinsing series in the order of chloroform, acetone, and methanol.
3. Perform a final rinse in 0.2-μm filtered DI or DW water.
4. Air-dry in a laminar-flow clean bench.
5. Polish coupons for biofilm development to an appropriate finish. Then place in a sonicator bath and critically clean in the same manner as described for the flow cells.
6. Determine surface topography using a profilometer.
7. Assemble the flow cells, wrap in two layers of Bioshield sterilization material (Baxter, Valencia, CA), and ethylene oxide gas-sterilize. Although the construction materials should be compatible with conditions encountered in a steam sterilizer (15 psi/$1.03 \times 10^5$ Pa, 121°C), the flow cells may become slightly warped during steam sterilization.

### *Colonization Studies*

T-type connections are made to 20 mm × 150 mm anaerobe tubes (Fischer Scientific) to eliminate the pulse flow that is associated with peristaltic systems. Throughout the feed system, tubing and connections are of silicone and polypropylene, respectively. Bacterial cultures in continuous culture systems or from natural environments may be pumped through the flow cells, with recirculation or in a single-pass mode.

## Shear Stress Measurement Module

A radial flow cell described by Fowler and McKay [4] can be used to study the influence of hydraulic shear on bacterial colonization and biofilm metabolic activity. A continuous shear gradient is generated across 100-mm diameter test substrata, and the resulting biofilm may be characterized by the method described. A schematic of the test system is shown in Figure 14-5. The cell adhesion measurement module (CAMM) is made up of a set of parallel plates, the uppermost of which is used as the test coupon. The system is constructed of 316 SS, and may be steam-sterilized. The CAMM is manufactured by LH Fermentation (Hayward, CA).

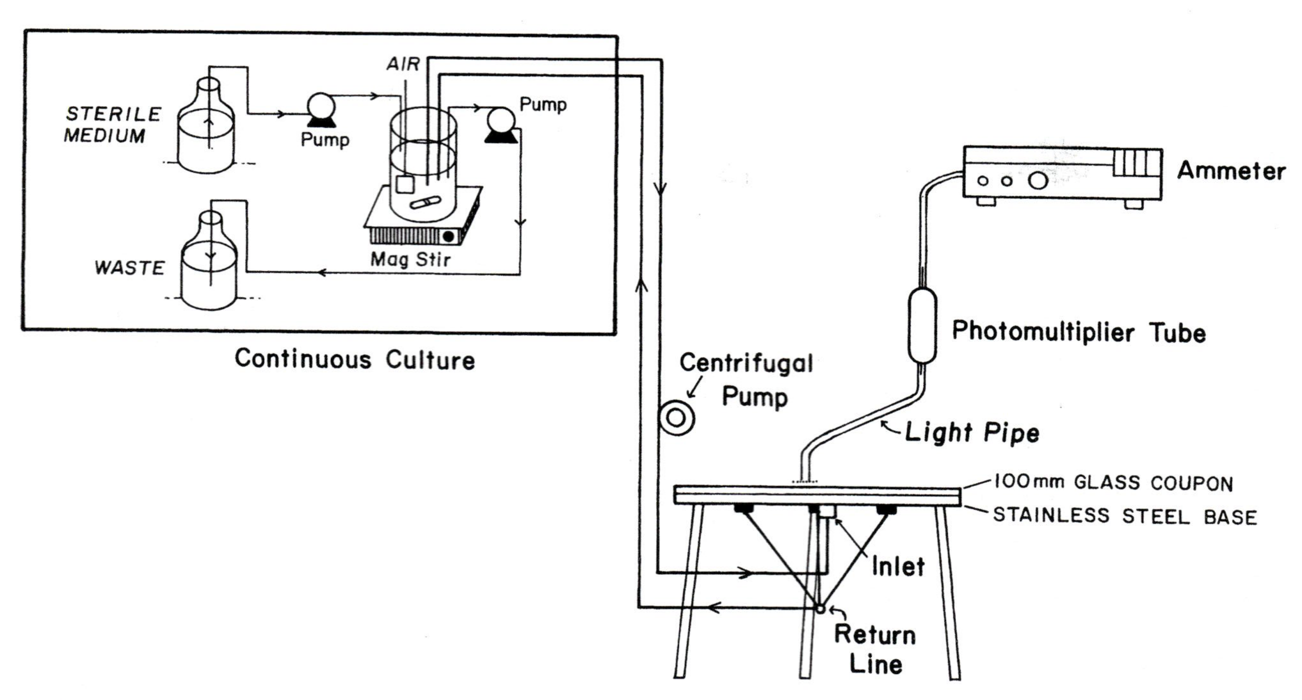


Figure 14-5 CAMM test system illustrating the direction of flow and shear generation. Shear force is inversely proportional to radius, with the greatest shear forces generated near the entry zone. The parallel plate separation distance, h, was maintained at 1 mm.

Procedure

1. Critically clean the CAMM before use as described for the laminar-flow adhesion cells.
2. Culture medium and suspended bacterial cells contact the center of the test coupon via a 3-mm diameter inlet hole in the lower parallel plate. Use a magnetically coupled drive pump or a peristaltic pump (Cole-Parmer, Chicago) to generate volumetric fluid velocities in the range of 60–1000 ml/min. Medium and cells exit the CAMM at the outer edge of the coupon, approximately 50 mm from the entry zone. Use pulseless pumps (or peristaltic pumps with pulse-dampening tubes) and silicone tubing throughout the system. Shear is generated radially, in inverse proportion to the radius:

$$T = \frac{3Q \times (\mu)}{(\pi r) \times h^2}$$

where T = shear force, dynes/cm$^2$, Q = flow rate, ml/s, $\mu$ = fluid viscosity, kg/m·s, r = radius from entry point, cm$^2$, and h = plate separation distance, cm.

## Coupon Systems for Natural Aquatic Systems

Several "baiting systems" have been described for recovering bacteria from natural waters and laboratory test loops. Henrici described a slide-baiting system for recovering stalked and appendaged organisms from natural stream waters [5]. One design for such a system is shown in Figure 14-6. It comprises a HDPE or Teflon slide holder, a series of glass microscope slides, coverslips or other desired substrate, and a fastening band. Colonized surfaces are removed with clean forceps, and biofilms are examined microscopically or recovered and analyzed as described in the next section. A device designed to sample Neuston-layer bacteria has also been described [2]. The device is made up of a Teflon sheet within a floating rack assembly, enabling the Teflon surface to contact the upper surface of fresh or

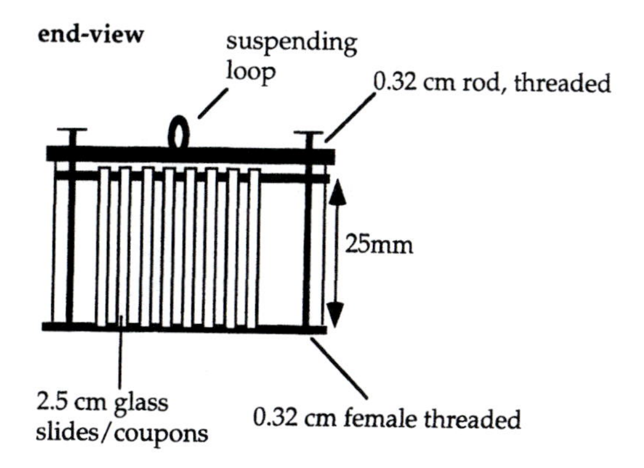


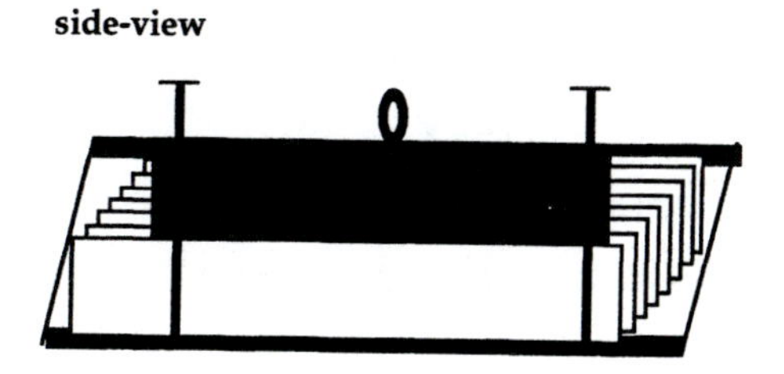


Figure 14-6 Coupon holder system.

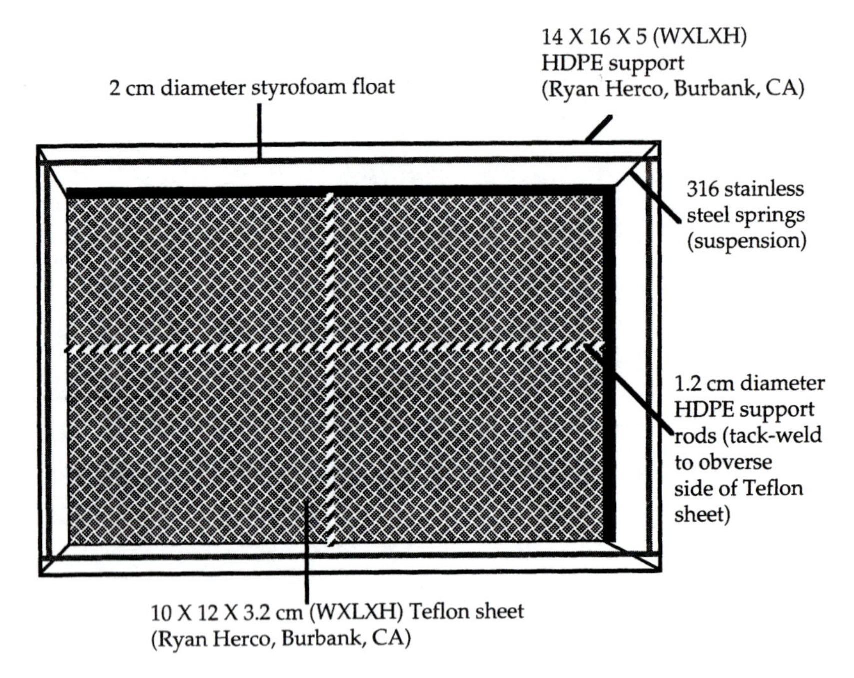


Figure 14-7 Teflon "baiting" system for studies of bacterial activity at the air-water interface.

seawaters. One design for such a device is shown in Figure 14-7. Laboratory studies of growth and metabolic activity at the air-water interface may be accomplished with this type of sampling device.

## QUANTITATIVE RECOVERY OF CELLS FROM SURFACES

Colonized surfaces are quantitatively extracted for microscopic direct cell counts using 1.2-cm diameter glass O-ring extractors (Kontes Glass, Vineland, NJ) equipped with N-buna type O-rings. The extraction area is 1.131 $cm^2$.

Procedure

1. Clamp the O-ring extractors onto the surface with C-clamps as shown in Figure 14-8.

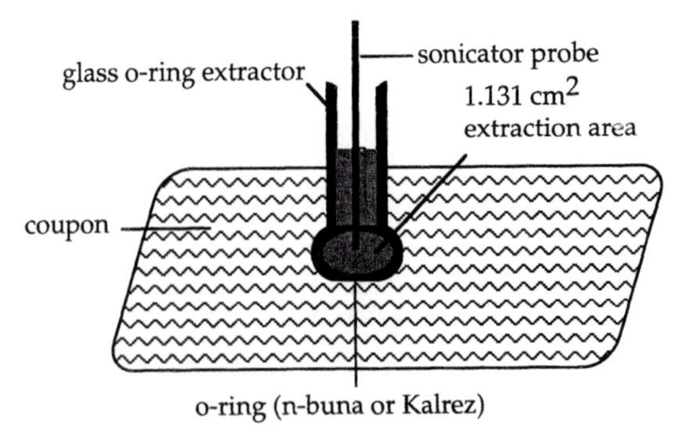


Figure 14-8 Glass O-ring extractors for biofilm sampling.

2. Add a 0.5–1.0 ml volume of ice-cold, sterile phosphate buffered saline, pH 7.2, or other appropriate medium to the extractor funnel.
3. Preclean a 3-mm diameter sonicator probe (Heat Systems, Plainview, NJ) with a isopropanol wipe, and then rinse in 0.2-μm filtered DI or DW.
4. Apply three 3-s pulses at 20% power to the extraction medium inside the o-ring extractors to remove viable cells and biomass constituents from surfaces. Validate sonication cycles for other power settings or instruments to ensure the quantitative recovery of cells from surfaces.
5. Remove the sonicate from the O-ring extractors, and add to sterile tubes. If the sample is to be processed for protein, carbohydrate, or lipids, the tubes should be washed first in Alconox and then cleaned with chloroform or hexane, followed by methanol. Do a final rinse in DI/DW before steam-sterilization of the tubes.
6. Rinse the interior surfaces of the O-ring extractors, and test coupons with an equivalent volume of extraction medium. Add this second aliquot to the sterile tubes containing the original sonicate.

Teflon scrapers can also be used to quantitatively recover attached cells and biomass constituents from surfaces. The scrapers are available from Fisher Scientific. Before use, they should be precleaned with Alconox, taken through a solvent series, and rinsed in DI/DW as described.

## EVALUATION OF BACTERIAL NUMBERS

### Viable Counts

#### Procedure

1. Add approximately 10 precleaned 2- to 3-mm diameter glass beads (Baxter Scientific) to the biofilm material recovered above (1–2 ml total solution volume).
2. Mix each tube for 6 s at high speed on a vortex mixer (Baxter Scientific).
3. Serially dilute and process the samples for viable plate counts.

### Total Direct Counts

Total direct counts of bacterial numbers may either be done directly on the coupon surface or on dilutions of biofilm extracts. It is often difficult, however, to obtain a reliable count directly from surfaces with large numbers of adherent cells.

#### Procedure: Direct Counts on Surfaces

1. Add a sufficient volume of 0.01 mg/ml acridine orange (AO; Sigma Chemical, St. Louis, MO) in 100 mM phosphate buffer, pH 7.2, to cover the area to be enumerated. Alternatively, use a solution of 1 μm/ml 4′,6-diamidino-2 phenylindole (DAPI) in 100 mM phosphate buffer, pH 7.2, for staining.
2. Let stand for 5 min at ambient temperature.
3. Gently pour off the stain from the coupon surface, and rinse with a 10-ml volume of sterile phosphate-buffered saline (PBS). The PBS should be gravity-delivered via a 10-ml pipette.
4. Allow the stained and rinsed coupon to air-dry.
5. Place on drop of low-fluorescing immersion oil (e.g., type B, Fisher Scientific) on the coupon surface.

6. Examine the coupon surface directly under epifluorescent illumination using the fluor clusters appropriate for AO or DAPI.
7. Use a calibrated eyepiece reticule for delineating the counting area. Count a sufficient number of fields such that a coefficient of variation (standard deviation/average) of less than 20% is obtained.
8. Calculate the number of cells/cm$^2$ by dividing the average count by the area of the counting reticule.

Procedure: Direct Counts of Bacteria Recovered from Surfaces

1. Add 0.5-ml aliquots of dilutions prepared for viable counts to 25-mm glass microfiltration units (Corning/Costar, Cambridge, MA) containing black, 0.2-μm polycarbonate membrane filters (Corning/Costar, Cambridge, MA). Note that the shiny side of the membranes should be oriented upward on the microfiltration funnel. Apply a minimal vacuum to the microfiltration unit.
2. Remove the vacuum source, and add a 0.5 ml volume of 0.01 mg/ml AO (Sigma Chemical, St. Louis, MO) in 100 mM phosphate buffer, pH 7.2, to the membrane surface. Alternatively, use a 1 μg/ml solution of DAPI in 100 mM phosphate buffer, pH 7.2, for staining.
3. Let stand 1 min.
4. Apply a minimal vacuum to remove the stain from the membrane surface, carefully remove the membrane, and let air-dry for approximately 2 min.
5. Place the dried membrane onto the surface of a clean glass microscope slide containing a small drop of low-fluorescing immersion oil. Allow the oil to partially diffuse through the membrane, and then add a coverslip to the surface of the membrane. Place a small drop of low-fluorescing immersion oil on the coverslip.
6. Examine the membrane surface under epifluorescent illumination as described in the procedure, Direct Count on Surfaces.
7. Calculate the number of cells/cm$^2$ of test surface by using the following formula:

$$a = \frac{X \times (b)}{(m) \times (v) \times (d) \times (e)}$$

where a = cells/cm$^2$, x = average number of cells/eyepiece reticule area, b = effective filtration area, cm$^2$, m = area of eyepiece reticule, cm$^2$, v = sample volume filtered, ml, d = dilution factor, 1/ml, and e = area extracted, cm$^2$.

### *Viable Direct Counts*

Counts of actively respiring biofilm bacteria (organisms with functioning electron transport chains) can be performed using the redox dye 5-cyano-2,3-ditolyl tetrazolium chloride (CTC Polyscience, Warrington, PA). Respiring organisms reduce the CTC to a fluorescent formazin salt that can be detected under epifluorescent illumination. Incubation of the CTC can be performed in situ; however, heavy accretions of bacteria and biofilm constituents may impede transport of CTC to cells at the biofilm-substratum interface.

Procedure

1. Using the microfiltration system and membrane filters described in earlier procedures, filter 0.5–1.0 ml of the biofilm extract from viable-count dilution tubes.

2. Remove the vacuum source, and add 0.5 ml of a 1.25 mM solution of CTC in pH 7.2, 100 mM phosphate buffer, to the membrane surface.
3. Let stand at the in situ temperature for 0.5–1.5 h. The optimal incubation time must be determined for each type of biofilm and suite of experimental conditions.
4. Apply a minimal vacuum to the microfiltration apparatus.
5. Remove the vacuum and then add 0.5 ml of the DAPI solution to counterstain.
6. Let stand 1 min and then filter.
7. Process the membrane for epifluorescence microscopy. The same fluor combination used for AO is appropriate for enumerating cells containing the reduced fluorescent formazon compound. Actively respiring cells fluoresce orange; inactive cells, blue-green.
8. Determine the total viable numbers of bacteria as described in the direct counts procedures, step 7.

## DETERMINATION OF BIOFILM BIOMASS CONSTITUENTS

### Proteins

A modified Lowry protein assay is performed on extracts from colonized surfaces. A test kit for performing this assay is also available from Sigma Chemical (St. Louis, MO).

#### Procedure

1. Prepare the following reagents for the assay:
   a. 2% $Na_2CO_3$ solution in 0.1 N NaOH
   b. 1% $CuSO_4$ solution in DW
   c. 2% sodium tartrate solution in DW
   d. Folin reagent diluted 1:1 with DW
   e. 10 mg of bovine serum albumin (Sigma Chemical, St. Louis, MO) dissolved in 100 ml of reagent a; prepare a series of protein standards in 500-μl volumes in clean 13 × 100 mm glass test tubes using DW as diluent: 0, 10, 25, 50, 75, and 100 μg/ml
2. Prepare reagent e fresh each day by mixing reagents a:b:c at a ratio of 100:1:1.
3. Prepare one tube containing 0.5 ml of biofilm extract.
4. Add 2.5 ml reagent e to each sample or standard tube, mix, and let stand 10 min.
5. Add 0.25 ml reagent d rapidly to each tube, and mix immediately.
6. Let stand 30 min.
7. Determine the optical density of the solutions at 550 nm.
8. Calculate the concentration of protein in the biofilm extract solutions from a standard curve developed with the protein standards. Determine the concentration of protein/unit area by the following formula:

$$P = \frac{C \times V}{A \times 0.5}$$

where P = concentration of protein/cm$^2$, C = measured protein concentration, μg/ml, V = volume of extraction solution, ml, and A = extraction area, cm$^2$.

## Carbohydrates

A phenol/sulfuric acid assay is performed on extracts from colonized surfaces.

### Procedure

1. Add 0.5 ml of the extracted biofilm solution to a precleaned 16 × 100 mm glass test tube. Tubes should be acid-washed or heated to 200°C overnight.
2. Prepare a series of D-glucose standards in DW: 0, 10, 25, 50, 75, and 100 μg/ml in DW.
3. Prepare a solution of 2.5 g hydrazine sulfate in 500 ml concentrated sulfuric acid. This reagent should be prepared in a clean, 1-liter beaker surrounded by ice (considerable heat is generated during mixing).
4. Add 0.5 ml of a 5% (vol/vol) solution of 2× distilled phenol in DW to each sample or standard tube.
5. Using a clean, glass 10-ml pipette, rapidly add 2.5 ml of the acid solution to each sample or standard tube. A rubber pipette bulb should be used to force the acid into the tubes in a rapid manner. *Appropriate safety equipment should be worn (face shield, laboratory coat, etc.). Considerable heat is generated during this step.*
6. Let stand 1 h at room temperature.
7. Measure the optical density of each solution at 490 nm.
8. Calculate the concentration of carbohydrates in the biofilm extract solutions from a standard curve developed with the glucose standards. Determine the concentration of carbohydrate/unit area using the same formula as for protein analysis.

## DETERMINATION OF BIOFILM METABOLIC ACTIVITY

The metabolic activity of bacteria within biofilms may be measured by a pulse-labeling procedure. Cells are incubated in situ under the desired experimental conditions with the appropriate $^{14}C$ label. The biofilm is then extracted and metabolic activity determined via measurement of $^{14}C$ incorporation into membrane phospholipids.

### Procedure

1. Use the laminar-flow adhesion cells, the CAMM, or other test system for these experiments.
2. Add the appropriate radiolabel to a minimal volume of sterile, C-free bulk-phase medium in a 125-ml Erlenmeyer flask. The volume to be used must be sufficient to include the test system void volume. The specific activity of the label, in μCi/mM, should be noted. In experiments with *Pseudomonas aeruginosa* biofilms, a 400-μl volume of 1,4-$^{14}C$ succinate (New England Nuclear, Boston) with a specific activity of 6 μCi/μM (100 μC/ml) was used successfully [6].
3. Recirculate the radiolabel through the colonized flow cells at the in situ flow rate used during the colonization experiment. A 60-min labeling period was previously found to provide optimal labeling; however, the optimal labeling period is a function of the test organisms and experimental systems employed.

4. After the recirculation period, replace the radiolabel-containing flask with a 125-ml Erlenmeyer flask containing 50–100 ml sterile, C- and label-free bulk-phase medium.
5. Pass the label-free medium through the adhesion system at the in situ flow rate, and collect the effluent in a waste container.
6. Carry out all manipulations (e.g., tubing and pump changes) performed during the flushing and labeling procedures in such a manner as to minimize perturbations of the biofilm. Keep the coupons wetted at all times during the labeling procedure, and pass through an air-medium interface only once, immediately before biofilm extraction.
7. Immediately after the rinsing step, remove the test coupons from the adhesion cells, and attach the O-ring extractors to the coupon surfaces. Kalrez-type O-rings (Kontes Glass, Vineland, NJ) should be used with the organic solvents used in this procedure.
8. Add 0.6 ml of Bligh-Dyer reagent [1] to each of the glass O-ring extractors. The Bligh-Dyer reagent is composed of a 1:2:0.9 vol/vol/vol mixture of chloroform:methanol:10 mM phosphate buffer, pH 6.8.
9. Allow to stand for 2 h at ambient temperature.
10. Remove the 0.6-ml extracts from the glass O-ring extractors to 13 × 100 mm test tubes. These tubes should be either precleaned with the Bligh-Dyer reagent or heated overnight at 200°C.
11. Rinse the coupon surfaces with two 0.6-ml aliquots of the Bligh-Dyer cocktail, and combine with the original extraction volume.
12. Transfer the contents of the 13 × 100 mm test tubes to 15-ml separatory funnels.
13. Add 1 ml of a 1:1 (vol/vol) chloroform:phosphate buffer solution to the 1.8 ml of extract. Shake well.
14. Allow to stand for 24 h at room temperature to effect phase separation.
15. Transfer organic phase (top phase) to 7-ml scintillation vials. Adjust the pH to between 8 and 9 by the addition of 1 M KOH.
16. Dry the solution under a gentle nitrogen stream. A small heat lamp may be used to speed drying.
17. Add 4.0 ml of scintillation cocktail (e.g., Ecolume, ICN, Irvine, CA) to each of the vials.
18. Count the samples on a β-liquid scintillation counter (e.g., LKB model 1212, Gaithersburg, MD), using quench correction to obtain corrected disintegrations per minute (DPM).
19. Calculate biofilm bacteria metabolic activity on a per unit area basis according to the following formula:

$$A = \frac{\text{DPM}}{1.1 \times 10^6 \times E \times S}$$

where A = metabolic activity in units of μM substrate assimilated/cm$^2$, E = extraction area, cm$^2$, and S = specific activity of labeled substrate, μCi/μM.

## ELECTRON MICROSCOPIC EXAMINATION OF BIOFILM ARCHITECTURE

Biofilm specimens are processed for transmission electron microscopy (EM) using a fixation, embedding, and thin sectioning procedure. This procedure is useful for examining biofilms associated with polymeric or metallic substrata.

Procedure

1. Fix colonized coupons in 2.5% (vol/vol in PBS, pH 7) EM-grade glutaraldehyde (Sigma Chemical, St. Louis) solution for 15 min at room temperature.
2. Dehydrate the specimens through an ethanol drying series with PBS, pH 7 as diluent: 25%, 5 min; 50%, 10 min; 75%, 10 min; 100%, 1 h.
3. Immerse the fixed and dehydrated coupons in a 30:70 mixture (wt/vol in 100% ethanol) of LR white medium-grade resin (Ted Pella Co., Redding, CA) for 3 min at room temperature. Disposable aluminum weighing dishes (Fisher Scientific) can serve as embedding containers.
4. Immerse the coupons in a 70:30 mixture of the LR white for 30 min at room temperature, followed by immersion in 100% LR white for 10 min at room temperature.
5. Remove the coupons to small glass jars, and sparge with nitrogen for 5 min. Glass Ball canning jars are useful containers for this procedure.
6. Immediately seal the jars, and place in a 65°C oven for 12 h.
7. At the end of the 12-h curing period, a hard resin forms on the surface of the coupon. Separate the embedded biofilm from the coupons by freezing at −80°C for approximately 5 min, followed by immersion in room-temperature deionized water.
8. Carefully pry off the resin from the coupons using a razor blade.
9. Thin-section (1.2-μm sections) the resin on a microtome.
10. Stain the sections in 0.01 M lead citrate and saturated uranyl acetate solutions for 30 and 2 min, respectively.
11. Gently wash the sections in deionized water; then, affix the EM grids and examine via transmission electron microscopy (e.g., Hitachi model H-600, Tokyo).

## REFERENCES


1. Bligh, E.G. and Dyer E.J. 1959. A rapid method of total lipid extraction and purification. Can. J. Biochem. Physiol. 37:911–917.
2. Dahlback, B., Hermannsson, M., Kjelleberg, S., and Norkrans, B. 1981. The hydrophobicity of bacteria—an important factor in the initial adhesion at the air-water interface. Arch. Microbiol. 128:267–270.
3. Davies, S.J. and White, C.M. 1928. An experimental study of the flow of water in pipes of rectangular section. Proc. Roy. Soc. London A 119:92–107.
4. Fowler, H.W. and McKay, A.J. 1980. The measurement of microbial adhesion. In Microbial Adhesion to Surfaces (Berkely, R. C. W., Lynch, J. M., Melling, J., and Rutter, P. R., Eds.), pp. 143–161. Vincent Harwood, Chichester, England.
5. Henrici, A.T. 1933. Studies of freshwater bacteria I. A direct microscopic technique. J. Bacteriol. 25:277–286.
6. Mittelman, M.W., Nivens, D.E., Low, C., and White, D.C. 1990. Differential adhesion, activity, and carbohydrate:protein ratios of *Pseudomonas atlantica* monocultures attaching to stainless steel in a linear shear gradient. Microb. Ecol. 19:269–278.


## BIBLIOGRAPHY


Angell, P., Arrage, A.A., Mittelman, M.W., and White, D.C. 1993. On-line, non-destructive biomass determination of bacterial biofilms by fluorometry. J. Microbiol. Meth. 18: 317–327.

Caldwell, D.R. 1990. Analysis of biofilm formation: confocal laser microscopy and computer image analysis. 77th Annual Meeting of the International Association of Milk Food Environmental Sanitarians, pp. 11–16.

Characklis, W.G. and Marshall, K.C. 1990. Biofilms. John Wiley and Sons, New York.

Kogure, K., Simidu, U., and Taga, N. 1979. A tentative direct microscopic method of counting living bacteria. Can. J. Microbiol. 25:415–420.

Lewandowski, Z., Altobelli, S.A., and Fukushima, E. 1993. NMR and microelectrode studies of hydrodynamics and kinetics in biofilms. Biotechnol. Prog. 9:40–45.

Marshall, K.C. 1986. Microscopic methods for the study of bacterial behaviour at inert surfaces. J. Microbiol. Meth. 4:217–227.

Mittelman, M.W. 1994. Emerging techniques for the evaluation of bacterial biofilm formation and metabolic activity in marine and freshwater environments. In Recent Developments in the Control of Biodeterioration (Morse, D., Ed.), pp. 49–56. Oxford University Press, London.

Mittelman, M.W., Kohring, L.L. and White, D.C. 1992. Multipurpose laminar-flow adhesion cells for the study of bacterial colonization and biofilm formation. Biofouling 6: 39–51.

Nichols, P.D., Henson, J.M., Guckert, J.B., Nivens, D.E., and White, D.C. 1985. Fourier transform-infrared spectroscopic methods for microbial ecology: analysis of bacteria, bacteria-polymer mixtures, and biofilms. J. Microbiol. Meth. 4:79–94.

Pedersen, K. 1982. Method for studying microbial biofilms in flowing-water systems. Appl. Environ. Microbiol. 43:6–13.

Peterson, G.L. 1977. A simplification of the protein assay of Lowry which is more generally applicable. Anal. Biochem. 83:346–351.

Strickland, J.D.H. and Parsons, J.R. 1968. A Practical Handbook of Seawater Analysis. Bulletin Fish. Res. Board Can., Vol. 167, pp. 174–175. Ottawa.

Tabor, P.S. and Neihof, R.A. 1982. Improved method for determination of respiring individual microorganisms in natural waters. Appl. Environ. Microbiol. 43:1249–1255.

Walker, J.T., Wagner, D., Fischer, W., and Keevil, C.W. 1994. Rapid detection of biofilm on corroded copper pipes. Biofouling 8:55–63.

Yu, F.P. and McFeters, G.A. 1994. Rapid in situ assessment of physiological activities in bacterial biofilms using fluorescent probes. J. Microbiol. Meth. 20:1–10.

Yu, F.P., Pyle, B.H., and McFeters, G.A. 1993. A direct viable count method for the enumeration of attached bacteria and assessment of biofilm disinfection. J. Microbiol. Meth. 17:167–180.

15

EUGENE L. MADSEN

# Theoretical and Applied Aspects of Bioremediation

## *The Influence of Microbiological Processes on Organic Contaminant Compounds in Field Sites*

This chapter is important because it demonstrates how the diverse techniques derived from the science of microbial ecology play a role in technology designed to eliminate environmental contaminants from field sites. Perhaps the first step in understanding bioremediation is acknowledging that it is inherently multidisciplinary. Bioremediation integrates the approaches, protocols, strategies, and analyses from microbiology, geochemistry, hydrology, soil science, physiology, molecular biology, analytical chemistry, and both civil and chemical engineering. Furthermore, each of these scientific disciples contributes a spectrum of expertise to bioremediation, ranging from theoretical and basic scientific issues to applied and practical ones. It is not surprising then that the term "bioremediation" appears in many contexts and is used by individuals of diverse perspectives and backgrounds. To dispel possible denotative and connotative ambiguities about bioremediation, this chapter begins with a series of conceptual and historical definitions that form a framework for the section on techniques that follows.

Material in this chapter addresses almost exclusively the fate of *organic* compounds in contaminated field sites. This emphasis reflects both the author's experience and a need to limit the scope of the chapter. It must be recognized, however, that microbiological processes affect *inorganic* environmental contaminants (i.e., heavy metals, acid mine drainage, and cyanides) as well [16, 18, 40, 51, 101, 123, 124, 141, 157, 180, 210]. Although conceptual portions of this chapter are readily applicable to microbial amelioration of all contamination problems, both organic and inorganic, laboratory and field protocols presented in the last section primarily address organic contaminants. For techniques pertinent to the bioremediation of inorganic contaminants, readers should consult other references [31, 75, 84, 141, 217, 223, 232].

## CONCEPTUAL OVERVIEW OF BIOREMEDIATION

### The Role of Microorganisms in the Biosphere's Carbon Cycle

When a tree falls in the forest, when crop residues are left in fields, when the oil tankers spill their cargo into pristine bays, and even when gasoline is inadvertently dribbled onto soil when a homeowner is trimming his or her lawn, microorganisms are there. We are in *their* midst. Just as humans eat meals to carry on our ways of life, so microorganisms carry on their ways of life using an astounding diversity of enzymes to digest materials in the biosphere. In an evolutionary sense, microorganisms (primarily heterotrophic microorganisms) are the recycling agents responsible for biosphere maintenance. These agents exploit thermodynamically favorable chemical reactions and, in so doing, derive carbon and energy from deceased biomass. As a result of microbial decay processes, essential nutrients (e.g., nitrogen and phosphorous) present in the biomass of one generation of biota are made available to the next.

Figure 15-1 emphasizes the importance of naturally occurring microbial processes in maintaining the biosphere. At the center of Figure 15-1 is a timeline that, simplistically but usefully, divides all of history into two periods. Before the 20th century, naturally occurring biodegradation processes were adequate in recycling the organic materials on the surface of the Earth. A myriad of microbial processes digested different types of biomass derived directly or indirectly from photosynthesis, so that organic substances seldom accumulated to cause environmental pollution. Of course, under certain circumstances the microbial precursors to petroleum have accumulated. In some portions of the earth's crust, tars and oils have formed and later were released into the biosphere. In the timeframe of the scenario described here, however, these accumulations and releases of organic compounds were transient or of minor toxicological impact. Therefore, few, if any, remedies were needed. In contrast, in the 20th century two major develop-

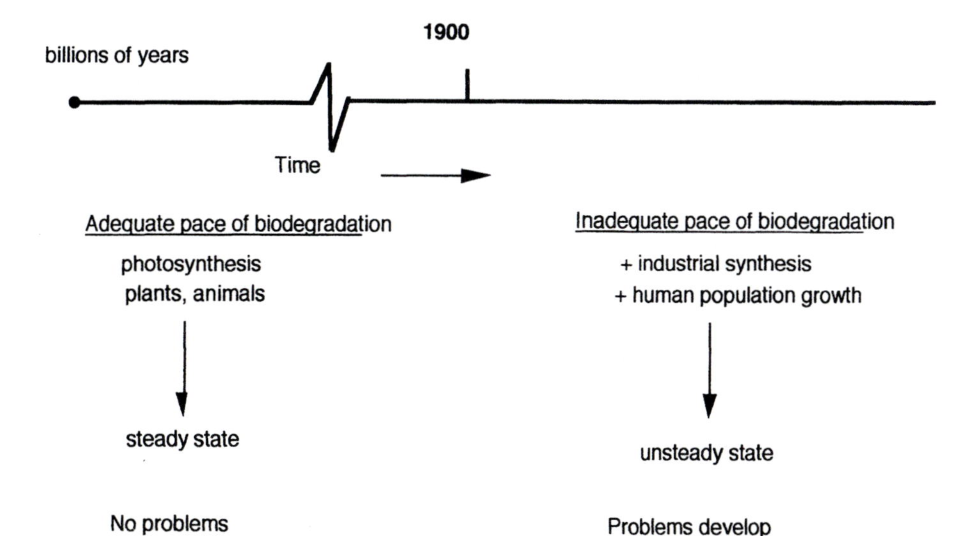


Figure 15-1 Historical perspective on the effectiveness of naturally occurring biodegradation processes in maintaining the biosphere.

ments occurred: (1) Human beings developed the ability to synthesize and disseminate large quantities of industrial chemicals, and (2) the human population increased dramatically. The environmental stresses implicit in natural resource exploitation by large human populations, in combination with the production and utilization of industrial products, have become apparent in the 20th century. Our awareness of environmental pollution problems has increased because of concomitant improvements in analytical chemistry, epidemiology, and toxicological expertise. Thus, developments in 20th-century civilization have caused rates of production and dispersal of industrial and other toxins to outpace naturally occurring biodegradation processes. An unsteady state has resulted, and threats to human health and ecosystem function are not uncommon in the latter part of the 20th century.

The term "bioremediation" itself provides rather profound insights into the state of industrialized societies and both their scientific and technological knowledge. Remedies to problems of environmental pollution are only necessary when the evolutionarily proven biodegradation capabilities of microorganisms somehow become inadequate. This inadequacy has been documented in numerous instances when humans or wildlife are threatened by toxic chemicals [32, 45, 158]. Industrialized societies face the challenge of using the same science, technology, and engineering tools that have created environmental pollution to eliminate it. In seeking technological remedies to environmental pollution, physical and chemical processes may be essential (Fig. 15-1), but microbiological processes also offer significant promise.

## Definitions and Relationships between Bioremediation and Biodegradation

*Biodegradation* is the partial simplification or complete destruction of the molecular structure of environmental pollutants by physiological reactions catalyzed by microorganisms [5, 6, 14, 15, 66, 129, 131, 227, 246]. It is routinely measured by applying chemical and physiological assays to laboratory incubations of flasks containing pure cultures of microorganisms, mixed cultures, or environmental samples (soil, water, or sediment).

*Bioremediation* is the intentional use of biodegradation processes to eliminate environmental pollutants from sites where they have been released either intentionally or inadvertently. Bioremediation technologies use the physiological potential of microorganisms, as documented most readily in laboratory assays, to eliminate or reduce the concentration of environmental pollutants in field sites to levels that are acceptable to site owners and regulatory agencies that may be involved [149, 191].

The fundamental divisions in approaches to implementing bioremediation technology are based on two questions: Where will the contaminants be metabolized?, and how aggressively will site remediation be approached? Regarding location (Fig. 15-2), microbial processes may destroy environmental contaminants in situ, where they are found in the landscape, or ex situ, which requires that contaminants be mobilized out of the landscape into some type of containment vessel (a bioreactor) for treatment. Regarding aggressiveness, **intrinsic bioremediation** is passive—it relies on the innate capacity of microorganisms present in field sites to respond to and metabolize the contaminants. Because intrinsic bioremediation occurs in the landscape where both indigenous microorganisms and contaminants reside, this type of bioremediation necessarily occurs in situ. Alternatively, **engineered bioremediation** takes an active role in modifying a site to encourage and enhance the biodegradative capabilities of microorganisms. Each of the two major engineered bioremediation approaches may exploit solid-, slurry-, or vapor-

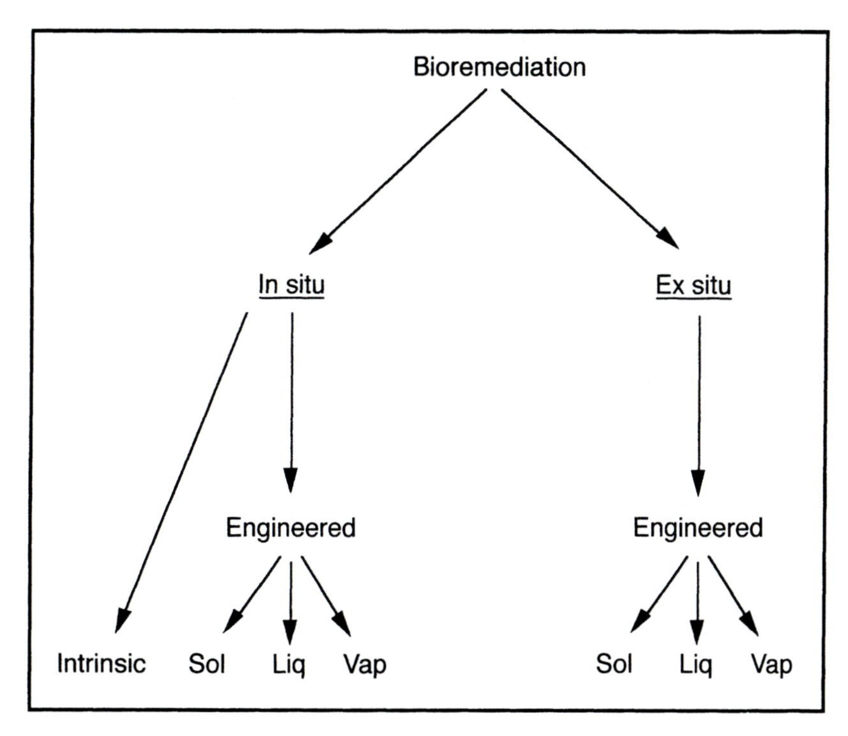


Figure 15-2 Overview of bioremediation approaches. Categories are based, respectively, on where mediation will occur (in situ vs. ex situ), on how aggressively remediation is pursued (engineered vs. intrinsic), and on the physical status of the treatment system—solid (Sol), liquid (Liq), or vapor-(Vap) phase treatment. (From Madsen, [131].)

phase systems for encouraging microorganisms to proliferate and metabolize the contaminant chemicals (Fig. 15-2).

Selection of the most effective bioremediation strategy is based on characteristics of the contaminants (toxicity, molecular structure, solubility, volatility, and susceptibility to microbial attack), the contaminated site (geology, hydrology, soil type, and climate and the legal, economic, and political pressures felt by the site owner), and the microbial process that will be exploited, such as pure culture, mixed cultures, and their respective growth conditions, and supplements [131, 218]). Engineered bioremediation relies on a variety of engineering procedures—control of water flow, aeration, chemical amendments, physical mixing, and the like—that influence both microbial populations and targeted contaminants [1, 56, 140, 175]. Furthermore, the efficacy of the remediation processes must be documented by chemical analysis of samples of water, air, and soil taken from the contaminated site. A full discussion of bioremediation technologies is beyond the scope of this chapter. However, readers may consult other references [6, 15, 24, 58, 83, 85–89, 109, 149, 152, 165, 167, 172] for additional details.

### *Intrinsic Bioremediation*

Intrinsic bioremediation is the management of contaminant biodegradation without taking any engineering steps to enhance the process. It uses the innate capabilities of naturally occurring microbial communities to metabolize environmental pollutants. The capacity of native microorganisms to carry out intrinsic bioremediation must be documented in laboratory biodegradation tests performed on site-specific samples. Furthermore, the effectiveness of intrinsic bioremediation must be proven with a site-monitoring regime that includes chemical analysis of contaminants, final electron acceptors, and other reactants or products indicative of biodegradation processes. This bioremediation strategy differs from no-action alternatives in that it requires adequate assessment of the existing biodegradation rates and potential and adequate monitoring of the process. It may be used alone or in conjunction with other remediation techniques. For intrinsic bioremediation to be effective, the rate of contaminant destruction must be faster than the rate of contaminant migration. These relative rates depend on the type of contaminant, the microbial community, and the site hydrogeochemical conditions. Intrinsic bioremediation has been documented for a variety of contaminants and habitats, including low molecular weight polycyclic aromatic compounds in groundwater [134, 147, 234], gasoline-related compounds in groundwater [176], crude oil in marine waters [30], and low molecular weight chlorinated solvents in groundwater [136, 189].

### *Engineered Bioremediation*

Engineered bioremediation either accelerates intrinsic bioremediation or replaces it completely through the use of site modification procedures, such as excavation, hydrological manipulations, and installation of bioreactors, that allow concentrations of nutrients, electron acceptors, or other materials to be managed in a manner that hastens biodegradation reactions. Engineered bioremediation is especially well suited for treating nonvolatile, sparingly soluble contaminants the properties of which impede successful treatment by other technologies. Engineered bioremediation may be chosen over intrinsic bioremediation because of considerations of time, cost, and liability. Because engineered bioremediation accelerates biodegradation reactions, this technology is appropriate for situations where time constraints for contaminant elimination are short or where transport processes are causing the contaminant plume to advance rapidly. The need for rapid pollutant removal may be driven by an impending property transfer or by the impact of the contamination on the local community. A shortened clean-up time means a correspondingly lower cost of maintaining the site.

Engineered ex situ bioremediation has been used in municipal sewage treatment systems for over a century [140]. In sewage treatment systems, wastewaters from municipalities are directed through an array of controlled environments that encourage microbial growth in filters, tanks, and digestors. Physical, chemical, and microbiological manipulations remove carbonaceous, nitrogenous, and other materials from water before it is discharged into rivers, lakes, or oceans. Engineered in situ bioremediation was implemented by R. Raymond and colleagues to clean up petroleum-contaminated groundwater over two decades ago [171]. In the pioneering version of this technology, a groundwater circulation system was established that enhanced the mixing of contaminants, microbial cells, and nutrients designed to encourage aerobic catabolic reactions. More recent engineered bioremediation case studies have been described [24, 58, 152, 172].

Table 15-1 compares in situ and ex situ approaches to engineered bioremediation. As much as possible, in situ approaches engineer the landscape to mimic conditions known to foster biodegradation reactions readily demonstrable in laboratory flasks. Thus, the major barrier to successful in situ engineered bioremediation is the impossibility of fully controlling the variety of intractable processes and heterogeneities characteristic of open field sites. Engineered bioremediation must contend with variability in properties of the site, the pollutant chemicals, the microorganisms, and the regulatory agency overseeing each clean-up effort. The bioreactor that may be central to an ex situ strategy to engineered bioremediation offers better control over biodegradation processes, as pollutant metabolism within the bioreactor can be verified and enhanced. When compared side by side it is

Table 15-1 Comparison of In situ and Ex situ Strategies for Engineered Bioremediation Systems

| | In Situ | Ex Situ |
|---|---|---|
| Location | In the landscape | In a controlled bioreactor |
| Requirements | Engineer the landscape to resemble a laboratory flask | Move contaminants from landscape to on-site bioreactors |
| Characteristics | Relatively poor control of biodegradation process | Greater control |
| Obstacles | Complexities of landscape that may prevent success | Complexities of landscape partially overcome |
| | Pollutant mixtures | Pollutant mixtures |
| | Unknown site histories | Unknown site histories |
| | Mass balances uncertain | Decent bioreactor mass balances |
| | Biotic versus abiotic processes | Biotic processes defined in bioreactor |
| | Incompatibility of site characteristics and microbiological processes | Incompatibility of site characteristics and microbiological processes |
| | Production of pollutants by microorganisms | Production of pollutants by microorganisms |
| | How clean is clean? | How clean is clean? |

clear that the two engineered bioremediation strategies share certain obstacles, but each also offers different advantages. If contaminants are strongly sorbed onto soil or sediment solids, the ex situ approach may be less appropriate. In either case, qualitative evidence must be obtained to prove that microbiological processes are responsible for pollutant loss. Quantifying the proportion of loss attributable to biotic versus abiotic processes remains a major challenge to implementation of site remediation technologies [30, 129, 237].

## The Spectrum of Compounds Susceptible to Biodegradation (Bioremediation)

Compounds that are most susceptible to microbial metabolism are naturally occurring, have a simple molecular structure, are water soluble, exhibit no sorptive tendencies, are non-toxic, and serve as a growth substrate for aerobic microorganisms. By contrast, compounds that are resistant to microbial metabolism exhibit such properties as a complex molecular structure, low water solubility, strong sorptive interactions, and toxicity or do not support the growth of microorganisms. Swoboda-Colberg [213] has recently prepared an insightful summary of environmental contaminants, their fates, and sources.

Table 15-2, modified from the National Research Council [149], provides an overview of classes of compounds, their inherent biodegradability, their aqueous mobility, frequency of occurrence in groundwater plumes, whether or not hydrophobic sorptive partitioning reactions are likely to impair biodegradation, and their overall amenability to bioremediation measures. Table 15-2 presents a broad perspective on how chemical and microbiological properties jointly affect prospects for bioremediation, and of course, there are exceptions to its classification scheme. However, the trends are clear: Bioremediation treatment technology has been established only for certain classes of petroleum hydrocarbons, whereas the technologies for treating all other classes in Table 15-2 are still emerging.

Petroleum hydrocarbons are naturally occurring chemicals featuring a variety of molecular weights and functional groups. Benzene, toluene, ethylbenzenes, and xylenes (BTEX) are components of gasoline that are distributed widely and have been the focus of substantial biodegradation and bioremediation research [26, 28]. BTEX, the low molecular weight fuels, alcohols, and ketones are readily miner-

Table 15–2 An Overview of Relationships between Chemicals, Their Properties, and Bioremediation Prospects

| Chemical Classes[a] | Biodegradability[b] (A, N, AN) | Mobility[c] | Frequency of Occurrence[d] | Partitioning Reactions[e] | Prospects for Bioremediation[f] |
|---|---|---|---|---|---|
| Hydrocarbons | | | | | |
| BTEX | A1, N2, AN2 | H | F | M | Es |
| Low MW, gasoline, #2 fuel oil | A1, N3, AN2 | M | F | M | Es |
| High MW. oil, PAH | A2, N4, AN4 | L | C | S | Em |
| Creosote | A1, N2, AN4 | L | I | S | Em |
| Oxygenated hydrocarbons | | | | | |
| Low MW alcohols, ketones, esters, ethers | A1, N5, AN3 | H | C | W | Es |
| Halogenated aliphatics | | | | | |
| Highly chlorinated | A4, A3, N5, AN2 | M | F | M | Em |
| Less chlorinated | A2, A3, N5, AN2 | H | F | M | Em |
| Halogenated aromatics | | | | | |
| Highly chlorinated | A4, A2, N5, AN2 | L | C | S | Em |
| Less chlorinated | A2, A3, N2, AN2 | M | C | M | Em |
| PCBs | | | | | |
| Highly chlorinated | A4, N5, AN2 | L | I | S | Em |
| Less chlorinated | A2, A1, N5, AN4 | L | I | S | Em |
| Nitroaromatics | A2, N5, AN2 | M | C | M | Em |

[a]BTEX = benzene, toluene, ethylbenzene, xylenes; MW = molecular weight; PAH = polycyclic aromatic hydrocarbon; PCBs = polychlorinated biphenyls.
[b]The three alphanumeric entries for each compound provide a biodegradability rating (1–5) under aerobic (A), nitrate-reducing (N), and other anaerobic (AN) conditions. 1 = readily mineralizable as growth substrate; 2 = biodegradable under narrow range of conditions; 3 = metabolized partially when second substrate is present (co-metabolized); 4 = resistant; 5 = insufficient information.
[c]H = highly mobile; M = moderately mobile; L = least mobile.
[d]Based on survey of groundwater contaminants. F = very frequent; C = common; I = Infrequent.
[e]S = strong sorptive characteristics; M = moderate characteristics; W = weak characteristics.
[f]Es = established; Em = emerging.

alizable aerobically; hence, these compounds have been removed successfully from contaminated sites via established bioremediation procedures. Recently, many BTEX components, traditionally considered resistant to anaerobic microbial attack, have also been found to be biodegradable under a variety of anaerobic conditions [44, 54, 125, 127, 128]. The high molecular weight petroleum components and creosotes are metabolized slowly, partly as a result of their structural complexity, low solubility, and strong sorptive characteristics. Thus, bioremediation techniques for these classes of petroleum hydrocarbons are still emerging [145].

Though halogenated organic compounds have been found in nature [71], they do not have commercial significance compared with the synthetic chemicals listed in the lower portion of Table 15-2. When halogen atoms (chlorine, bromine, fluorine) are introduced into organic molecules, many properties, such as solubility, volatility, density, hydrophobicity, and toxicity, change markedly. These changes confer improvements that are valuable for commercial products, but also have serious implications for microbial metabolism. The susceptibility of the chemicals to enzymatic attack is sometimes drastically altered by halogenation, and environmentally persistent compounds often result.

Halogenated aliphatic compounds are straight chain hydrocarbons in which varying numbers of hydrogen atoms have been replaced by halogen atoms. They are effective solvents and degreasers that have been widely used in many manufacturing and service industries. Some highly chlorinated representatives of this class, such as tetrachloroethene, are completely resistant to aerobic microbial attack while being susceptible to anaerobic reductive dehalogenation under methanogenic and other anaerobic conditions (Table 15-2; [1]). In fact, recent laboratory and field evidence shows that complete reductive dechlorination from tetrachloroethene to the relatively non-toxic compound ethene can occur [62, 94, 110, 139, 139a, 189, 215, 243a]. Furthermore, as the degree of halogenation in aliphatics diminishes, their susceptibility to aerobic metabolism increases. This is especially so when the active microorganisms possess one of a variety of nonspecific oxygenase enzyme systems [224]. Oxygen atoms can be fortuitously inserted into the less halogenated ethenes by a process known as *co-metabolism* when microorganisms are supplied such substrates as methane, toluene, or phenol that activate the oxygenase enzymes. Co-metabolism radically alters both the toxicity and biodegradability of the halogenated ethenes. Thus, a common treatment rationale for the chlorinated aliphatics is to remove the chlorine atoms anaerobically and then complete the biodegradation process using aerobic co-metabolism. Procedures for bioremediating sites contaminated with chlorinated aliphatics are developing rapidly. Both anaerobic treatment (driven by supplementing field-site waters with an electron source) and aerobic co-metabolic treatment (driven by the additions of methane or aromatic substrates) are being field-tested [23, 78, 95, 161a, 187].

Halogenated aromatics, such as phenoxy acetic acid pesticides, pentachlorophenol, polychlorinated biphenyls (PCBs) and others, consist of one or more benzene rings that bear halogens as well as other chemical functional groups (i.e., hydroxyls, carboxyls, etc.). The aromatic benzene nucleus is susceptible to both aerobic and anaerobic metabolism, though the latter process occurs relatively slowly. Overall, however, the presence of halogen atoms on the aromatic ring, their position, and their interaction with functional groups are what govern biodegradability. A high degree of halogenation may prevent aromatic compounds from being oxidatively (aerobically) metabolized, as is the case for PCBs [21]. However, as discussed for the aliphatic compounds, the highly halogenated aromatics are subject to anaerobic reductive dehalogenation. As the halogen atoms are replaced by hydrogens, the molecules become susceptible to aerobic attack. Thus, a common bioremediation scenario for treating soils, sediments, or water contaminated with halogenated aromatic chemicals is anaerobic dehalogenation followed by aerobic mineralization of the residual compounds [61]. However, when the proper substituent group accompanies the halogens on the aromatic ring, aerobic metabolism may proceed rapidly, as is the case for pentachlorophenol. Advances in our understanding of how microorganisms metabolize nitroaromatic compounds, common components of explosives and pesticides, have also been made [63, 151, 198].

## Verification of Bioremediation

There are three reasons to establish sound scientific criteria for microbiological involvement in contaminant loss. First, biodegradation processes are often unique in their capacity to break intramolecular bonds of contaminant compounds; thus, contaminants can be destroyed and not simply transferred from one location to another, as is the case for many other pollution control technologies. Second, when the mechanism of pollutant destruction is certain, key site management decisions about process enhancement can be made. Finally, for bioremediation to meet society's pollution control needs, the industry must adopt some standards for uniformity and quality control so that credibility and reliability can be attained [149].

But the question remains: What is adequate proof of bioremediation? The U.S. legal system provides several categories of certainty in interpreting evidence, depending on the type of case and the significance of the issues. Among the different burdens of proof are proof beyond a reasonable doubt, proof in a clear and convincing manner, or proof beyond a preponderance of doubt. This chapter neither intends, nor is able, to dictate to regulatory or legal agencies which level of proof should be deemed adequate for bioremediation technology practitioners. Nonetheless, this section introduces approaches that can be used to distinguish biotic from abiotic reactions affecting contaminants at field sites where bioremediation technology is being applied (see [129–131, 190, 212] for additional discussion).

One must recognize that only under relatively rare circumstances is a proof of bioremediation unequivocal when relying on a single piece of evidence. In the majority of cases, the complexities of contaminant mixtures, their hydrogeochemical settings, and competing abiotic mechanisms of contaminant loss make it a challenge to identify biodegradation processes. Unlike controlled laboratory experimentation where measurements can usually be interpreted easily, cause-and-effect relationships are often very difficult to establish at field sites. Furthermore, certain bioremediation data that may be convincing for some authorities may not be convincing for others. Thus, in seeking proof of bioremediation, several approaches described below should be pursued independently. The strategy should be to build a consistent, logical case relying on convergent lines of independent evidence.

The consensus of a recent National Research Council [149] committee in recommending criteria proving in situ bioremediation is as follows:

1. Develop historical records documenting the loss of contaminants from field sites.
2. Perform laboratory assays unequivocally showing that microorganisms in site-derived samples have the potential to metabolize the contaminants under expected site conditions.
3. Demonstrate that the metabolic potential measured under Step 2 is actually expressed in the field. Therefore, microbiological mechanisms of contaminant attenuation must be distinguished from abiotic ones. Evidence deemed suitable for these purposes will vary according to the contaminants and conditions found at each site.

Table 15–3 Overview of Steps Required for Demonstrating Successful Bioremediation and Corresponding Methodological Protocols

| Step | Protocols |
|---|---|
| 1. Develop historical records documenting the loss of contaminants from sites. | 1. Determine contaminant behavior in the field via a series of chemical analytical procedures performed in the field and in the laboratory on environmental samples. |
| 2. Perform laboratory assays showing that microorganisms in site samples have the potential to transform the contaminants under expected site conditions. | 2. Measure biodegradation activity in field samples incubated in the laboratory. Support these data with information from pure cultures, cell-free preparations, and molecular procedures. |
| 3. Obtain evidence that biodegradation potential is actually expressed in the field. | 3. Demonstrate expression of biodegradation in situ (1) through the use of conservative tracers, (2) by field detection of intermediary metabolites, (3) through the use of replicate field plots to distinguish biotic from abiotic contaminant loss, (4) using emerging molecular procedures for detecting mRNA and key enzymes indicative of expressed catabolic genes, (5) by documenting the adaptation of indigenous microorganisms to contaminants using field gradients of specific contaminant-degrading populations and metabolic co-reactants over time and space, or (6) via rigorous computer modeling strategies. |

From Madsen [130, 131] and NRC [149].

Table 15-3 provides an overview of the three steps, as well as the supporting protocols.

## TECHNIQUES OF BIOREMEDIATION AND BIODEGRADATION

### Overview and Prioritization of Measurements Required for Documenting Biodegradation and Bioremediation

Because bioremediation and biodegradation aim to understand and exploit the activities of microorganisms in field sites, relevant protocols can conceivably draw on virtually all of the techniques of microbial ecology. Several of the techniques in microbial ecology that are most pertinent to biodegradation and bioredmediation processes are listed in Table 15-4. In addition, a variety of chemical analytical protocols (Table 15-5) are essential for understanding on-site geochemistry. Of course, engineering procedures [58, 109, 149, 152, 172] are also important. A comprehensive treatment of every aspect of bioremediation is beyond the scope of this chapter. What is presented, however, are the core principles, logic, and basic protocols useful in assembling convergent lines of independent evidence that are required for establishing, determining, and verifying bioremediation processes. A generalized strategy for harnessing microbial processes for the purpose of bioremediation has been presented by Madsen [129]. This strategy appears in diagrammatic form in Figure 15-3. Steps 1–3 of Figure 15-3 are aimed at the following:

Understanding the context: What are the hydrological, climatic, and biogeochemical characteristics that dominate a contaminated field site?

Understanding the nature of the environmental problem that needs to be remedied: Which compounds are present and at what concentrations? Is the

Table 15–4 Techniques of Microbial Ecology that are Particularly Relevant to Biodegradation and Bioremediation

| Technique | Relevance | Reference |
|---|---|---|
| Microscopy of cells in fixed site-derived samples | A variety of general nucleic acid stains or fluorescently- labeled oligonucleotide or immunological probes reveal the abundance of total cells or specific genes or populations. | 9, 27, 117, 220 |
| Microcopy of cells in site-derived samples after brief incubation with substrates that reveal the general physiological status of the organisms | A variety of respiratory stains or cell growth assays distinguish active from inactive cells. | 4, 102, 138, 203, 235 |
| Viable plate counts on media containing general or specific carbon sources | Depending on the agar medium employed, growth from dilutions of site-derived microorganisms can quantitatively measure both general heterotrophs and populations capable of metabolizing specific site contaminants. | 4, 216, 231, 240 |
| Biodegradation assays performed on environmental samples, mixed cultures, or pure cultures | Biodegradation potential is demonstrated in laboratory flasks by a net loss of parent contaminant compound in live versus abiotic treatments. These physiological assays document contaminant metabolism and also tie it to consumption of specific final electron acceptors (e.g., $O_2$, $NO_3^-$) and formation of metabolic end products, (e.g., $CO_2$, $CH_4$). | 11, 60, 208, 209, 248 |
| Biochemical assays demonstrating enzymatic activities and specific metabolic pathways | Assays confirm biodegradation potential and explore details of metabolic pathways. They provide information on metabolites that may be extracted from site samples to demonstrate in situ metabolism of contaminants. | 60, 65, 192, 211, 224, 225, 234 |
| Molecular probing (using PCR, RT-PCR, or specific oligonucleotides) of field site-derived nucleic acids to demonstrate the presence of particular organisms, genes, or gene transcripts | Probing confirms biodegradation potential, explores metabolic diversity and both gene exchange and expression issues, and documents in situ gene expression if mRNA and enzyme detection procedures are conducted in properly fixed field samples. | Chapter 13; 3, 19, 34, 35, 90, 92, 120, 154–156, 160, 161, 179, 181, 184, 200, 201 |
| Enumeration of protozoa and assays for predator-prey interactions | High numbers of protozoa in field sites indicate that actively growing bacteria are present. If this growth is caused by the contaminants, in situ biodegradation can be established. | Chapter 8; 68, 96, 134 |

Table 15–5 Organic and Inorganic Chemical Assays Pertinent to Understanding the Biogeochemistry of Contaminated Field Sites

| Analytical Approach | Sample Preparation | Information | Reference |
|---|---|---|---|
| Portable field meters | None is required if probes can be immersed in soil, sediment, or waters in situ. | Temperature, $O_2$, conductivity, and other measures pending availability of specific-ion electrodes and other analytical probes and standards | 10, and a variety of commercial manufacturers (e.g., Yellow Springs Instruments, Inc., Yellow Springs, OH) |
| Instruments in mobile field laboratory (e.g., spectrophotometer, GC, and others) and instruments in base analytical laboratory (e.g., generally higher precision spectrophotometer, GC, GC/MS, HPLC, HPLC/MS, ion chromatograph, and others) | Sample gathering and fixation protocols vary with each specific assay. Fixation is designed to avoid chemical artifacts that may develop between the time that samples are removed from the field site and assays are completed. | Inorganic nutrients and electron acceptors (e.g., $O_2$, $NO_3^-$, $Fe^{3+}$/$Fe^{2+}$, $Mn^{4+}$/$Mn^{2+}$, $NH_4^+$, $S^{2-}$, $SO_4^{2-}$, $H_2$, $CO_2$); organic constituents (e.g., the contaminants, co-contaminants, metabolites, dissolved and total organic carbon) | 10, EPA methods as outlined by 103–105, 146, 226, 228, and a variety of commercial manufacturers (e.g., Hach Co., Loveland, CO). See also the new journal, *Field Analytical Chemistry Technology.* |

concern due to synthetic organic compounds (such as pesticides), to naturally occurring organic compounds (such as crude oil), or to inorganic materials (such as metals or acids)?

Understanding rules of microbial ecology, physiology, biochemistry, genetics, and thermodynamics that tell us if and when the contaminant materials can be resources that are exploitable by microbial processes: Our understanding of these rules changes as research in each of the relevant areas progresses. Nonetheless, a current synthesis of all the pertinent disciplines greatly aids the development of successful bioremediation strategies.

Step 4 of Figure 15-3 integrates information with decision making, aiming at:

Understanding to what degree resource exploitation has already begun: Is the naturally occurring microbial community at the field site already actively engaged in contaminant elimination? If so, then this intrinsic decontamination capability of on-site microorganisms may be adequate for achieving clean-up goals or may be enhanced easily. If intrinsic bioremediation is not occurring, then impediments—including absence of the appropriate microorganisms, nutrient limitations, the nature of the contaminant themselves, possible toxic effects of co-contaminants, and other site-related characteristics—should be identified.

Assessing the severity of the problem, the nature of the contaminants, the nature of contaminant-specific microbial processes, available engineering tools, and cost considerations so that the most appropriate engineered or intrinsic remediation procedure for the field site can be identified.

Steps 5 to 8 of Figure 15-3 have these objectives:

Implementing the bioremediation strategy,
Monitoring its effectiveness,

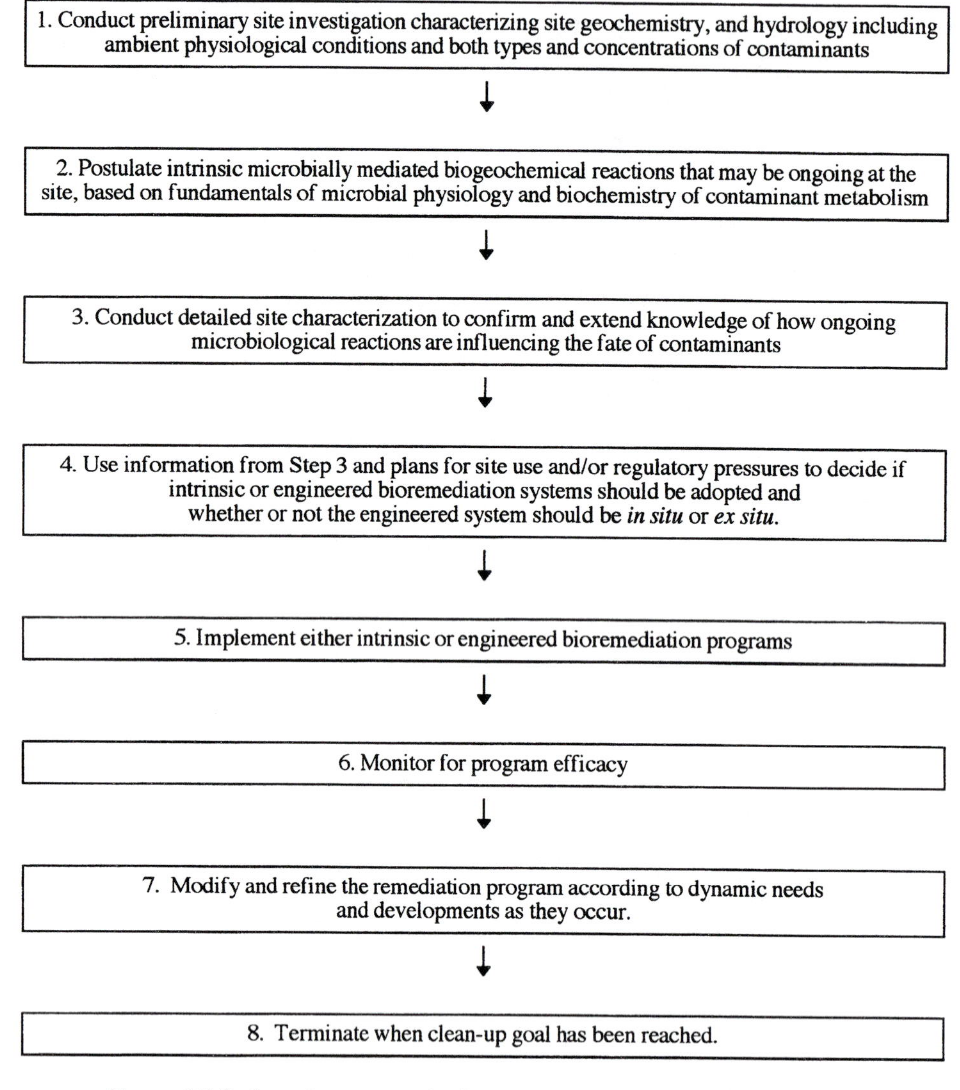


Figure 15-3 Stepwise strategy for implementing a bioremediation project.

Modifying the remediation program according to virtually inevitable, unforeseen developments that occur during clean-up efforts at each site, and
Terminating the program when its goals have been achieved.

## Step 1. Site Characterization Protocols: Understanding Site Biogeochemistry and Establishing Historical Trends of Contaminant Behavior

There is a critical need to relate results of geochemical measurements, performed on field samples, directly to processes and conditions in the field (see Chapter 10). Therefore, the utmost care must be taken to avoid artifacts that may be imposed on analytical results by imprudent delays in analysis completion, improper sample fixation, or laboratory incubations. Whenever possible, portable field instruments should be used. In this regard, cone penetrometry [42] and a variety of in situ water analyses and sample gathering and fixation procedures are relevant to understanding on-site biogeochemical processes. As is suggested in

Table 15-5, however, many measurements cannot be completed in the field. When sample removal from field sites cannot be avoided, a variety of crucial decisions must be made. Selecting the locations within field sites for obtaining representative samples is no simple matter [159, 239]. Furthermore, deciding how to gather samples (with a shovel, spoon, pump, or drilling rig with or without aseptic technique) is also a critical issue [103–105; see Chapter 10]. In addition, the vessels for containing environmental samples must be clean, leak-proof, and compatible with all intended uses. Once a field sample has been transferred to a container, the sample fixation protocol must be selected carefully to avoid artifacts. Obviously, fixation must prevent the parameter(s) of interest from changing and be compatible with the intended measurement procedures. Fixation may be accomplished by freezing (in liquid nitrogen or by placing samples on dry ice) or by adding biological inhibitors (e.g., formalin) and chemical fixatives (e.g., acid) [see Chapter 10 and [10] for details].

Final electron acceptors that dominate the physiological reactions of field sites or discrete zones therein provide useful criteria for categorizing biogeochemical regimes. For investigations of intrinsic bioremediation, understanding these ambient conditions that control and respond to in situ physiological processes is essential. The amount of effort required for defining which of the many possible biogeochemical regimes—aerobic, denitrifying, iron-reducing, manganese-reducing, fermentative, dehalogenating, sulfate-reducing, or methanogenic; [79, 143, 206, 247]—actually occur in the field is site specific. And, of course, spatial heterogeneity is the key impediment for successful site characterization. For example, in uniform well-mixed aquatic habitats, consistent readings from an oxygen probe at a variety of locations can be interpreted accurately and extrapolated to the site as a whole. However, in soils and saturated sediments that are spatially heterogeneous, it is very difficult to know precisely where and when particular physiological regimes are established.

All site characterization data must be interpreted in terms of the physiological processes that produce and consume geochemical constituents [25, 197]. Key insights into in situ microbial physiology can be provided by field measurements of co-reactants and end products of microbial metabolism (e.g., $CO_2$, $Fe^{2+}$, $Mn^{2+}$, $S^{2-}$, $N_2O$, $NH_4^+$, organic acids, reductive dehalogenation daughter products, and methane), as well as by concentration gradients of final electron acceptors (e.g., $O_2$, $NO_3^-$, $Fe^{3+}$, $Mn^{4+}$, $SO_4^{2-}$, halogenated compounds) along site transects. In this regard, Lovley et al. [126] have devised a gas sampling bulb protocol for anaerobic groundwaters in the field that, in combination with hydrogen gas determinations and Winkler titrations for $O_2$ [10], provides definitive information on dominant anaerobic redox couples. Yet, in interpreting such field measurements, one must be mindful of the presence of microenvironments, which may allow localized pockets of anaerobiosis to occur in seemingly aerobic habitats. Furthermore, many of the reduced end products may diffuse away or be transported from the location where they were produced. For instance, detection of methane in field samples (from a natural gas-free locality) indicates that the highly reducing biogeochemical conditions associated with methanogenesis are operative in the vicinity of the sampling point. However, because methane is a volatile and mobile gas, its detection does not necessarily define the physiological activities in progress at the time and location of sample removal.

Establishing site-specific historical records of the behavior of contaminants, co-reactants, and metabolic products may initially appear simple. Theoretically, compilation of such field data documents the effects of contaminant-attenuating processes over time. In conjunction with other assays, the field data can be interpreted in ways that may implicate biodegradation as a cause of pollutant losses. However, as with understanding in situ physiological regimes, obtaining robust, interpretable field monitoring data may be an elusive goal if contaminant characteristics and site conditions are complex. The overall objective is to establish a site-monitoring regime using a network of consistent sampling locations that afford the acquisition of contaminant concentration and other measurements that are comparable over time. If the distribution of contaminants at the site and factors influencing contaminant transport (e.g., climate, hydrology, commercial, or industrial activities) are erratic, then the pertinent database on contaminant behavior may be so noisy as to mask any trends. However, in many sites the type of contaminant monitoring protocols required by concerned regulatory agencies can be integrated over time and sometimes produce a clear historical record of diminishing contaminant concentrations from year to year. When such data exist, they assist in meeting the first criterion for proving in situ bioremediation (see Table 15-3).

## Step 2: Microbiological Protocols for Measuring Biodegradation Activity

Microbiological protocols, the heart of this chapter, are designed to confirm, demonstrate, and explore both the net chemical changes and the associated intracellular details pertinent to how microorganisms influence the fate of organic contaminants. This type of information is necessary for meeting the second criterion for demonstrating in situ bioremediation. As outlined in Tables 15-3 and 15-4, a variety of microbiological procedures spanning a broad range of disciplines are pertinent to biodegradation and bioremediation. Figure 15-4 provides a comprehensive overview of the variety of objectives, disciplines, and protocols that play key roles in biodegradation research and the practice of bioremediation technology. The two phases that serve as main divisions in Figure 15-4 are distinguished by the degree to which scientific detail is pursued. Phase 1 treats samples of soil, sediments, or water simply as "black boxes" that do or do not make contaminant compounds disappear as judged by analytical chemical criteria. Phase 2 begins with the isolation of pure cultures of contaminant-degrading microorganisms. Once these pure cultures have been obtained, then refined physiological and biochemical investigations may be performed. Furthermore, as DNA sequences of genes that code for metabolic pathways become increasingly available, molecular procedures (Table 15-4) will continue to gain prominence in biodegradation protocols. Among the final goals of the procedures shown in Phase 2 is understanding the molecular basis for gene expression and regulation.

### *Design and Implementation of Biodegradation Assays Using Environmental Samples*

The traditional black box approach to biodegradation assays asks the question, Are microorganisms within this complex soil, water, or sediment microbial community able to metabolize the compound of interest? To answer this question, one aseptically gathers samples from a given field location, dispenses known weights or volumes of the samples to replicated vessels, handles the samples in a variety of ways that include a treatment that has either been sterilized or poisoned, incubates the test samples under laboratory conditions, and employs both chemical and physiological assays that monitor the fate of the test compound within experimental vessels over time (Fig. 15-4, Phase 1). The objective of this experimental design for biodegradation is remarkably simple. Yet a series of substantial obstacles must be overcome before obtaining clear data that truly test a given set of specific hypotheses. Every design parameter selected for inclusion in a biodegradation assay can influence the resultant data. Therefore, decisions made in implementing biodegradation assays should be well reasoned. Table 15-6 sum-

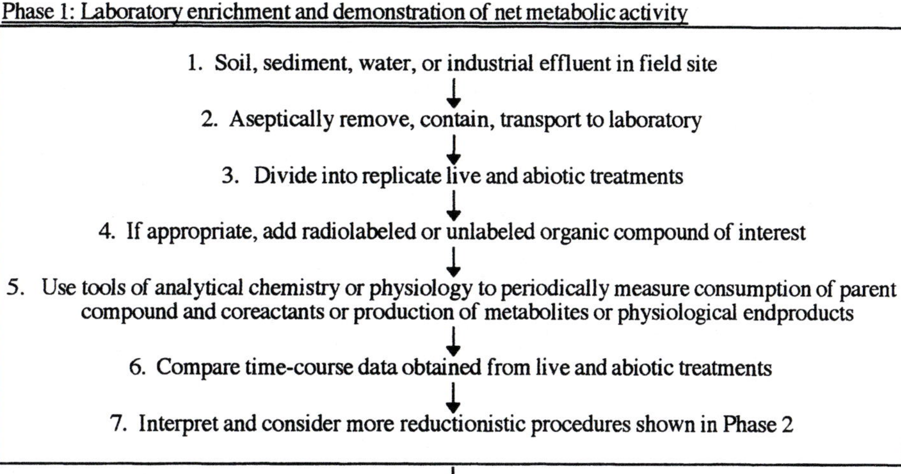


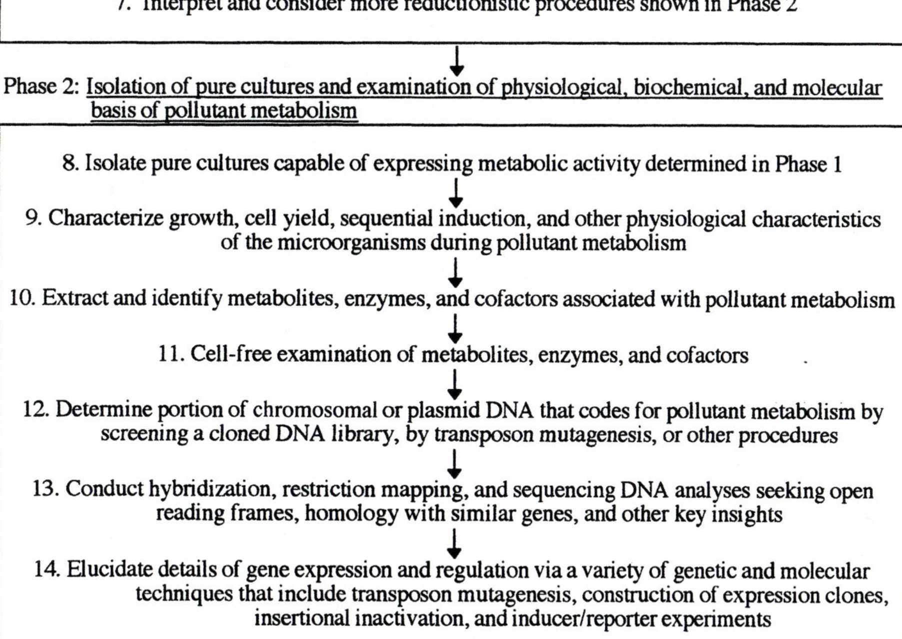


Figure 15-4 Two phases of procedures for understanding biodegradation processes. Phase 1 begins with environmental samples. Phase 2 proceeds through biochemical and molecular aspects of pollutant metabolism by single microorganisms. (From Madsen [131].)

marizes many of the practical and theoretical decisions that must be made in developing biodegradation protocols using environmental samples. Step 1 in Table 15-6, a background issue considering information use, is fundamental to all subsequent experimental decisions. The degree to which experimental minutiae of a given testing protocol must be initially considered is commensurate with the scrutiny the final data will undergo. Artifacts and biases in data are virtually unavoidable in biodegradation assays; thus, it may be wise to simply accept methodological limitations, rather than worry about initial potential technical design flaws that may later have no practical impact.

Once the reason for conducting the biodegradation assay has been put in perspective (Step 1, Table 15-6), then another background issue–that of physiological conditions—should be confronted. Step 2 acknowledges the fact that biodegradation is only a small portion of the perhaps thousands of physiological reactions simultaneously occurring when the mixed microbial populations in environmental

Table 15–6 Steps and Decisions Essential for Implementing Biodegradation Assays Using Environmental Samples

| Step | Decisions |
|---|---|
| 1. *Background*: Determine how the resultant data will be interpreted and used. | Objectives range from information about crude "biodegradation potential" to tests of specific hypotheses about physiological or biochemical factors governing biodegradation processes. |
| 2. *Background*: Select the physiological conditions under which pollutant metabolism is to be measured. | The pivotal physiological concern is defining the mechanisms by which the compound is metabolized. Of primary importance is discerning between such possibilities as co-metabolic reactions, use as an electron acceptor, and use as a carbon and energy source. Other concerns address conditions in experimental flasks, such as nutrient sufficiency, which final electron acceptor regime should dominate, what pollutant concentration ranges should be examined, and if conditions should be constantly changing (batch culture) or constant (continuous culture) during the assay. |
| 3. *Practice*: Select and aseptically sample a source of microorganisms with physiological activity that is of interest. | Aseptic techniques involve use of such tools as flame-sterilized scoops, spatulas, and knives and sample placement within sterilized glass or plastic containers. |
| 4. *Practice*: Select the physical apparatus, hence the physiological setting, for biodegradation reactions to occur. | Glass (or plastic) vessels must be assembled. These contain the microbial community being studied, as well as any accompanying components of soils, sediments, and water in varying ratios. The experimental hardware may be fitted with a variety of gas and water exchange assemblies for maintaining physiological conditions and assaying reaction progress. |
| 5. *Practice*: Select a metabolic activity assay that is sensitive, effective, convenient, inexpensive, and compatible with experimental objectives. | The general assay categories are physiological assays (e.g., respirometry or growth) and chemical assays, which include GC, GC/MS, HPLC, and radiotracer techniques. |
| 6. *Practice*: Aseptically prepare stock solutions of $^{14}C$-labeled organic compounds. Verify their radiopurity. | The validity of the results from biodegradation assays using $^{14}C$-labeled substrates is dependent on substrate radiopurity and aseptic preparation of stock solutions. |
| 7. *Practice*: Complete the experimental design parameters for the assay vessels and the assays themselves. | Design parameters incude the following:<br>a. concentration of the test substrate(s)<br>b. number of replicated flasks<br>c. whether flasks can be sampled repeatedly or if they require sacrifice<br>d. frequency of sampling<br>e. method of preparing abiotic controls<br>f. method of separating radioactive parent and product compounds from one another |

Modified from Madsen [131].

samples are incubated in the laboratory. These physiological processes feed one another, interact in complex ways, and can be governed by many of the sometimes inadvertent physical and chemical manipulations made while preparing, incubating, and sampling assay vessels. Uncertainties become particularly striking when attempting to trouble-shoot failed attempts at biodegradation assays. The interplay between fundamental knowledge of physiology and experimental design parameters demands that a variety of issues be confronted, including (1) the mechanism by which the compound is metabolized (e.g., as a carbon source, a nitrogen source, or as a co-metabolic substrate the transformation of which will occur only when other compounds are supplied); (2) inclusion versus exclusion of potential growth-limiting vitamins and minerals; (3) inclusion versus exclusion of air as headspace in the reaction vessel; (4) the solid-to-liquid ratios used in each test vessel; (5) the multiple roles of compounds in physiological reactions (for instance, nitrate can serve as both a nutrient and a final electron acceptor); and (6) the fact that the compound whose biodegradation is being tested may be toxic at high concentrations or may fall below some minimum threshold value for uptake and cell growth at low concentrations. Background considerations raised under Steps 1 and 2 of Table 15-6 guide most of the practical steps needed for completing the implementation of the biodegradation assays (Steps 3–7).

Proper sampling of the source of microorganisms of interest (Step 3) is critical for achieving valid biodegradation data. All of the physiological information generated late in biodegradation protocols is only as sound as were the investigators' hands in gathering microorganisms that truly represent the sampling site. The same care described for geochemical site characterization must be used during microbiological sampling. In addition, aseptic techniques (such as the use of flame-sterilized implements and the enclosure of samples within previously sterilized vessels) must be employed. Because microbiological characteristics of environmental samples are at least as prone to postsampling changes as chemical characteristics, the fixation procedures recommended for chemical sampling procedures might initially seem applicable to microbiological sampling. This is the case for a variety of microbiological characteristics, such as extraction of nucleic acids or phospholipid fatty acids, that do not require that cell viability be maintained. However, biodegradation assays are, by definition, measures of dynamic chemical changes effected over time by live microorganisms. Therefore, fixed cells cannot be used in biodegradation assays. Cooling of the samples on ice until laboratory processing is the most widely recommended sample-holding procedure. The microbial populations present in soil and water samples may begin to shift and change the moment an environmental sample is removed from a field site [130]. These changes continue through cold storage, distribution of the samples to biodegradation vessels, and continued laboratory incubation during the biodegradation assays. It is this inevitable, intractable set of microbiological changes (as well as our inability to match laboratory to in situ field conditions) that make it virtually impossible to extrapolate the results of laboratory biodegradation assays directly to field sites.

Depending on one's objectives, selecting the physical apparatus and hence the physical/chemical setting of the biodegradation assay (Step 4, Table 15-6) can be a trivial matter or present severe logistical problems. Despite the fact that all biodegradation equipment has the same general goal—to contain both microorganisms and test organic compounds in ways that allow physiological modification of the organic compound to be measured—this goal has been manifest in hundreds of different ways in the published literature. The physical equipment used by a given investigator reflects a variety of factors that include the chemical properties of the test component; physiological conditions desired; the sensitivity and accuracy needed to document the particular biodegradation reaction(s); the biochemical mechanism of biodegradation; the fastidiousness of the microorganism(s) carrying out the reaction; the amount of control sought by investigators over chemical, nutritional, or microscale gradients within experimental vessels; the degree to which the investigator desires conditions during the biodegradation assay to match those in the field site of interest; and the availability of personnel, hardware, or other resources.

Selection of the physical apparatus is often intricately linked to how the metabolism of the organic compound will be documented. For instance, if the historically established biological oxygen demand (BOD) assay (traditionally used for sewage treatment purposes) is the assay of choice, then the widely used BOD bottle is likely to be the accompanying apparatus of choice, and biodegradation will be gauged using an indirect metabolic activity assay, oxygen consumption. This and other indirect physiological biodegradation tests infer chemical loss through a related vital microbiological parameter—cell growth, respiration, calorimetry, or the like) (Step 5, Table 15-6). In contrast, methods that focus directly on demonstrating diminished mass of the contaminant or the production of contaminant-specific metabolites are considered more rigorous. These direct procedures use specific chemical purification and detection hardware (including gas chromatography (GC), high-performance liquid chromatography (HPLC), gas chromatography/mass spectrometry (GC/MS), spectrophotometry, and radioisotopic-tracers) to monitor the abundance of test chemicals in experimental vessels. If a $^{14}C$-labeled version of the test compound is added to the test apparatus and later steadily increasing amounts of $^{14}C$ are recovered from the assay vessel as $^{14}CO_2$, then microbial mineralization has been documented. Such radiotracer assays are often both elegantly rigorous and procedurally facile to complete.

The sixth entry in Table 15-6, the selection, preparation, and quality control for sterile $^{14}C$-labeled stock solutions—is a critical step in implementing biodegradation studies involving radiotracers. The location of the $^{14}C$-labeled atom(s) within the test molecule has a major bearing on experimental objectives and data interpretation. If nuances of biochemical pathways are of interest, many refined hypotheses can be explained by varying positions of the $^{14}C$-labeled atoms. However, if complete mineralization is of interest, then the central structural elements of the molecule (such as a benzene ring) should be labeled uniformly with $^{14}C$. Sterility is required both for maintaining the expensive stock solutions and for allowing clear, confident interpretations of data from experiments in which the stock solutions are used. If microorganisms able to transform the $^{14}C$-labeled substrate are present in the stock solutions, not only may the compound be altered before its intended use but also the source of biodegradation activity found in experimental treatments may be uncertain. Both of these possibilities are disastrous for biodegradation testing. Thus, extreme care in ensuring aseptic handling of the vessels, solutions, and dilutions of $^{14}C$-stock solutions is essential. Furthermore, repeated tests should be carried out that ensure the lack of microbial contamination of stock solutions and that assess the radiopurity of the substrate. Under most circumstances, an uninoculated, "reagent only" control treatment serves to test for the presence of biodegradation activity in the stock solutions.

The radiopurity of $^{14}C$-labeled substrates is readily assayed using HPLC separation techniques and a fraction collector (thin-layer chromatography is suitable if the proper solvent mix in the mobile phase is used). The HPLC analysis system must be tested on unlabeled standards so that the precise elution time for given flow rates of the mobile phase is known. The total $^{14}C$ activity contained within a given volume of stock solution (determined via scintillation counting) is compared with the proportion of $^{14}C$ eluting at the expected time, thereby discerning the radiopurity of the substrate standards. In subsequent biodegradation tests, the proportion of transformed substrate (measured via either loss of parent or conversion to $^{14}CO_2$ or daughter products) must far exceed (i.e., at least by a factor of 2 to 4) the level of $^{14}C$ impurities present in the stock solution. The manufacturer's specifications on product radiopurity (e.g., over 98%) can generally be trusted for newly purchased materials.

When received from the manufacturer, the $^{14}C$-labeled compound may be carrier-free (i.e., as vanishingly small quantities of pure liquid, solid, or gas) or in a carrier solvent, such as ethanol or acetone. In general it is best to request the carrier-free form of substrate from the manufacturer, which avoids issues of solvent effects or co-substrate influences on the results of microbial activity assays. The precise details of preparing and maintaining small volumes of sterile concentrated primary stock solutions are beyond the scope of this chapter. However, stock preparation approaches include the addition of sterile water to the manufacturer's vessel (for solubilization and repeated collection of rinsates); dilution and washing of volatile $^{14}C$-labeled solvents (originally supplied in sealed glass ampules) with their unlabeled forms; preparation of saturator vessels that allow sparingly soluble substrates simply to float in sterile deionized water so that water drawn off contains fully dissolved aqueous-phase substrate at its saturated solubility; and filter-sterilization using syringe-fitted units with housings and filtration membranes that fail to bind the radiolabeled substrate of interest. Other working stock solutions, derived from dilutions of the primary stock, can be readily prepared by aseptic transfer of known volumes of the primary stock to known volumes of suitable sterile solvents (often water). All stocks should be stored in a manner that stabilizes the material—usually in a glass vessel, sealed with a Teflon-lined cap, and kept cold or frozen.

The seventh and final step in Table 15-6 is necessary for completing the logistical considerations implicit in all previous steps. Major issues in this last step include

**Concentration of the Test Substrate(s)** Pollutant compounds should be added to test vessels at environmentally relevant concentrations. Ideally, the substrates should be free of carrier solvents. It is essential to understand a variety of properties of test compounds, such as aqueous solubility, volatility, and toxicity (to the microorganisms and to the microbiologist), before initiating biodegradation assays and the elective enrichment, isolation, and enumeration protocols that may follow. Table 15-7 provides selected guidelines for selected pollutants that are frequently examined for their biodegradability.

**Number of Replicated Experimental Flasks per Treatment** Three replicates is generally considered the minimum, but the number may go above 10, if experimental aims involve fine discrimination among treatment types. (You and your statistician should debate and agree upon replication issues.)

**Whether Flasks Can Be Sampled Repeatedly or Require Sacrifice** This issue is largely determined by the physical and chemical status of the test compound within the experimental apparatus. If the reaction mixture of cells, water, particles, gases, and the like can be subsampled in a reproducible manner while maintaining both the composition of the mixture and the integrity of the biodegradation process, then the repeated sampling strategy should be used. This approach obviates the need for the preparation of large numbers of replicate assay vessels at the beginning of the testing procedures. Furthermore, repeated sampling of the same vessels also ensures that the same microbiological populations (or their progeny) are sampled throughout an experiment, thus diminishing inoculum variability as a source of noise in the data.

It becomes necessary to sacrifice replicate vessels when subsampling cannot accurately quantify the compound of interest within test vessels. For example, in many studies of the metabolism of hydrophobic pesticides, not only may the compound's molecular structure be resistant to complete mineralization but the pesticide may also partition itself in complex patterns among a variety of soil components, such as clays and humus. Uniformly sampling each of these components within test vessels is usually impossible. Thus, the analytical approach

Table 15–7 Characteristics of Selected Environmental Contaminants and Considerations for Carrying out Biodegradation Assays and Strain Isolation Procedures

| Compound | Aqueous Solubility (ppm) | Recommended Starting Concentrations in Aqueous Enrichment Culture (ppm) | Mode of Compound Delivery in Media | Major Considerations | References |
|---|---|---|---|---|---|
| Benzene | 1,780 | 5[a] to saturation | Vapor phase or embedded in wax | Toxicity, volatility; used aerobically and anaerobically as carbon and energy source | 60, 67, 128 |
| Toluene | 535 | 5[a] to saturation | Vapor phase | Toxicity, volatility; used aerobically and anaerobically as carbon and energy source | 50, 54, 125, 192, 193 |
| Naphthalene | 30 | 30 | Vapor phase | Solubility, volatility; used aerobically as carbon and energy source; anaerobic metabolism may also occur | 43, 52, 178 |
| Phenanthrene | 1.3 | 1.3 | Spray plate | Solubility; used aerobically as carbon and energy source; anaerobic metabolism may also occur | 43, 108, 145, 194, 229 |
| Phenol | 82,000 | 50 | Dissolved in medium | Toxicity, volatility; used aerobically and anaerobically as carbon and energy source | 17, 112 |
| Tetrachloroethylene | 150–200 | 0.5<br>$3 \times 10^4$ | Dissolved in medium and vapor phase; dissolved in hexadecane | Toxicity, volatility; used anaerobically as final electron acceptor, not attacked aerobically; electron donors include $H_2$, methanol, butyrate | 62, 94, 110, 139, 139a, 215 |
| Trichloroethylene | 1,100 | 0.5 | Dissolved in medium and vapor phase | Toxicity, volatility; used anaerobically as a final electron acceptor and aerobically co-metabolized by oxygenase enzymes | 62, 122 |

[a]5 ppm is designed to avoid toxicity, especially under anaerobic conditions. Saturated concentrations may be toxic to some microorganisms, but realistically simulate field conditions in contaminated sites.

used for each sampling point in time may be to remove three or more replicate vessels, subjecting the entire contents of each to an array of dissection, extraction, chromatographic, and detection procedures that quantify the disposition of $^{14}C$-labeled parent and daughter compounds.

**Frequency of Sampling** Determining how often replicate biodegradation vessels should be sampled is no trivial issue. Plots of data showing the loss of contaminants versus time generally convey the most information if several early data points establish a stable initial concentration of the substrate and then gradual loss begins and plateaus at some low level. Thus, three phases in a biodegradation curve are typically seen: lag, active biodegradation, and cessation. During lag and cessation periods, the sampling interval is often immaterial, unless specific experiments aim at quantifying subtle kinetic aspects of lag and cessation phases. The key issue about the active biodegradation phase is that it should not be missed. During the active biodegradation phase, the sampling intervals should be adjusted to one third or one quarter that of the active period. For instance, if the substrate concentration drops from 100% to 20% in a 24-h period, the sampling interval should be 6–8 h. That sampling regimen produces clean, unambiguous substrate disappearance curves. However, depending on the statistical needs of the investigator, sampling intervals may be one tenth or one twentieth of the active period [182].

With no prior knowledge of how long the lag time or active periods will be, all that an investigator can do is be pragmatic: Sample at convenient intervals that gradually increase with the duration of the test; for example, first daily and then every 2, 4, and then 7 days; for some anaerobic incubations (see Biodegradation Protocol 3) the sampling interval may be monthly. In addition, expect to repeat the experiment several times using data from each successive assay to refine the experimental design (especially sampling-interval) parameters. It is the relationship between the mass of the substrate and the mass and activity of the biodegrading populations that governs rates of substrate loss. If the substrate of interest is a carbon and energy source (e.g., phenol) added to soil at a concentration of 1 ppm (1 $\mu$g/g soil) and the initial phenol-degrading population is $10^4$ cells/g soil (or $10^{-8}$ g cells/g soil, presuming that cells weigh 1 pg each), then a substantial time lag is inevitable because an initial doubling of biomass will only consume 1% ($10^{-8}$ g/$10^{-6}$ g) of the phenol. However, after several generations, as the mass of the active cells approaches 1 ppm ($10^6$ g cells/g soil), then in a single generation period nearly all of the phenol can theoretically be metabolized. This is a simple but insightful illustration of how both lag time and active biodegradation phases are governed by relative masses of test substrate and the active cells present in test vessels.

**Methods of Preparing Abiotic Controls** These methods include sterilization, use of biological poisons, and the uninoculated control. These treatments are necessary so that the contrasting behavior of the test compound in the live versus abiotic treatments can be attributed to microbial activity. The ideal abiotic control treatment would simply eliminate all biological function while keeping the physical and chemical properties of environmental samples (composition and structure of everything from the test compound to microbial cells to the surfaces of clay minerals and humus polymers) completely unaltered. This is an impossibility.

1. *Sterilization*: Unquestionably the best sterilization procedure is by filtration (0.22-$\mu$m membrane filter pore size). Filtration removes all cells, retains dissolved chemical constituents in environmental samples, and is only appropriate for relatively pristine aquatic samples. High-solid samples such as

soil cannot be filter-sterilized, so alternative treatments must be used. Autoclaving soil for 1-h periods on 3 successive days effectively eliminates biological activity, as does gamma irradiation (2 to 3 Mrads) and gas-phase sterilization (i.e., by ethylene oxide treatment). Soil sterilization techniques have recently been reviewed by Wolf and Skipper [236].

2. *Use of Biological Poisons*: Chemical agents (poisons) can be introduced easily and are highly effective in aquatic samples where the test medium is uniform and homogeneous and good cell-poison contact can be ensured. Because soil is a complex mixture of reactive components, the activity of chemical agents as metabolic poisons is often diminished greatly. Thus, high concentrations and multiple agents are often recommended when attempting to curtail microbial processes in soil. The chemical inhibitors of microbial activity that have been tested include $HgCl_2$, sodium azide, a variety of acids, strong base (NaOH), sodium cyanide, formaldehyde, and various antibiotics [33, 236]. The effectiveness of inhibitors varies with dosage and each specific environmental sample. Therefore, preliminary tests verifying process inhibition should be conducted while refining the design of biodegradation protocols.
3. *Uninoculated control*: These treatments, whenever possible, should be included in experimental designs. They verify that the reactants of interest are stable in the absence of microorganisms and that procedures taken to exclude microorganisms were effective. In many instances, uninoculated controls can only be prepared when a substantial portion of the biodegradation reaction medium is synthetic (e.g., a mixture of water mineral salts and the test compound) and has been sterilized. When the reaction medium consists primarily of the environmental sample itself (e.g., soil or lake water), then clearly the inoculum is already present. In such cases, the uninoculated control cannot be prepared; instead, sterilization or poisoned treatments must be pursued.

**Methods of Separating Radioactive Parent and Product Compounds from One Another** In biodegradation procedures using radiolabeled test compounds, it is generally unnecessary to use chromatographic separation measures for distinguishing the compounds of interest from others present in environmental samples. Instead, documentation of microbial attack on the parent labeled compound usually relies on changes in properties (volatility, molecular weight, solubility, hydrophobicity, etc.) of the radiotracer. These changes generally allow investigators to use simple partitioning procedures that quantify the radioactivity in different pools of the initial parent and intermediary or final daughter products. The simplest example of such a procedure is the conversion of a water-soluble, nonvolatile parent compound, such as nitrophenol, to $CO_2$. The parent remains in aqueous solution always, while $CO_2$ will leave aqueous solution at low pH and can be partitioned into a separate headspace-located reservoir of base. Not all combinations of radiolabeled parent and daughter products differ unambiguously in their partitioning behavior. For instance, $^{14}C$ benzene is both volatile and reasonably soluble in aqueous media. Thus, it may be erroneous to presume at the end of a mineralization assay that 100% of the $^{14}C$ detected in a separate reservoir of base represents $^{14}CO_2$. A variety of checks have been devised for ensuring that $^{14}C$-labeled protocols are partitioned unambiguously. These checks take the form of additional procedures or control flasks that test alternative interpretations of the data. Biodegradation Protocol 2, Step 7, provides examples of methods for separating parent and daughter compounds. Whenever one is uncertain about the partitioning behavior of $^{14}C$-labeled pools of parent and daughter compounds in a given experimental apparatus, the trapping efficiencies of authentic standards must be tested after being added to test vessels, alone and in mixtures, in the absence of microorganisms.

### *Biodegradation Protocol 1. Loss of a Soluble Non-Volatile $^{14}$C-labeled Compound from Batch-Incubated Aerobic Water Samples*

Procedure

1. Aseptically gather the aqueous sample of interest as described in Table 15-6. Maintain at 4°C on ice or in a refrigerator. Aseptically dispense 200 ml to four sterile 500-ml flasks. Add 10 ml of acid mercuric chloride solution (0.25 M $HgCl_2$ in 5% HCl) to one of the flasks to eliminate microbial activity. Alternatively, sterilize the abiotic control flask via filtration or autoclaving (Table 15-6).
2. Aseptically dispense at least $2.0 \times 10^6$ dpm of radiolabeled compound from a sterile stock solution. This will yield at least $10^4$ dpm/ml. Add sufficient non-radiolabeled compound, which is ideally carrier-free, to attain a concentration of 1 ppm (if the water sample is pristine, addition of 200 μg is required). Decisions about the amount of non-radiolabeled compound should be based on experimental objectives and chemical analyses performed on the sample (see Table 15-6). The mass of $^{14}$C-labeled compound added, as calculated based on the manufacturers' specifications of specific activity (mCi/mmole), is often negligible.
3. Plug 500-ml flasks with snug-fitting cotton gauze or foam stoppers. Cover these with loose-fitting aluminum foil or metal caps that extend beyond the lip of each flask. Place on a rotary shaker, and adjust both aeration levels and temperature to conform with overall experimental objectives and to the degree to which laboratory conditions are intended to match those of field sites.

   *Note*: Microbial respiration will produce volatile $^{14}CO_2$. To prevent inhalation and other exposure to individuals nearby, incubate the flasks in a manner that either contains possible $^{14}CO_2$ within a vented hood or scavenges the $^{14}CO_2$ within each flask with base (see Protocol 2).
4. Periodically remove 2 to 4 ml from each flask using sterile pipette tips attached to a hand-controlled manual pipettor (e.g., Gilson P5000, Rainin Instruments). Transfer to a 10-ml glass test tube. Place in a vented hood. Acidify to pH 2 with 1 drop of concentrated $H_2SO_4$. Flush the acidified sample with air bubbles for 10 min in a hood.

   *Note*: The air bubble distribution manifold can be assembled easily from a 0.5 m length of 0.5 cm Tygon tubing (the main "feedline") that is folded, clasped, and sealed at the distal end. This tubing is pressurized with air (via a house line or aquarium-type pump) at the proximal end, and multiple "bubbling lines" are used to direct the air into the bottom of each test tube. Each bubbling line consists of a double-pointed 18-gauge needle (Vacutainer needle holder, Becton Dickinson), one end of which is inserted into a ≈15-cm length of tightly fitting narrow Tygon tubing. The other end of the narrow tubing is fitted snugly over a 10-μl glass capillary tube. To use the bubbling line, connect it to the pressurized air by inserting the unused end of the double-pointed needle through one wall of the feedline, and drop the glass capillary tube end of the bubbling line into the bottom of the $^{14}$C solution contained by the test tube. Ten minutes is sufficient to drive off any $^{14}CO_2$ that may be present.

5. Measure residual $^{14}C$ present in the water samples by pipetting a precise volume (1.0 to 3.0 ml) of fluid from each bubbled sample to the appropriate volumes of scintillation cocktail (see the manufacturer's specifications) and counting in a scintillation counter. Differences in the quantity of $^{14}C$ recovered from live versus sterile and initial versus final flasks are indicative of substrate loss via mineralization (i.e., conversion to $^{14}CO_2$).

   *Note*: This procedure is suited to the documentation of the aerobic conversion of non-volatile organic compounds readily respired to $CO_2$. Because of the presumption that the lost $^{14}C$ is $^{14}CO_2$, this procedure is not recommended for compounds with a doubtful susceptibility to aerobic mineralization.
6. After each 10-min bubbling period, rinse the glass capillary tubes with two washes of deionized water to eliminate carryover from one sampling time to the next. Properly dispose of and store the residual aqueous $^{14}C$ present in the air-purged test tubes and the glass capillary rinsate. Because the flask-incubated water samples can be subsampled repeatedly and because more than 100 bubbling lines can simultaneously purge water subsamples, this procedure is perhaps the simplest and least labor intensive of all used in biodegradation studies [207].

### *Biodegradation Protocol 2: Production of $^{14}CO_2$ from a $^{14}C$-labeled Compound in Batch-Incubated Soil*

Possible drawbacks of the methodology described in Protocol 1 are its demand for a homogeneous aqueous sample matrix and its presumption that $^{14}C$ lost from solution is synonymous with conversion of parent compound to $^{14}CO_2$. If additional information is desired, a modification in the experimental apparatus is implemented so that a reservoir of inorganic (e.g., NaOH) or organic (e.g., phenethylamine) base is present within each $^{14}C$-amended test vessel. By monitoring $^{14}CO_2$ evolution in headspace above the environmental sample, microbial biodegradation activity in heterogeneous samples, such as soil, can be measured.

Procedure

1. Aseptically gather soil samples of interest (see Table 15-6). Maintain at 4°C on ice or in a refrigerator. Sieve or mix to uniformity if the assay is aimed at ascertaining general biodegradation potential. Avoid sieving or keep as an intact core and incubate at field temperatures if the simulation of field conditions is desired. Aseptically dispense 10 to 30 g of soil to each of four 125-ml flasks. Criteria for selecting the size and type of flask, stopper, and $^{14}CO_2$ trap are discussed under Step 3 (see also Table 15-6). Eliminate metabolic activity in one of the flasks either by autoclaving (for 1 h on each of 3 successive days) or by adding $\approx$ 5 ml of the $HgCl_2$/HCl solution described in Protocol 1.
2. Aseptically dispense $10^5$ dpm of radiolabeled substrate to the soil surface in each flask. Mix to uniformity with a sterile stainless steel spatula or glass rod. Dispense aqueous phase or neat non-labeled chemicals to soil to achieve the final desired total concentration of test compound (see Table 15-6). If intact soil cores are used, inject the substrates at a variety of depths in an attempt to achieve uniform distribution.
3. Configure the flask that contains the soil and both labeled and unlabeled substrates to contain an internal reservoir of base that can be emptied periodically, counted to ascertain $^{14}C$ activity, and replenished with fresh base. Many possible hardware arrangements can meet these simple logistical re-

quirements; two are described here. In the first, a "biometer flask" (Bellco, Inc.) consists of a 250-ml Erlenmeyer flask with a 50-ml glass side arm attached. The Erlenmeyer is sealed with a rubber stopper containing a glass tube that extends into the headspace and is fitted with a stop cock and an ascarite-filled column that allows $O_2$ to be replenished within the flask while preventing the entry of atmospheric $CO_2$. The attached side arm compartment is sealed with a rubber stopper pierced with a 15-cm long 15-gauge cannula. The top of the cannula is sealed with either a rubber policeman or a septum at all times, except when withdrawing or replenishing 10 ml of 0.2 N KOH used to trap $^{14}C$- (and unlabeled) $CO_2$ [20]. The second hardware arrangement consists of a 25-ml glass vial containing 4 to 8 g of soil, a sterile glass marble, and a sterile glass shell vial (15 × 45 mm, Kimble, Vineland, NJ). The marble supports the shell vial so that it is accessible to the sampling syringe through a septum. The entire assembly is sealed with a rubber septum held in place with a screw-cap ring (e.g., Pierce, Rockford, IL; many different manufacturers offer different combinations of glass vials, caps, septa, and sampling valves). A 0.4-ml volume of a 0.5 N NaOH solution is removed periodically and replenished from the shell vial using a 4-cm 20-gauge needle with a 1-ml syringe [135].

4. Seal the test vessels to allow complete recovery of $^{14}CO_2$ but also provide adequate final electron acceptor (often $O_2$) so as not to limit metabolic activity. Assurance of adequate $O_2$ can be achieved via passive replenishment, such as by using the ascarite trap above [11, 20] or by adjusting the ratio of headspace to soil so that it is relatively large ($\approx$ 10:1; headspace $O_2$ should meet the total theoretical $O_2$ demand based on the organic carbon content of the enclosed soil). The $O_2$ supply can also be actively replenished via continuous purging [248].
5. Determine the conversion of radiolabeled substrate to $^{14}CO_2$ by mixing a known proportion of the basic trapping solution (taken from within each test vessel) with an appropriate volume of scintillation cocktail (see the manufacturer's guidelines) and counting in a scintillation counter. Control assays that should also be counted include reagent blanks (trapping solution only) and base retrieved from the sterile or poisoned treatments. Data are presented as time-course plots of the cumulative % $^{14}CO_2$ produced from the total initial $^{14}C$ added to each flask. To terminate an experiment, saturate the contents of each test vessel with 0.1 N HCl. This converts any residual bicarbonate $^{14}C$ species to $^{14}CO_2$ that, after 12–24 h, should be trapped and counted. Resultant data may be interpreted using a variety of kinetic modeling approaches [8, 116, 195].
6. Make modifications to Biodegradation Protocol 2 as necessary to accommodate volatile compounds. As mentioned in Table 15-6, the apparatus selected to assess biodegradation must allow mass balances to be assembled. Yet, some pollutant compounds exhibit volatility, sorption, and solubility traits that make them difficult to contain, cause them to partition into rubber stoppers, or hinder distinguishing between $^{14}CO_2$ and parent compound in the trapping fluid. Under these circumstances, special steps need to be taken.

   To prevent sorption or partitioning of volatile hydrophobic test chemicals into rubber stoppers, use Teflon-faced stoppers. For some stoppers, the Teflon coating may not extend to the perimeter, necessitating additional application from an aerosol-based Teflon product. Furthermore, Teflon may prove to be ineffective in preventing sorptive losses; under such circumstances all-glass or all-metal vessels may be required.
7. There is a possibility that $^{14}C$ recovered in the basic trapping solutions may not be $CO_2$. If this is the case, process the base further. At least two pro-

cedures are available. The simplest, which applies to aqueous (i.e., NaOH) trapping solutions, prescribes adding 1 ml of a 3 N $BaCl_2$ [11, 121]. Barium forms an insoluble precipitate with carbonate, which is five orders of magnitude less soluble than barium hydroxide. Therefore, after centrifugation, all $^{14}CO_2$ should be removed from solution, whereas non-carbonate $^{14}C$ should remain in solution. Alternatively, transfer the radioactive basic solution (both organic and inorganic) to a purging manifold, which delivers a train of bubbles through a series of six sequentially linked glass scintillation vials containing air (one vial), a hydrophobic scintillation fluid (toluene- or xylene-based or a less toxic equivalent; two vials), air (one vial), and either an organic (phenethylamine or methylamine) or inorganic (0.1 N KOH or 0.5 N NaOH; two vials) base [137]; the vials with only air serve to guard against liquid-transfer between vials. Once the purging train has been assembled and the flow of bubbles established (vacuum-based systems are superior to pressure-based ones), enough concentrated HCl is added to the original $^{14}C$-laden solution to achieve pH 2. The acid converts all $^{14}C$ carbonate species to volatile $^{14}CO_2$, which will pass through the first four traps, but be retained in the last two. However, $^{14}C$ as parent or volatile daughter products should be retained in the hydrophilic scintillation cocktail. By counting $^{14}C$ in each trap, $^{14}CO_2$ is distinguished from other $^{14}C$ compounds. Related procedures for separating $^{14}C$ parent from daughter compounds have been described by Anderson [11] and Zibilske [248].

### *Biodegradation Protocol 3: Anaerobic Metabolism of Soluble Organic Compounds in Batch-Incubated Slurries*

Anaerobic metabolism of soluble pollutant compounds can be readily examined using narrow-necked serum bottles sealed with butyl-rubber stoppers. The basic idea is to prepare a slurry-phase mixture of a mineral salts medium, the organic compound (or compounds) of interest, and the inoculum of interest. The butyl-rubber stoppers are impermeable to $O_2$ and reseal around the perforations made by needles of sampling syringes. Thus, long (over 1 year) anaerobic incubations can be carried out in which the same bottles are sampled repeatedly. Samples are subject to analysis (e.g., HPLC or GC, see Table 15-5) so that time-course data describing the loss of parent compound or the production of intermediary metabolites or both can be assembled. Appropriate abiotic controls (uninoculated and inoculated but poisoned or autoclaved) are prepared, as is an inoculated control without added substrate. The latter is used to verify that daughter compounds, suspected to be derived from the added substrate, are not produced from materials carried by the inoculum only. Components of the inorganic medium ensure a lack of mineral nutrient limitation and also allow the oxidation-reduction status to be adjusted so that conditions will favor one of the variety of possible anaerobic physiological regimes (e.g., dissimilatory nitrate reduction, Mn reduction, Fe reduction, sulfate reduction, or methanogenesis). Procedures below are based on literature references [29, 128, 133, 209, 245].

Procedure

1. Aseptically gather soil, sewage, water, or sediment samples of interest. Maintain at 4°C on ice or in a refrigerator. Sieve or mix to uniformity (see Table 15-6).
2. Prepare anaerobic inorganic salts medium containing (per liter) 0.27 g $KH_2PO_4$, 0.35 g $K_2HPO_4$, 0.53 g $NH_4Cl$, 0.10 g $MgCl_2 \cdot 6H_2O$, 73 mg $CaCl_2 \cdot 2H_2O$, 20 mg $FeCl_2 \cdot 4H_2O$, and 1 ml of a trace metal solution. One

such solution, "metals 44" [204], contains (per liter) ethylenediaminetetraacetic acid (EDTA) 250 mg; $ZnSO_4 \cdot 7H_2O$, 1095 mg; $FeSO_4 \cdot 7H_2O$, 500 mg; $MnSO_4 \cdot H_2O$, 154 mg; $CuSO_4 \cdot 2H_2O$, 39.2 mg; $Co(NO_3)_2 \cdot 6H_2O$, 24.8 mg; $Na_2B_4O_7 \cdot 10H_2O$, 17 mg, and a few drops of sulfuric acid added to retard precipitation.

3. After being autoclaved for sterilization and $O_2$ removal, maintain the medium under a positive pressure of $N_2$ gas that was previously passed through fine copper ribbons at 300°C to remove traces of $O_2$ (see Chapter 5). When the medium has reached room temperature, add the test organic substrate and reagents designed to encourage specific physiological regimes. These include 1.2 mg $Na_2HCO_3$ and 0.12 mg $Na_2S$ for methanogenesis; 0.12 mg $Na_2S$ and 200 mM $Na_2SO_4$ for sulfate reduction; a 10% (vol/vol) slurry of Fe(III) oxide for iron-reducing conditions; a 5% (vol/vol) slurry of Mn(IV) oxide for manganese-reducing conditions, or 200 mM $KNO_3$ for dissimilatory nitrate reduction. The Fe(III) oxide slurry is prepared by dissolving 108 g of $FeCl_3$ in 1 liter of distilled water and slowly raising the pH to 7 with drops of 10 N NaOH. Wash the resultant floc by multiple centrifugations to remove excess salt. Store concentrated suspension in distilled water at 4°C [64]. The Mn(IV) oxide slurry is prepared by dissolving 3.16 g of $KMnO_4$ and 3.2 g of NaOH in 1 liter of distilled water. Dissolve 5.94 g of $MnCl_2$ in another liter of water. Add the $MnCl_2$ solution slowly to the basic $KMnO_4$ solution while mixing constantly with a magnetic stir bar. Wash and store the resultant floc as described for Fe(III) oxide [64]. Maintain the medium under positive pressure of $N_2$ gas until dispensed to serum bottles.
4. Add the soil, sediment, sewage, or water to the medium. The medium is anaerobically dispensed to serum bottles either within an anaerobic hood or via a gas distribution system that prevents the intrusion of atmospheric $O_2$ (see Chapter 5). Experimental objectives may call for the addition of inocula or other medium components after the bulk of the medium has been dispensed to individual serum bottles. For each experimental treatment (i.e., particular combination of inoculum, test compound, and physiological regime), usually prepare three replicate bottles and one poisoned, autoclaved, or uninoculated control. Bottles containing medium and inoculum but no test compound should be prepared in triplicate.

   *Note*: The inoculum is usually added as a 1:9 dilution with the medium to achieve a 10% slurry. The key issue is creation and maintenance of a homogeneous suspension with a constant composition as subsamples are removed while incubations proceed. Sampling the slurry with needle and syringe should not cause the selective removal of fine or soluble components, thereby changing the solid-to-liquid ratios.
5. To document biodegradation, periodically (for sampling intervals, see Table 15-6) invert the bottles and use a needle and syringe to withdraw 0.5 to 1.0 ml from the slurry. Analyze immediately (via GC, HPLC, or the like; see Table 15-5) or stabilize by freezing or adding chemical inhibitors compatible with the analytical procedures used for later analysis.
6. Do not tacitly assume that the microbial communities used as inocula successfully establish the metabolic regimes sought in the experimental treatments. This must be verified by independent analyses as follows. Dissimilatory nitrate reduction is documented by measuring the loss of nitrate (via ion chromatography or a nitrate-specific electrode), the absence of methane (via GC analysis of headspace gases), and the transient occurrence of $N_2O$ (via GC analysis of headspace gases) and $NO_2^-$ (via ion chromatography). Mn reduction is documented by the visual inspection of bottles—the dark brown Mn(IV) oxide is converted to whitish-colored rhodochrosite ($MnCO_3$)

precipitate. Fe reduction converts the non-magnetic orange-red Fe(III) oxide to a black magnetic precipitate, magnetite ($Fe_3O_4$). Sulfate reduction is documented by measuring the loss of sulfate (via ion chromatography) and an absence of GC-determined methane in headspace gases. Methanogenesis is determined again by GC analysis of headspace gases.

### *Enrichment, Isolation, and Enumeration of Pollutant-Metabolizing Microorganisms from Environmental Samples*

Once procedures described in the previous section demonstrate successfully that biodegradation activity exists somewhere within the complex mixture of laboratory-incubated cells, water, and particulate materials present in an environmental sample, it may be desirable to enter the next phase of biodegradation study—that of isolating pure cultures capable of catalyzing pollutant metabolism (Fig. 15-4, phase 2). Many of the steps and decisions required for implementing enumeration, enrichment, and isolation procedures are identical to those used in implementing biodegradation assays (see Tables 15-6 and 15-7). The successful cultivation of microorganisms able to catalyze particular chemical reactions is based on a mixture of scientific rigor, intuition, experience, and art. Enrichment, isolation, and enumeration procedures rely upon contrived conditions designed to establish resources in laboratory vessels and growth media that allow for the selective growth of an often minuscule proportion of the populations initially present in inocula obtained from soil, water, or sediment in field sites. Investigators must be fully aware of the physiological ramifications of every decision made when preparing enrichment and isolation media [69, 131, 249]. It is the attention to details of media design that usually determines the success or failure of attempts to isolate microorganisms. The subtle microbial physiology issues discussed at length in the overview of techniques and in Tables 15-6 and 15-7 (i.e., What is (are) the mechanism(s) by which microorganisms metabolize the test compound? What physiological rules must be observed? Are the sought physical, chemical, and physiological conditions actually achieved in the experimental apparatus? and so on) apply fully to strategies used to enrich and isolate microorganisms capable of catalyzing particular biodegradation reactions. In general, one must be aware of the fact that every nuance of cultivation procedures necessarily selects for and enriches particular populations from within the array of populations that comprise the sampled microbial community. In addition, both inclusion and exclusion of medium components can sway the results. For instance, omission of a nitrogen source favors microorganisms capable of $N_2$-fixation whereas inclusion of a vitamin mixture allows the growth of organisms lacking vitamin biosynthesis capabilities.

**Preparation of Stanier's Basal Salts Medium** This medium is used for isolating organisms able to use pollutants as carbon and energy sources [204].

*Solution A*

1. Add 67 g $Na_2HPO_4 \cdot 6H_2O$ and 34 g $KH_2PO_4$ to 500 ml distilled water.
2. Adjust pH to 7.25 with 1 M KOH solution and/or KOH pellets.
3. Sterilize by autoclaving, seal, and store at 4°C.

*Solution B*

1. Dissolve 20 g nitrilotriacetic acid in 1600 ml distilled water.
2. Neutralize by adding 14.6 g of KOH.

3. Add 28.9 g $MgSO_4 \cdot 7H_2O$, 6.68 g $CaCl_2 \cdot 2H_2O$, 0.0185 g $(NH_4)_6Mo_7 O_{24} \cdot 4H_2O$, 0.698 g $FeSO_4 \cdot 7H_2O$, 0.25 g $Na_2$ EDTA, 1.1 g $ZnSO_4 \cdot 7H_2O$, 0.154 g $MnSO_4 \cdot H_2O$, 0.04 g $CaSO_4 \cdot 5H_2O$, 0.025 g $Co(NO_3) \cdot 6H_2O$, and 0.018 g $Na_2B_4 O_7 \cdot 10H_2O$. The pH should be $\approx$ 4.6.
4. Adjust to pH 6.6–6.9 by slowly adding about 100 ml 1 M KOH solution.
5. When the proper pH has been reached, filter-sterilize the solution.
6. Seal and store at 4°C.

*Solution C*

1. Add 10 g $(NH_4)_2SO_4$ to 100 ml distilled water.
2. Sterilize by autoclaving.
3. Seal and store at 4°C.

Procedure

1. To prepare 1 liter of sterile medium, autoclave 1 liter of distilled water and aseptically add 40 ml of solution A, 20 ml of Solution B, and 10 ml of Solution C.
2. If deemed appropriate, aseptically add a vitamin solution (see Chapter 5; 238).
3. If agar plates are desired for growth and isolation of colonies, add 15 g of purified agar (e.g., Noble Agar) to the liter of distilled water before autoclaving.

Procedure: Preparation of Enrichment Cultures

1. Dispense 100 ml of medium to a 250-ml Erlenmeyer flask.
2. Inoculate with the environmental sample (e.g., 10 ml of lake water or 1 to 5 g soil or sediment).
3. Add the test substrate to be used as carbon source to the flask. The need to sterilize, the mode of sterilization, the mode of substrate delivery, and its concentration are dependent on toxicity, solubility, and volatility properties of the substrate. Guidelines for common substrates appear in Table 15-7. As a specific example, such volatile solvents as benzene and toluene can be added directly to a small reservoir within the enrichment flask. The internal reservoir may be a cotton-plugged glass test tube that is so tall that it is kept vertical by the sides of the enrichment flask, or it may be a plastic Eppendorf tube suspended beneath the vessel's cap. When using volatile substrates, the flask is sealed (i.e., with a screw cap) to prevent exposure to the benzene and toluene). Soluble, non-volatile substrates can simply be prepared as concentrated stock solutions, filter-sterilized, and added to the medium to reach desired concentrations.
4. Place on a shaker at aeration and temperatures appropriate to experimental objectives.
5. Watch for turbidity in the medium.
6. If appropriate, prepare uninoculated medium to verify that growth is from the inoculum source.

**Preparation of Dilutions** This step may be performed when turbidity is observed in the enrichment flasks (Preparation of Enrichment Cultures, Step 5) or on aseptically prepared dilutions of environmental samples without enrichment culture. Diluent consists of 9- and 90-ml aliquots of phosphate-buffered saline (PBS; 1.18 g $Na_2HPO_4$, 0.22 g $NaH_2PO_4 \cdot H_2O$, 8.5 g NaCl in 1000 ml distilled water, adjust pH to 7.2; see Zuberer [249] for several alternative diluents) autoclaved in screw-capped test tubes or bottles.

Procedure

1. Prepare a $10^{-1}$ dilution by transferring 1 or 10 ml of enrichment culture to 9 or 90 ml of PBS, respectively. When environmental samples (water, soil or sediment) are being processed, known volume or dry or wet weights should be used.
2. Vortex mix to disperse cells and particles.
3. Sequentially repeat this dilution step until a $10^{-6}$- or $10^{-7}$-fold dilution series has been prepared.
4. Use spread or pour-plate procedures to dispense 0.1 ml or 1.0 ml, respectively, of each dilution to the substrate-ordered agar growth medium in two to six replicated plates per dilution.
5. Invert plates and prevent desiccation during incubation at a temperature appropriate for experimental objectives.
6. If the organic substrate is volatile enough to be delivered to cells in the vapor phase, divide the plates from each dilution between two vessels (ideally glass ones, though plastic tubes have been found suitable) either with or without the substrate, for example, naphthalene crystals in an open dish or benzene liquid in a cotton-plugged test tube.
7. Periodically inspect plates for colony formation. Control plates that are inoculated but lack the added growth substrate should contain colonies (growing on traces of organics in the agar or scavenged from the air); however, cells using the provided growth substrate should be large and robust and thus easily distinguished from the background of small pinpoint colonies.

Procedure: Calculation of the Number of Biodegrading Microorganisms

1. Inspect the plates from the dilution series and select the dilution containing between 30 and 300 colonies for quantifying the number of biodegrading microorganisms. In deciding whether or not to count a given colony, only select those that are larger and have grown more quickly than the colonies on the no-added-carbon control plates.
2. To calculate the abundance of biodegradative organisms, take the average and standard deviation from the dilution that falls between 30 and 300 cells per plate and then multiply by the inverse of the dilution while accounting for the volume of plated diluent. Thus, if a dilution series was performed on contaminated groundwater and 0.1 ml of diluent at the $10^{-2}$ dilution contained 120 colonies, the reported biodegrading cell abundance is $1.2 \times 10^5$ per ml.

Procedure: Isolation of Biodegrading Cultures

1. Confirm colony purity and the cultures ability to utilize the substrate by picking them and restreaking them on the medium, with and without a carbon source.
2. During this purification and isolation stage, include a transfer to a general heterotrophic medium, such as 5% PTYG (0.6 g $MgSO_4 \cdot 7H_2O$, 0.07 g $CaCl_2 \cdot 2H_2O$, 0.25 g peptone, 0.25 g Trypticase, 0.5 g yeast extract, and 0.5 g glucose, 15 g agar). However, prolonged growth on rich media is not recommended because this may result in the loss of catabolic plasmids. In this way, cultures may lose their ability to metabolize the carbon source of interest.
3. Further verify the ability of isolated microorganisms to metabolize the pollutant compound(s) in aqueous media using radiotracer or analytical chemistry assays described in the overview section.

### *Cell-Free Enzyme Activity Assays for Key Biodegradation Steps*

As outlined by Dagley and Chapman [47], a variety of methodologies and strategies must be tapped when determining the metabolic pathways used by microorganisms during pollutant biodegradation. The methodologies include isolation of chemical intermediates from culture fluids, the use of metabolic inhibitors, partial catabolism of substrates and their analogs, simultaneous adaptation (sequential induction) experiments, the use of radiotracers, and the use of cell-free extracts. Key to interpreting the results of this constellation of approaches is a thorough understanding of the strengths and pitfalls of each method. Also essential is a rigorous conceptual view of microbial physiology that considers alternative hypotheses for explaining the results of each method and seeks consistency in their respective data. A thorough discussion of these issues is beyond the scope of this chapter, but readers are referred to Dagley and Chapman [47], Gibson [65], and Focht [60] for guidance. For the purposes of this section, it is necessary to acknowledge that examining transformations of pollutant compounds in the absence of whole microbial cells leads toward a reductionistic biochemical understanding of the mechanisms by which microorganisms degrade pollutant compounds (Fig. 15-4). Refined biochemical information is derived from cells that are broken or lyzed so that their interior contents (especially soluble, inducible enzymes) can be separated from other cellular constituents. By progressing from experiments using crude cell extracts to those using highly purified enzymes [174], the detailed mechanisms of enzyme-substrate binding, co-factor interactions, and metabolic regulation can be pieced together.

The enzyme assay using crude cell-free extracts is included here because such information adds a level of redundancy and clarity to descriptions of biodegradation processes. After pure cultures capable of metabolizing environmental pollutants have been obtained, an investigator is best served by tying the capabilities of new site-derived microorganisms to the existing pool of knowledge about the process of interest (see Fig. 15-4, Table 15-8, and Chapter 13 for corresponding information about nucleic acid sequences).

As indicated by accepted texts in biochemical methodology, the essential aim of enzyme assays is to create an in vitro context where catalysis can be discerned through simple chemical measurements, such as spectrophotometry. Control assays, in which the crude or highly purified cell extract is boiled to achieve enzyme denaturation, are performed to demonstrate that the catalytic agent is heat labile, a trait common to enzymes derived from non-thermophilic microorganisms. Additional control assays often include preparations in which the enzyme or its substrate is omitted from the reaction mixture. The assay described here is a general example of cell-free enzyme assays and is based on procedures from Gibson [65], Focht [60], and Shields et al. [192].

The enzymes that are best characterized and most pertinent to biodegradation reactions are those involved in the catabolism of aromatic compounds. As elaborated by Gibson [65], not only is the reaction between molecular oxygen and aromatic compounds of major importance to the biogeochemical cycling of carbon and oxygen but also the mechanism of aromatic ring fission (either "ortho" between hydroxyl groups; or "meta" at the bond adjacent to one of the hydroxyl groups) has significant taxonomic implications for the *Pseudomonads*.

Procedure: Cell-Free Enzyme Assay for Catechol-2,3-dioxygenase (meta-pyrocatechase; EC 1.13.112)

1. Grow cells in suitable liquid medium containing a carbon source selected for sound physiological reasons (i.e., designed to regulate catechol-2,3-diox-

ygenase activity). One liter of medium (such as that described in the protocol for the isolation of aerobic pollutant-degrading microorganisms) shaken in a 2-liter flask should provide a sufficient quantity of cells. Incubate under appropriate aeration and temperature conditions.

2. Harvest, wash, and lyse cells from the late exponential growth phase. Using a refrigerated centrifuge, spin down the cells (6000 *g* for 10 min), and discard the supernatant. Resuspend in 300 ml potassium dihydrogen phosphate buffer (0.05 M, pH 7.5). Centrifuge, decant, and resuspend twice again in 300 ml of buffer. After the fourth and final centrifugation, resuspend in 10 ml of buffer. Cool the turbid cell suspension in an ice bath, and disrupt the cells with a sonic probe. The duration of sonic bursts should be 15 s alternating with equal periods of swirling the suspension in the ice bath. Repeat 8 to 12 times to facilitate the lysis of Gram-negative bacteria. The tendency of Gram-positive bacteria to resist cell lysis may be overcome by adding an equal volume of glass beads (less than 10-μm diameter) to the cell suspension before sonication. Other more rigorous procedures, such as a French pressurized cell (25 Mpa), may also be employed. Successful cell lysis is suggested by an increased uniform viscosity in the sonicated or pressurized fluids.
3. To separate cell fragments from aqueous-phase intracellular components, centrifuge at 40,000 *g* at 4°C for 40 min. Assay a small aliquot of the supernatant for protein content (e.g., the Bradford Coomassie Brilliant Blue dye binding procedure; Bio-Rad protein assay kit, Melville, NY). The protein concentration should be approximately 10 $mg \cdot ml^{-1}$. Such cell-free preparations can be further purified for detailed enzyme characterization [174] or used as crude extracts via the spectrophotometric enzyme activity assay.
4. Assay for catechol-2,3-dioxygenase by diluting (0.05 M potassium dihydrogen phosphate, pH 7.5) the cell-free enzyme extract to achieve a concentration of approximately 0.1 mg protein per ml. To a cuvette with a 1-cm light path, add 2.8 ml of phosphate buffer and 0.1 to 1.0 ml of the cell-free enzyme extract, depending on the achieved enzyme concentration. Mix by inverting, and hold for up to 2 min to allow temperature equilibration. Add 0.1 ml of a freshly prepared catechol solution (0.01 M in deionized water, methanol, or N, N-dimethyl formamide). Immediately monitor optical density in a UV-visible spectrophotometer at 375 nm. Formation of the yellow ring fission product, 2-hydroxymuconic semialdehyde, is rapid. The rate may diminish at optical density readings above 0.1. The activity can be determined from the slope of a graph of product formation versus time. One unit of enzyme is defined as the amount that oxidizes 1 μmole of catechol per minute. A molar extinction coefficient for 2-hydroxymuconic semialdehyde of 33,000 $cm^{-1} \cdot M^{-1}$ (at pH 7.5) is used for quantifying its concentrations. After protein determination (Step 3), compare the specific activity in enzyme units per mg cell protein. Control assays (keeping reaction volumes constant between treatment and controls) consist of preparations using boiled cell-free enzyme extract and that substitute buffer for both the cell-free extract and the catechol solution.

### Step 3: Strategies for Demonstrating the Expression of Biodegradation Activity in the Field

Three sources of uncertainty must be confronted and overcome when demonstrating that microorganisms are the active agents of pollutant loss in bioremediation projects. First, we must acknowledge that extrapolation from laboratory-based metabolic activity assays to the field is usually impossible because of the propensity of microorganisms in field samples to respond to laboratory-imposed physiological conditions that are unlikely to perfectly match those in the field [130]. Second, the spatial heterogeneity of field sites may impede or completely prevent trends in the behavior of environmental contaminants from being discerned. Third, the action of a multitude of both abiotic and biotic processes may simultaneously contribute to pollutant attenuation [129]. To contend with these challenges, a variety of strategies have been developed for verifying the success of bioremediation efforts in truly activating pollutant-destroying microbial processes in field sites. Perhaps the simplest of the approaches is to release $^{14}C$-labeled contaminants into field sites and then conduct assays analogous to laboratory ones described earlier. Despite obvious obstacles from regulatory agencies (spurred by human health concerns), such field releases have occasionally been reported (e.g., [119]). In general, however, less hazardous approaches have been implemented, as discussed in this section.

### *Use of Conservative Tracers*

Conservative tracers are materials with chemical and transport properties that match those of microbiologically reactive chemicals (contaminants, electron acceptors, metabolites, respiratory products), but the tracers themselves are not microbiologically reactive. Conservative tracers may be intentionally released by the bioremediation practitioner or may be fortuitously present in the contaminants. By measuring concentrations of the tracer and microbiologically active chemicals, their ratios can be computed. Because tracers account for transport-related abiotic attenuation processes, changes in contaminants, reactants, or products relative to the tracer indicate in situ biodegradation for otherwise stable compounds.

Internal tracers are materials fortuitously present in contaminant mixtures; their presence in field site samples can be used to deduce the loss of biodegradable materials. Analyses of contaminants in site-derived samples can document diminishing contaminant concentrations, but not directly identify the cause. However, if mixtures of contaminants are present, then microbiological causality can sometimes be established by measuring the ratios of biodegradable to nonbiodegradable organic compounds. Phytane, a branched alkane molecule 20 carbon atoms long that occurs in crude oil, is an example of an organic compound that can serve as a conservative tracer [13, 14, 168]. Because of its branched-chain structure, phytane is resistant to microbial attack compared to the straight-chain compound, octadecane. These two molecules have very similar molecular weights, have shared volatility and transport characteristics, and are likely to undergo nearly identical abiotic reactions. Therefore, phytane is a valid conservative tracer for in situ octadecane biodegradation. A decrease in the ratio of straight to branched-chain members of this pair of molecules is proof of in situ microbial attack of the octadecane. A slight drawback of this approach is the fact that naturally occurring microbial populations often eventually attack phytane, thus causing the octadecane:phytane ratio to underestimate in situ biodegradation rates.

Other strategies for using the ratio of degradable to "nondegradable" contaminants have been implemented successfully. These include ratios of heptadecane to pristane [13, 14, 168], ratios of aerobically metabolizable to resistant PCB congeners [76], and ratios of crude oil components to hopanes [30, 166, 212]. In each case, the specific biochemistry of the reactions, the chemistry of contaminant mobility, and the analytical chemistry of field site-derived samples must be well understood.

Recent metabolic evidence has shown that, in addition to being isomer-specific, microbial metabolism can be stereospecific. Some organic contaminants occur as racemic mixtures of stereoisomers. Hexachlorocyclohexane, for example, exists in two different chiral forms, only one of which is readily metabolized by microorganisms [38, 55, 248a]. Thus, chemical analyses documenting the selective disappearance of α-hexachlorohexane relative to its more resistant enantiomer is proof of in situ biodegradation. This type of rationale is contaminant-specific and requires substantial prior biochemical and physiological knowledge, but it illustrates an important principle of potential practical importance for bioremediation projects.

The use of external tracers, conservative compounds added to field sites after a pollution event has occurred, represents another approach for documenting in situ biodegradation activity. In a variety of engineered bioremediation approaches, it is common to supplement circulating waters or bioreactors or gas mixtures with oxygen. Such circumstances lend themselves well to the inclusion of a gaseous conservative tracer, He, along with the oxygen source. Monitoring on-site concentrations of He provides a basis for understanding in situ transport and dilution processes. By simultaneously measuring in situ concentrations of $O_2$ and $CO_2$, changes relative to He can be computed. Over gradients of time and space, in situ contaminant respiration is indicated by a diminishing $O_2$:He ratio and an increasing $CO_2$:He ratio [149].

In circumstances where reactants or products of biodegradation are non-gaseous, non-gaseous conservative tracers can be used. As an example, if nitrate is the electron acceptor in a biodegradation reaction, the conservative anion, bromide, may be fortuitously present or be intentionally released so that nitrate: bromide ratios can be traced over time and space at the site. These data can be interpreted in a manner similar to $O_2$:He ratios and may indicate in situ respiration of contaminants. However, because nitrate can be metabolized as a nitrogen source and as a final electron acceptor, this approach is of dubious applicability under N-limited site conditions. Nonetheless, the principle of using non-gaseous conservative tracers to demonstrate the relative disappearance of biodegradation reactants remains valid.

In situ toluene biodegradation has recently been documented by injecting both gaseous (propane) and non-gaseous (NaCl) tracers into a toluene-contaminated stream [106]. Actual stream concentrations of toluene were always significantly lower than those predicted when only dilution and volatilization were considered. The difference was attributed to biodegradation.

### *Stable Isotope Fractionation*

Stable isotope fractionation patterns can also provide convincing evidence of in situ biodegradation. This strategy relies on signature isotopic ratios and mass spectrometric analyses to distinguish biotic from abiotic processes that affect contaminants and their metabolic products. Unlike most abiotic processes, enzymatic processes often selectively act on compounds bearing the lighter isotopes. Thus, reactants become selectively depleted in lighter isotopes, and products are correspondingly enriched [72, 91]. The analyses are elaborate, expensive, and only pertinent if the contaminant source bears a characteristic ratio of stable isotopes (e.g., $^{13}C/^{12}C$). However, given the proper circumstances, isotopic fractionation can document the in situ fate of contaminants through metabolites, $CH_4$ and $CO_2$ [2]. Under special circumstances, stable isotopic tracers should theoretically be useful in documenting the fate of contaminants in food chains. These approaches are particularly advantageous because the results are self-evident, and there may be no need to sample adjacent uncontaminated areas to evaluate relative responses.

A research-oriented version of the use of stable isotopic ratio for verifying bioremediation would use the analytical procedures and rationale required for documenting $^{13}C/^{12}C$ (or other) ratios, but impose an isotopic signature on site contaminants by amending the site with isotopically labeled forms of the contaminant. For these purposes, the amended contaminants can be synthesized using stable isotopes, such as deuterium or $^{13}C$. The imposition of isotopic signatures may serve two purposes. If expected transformation products (metabolic intermediates or $CO_2$) are found to carry the isotopic signature, then the reaction has actually occurred in the field site, and in situ biodegradation has been proven. Alternatively, predictable enzyme-selective isotopic fractionation patterns may also be discerned. Among the major drawbacks to adding isotopically labeled pollutant compounds to contaminated sites are that pollutant burden is increased and time-dependent transport, mixing, and sequestration reactions may prevent the behavior of the freshly added contaminant from being indicative of the unlabeled contaminants. These aging, sorption, and sequestration issues have been raised [7, 46, 77, 135, 162, 163, 170, 177, 183, 205].

### *Detection of Intermediary Metabolites*

GC/MS determination of compounds that are produced uniquely during microbial metabolism of particular contaminants is an insightful and elegant procedure for demonstrating in situ biodegradation activity. In using organic contaminant-specific analyses to document in situ bioremediation, the sampling and analytical strategies are guided by biochemical knowledge pertinent to the metabolism of specific contaminants. By understanding metabolic pathways and recognizing that microbiological processes may transform contaminant compounds into unique intermediary metabolites, evidence for in situ biodegradation can be obtained. For instance, during co-metabolic microbial attack of trichloroethylene (TCE), trans-dichloro-ethylene oxide may be produced [187]. Similarly, under anaerobic conditions chlorinated ethenes may undergo reductive dehalogenation reactions to yield vinyl chloride and ethylene [23, 189, 243a]. Detection of these compounds, usually of unique microbiological origin, in field site samples serves to indicate in situ biodegradation. To ensure the validity of these approaches, the intermediary metabolites cannot have been present in the contaminants originally released, and they should be absent in adjacent uncontaminated areas. The uncontaminated "control" site is essential for the rationale of this approach. If site heterogeneity or contaminant distributions preclude measurements in a control area that is virtually identical to the contaminated area, except for the contamination, this approach is not valid.

Whenever unique intermediary metabolites are targeted for analysis in field samples, knowing their stability is essential for valid interpretation of the data. When microbiologically stable "dead end" metabolites are found in field samples, one can conclude that metabolic attack has occurred in situ, but at uncertain times and locations. By contrast, Wilson and Madsen [234] have proposed that detection of unstable transient metabolites in field samples, handled in ways designed to prevent artifacts, can serve as evidence for in situ biodegradation in real time. Field extraction of intermediary metabolites has been applied to the biodegradation of PCBs [57], TCE [187], naphthalene [234], and ankylbenzines [23a, b] among others. It is appealing to speculate that in situ rates of biodegradation reactions would be proportional to field concentrations of intermediary metabolites. However, rigorous field calibration verifying such relationships have not been completed.

### *About Molecular Procedures for Detecting Biodegradation Genes and Gene Expression*

The science and technology of bioremediation are influenced by a broad diversity of disciplines. Molecular biology, seemingly distant from contaminated field sites, is having an increasingly important impact on the theory of bioremediation. Chapter 13 provides a detailed description of the variety of molecular procedures that are relevant to environmental microbiology [see also 3, 19, 34, 35, 181, 200, 201]. Molecular biology provides highly refined insights into the regulation and expression of biodegradation genes (see Fig. 15-4). This fundamental knowledge has the potential to explain how microorganisms may react to complex mixtures of environmental pollutants and their metabolites, which may serve as inducers or repressors of gene expression. Molecular aspects of pollutant metabolism can also provide insights into the evolution of a catabolic function via comparative phylogenetic studies [12, 74, 221, 233, 243]. Molecular biology also provides procedures both to design new biodegradation pathways and demonstrate in situ gene expression via mRNA analysis [59, 90, 115, 156, 160, 179, 184, 234a]. A brief overview of compounds, enzyme systems, bacterial strains, and investigator's contributions to the elucidation of biochemical and genetic aspects of pollutant metabolism is presented in Table 15-8.

Despite its elegance and powerful tools [34], molecular biology currently has a limited impact on the practice of bioremediation. The limitations are caused by (1) the relatively small database of genetic sequences pertinent to pollutant metabolism and (2) uncertainties about the real-world relevance of those genetic sequences because they were obtained from a small number of domesticated laboratory microorganisms. At issue here is metabolic diversity and whether or not genes of unrelated lineage have converged on the same metabolic function. Physical/chemical impediments to efficient unbiased extraction, amplification, and analysis of nucleic acids from environmental samples also limit molecular biology's impact [80, 92, 142, 155; See [130] for a general discussion of these issues].

Regardless of those current limitations, future prospects are quite bright for gaining insights into the genetic and molecular aspects of field biodegradation processes. As indicated above and in Table 15-3, molecular procedures measuring mRNA and enzymes involved in gene expression in situ may soon be added to the criteria for documenting field biodegradation of environmental contaminants. Major forays using molecular procedures on model biodegradation systems and in field situations have already been completed. These include work on 2,4-D metabolism in soil (examining plasmid pJP4) [93, 98–100, 219]; on chlorobenzoate metabolism (examining transposon Tn5271 in a groundwater treatment system) [242]; on naphthalene catabolism in coal-tar contaminated sediments or soil [59, 80, 80a, 179]; in TCE-contaminated sediments undergoing field co-metabolic treatments by methanotrophs [78, 161a]; and on the use of the light-producing lux cassette in combination with catabolic genes to report both the presence of environmental contaminants and the physiological activity of degrading populations [36, 37, 107, 118, 179].

### *Replicate Field Plot Responses for Distinguishing Biotic and Abiotic Contaminant Loss*

Another approach for demonstrating in situ biodegradation relies on the fact that rates of microbial processes can be enhanced by relieving nutrient or other limitations in field plots. Abiotic processes operating in comparable field plots are not likely to respond to the additions of nutrients or microorganisms. Thus, when nutrients or inocula or both are added to one set of field plots, but not another, relative enhancement of in situ biodegradation should result. The contrast in contaminant loss between enhanced and unenhanced treatments can be attributed to in situ biodegradation. To apply this approach to field sites requires a hydrogeological setting uniform enough for the preparation of comparable replicated test plots. A variety of studies have shown that contaminant metabolism can be ac-

**Table 15–8** Selected Compounds, Enzyme Systems, and Microorganisms That Have Served to Elucidate Biochemical and Genetic Aspects of Contaminant Metabolism

| Substrate | Final Electron Acceptor | Enzyme System | Strain Designations | References[a] |
|---|---|---|---|---|
| Benzene | $O_2$ | Benzene dioxygenase | *Pseudomonas putida* | 97 |
| Benzoate | $O_2$ | Benzoate dioxygenase | *Acinetobacter calcoaceticus* | 150 |
| Toluene | $O_2$ | Toluene dioxygenase | *P. putida* F1 | 250 |
| | $O_2$ | Toluene methyl group monooxygenase | *P. putida* mt-2 | 241 |
| | $O_2$ | Toluene *para* monooxygenase | *P. mendocina* KR1 | 244 |
| | $O_2$ | Toluene *ortho* monooxygenase | *P. cepacia* G4 (*Burkholderia*) | 192, 193 |
| | $O_2$ | Toluene *meta* monooxygenase | *P. picketti* PK01 | 112 |
| | $NO_3^-$ | Biochemistry and genetics uncertain | *Pseudomonas* strain T and strain T1 | 50, 54 |
| | $Fe^{3+}$ | Biochemistry and genetics uncertain | Strain GS15 (*Geobacter metallireducens*) | 125 |
| | $SO_4^-$ | Biochemistry and genetics uncertain | Strain Tol2 | 169 |
| Biphenyl | $O_2$ | Biphenyl dioxygenase | *Pseudomonas* strain LB400 | 53, 73 |
| Phenol | $O_2$ | Phenol monooxygenase | *Pseudomonas* sp. EST1001 | 153 |
| Naphthalene | $O_2$ | Naphthalene dioxygenase | *P. putida* strains G7, NCIB 9816-4, OUS82, and *Comamonas testosteroni* strain GZ39 | 49, 70, 113, 196, 214 |
| Methane | $O_2$ | Methane monooxygenase | *Methylococcus capsulatus* Bath | 202 |

[a]For appropriate DNA probes, sequences, plasmids, or metabolic pathways.

celerated in amended field plots [30, 114, 132, 168, 171, 199, 230] through (1) nutrient amendments (N, P), (2) final electron acceptor amendments (e.g., $O_2$, $SO_4^-$, $NO_3^-$), (3) co-metabolite amendments (e.g., biphenyl for enhancing PCB utilization and $CH_4$ or phenol for enhancing TCE oxidation), and (4) the inoculation of microorganisms.

### *Documenting Adaptation of Indigenous Microorganisms to Contaminants Using Profiles of Contaminant-Specific Populations and Consumption of Co-Reactants over Time or Space*

Protozoan predation of actively growing bacteria and metabolic adaptation are potentially valuable field microbiological measures of in situ biodegradation. If both of these two phenomena can be documented at the same site, then in situ biodegradation is indicated [134]. To document these phenomena, cores at a relatively uniform site, from inside and outside the contaminant plume, need to be divided so that each can be quantitatively assayed for numbers of vegetative protozoa and biodegradation potential. If relatively high numbers of vegetative protozoa are found inside the contaminant plume, in situ growth of bacterial prey is indicated. Then if the microbial community inside the contaminant plume is also found to metabolize the contaminant relatively rapidly, in situ bacterial growth can be attributed to the contaminants. The uncontaminated "control" site is essential for the rationale of this approach. If site heterogeneity or contaminant distributions preclude measurements in a control area (that is virtually identical to the contaminated area, except for the contamination) this approach is not valid.

In an approach related to the documentation of metabolic adaptation, contrasts between microbial population and activity parameters before and after bioremediation can also demonstrate in situ biodegradation. The rationale for this type of proof relies on predictable responses of microbial communities to specific site manipulations aimed at stimulating in situ biodegradation processes. If baseline measures of site chemistry and microbiology can be established before implementing the bioremediation plan, then departures from the baseline data should be consistent with the amounts of contaminant thought to be metabolized. Among key parameters to be measured are total microbial numbers and biomass; numbers of microorganisms capable of degrading the contaminant(s); biodegradation potential; activity levels of biodegradative enzymes, mRNA transcripts, and genes; numbers of protozoans; and masses of co-reactants (nutrients, electron donors, electron acceptors, and the like) that were consumed during the bioremediation period. This approach was used in a very thorough manner in a large-scale field demonstration of methanotrophic cleanup of TCE-contaminated aquifer material [78, 161a].

Field concentrations of co-reactants and products of microbial activity ($CO_2$, $Fe^{3+}/Fe^{2+}$, $NO_3^-/NO_2^-$, $CH_4$, $O_2$, $Cl^-$) provide key insights into the metabolic status of native microbial communities in field sites. Co-reactant profiles can reveal in situ biodegradation activity in both intrinsic and engineered bioremediation projects. For instance, contaminant-stimulated in situ respiration can be documented when measurements show a depletion of final electron acceptor along site profiles containing the contaminant and when the depletion is absent in adjacent uncontaminated profiles. When data showing in situ respiration can be coupled to a historical trend of contaminant attenuation (based on sequential chemical analyses of samples taken from fixed locations within the respiration zone), then in situ biodegradation can be inferred. In addition to spatial gradients of final electron acceptors, dynamic in situ time-course respiration assays can also be sought. These assays are perhaps best suited to engineered bioremediation projects undergoing air sparging [81–83, 149]. Immediately after terminating the flow of sparged gases, an oxygen probe is lowered into groundwater wells to measure rates of oxygen consumption. To distinguish the respiratory demand of the contaminants from ambient background respiration rates, comparable in situ respiration should be measured in adjacent uncontaminated wells. Relatively rapid oxygen loss in the contaminated area can be attributed to in situ microbial metabolism of the pollutant, when analytical data confirm disappearance of the contaminant(s). An innovative improvement in the in situ time-course respiration assay uses helium gas as a conservative tracer so that abiotic mechanisms of $O_2$ depletion (i.e., volatilization, dilution, and reaction with reduced substances) do not confound data interpretation. The helium-tracer procedure incorporates a known concentration of helium gas into the air used to sparge the contaminated zone. Instead of determining time courses that simply show the rate of $O_2$ loss, the rate of oxygen depletion relative to the helium tracer is used to indicate in situ microbial respiration. A corresponding determination is performed in adjacent uncontaminated wells. On-site analysis of helium concentrations in the water samples is performed using a portable GC, and analytical data from monitoring points are needed to confirm contaminant loss.

In each instance where the measurement of oxygen disappearance has been recommended above, measurement of $CO_2$ production could theoretically also be used. As microbial communities consume oxygen and mineralize contaminants, carbon dioxide is produced. Static vertical profiles or dynamic time courses of carbon dioxide enrichment indicate in situ respiration. A link tying the $CO_2$ production to the loss of contaminant is established by finding enhanced respiration rates in wells inside the contaminant plume compared with outside and by finding diminishing concentrations of contaminants on site. Gas chromatography is the method of choice for determining $CO_2$ concentrations in field samples. However, this approach is inappropriate where the bicarbonate buffering capacity of the site completely masks respiratory $CO_2$ production. In interpreting all in situ respiration and other measurements designed to document microbiological reactions, it must be acknowledged that alternative, abiotic mechanisms of substrate depletion (e.g., $O_2$ depletion via reactions with $Fe^{2+}$) may occur. This fact emphasizes the need to seek multiple lines of independent evidence in interpreting the behavior of contaminants in field sites. Furthermore, site-specific spatial heterogeneity may prevent easily interpretable patterns of oxygen (and other co-reactants) from being discerned.

Finally, using logic analogous to the type that links field gradients of depleted final electron acceptors to the loss of contaminants that serve as microbial carbon and electron sources, in situ co-metabolism can be documented by showing a mechanistic link between the disappearance of co-metabolizable contaminants and the pulses of both electron donor and electron acceptor delivered to the site. This approach is typically accomplished in an engineered bioremediation setting. The data must be derived from contaminated zones where analyses prove that contaminants persist when electron donor or acceptor is supplied singly, but concentrations of electron donor, acceptor, and contaminant(s) diminish when electron donor and acceptor are supplied simultaneously [78, 95, 161a, 184a–188].

## Confirming Biodegradation and Bioremediation Using Computer Modeling Strategies

The essential goal of the computer modeling strategies for proving in situ biodegradation is to show that all known abiotic processes are unable to account for empirically observed rates of contaminant disappearance. A microbiological component is therefore implicated. Computer models provide a qualitative and quantitative framework for integrating relevant site and contaminant characteristics so that rates and extents of contaminant transport can be predicted. The response of transport models to a range of values for key modeling parameters can be compared to actual on-site contaminant distributions. In situ biodegradation is implicated by the model when the field data make sense only after contaminant losses to biodegradation processes have been included in the model. A drawback of this approach is that the validity of each model must be evaluated on a site-by-site basis. Determining each of the many modeling parameters (i.e., hydraulic conductivity, hydraulic gradient, contaminant sorption, etc.) may be as demanding as meeting the more direct criteria of the other strategies for demonstrating bioremediation. Nonetheless, computer modeling of contaminant masses and their transport and fate has played and will continue to play a key role in enhancing both our understanding of contaminant behavior in field sites and the role of microbial processes in that behavior [22, 39, 41, 48, 148, 164, 222]. Of particular interest is the prospect that quantitative modeling will provide insights into rates of in situ biodegradation.

## CONCLUDING REMARKS

As is evident from introductory remarks in the first two sections of this chapter, bioremediation is a technology that is reliant on the fundamentals of microbiology and biochemistry that govern microbial metabolic processes. However, advances in the use and applications of bioremediation technology are also tied to such diverse disciplines as chemical engineering, analytical chemistry, hydrogeology, and environmental law. This chapter has attempted to provide the reader with step-by-step protocols for practicing techniques in the science of biodegradation while simultaneously conveying an understanding of the broad scope of information that influences the technology of bioremediation.

*Acknowledgments* During preparation of this manuscript, the author's research laboratory was supported by a grant from the Cornell Biotechnology Program, which is sponsored by the New York State Science and Technology Foundation (grant NYS CAT 92054), a consortium of industries, and the National Science Foundation. Additional support was provided by the Air Force Office of Scientific Research (grants AOFSR-91-0436, F49620-93-1-0414, and F49620-95-1-0346); the National Institute of Environmental Health Sciences (grant 08-P2E505950A); Merck and Co., Inc. (grant PY-641308); and the New York State Center for Hazardous Waste Management. Expert manuscript preparation from P. Lisk is greatly appreciated.

## REFERENCES

1. Adriaens, P. and Vogel, T.M. 1995. Biological treatment of chlorinated organics. In Microbial Transformation and Degradation of Toxic Organic Chemicals (Young, L.Y. and Cerniglia, C.E., Eds.), pp. 435–486. Wiley-Liss, New York.
2. Aggarwal, P.K. and Hinchee, R.E. 1991. Monitoring in situ biodegradation of hydrocarbons by using stable carbon isotopes. Environ. Sci. Technol. 25:1178–1180.
3. Akkermans, A.D.L., van Elsas, J.D., and de Bruijn, F.J. 1995. Molecular Ecology Manual. Kluwer Academic Publishers, Norwell, MA.

4. Alef, K. and Nannipieri, P., Eds. 1995. Methods in Applied Soil Microbiology and Biochemistry. Academic Press, San Diego.
5. Alexander, M. 1981. Biodegradation of chemicals of environmental concern. Science 211:132–138.
6. Alexander, M. 1994. Biodegradation and Bioremediation. Academic Press, New York.
7. Alexander, M. 1995. How toxic are toxic chemicals in soil? Environ. Sci. Technol. 29:2713–2717.
8. Alexander, M. and Scow, K.M. 1989. Kinetics of biodegradation in soil. In Reactions and Movement of Organic Chemicals in Soils (Sawhney, B.L. and Brown, K., Eds.), pp. 243–269. Soil Science Society of America, Madison, WI.
9. Amann, R.I., Ludwig, W., and Schleifer, K.-H. 1995. Phylogenetic identification and in situ detection of individual microbial cells without cultivation. Microbial. Rev. 59:143–169.
10. American Public Health Association. 1992. Standard Methods for the Examination of Water and Wastewater, 18th ed. APHA, Washington, DC.
11. Anderson, J.P.E. 1982. Soil respiration. In Methods of Soil Analysis, Part 2 (Page, A.L., Miller, R.H., and Keeney, D.R., Eds.), pp. 831–871. Soil Science Society of America, Madison, WI.
12. Asturias, J.A., Diaz, E., and Timmis, K.N. 1995. The evolutionary relationship of biphenyl dioxygenase from Gram-positive *Rhodococcus globerulus* P6 to multicomponent dioxygenases from Gram-negative bacteria. Gene 156:11–18.
13. Atlas, R.M. 1981. Microbial degradation of petroleum hydrocarbons: an environmental perspective. Microbiol. Rev. 45:180–209.
14. Atlas, R.M. 1992. Oil spills: regulation and biotechnology. Curr. Opin. Biotechnol. 3:220–223.
15. Atlas, R.M. and Cerniglia, C.E. 1995. Bioremediation of petroleum pollutants. Bioscience 45:332–338.
16. Babu, G.R.V., Wolfram, J.H., and Chapatwala, K.D. 1992. Conversion of sodium cyanide to carbon dioxide and ammonia by immobilized cells of *Pseudomonas putida*. J. Indust. Microbiol. 9:235–238.
17. Bak, F. and Widdel, F. 1986. Anaerobic degradation of phenol and phenol derivatives by *Desulfobacterium phenolicum* new species. Arch. Microbiol. 146:177–180.
18. Barkay, T., Liebert, C., and Gillman, M. 1989. Environmental significance of the potential for mer(Tn21)-mediated reduction of $Hg^{2+}$ to $Hg^0$ in natural waters. Appl. Environ. Microbiol. 55:1196–1202.
19. Barkay, T., Nazaret, S., and Jeffrey, W. 1995. Degradative genes in the environment. In Microbial Transformation and Degradation of Toxic Organic Chemicals (Young, L.Y. and Cerniglia, C.E., Eds.), pp. 545–578. Wiley-Liss, New York.
20. Bartha, R. and Pramer, D. 1965. Features of a flask and method for measuring the persistence and biological effects of pesticides in soil. Soil Sci. 100:68–70.
21. Bedard, D.L. and Quensen, III, J.F. Microbial reductive dechlorination of polychlorinated biphenyls. In Microbial Transformation and Degradation of Toxic Organic Chemicals (Young, L.Y. and Cerniglia, C.E., Eds.), pp. 127–216. Wiley-Liss, New York.
22. Bedient, P.B. and Rifai, H.S. 1993. Modeling in situ bioremediation. In In situ Bioremediation: When Does It Work?, pp. 153–159. National Academy Press, Washington, DC.
23. Beeman, R.E., Howell, J.E., Shoemaker, S.H., Salazar, E.A., and Buttram, J.R. 1994. A field evaluation of in situ microbial reductive dehalogenation by the biotransformation of chlorinated ethenes. In Bioremediation of Chlorinated and Polycyclic Aromatic Hydrocarbon Compounds (Hinchee, R.E., Leesen, A., Semprini, L., and Ong, S.K., Eds.), pp. 14–27. Lewis Publishers, Boca Raton, FL.
23a. Beller H.R., Ding, W.H., and Reinhard, M. 1995. Byproducts of anaerobic alkylbenzene metabolism useful as indicators of in situ bioremediation. Environ. Sci. Technol. 29:2864–2470.
23b. Beller, H.R. and Spormann, A.M. 1997. Anaerobic activation of toluene and *o*-xylene by addition to fumanate in identifying strain T. J. Bacteriol. 179:670–676.

24. Bewley, R.J.F. 1996. Field implementation of in situ bioremediation: key physicochemical and biological factors. In Soil Biochemistry (Stotzky, G. and Bollag, J.-M., Eds.), pp. 473–541. Marcel Dekker, New York.
25. Borden, R.C., Gomez, C.A., and Becker, M.T. 1995. Geochemical indicators of intrinsic bioremediation. Ground Water 33:180–189.
26. Bossert, I.D. and Compeau, G.C. 1995. Cleanup of petroleum hydrocarbon contamination in soil. In Microbial Transformation and Degradation of Toxic Organic Chemicals (Young, L.Y. and Cerniglia, C.E., Eds.), pp. 77–126. Wiley-Liss, New York.
27. Bottomley, P.J. 1994. Light microscopic methods for studying soil microorganisms. In Methods of Soil Analysis, Part 2 (Weaver, R.W., Angle, S., Bottomley, P., Bezdicek, D., Smith, S., Tabatabai, A., and Wollum, A., Eds.), pp. 81–105. Soil Science Society of America, Madison, WI.
28. Bowlen, G.F. and Kosson, D.S. 1995. *In situ* process for bioremediation of BTEX and petroleum fuel products. In Microbial Transformation and Degradation of Toxic Organic Chemicals (Young, L.Y. and Cerniglia, C.E., Eds.), pp. 515–544. Wiley-Liss, New York.
29. Boyd, S.A., Shelton, D.R., Berry, D., and Tiedje, J.M. 1983. Anaerobic biodegradation of phenolic compounds in digested sludge. Appl. Environ. Microbiol. 46:50–54.
30. Bragg, J.R., Prince, R.C., Harner, E.J., and Atlas, R.M. 1994. Effectiveness of bioremediation for the Exxon Valdez oil spill. Nature 368:413–418.
31. Brierley, C.L. 1990. Bioremediation of metal-contaminated surface and groundwaters. Geomicrobiology 8:201–224.
32. British Medical Association. 1991. Hazardous Wastes and Human Health. Oxford University Press. New York.
33. Brock, T.D. 1978. The poisoned control in biogeochemical investigations. In Environmental Biogeochemistry and Geomicrobiology, Vol. 3 (Krumbein, W.V., Ed.), pp. 717–726. Ann Arbor Science Publishers, Ann Arbor, MI.
34. Brockman, F.J. 1995. Nucleic-acid-based methods for monitoring the performance of *in situ* bioremediation. Mol. Ecol. 4:567–578.
35. Burlage, R.S. 1997. Emerging technologies: bioreporters, biosensors, and microprobes. In Manual of Environmental Microbiology (Hurst, C.J., Knudsen, G.R., McInerney, M.J., Stetzenbach, L.D., and Walter, M.V., Eds.), pp. 115–123. ASM Press, Washington, DC.
36. Burlage, R.S., Sayler, G.S., and Larimer, F. 1990. Monitoring of naphthalene catabolism by bioluminescence with nah-lux transcriptional fusions. J. Bacteriol. 172: 4749–4757.
37. Burlage, R.S., Palumbo, A.V., Heitzer, A., and Sayler, G. 1994. Bioluminescent reporter bacteria detect contaminants in soil samples. Appl. Biochem. Biotechnol. 45–46:731–740.
38. Buser, H.-R. and Miller, M.D. 1995. Isomer and enantioselective degradation of hexachlorocyclohexane isomers in sewage sludge under anaerobic conditions. Environ. Sci. Technol. 29:664–672.
39. Celia, M.A., Kindred, J.S., and Herrera, I. 1989. Contaminant transport and biodegradation. 1. A numerical model for reactive transport in porous media. Water Resources Res. 25:1141–1148.
40. Chapatwala, K.D., Babu, G.R.V., Armstead, E.R., White, E.M., and Wolfram, J.H. 1995. A kinetic study on the bioremediation of sodium cyanide and acetonitrile by free and immobilized cells of *Pseudomonas putida*. Appl. Biochem. Biotechnol. 51–52:717–726.
41. Chiang, C.Y., Salanitro, J.P., Chai, E.Y., Colthart, J.D., and Klein, C.L. 1989. Aerobic biodegradation of benzene, toluene, and xylene in a sandy aquifer—data analysis and computer modeling. Ground Water 27:823–834.
42. Chiang, C.Y., Loos, K.R., and Klopp, R.A. 1992. Field determination of geological-chemical properties of an aquifer by cone penetrometry and headspace analysis. Ground Water 30:428–436.
43. Coates, J.D., Anderson, R.T., and Lovley, D.R. 1996. Oxidation of polycyclic aromatic hydrocarbons under sulfate-reducing conditions. Appl. Environ. Microbiol. 62: 1099–1101.

44. Colberg, P.J.S. and Young, L.Y. 1995. Anaerobic degradation of nonhalogenated homocyclic aromatic compounds coupled with nitrate, iron, or sulfate reduction. In Microbial Transformation and Degradation of Toxic Organic Chemicals (Young, L.Y. and Cerniglia, C.E., Eds.), pp. 307–330. Wiley-Liss, New York.
45. Contu, A., Sarritzu, G., Breittmayer, J.P., and Schintu, M. 1986. Considerations on continuous pollution by mercury in a lagoon in Sarcinia, Italy: recent research on possible repercussions along the trophic chain. Water Sci. Technol. 18:307.
46. Crocker, F.H., Guerin, W.F., and Boyd, S.A. 1995. Bioavailability of naphthalene sorbed to cationic surfactant-modified smectite clay. Environ. Sci. Technol. 29: 2953–2958.
47. Dagley, S. and Chapman, P.J. 1971. Evaluation of methods used to determine metabolic pathways. In Methods in Microbiology, Vol. 6A (Norris, J.R. and Ribbons, D.W., Eds.), pp. 217–268. Academic Press, London.
48. Davis, J.W., Klier, N.J., and Carpenter, C.L. 1994. Natural biological attenuation of benzene in ground water beneath a manufacturing facility. Ground Water 32: 215–226.
49. Denome, S.A., Stanley, D.C., Olson, E.S., and Young, K.D. 1993. Metabolism of dibenzothiophene and naphthalene in *Pseudomonas* strains: complete DNA sequence of an upper naphthalene catabolic pathway. J. Bacteriol. 175:6890–6901.
50. Dolfing, J., Zeyer, J., Binder-Eicher, P., and Schwarzenbach, R.P. 1990. Isolation and characterization of a bacterium that mineralizes toluene in the absence of molecular oxygen. Arch. Microbiol. 154:336–341.
51. Dowling, N.J.E. and Guezennec, J. 1997. Microbiological-influenced corrosion. In Manual of Environmental Microbiology (Hurst, C.J., Knudsen, G., McInerney, M., Stetzenbach, L.D., and Walter, M., Eds.), pp. 842–855. ASM Press, Washington, DC.
52. Eaton, R.W. and Chapman, P.J. 1992. Bacterial metabolism of naphthalene: construction and use of recombinant bacteria to study ring cleavage of 1,2-dihydroxynaphthalene and subsequent reactions. J. Bacteriol. 174:7542–7554.
53. Erickson, B.D. and Mondello, F.J. 1992. Nucleotide sequencing and transcriptional mapping of the genes encoding biphenyl dioxygenase, a multicomponent polychlorinated biphenyl degrading enzyme in *Pseudomonas* strain LB 400. J. Bacteriol. 174: 2903–2912.
54. Evans, P.J., Mang, D.T., Kim, K.S., and Young, L.Y. 1991. Anaerobic degradation of toluene by a denitrifying bacterium. Appl. Environ. Microbiol. 57:1139–1145.
55. Falconer, R.L., Bidleman, T.F., Gregor, D.J., Semkin, R., and Teixeira, C. 1995. Enantioselective breakdown of a-hexachlorocyclohexane in a small Arctic lake and its watershed. Environ. Sci. Technol. 29:1297–1302.
56. Fiorenza, S., Duston, K.L., and Ward, C.H. 1991. Decision making—Is bioremediation a viable option. J. Hazard Mats. 28:171–183.
57. Flanagan, W.P. and May, R.J. 1993. Metabolite detection as evidence for naturally occurring aerobic PCB biodegradation in Hudson River sediment. Environ. Sci. Technol. 27:2207–2212.
58. Flathman, P.E., Jerger, D.E., and Exner, J.H. 1994. Bioremediation: Field Experience. Lewis Publishers, Boca Raton, FL.
59. Fleming, J.T., Sanseverino, J., and Sayler, G.S. 1993. Quantitative relationship between naphthalene catabolic gene frequency and expression in predicting PAH degradation in soils. Environ. Sci. Technol. 27:1068–1074.
60. Focht, D.D. 1994. Microbiological procedures for biodegradation research. In Methods of Soil Analysis, Part 2 (Weaver, R.W., Angle, S., Bottomley, P., Bezdicek, D., Smith, S., Tabatabai, A., and Wollum, A., Eds.), pp. 407–425. Soil Science Society of America, Madison, WI.
61. Focht, D.D. 1997. Aerobic biotransformations of polychlorinated biphenyls. In Manual of Environmental Microbiology (Hurst, C.J., Knudsen, G. McInerney, M., Stetzenbach, L.D., and Walter, M., Eds.), pp. 811–814. ASM Press, Washington, DC.
62. Freedman, D.L. and Gossett, J.M. 1989. Biological reductive dechlorination of tetrachloroethylene and trichloroethylene to ethylene under methanogenic conditions. Appl. Environ. Microbiol. 55:2144–2151.

62a. Führ, F. 1991. Lysimeter experiments with $^{14}$C-labelled pesticides—An agroecosystem approach. In Pesticide Chemistry (H. Frehse, Ed.), pp. 37–48. VCH, Weinheim, FRG.
63. Funk, S.B., Crawford, D.L., Crawford, R.L., Mead, G., and Davis-Hoover, W. 1995. Full-scale anaerobic bioremediation of trinitrotoluene (TNT) contaminated soil. Appl. Biochem. Biotechnol. 51–52:625–633.
64. Ghiorse, W.C. 1994. Iron and manganese oxidation and reduction. In Methods of Soil Analysis, Part 2 (Weaver, R.W., Angle, S., Bottomley, P., Bezdicek, D., Smith, S., Tabatabai, A., and Wollum, A., Eds.), pp. 1079–1096. Soil Science Society of America, Madison, WI.
65. Gibson, D.T. 1971. Assay of enzymes of aromatic metabolism. In Methods in Microbiology, Vol. 6A (Norris, J.R. and Ribbons, D.W., Eds.), pp. 463–478. Academic Press, New York.
66. Gibson, D.T. 1984. Microbial Degradation of Organic Compounds. Marcel Dekker, New York.
67. Gibson, D.T., Cardini, G.E., Maseles, F.C., and Kallio, R.E. 1970. Incorporation of oxygen-18 into benzene by *Pseudomonas putida*. Biochemistry 9:1631–1635.
68. Gonzalez, J.M., Sherr, B.F., and Sherr, E.B. 1993. Digestive enzyme activity as a quantitative measure of protistan grazing: the acid lysozyme assay for bacterivory. Mar. Ecol. Prog. Ser. 100:197–206.
69. Gottschal, J.C., Harder, W., and Prins, R.A. 1992. Principles of enrichment, isolation, cultivation, and preservation of bacteria. In The Prokaryotes, Vol I (Balows, A., Trüper, H.G., Dworkin, M., Harder, W., and Schieifer, K.-H., Eds.), pp. 149–196. Springer-Verlag, New York.
70. Goyal, A.K. and Zylstra, G.J. 1996. Molecular characterization of novel genes for polycyclic aromatic hydrocarbon degradation from Comamonas testosteroni GZ39. Appl. Environ. Microbiol. 62:230–236.
71. Gribble, G.W. 1994. The natural production of chlorinated compounds. Environ. Sci. Technol. 38:310A–319A.
72. Grossman, E.L. 1997. Stable carbon isotopes as indicators of microbial activity in aquifers. In Manual of Environmental Microbiology (Hurst, C.J., Knudsen, G.R., McInerney, M.J., Stetzenbach, L.D., and Walter, M.V., Eds.), pp. 565–576. ASM Press, Washington, DC.
73. Haddock, J.D., Horton, J.R., and Gibson, D.T. 1995. Dihydroxylation and dechlorination of chlorinated biphenyls by purified biphenyl 2,3-dioxygenase from *Pseudomonas* strain LB400. J. Bacteriol. 177:20–26.
74. Harayama, S., Kok, M., and Neidle, E.L. 1992. Functional and evolutionary relationships among diverse oxygenases. Annu Rev. Microbiol. 46:565–601.
75. Harding, G.L. 1996. Bioremediation and the dissimilatory reduction of metals. In Manual of Environmental Microbiology (Hurst, C.J., Knudsen, G., McInerney, M., Stetzenbach, L.D., and Walter, M., Eds.), pp. 806–810. ASM Press, Washington, DC.
76. Harkness, M.R., McDermott, J.B., Abramowicz, D.A., Salvo, J.J., Flanagan, W.P., Stephens, M.L., Mondello, F.J., May, R.J., Lobos, J.H., Carroll, K.M., Brennan, M.J., Bracco, A.A., Fish, K.M., Warner, G.L., Wilson, P.R., Dietrich, D.K., Lin, D.T., Morgan, C.B., and Gately, W.L. 1993. In situ stimulation of aerobic PCB biodegradation in Hudson River sediments. Science 259:503–507.
77. Hatzinger, P.B. and Alexander, M. 1995. Effect of aging of chemicals in soil on their biodegradability and extractability. Environ. Sci. Technol. 29:537–545.
78. Hazen, T.D., Looney, B.B., Enzien, M., Dougherty, D.M., Wear, J., Fliermans, C.B., and Eddy, C.A. 1994. In situ bioremediation of chlorinated-solvents via horizontal wells. Abstr. Annu. Meet. Am. Soc. Microbiol., p. 329.
79. Hemond, H.F. and Fechner, E.J. 1994. Chemical Fate and Transport in the Environment. Academic Press, New York.
80. Herrick, J.B., Madsen, E.L., Batt, C.A., and Ghiorse, W.C. 1993. Polymerase chain reaction amplification of naphthalene catabolic and 16S rRNA gene sequences from indigenous sediment bacteria. Appl. Environ. Microbiol. 59:687–694.
80a. Herrick, J.B., Stuart-Keil, K.G., Ghiorse, W.C., and Madsen, E.L. 1997. Natural horizontal transfer of a naphthalene dioxygenase gene between bacteria native to a coal tar-contaminated field site. Appl. Environ. Microbiol. 63:2330–2337.

81. Hickey, W.J. 1995. In situ respirometry: field methods and implications for hydrocarbon biodegradation in subsurface soils. J. Environ. Qual. 24:583–588.
82. Hickey, W.J. 1995. Soil ventilation: effects on microbial populations in gasoline-contaminated subsurface soils. J. Environ. Qual. 24:571–582.
83. Hinchee, R.E, Ed. 1994. Air Sparging. Lewis Publishers, Boca Raton, FL.
84. Hinchee, R.E., Means, J.L., and Burris, D.R., Eds. 1995. Bioremediation of Inorganics. Battelle Press, Columbus, OH.
85. Hinchee, R.E. and Olfenbuttel, R.F. 1991. In situ Bioreclamation. Butterworth-Heinemann, Boston.
86. Hinchee, R.E. and Olfenbuttel, R.F. 1991. On-Site Bioreclamation. Butterworth-Heinemann, Boston.
87. Hinchee, R.E. and Olfenbuttel, R.F. 1994. On-site Bioreclamation: Processes for Xenobiotic and Hydrocarbon Treatment. Battelle Press, Columbus, OH.
88. Hinchee, R.E. and Olfenbuttel, R.F. 1994. In situ Bioreclamation: Applications and Investigations for Hydrocarbon and Contaminated Site Remediation. Battelle Press, Columbus, OH.
89. Hinchee, R.E., Wilson, J.T., and Downey, D.C., Eds. 1995. Intrinsic Bioremediation. Battelle Press, Columbus, OH.
90. Hodson, R.E., Dustman, W.A., Garg, R.P., and Moran, M.A. 1995. In situ PCR for visualization of microscale distribution of specific genes and gene products in prokaryotic communities. Appl. Environ. Microbiol. 61:4074–4082.
91. Hoefs, J. 1987. Stable Isotope Geochemistry, 3rd ed. Springer-Verlag, New York.
92. Holben, W.E. 1997. Isolation and purification of bacterial community DNA from environmental samples. In Manual of Environmental Microbiology (Hurst, C.J., Knudsen, G., McInerney, M., Stetzenbach, L.D., and Walter, M., Eds.), pp. 431–436. ASM Press, Washington, DC.
93. Holben, W.E., Schroeter, B.M., Calabrese, V.G.M., Olsen, R.H., Kukor, J.K., Biederbeck, V.O., Smith, A.E., and Tiedje, J.M. 1992. Gene probe analysis of soil microbial populations selected by amendment with 2,4-dichlorophenoxyacetic acid. Appl. Environ. Microbiol. 58:3941–3948.
94. Holliger, C., Schraa, G., Stams, A.J.M., and Zehnder, A.J.B. 1993. A highly purified enrichment culture couples the reductive dechlorination of tetrachloroethene to growth. Appl. Environ. Microbiol. 59:2991–2997.
95. Hopkins, G.D., Munakata, J., Semprini, L., and McCarty, P.L. 1993. Trichloroethylene concentration effects on pilot field-scale in-situ groundwater bioremediation by phenol-oxidizing microorganisms. Environ. Sci. Technol. 27:2542–2547.
96. Ingham, E.R. 1994. Protozoa. In Methods of Soil Analysis, Part 2 (Weaver, R.W., Angle, S., Bottomley, P., Bezdicek, D., Smith, S., Tabatabai, A., and Wollum, A., Eds.), pp. 491–515. Soil Science Society of America, Madison, WI.
97. Irie, S., Doi, S., Yorifuji, T., Takagi, M., and Yano, K. 1987. Nucleotide sequencing and characterization of the genes encoding benzene oxidation enzymes of *Pseudomonas putida*. J. Bacteriol. 169:5174–5179.
98. Ka, J.O., Holben, W.E., and Tiedje, J.M. 1994. Genetic and phenotypic diversity of 2,4-dichlorophenoxyacetic acid (2,4-D)-degrading bacteria isolated from 2,4-D-treated field soils. Appl. Environ. Microbiol. 60:1106–1115.
99. Ka, J.O., Holben, W.E., and Tiedje, J.M. 1994. Use of gene probes to aid in recovery and identification of functionally dominant 2,4-dichlorophenoxyacetic acid-degrading populations in soil. Appl. Environ. Microbiol. 60:1116–1120.
100. Ka, J.O., Holben, W.E., and Tiedje, J.M. 1994. Analysis of competition in soil among 2,4-dichlorophenoxyacetic acid-degrading bacteria. Appl. Environ. Microbiol. 60:1121–1128.
101. Kalin, M., Cairns, J., and McCready, R.M. 1991. Ecological engineering methods for acid mine drainage treatment of coal wastes. Resources Conserv. Recycl. 5: 265–276.
102. Karl, D.M. 1986. Determination of in situ microbial biomass, viability, metabolism, and growth. In Bacteria in Nature Vol. 2 (Poindexter, J.S. and Leadbetter, E.R., Eds.), pp. 85–176. Plenum Press, New York.

103. Keith, L.H., Ed. 1996. Principles of Environmental Sampling, 2nd ed. ACS Professional Reference Book, American Chemical Society, Washington, D.C.
104. Keith, L.H. 1991. Environmental Sampling and Analysis: A Practical Guide. Lewis Publishers, Boca Raton, FL.
105. Keith, L.H., Ed. 1996. Compilation of EPA's Sampling and Analysis Methods, 2nd ed. Lewis Publishers, Boca Raton, FL.
106. Kim, H., Hemond, H.F., Krumholz, L.R., and Cohen, B.A. 1995. In-situ biodegradation of toluene in a contaminated stream. 1. Field studies. Environ. Sci. Technol. 29:108–116.
107. King, J.M.H., Digrazia, P.M., Applegate, B., Burlage, R., Sanseverino, J., Dunbar, P., Larimer, F., and Sayler, G.S. 1990. Rapid sensitive bioluminescent reporter technology for naphthalene exposure and biodegradation. Science 249:778–781.
108. Kiyohara, H., Ngao, N., and Yana, K. 1982. Rapid screen for bacteria degrading water-insoluble, solid hydrocarbons on agar plates. Appl. Environ. Microbiol. 43: 454–457.
109. Korus, R.A. 1997. Scale-up of processes for bioremediation. In Manual of Environmental Microbiology (Hurst, C.J., Knudsen, G.R., McInerney, M.J., Stetzenbach, L.D., and Walter, M.V., Eds.), pp. 856–864. ASM Press, Washington, DC.
110. Krumholz, L.R., Sharp, R., and Fishbain, S.S. 1996. A freshwater anaerobe coupling acetate oxidation to tetrachloroethylene dehalogenation. Appl. Environ. Microbiol. 62:4108–4113.
111. Kukor, J.J. and Olsen, R.H. 1990. Diversity of toluene degradation following long term exposure to BTEX In situ. In Biotechnology and Biodegradation (Kamely, D., Chakrabarty, A., and Omenn, G., Eds.), pp. 405–421. Gulf. Publishing, Houston.
112. Kukor, J.J., and Olsen, R.H. 1992. Complete nucleotide sequence of *tbuD*, the gene encoding phenol-cresol hydroxylase from *Pseudomonas picketti* PK01 and functional analysis of the encoded enzyme. J. Bacteriol. 174:6518–6526.
113. Kurkela, S., Lehvaslaiho, H., Palva, E.T., and Teeri, T.H. 1988. Cloning, nucleotide sequence and characterization of genes encoding naphthalene dioxygenase of *Pseudomonas putida* strain NCIB9816. Gene 73:355–362.
114. Lamar, R.T. and Dietrich, D.M. 1990. In situ depletion of pentachlorophenol from contaminated soil by *Phanerochaete* spp. Appl. Environ. Microbiol. 56:3093–3100.
115. Lamar, R.T., Schoenike, B., vanden Wymelenberg, A., Stewart, P., Dietrich, D.M., and Cullen, D. 1995. Quantitation of fungal mRNAs in complex substrates by reverse transcription PCR and its application to *Phanerochaete chrysosporium*-colonized soil. Appl. Environ. Microbiol. 61:2122–2126.
116. Larson, R.J. 1983. Kinetic and ecological approaches for predicting biodegradation rates of xenobiotic organic chemicals in natural ecosystems. In Current Perspectives in Microbial Ecology (Klug, M.J. and Reddy, C.A., Eds.), pp. 677–686. ASM Press, Washington, DC.
117. Lawrence, J.R., Korber, D.R., Wolfaardt, G.M., and Caldwell, D.E. 1997. Analytical imaging and microscopy techniques. In Manual of Environmental Microbiology (Hurst, C.J., Knudsen, G.R., McInerney, M.J., Stetzenbach, L.D., and Walter, M.V., Eds.), pp. 29–51. ASM Press, Washington, DC.
118. Layton, A.C., Lajoie, C.A., Easter, J.P., Jernigan, R., Sanseverino, J., and Sayler, G.S. 1994. Molecular diagnostics and chemical analysis for assessing biodegradation of polychlorinated biphenyls in contaminated soils. J. Indust. Microbiol. 13:392–401.
119. Lee, K., Wong, C.S., Cretney, W.J., Whitney, F.A., Parsons, T.R., Lalli, C.M., and Wu, J. 1985. Microbiol response to crude oil and corexit 9527: SEAFLUXES enclosure study. Microb. Ecol. 11:337–351.
120. Lee, S. and Fuhrman, J.A. 1990. DNA hybridization to compare species compositions of natural bacterioplankton assemblages. Appl. Environ. Microbiol. 56:739–746.
121. Lehmann, R.G., Varaprath, S., and Frye, C.L. 1994. Degradation of silicone polymers in soil. Environ. Toxicol. Chem. 13:1061–1064.
122. Little, C.D., Palumbo, A.V., Herbes, S.E., Lidstrom, M.E., Tyndall, R.L., and Gilmer, P.J. 1988. Trichloroethylene biodegradation by a methane-oxidizing bacterium. Appl. Environ. Microbiol. 54:951–956.

123. Lovley, D.R. 1993. Dissimilatory metal reduction. Annu. Rev. Microbiol. 47:263–290.
124. Lovley, D.R. 1995. Bioremediation of organic and metal contaminants with dissimilatory metal reduction. J. Indust. Microbiol. 14:85–93.
125. Lovley, D.R. and Lonergan, D.J. 1990. Anaerobic oxidation of toluene, phenol, and p-cresol by the dissimilatory iron-reducing organism, GS-15. Appl. Environ. Microbiol. 56:1858–1864.
126. Lovley, D.R., Chapelle, F.H., and Woodward, J.C. 1994. Use of dissolved $H_2$ concentrations to determine distribution of microbially catalyzed redox reactions in anoxic groundwater. Environ. Sci. Tech. 28:1205–1210.
127. Lovley, D.R., Woodard, J.C., and Chapelle, F.H. 1994. Stimulated anoxic biodegradation of aromatic hydrocarbons using Fe(III) ligands. Nature 370:128–131.
128. Lovley, D.R., Coates, J.D., Woodward, J.C., and Phillips, E.J.P. 1995. Benzene oxidation coupled to sulfate reduction. Appl. Environ. Microbiol. 61:953–958.
129. Madsen, E.L. 1991. Determining in situ biodegradation: facts and challenges. Environ. Sci. Technol. 25:1662–1673.
130. Madsen, E.L. 1996. A critical analysis of methods for determining the composition and biogeochemical activities of soil microbial communities in situ. In Soil Biochemistry (Stotzky, G. and Bollag, J.-M., Eds.), Vol. 9, pp. 287–370. Marcel Dekker, New York.
131. Madsen, E.L. 1997. Methods for determining biodegradability. In Manual of Methods in Environmental Microbiology (Hurst, C.J., Knudsen, G.R., McInerney, M.J., Stetzenbach, L.D., and Walter, M.V., Eds.), pp. 709–720. ASM Press, Washington, DC.
132. Madsen, E.L., Manin, C.L., and Bilotta-Best, S. 1996. Oxygen limitations and aging as explanation for the persistence of naphthalene in coal-tar contaminated surface sediments. Environ. Toxicol. Chem. 15:1876–1882.
133. Madsen, E.L., Francis, A.J., and Bollag, J.-M. 1988. Environmental factors affecting indole metabolism under anaerobic conditions. Appl. Environ. Microbiol. 54:74–78.
134. Madsen, E.L., Sinclair, J.L., and Ghiorse, W.C. 1991. In situ biodegradation: microbiological patterns in a contaminated aquifer. Science 252:830–833.
135. Madsen, E.L., Bilotta-Best, S.E., and Giorse, W.C. 1995. Development of a rapid $^{14}C$-based field method for assessing potential biodegradation of organic compounds in soil and sediment samples. J. Microbiol. Meth. 21:317–327.
136. Major, D.W., Hodgins, E.W., and Butler, B.J. 1991. Field and laboratory evidence of in situ biotransformation of tetrachloroethene to ethene and ethane at a chemical transfer facility in North Toronto. In On-Site Bioreclamation (Hinchee, R.E. and Olfenbuttel, R.F., Eds.), pp. 147–172. Butterworth-Heinemann, Stoneham, MA.
137. Marinucci, A.C. and Bartha, R. 1979. Apparatus for monitoring the mineralization of volatile $^{14}C$-labeled compounds. Appl. Environ. Microbiol. 38:1020–1022.
138. Marshall, K.C. 1986. Microscopic methods for the study of bacterial behavior at inert surfaces. J. Microbiol. Meth. 4:217–227.
139. Maymó-Gatell, X., Tandoi, V., Gossett, J.M., and Zinder, S.H. 1995. Characterization of an $H_2$-utilizing enrichment culture that reductively dechlorinates tetrachoroethenes to vinyl chloride and ethene in the absence of methanogenesis and acetogenesis. Appl. Environ. Microbiol. 61:3928–3933.
139a. Maymo-Gatell, X., Chien, Y-T., Gossett J.M., and Zinder, S.H. 1997. Isolation of a bacterium that reductively dechlorinates tetrachlorethylene to ethene. Science 267: 1568–1571.
140. McCarty, P.L. 1991. Engineering concepts for in situ bioremediation. J. Hazard. Mats. 28:1–11.
141. McHale, A.P. and McHale, S. 1994. Microbiol biosorption of metals: potential in the treatment of metal pollution. Biotechnol. Adv. 12:647–652.
142. More', M.I., Herrick, J.B., Silva, M.C., Ghiorse, W.C., and Madsen, E.L. 1994. Quantitative cell lysis of indigenous microorganisms and rapid extraction of microbial DNA from sediment. Appl. Environ. Microbiol. 60:1572–1580.
143. Morel, F.M.M. and Hering, J.G. 1993. Principles and Applications of Aquatic Chemistry. John Wiley & Sons, New York.

144. Mueller, J.G., Chapman, P.J., Blattmann, B.O., and Pritchard, P.H. 1990. Isolation and characterization of a fluoranthene-utilizing strain of *Pseudomonas paucimobilis*. Appl. Environ. Microbiol. 56:1079–1086.
145. Mueller, J.B., Lantz, S.E., Ross, D., Colvin, R.J., Middaugh, D.G., and Pritchard, P.H. 1993. Strategy using bioreactors and specially selected microorganisms for bioremediation of groundwater contaminated with creosote and pentachlorophenol. Environ. Sci. Technol. 27:691–698.
146. Mueller, W., Smith, D.L., and Keith, L.H. 1991. Compilation of EPA's Sampling and Analysis Methods. Lewis Publishers, Boca Raton, FL.
147. Murarka, I., Neuhauser, E., Sherman, M., Taylor, B.B., Mauro, D.M., Ripp, J., and Taylor, T. 1992. Organic substances in the subsurface: delineation, migration, and remediation. J. Haz. Mats. 32:245–261.
148. National Research Council (NRC). 1990. Ground Water Models: Scientific and Regulatory Applications. National Academy Press, Washington, DC.
149. National Research Council (NRC). 1993. In situ Bioremediation: When Does It Work? National Academy Press, Washington, DC.
150. Neidle, E.L., Hartnett, C., Ornston, L.N., Bairoch, A., Rekik, M., and Harayama, S. 1991. Nucelotide sequences of the *Acinetobacter calcoaceticus* benABC genes for benzoate 1,2-dioxygenase reveal evolutionary relationships among multicomponent oxygenases. J. Bacteriol. 173:5385–5395.
151. Nishino, S.F. and Spain, J.C. 1997. Biodegradation and transformation of nitroaromatic compounds. In Manual of Environmental Microbiology (Hurst, C.J., Knudsen, G., McInerney, M., Stetzenbach, L.D., and Walter, M., Eds.), pp. 776–783. ASM Press, Washington, DC.
152. Norris, R.D., Hinchee, R.E., Brown, R., McCarty, P.L., Semprini, L., Wilson, J.T., Kampbell, D.G., Reinhard, M., Bouwer, D.J., Borden, R.C., Vogel, T.M., Thomas, J.M., and Ward, C.H. 1994. Handbook of Bioremediation. Lewis Publishers, Boca Raton, FL.
153. Nurk, A., Kasak, L., and Kivisaar, M. 1991. Sequence of the gene (pheA) encoding phenol monooxygenase from *Pseudomonas* sp. EST1001: expression in *Escherichia coli* and *Pseudomonas putida*. Gene 102:13–18.
154. Ogram, A.V. and Bezdicek, D.F. 1994. Nucleic acid probes. In Methods of Soil Analysis, Part 2 (Weaver, R.W., Angle, S., Bottomley, P., Bezdicek, D., Smith, S., Tabatabai, A., and Wollum A., Eds.), pp. 665–687. Soil Science Society of America, Madison, WI.
155. Ogram, A. and Feng, X. 1997. Methods of soil microbial community analysis. In Manual of Environmental Microbiology (Hurst, C.J., Knudsen, G.R., McInerney, M.J., Stetzenbach, L.D., and Walter, M.V., Eds.), pp. 422–430. ASM Press, Washington, DC.
156. Ogram, A., Sun, W., Brockman, F.J., and Fredrickson, J.K. 1995. Isolation and characterization of RNA from low-biomass deep-subsurface sediments. Appl. Environ. Microbiol. 61:763–768.
157. Oremland, R.S., Culbertson, C.W., and Winfrey, M.R. 1991. Methylmercury decomposition in sediments and bacterial cultures: involvement of methanogens and sulfate reducers in oxidative demethylation. Appl. Environ. Microbiol. 57:130–137.
158. Paigen, B. and Goldman, L.R. 1987. Lessons from Love Canal New York USA: the role of the public and the use of birth weight growth and indigenous wildlife to evaluate health risk. In Health Effects from Hazardous Waste Sites (Andelman, J.B. and Underhill, D.W., Eds.), pp. 177–192. Lewis Publishers, Chelsea, MI.
159. Parkin, T.B. and Robinson, J.A. 1994. Statistical treatment of microbial data. In Methods of Soil Analysis, Part 2 (Weaver, R.W., Angle, S., Bottomley, P., Bezdicek, D., Smith, S., Tabatabai, A., and Wollum, A., Eds.), pp. 15–40. Soil Science Society of America, Madison, WI.
160. Paul, J.H. and Pichard, S.L. 1995. Extraction of DNA and RNA from aquatic environments. In Nucleic Acids in the Environment. Methods and Applications (Trevors, J.T. and VanElsas, J.D., Eds.), pp. 170–173. Springer Verlag, New York.
161. Pepper, I.L. and Pillai, S.D. 1994. Detection of specific DNA sequences in environmental samples via polymerase chain reaction. In Methods of Soil Analysis, Part 2

(Weaver, R.W., Angle, S., Pottomley, P., Bezdicek, D., Smith, S., Tabatabai, A., and Wollum, A., Eds.), pp. 707–725. Soil Science Society of America, Madison, WI.

161a. Pfiffner, S.M., Palumbo, A.V., Phelps, T.J., and Hazen, T.C. 1997. Effects of nutrient dosing on subsurface methanotrophic populations and trichloroethylene degradation. J. Ind. Microbiol. Biotechnol. 18:204–212.

162. Pignatello, J.J., Ferrandino, F.J., and Huang, L.Q. 1993. Elution of aged and freshly added herbicides from a soil. Environ. Sci. Technol. 27:1563–1571.

163. Pignatello, J.J. and Xing, B. 1996. Mechanisms of slow sorption of organic chemicals to natural particles. Environ. Sci. Technol. 30:1–11.

164. Pinder, G.F. and Abriola, L.M. 1986. On the simulation of nonaqueous phase organic compounds in the subsurface. Water Resources Res. 22:109S–119S.

165. Pollard, S.J.T., Hurdey, S.E., and Fedorak, P.M. 1994. Bioremediation of petroleum- and creosote-contaminated soils: a reivew of constraints. Waste Mgt. Res. 12:173–194.

166. Prince, R.C., Elmendorf, D.L., Lute, J.R., Hsu, C.S., Halth, C.E., Senius, J.D., Dechert, G.J., Douglas, G.S., and Butler, E.L. 1994. 17α(H),21β(H)-hopane as a conserved internal marker for estimating the biodegradation of crude oil. Environ. Sci. Technol. 28:142–145.

167. Pritchard, P.H. 1992. Use of inoculation of bioremediation. Curr. Opinion Biotechnol. 3:232–243.

168. Pritchard, P.H. and Costa, C.F. 1991. EPA's Alaska oil spill bioremediation project. Environ. Sci. Technol. 25:372–379.

169. Rabus, R., Nordhaus, R., Ludwig, W., and Widdel, F. 1993. Complete oxidation of toluene under strictly anoxic conditions by a new sulfate-reducing bacterium. Appl. Environ. Microbiol. 59:1444–1451.

170. Ramaswami, A. and Luthy, R.G. 1997. Measuring and modeling physiochemical limitations to bioavailability and biodegradation. In Manual of Environmental Microbiology (Hurst, C.J., Knudsen, G., McInerney, M., Stetzenbach, L.D., and Walter, M., Eds.), pp. 721–730. ASM Press, Washington, DC.

171. Raymond, R.L., Hudson, J.O., and Jamison, V.W. 1976. Oil degradation in soil. Appl. Environ. Microbiol. 31:522–535.

172. Rittmann, B.E., Seagren, E., Wrenn, B.A., Valocchi, A.J., Ray, C., and Raskin, L. 1994. In situ Bioremediation, 2nd ed. Noyes Publications, Park Ridge, NJ.

173. Robyt, J.F. and White, B.J. 1990. Biochemical Techniques Theory and Practice. Waveland Press, Prospect Heights, IL.

174. Rogers, J.E. and Gibson, D.T. 1977. Purification and properties of cis-toluene dihydrodiol dehydrogenase from *Pseudomonas putida*. J. Bacteriol. 130:1117–1124.

175. Ryan, J.R., Loehr, R.C., and Rucker, E. 1991. Bioremediation of organic contaminated soils. J. Hazard. Mats. 28:159–169.

176. Salanitro, J.P. 1993. The role of bioattenuation in the management of aromatic hydrocarbon plumes in aquifers. Ground Water Monitor. Remed. 13:150–161.

177. Sandoli, R.S., Ghiorse, W.C., and Madsen, E.L. Regulation of microbial phenanthrene mineralization in sediments by sorbent-sorbate contact time, inoculum, and gamma irradiation-induced sterilization artifacts. Environ. Toxicol. Chem. 15: 1901–1907.

178. Sanseverino, J., Applegate, B.M., King, J.M.H., and Sayler, G.S. 1993. Plasmid-mediated mineralization of naphthalene, phenanthrene and anthracene. Appl. Environ. Microbiol. 59:1931–1937.

179. Sanseverino, J., Werner, C., Fleming, J., Applegate, B., King, J.M.H., and Sayler, G.S. 1993. Molecular diagnostics of polycyclic aromatic hydrocarbon biodegradation in manufactured gas plant soil. Biodegradation 4:303–321.

180. Saouter, E., Gillman, M., and Barkay, T. 1995. An evaluation of mer-specified reduction of ionic mercury as a remedial tool of a mercury-contaminated freshwater pond. J. Indust. Microbiol. 14:343–348.

181. Sayler, G.S., Nikbakht, K., Fleming, J.T., and Packard, J. 1992. Application and molecular techniques to soil biochemistry. In Soil Biochemistry, Vol. 7 (Stotzky, G. and Bollag, J.-M., Eds.), pp. 131–172. Marcel Dekker, New York.

182. Scow, K.M. and Alexander, M.A. 1992. Effect of diffusion on the kinetics of biodegradation experimental results with synthetic aggregates. Soil Sci. Soc. Am. J. 56: 128–134.
183. Scribner, S.L., Benzing, T.R., Sun, S., and Boyd, S.A. 1992. Desorption and bioavailability of aged simazine residues in soil from a continuous corn field. J. Environ. Qual. 21:115–120.
184. Selvaratnam, S., Schoedel, B.A., McFarland, B.L., and Kulpa, C.F. 1995. Application of reverse transcriptase PCR for monitoring expression of the catabolic *dmpN* gene in a phenol-degrading sequencing batch reactor. Appl. Environ. Microbiol. 61: 3981–3985.
184a. Senprini, L. 1997. Strategies for the aerobic cometabolism of chlorinated solvents. Curr. Op. Biotechnol. 8:296–308.
185. Semprini, L. and McCarty, P.L. 1991. Comparison between model similations and field results for in-situ biorestoration of chlorinated aliphatics. 1. Biostimulation of methanotrophic bacteria. Ground Water 29:365–374.
186. Semprini, L. and McCarty, P.L. 1992. Comparison between model simulations and field results for in-situ biorestoration of chlorinated alphatics. 2. Cometabolic transformations. Ground Water 29:37–44.
187. Semprini, L., Roberts, P.V., Hopkins, G.D., and McCarty, P.L. 1990. A field evaluation of in-situ biodegradation of chlorinated ethenes: Part 2, results of biostimulation and biotransformation experiments. Ground Water 28:715–727.
188. Semprini, L., Hopkins, G.D., Roberts, P.V., Grbic-Galic, D., and McCarty, P.L. 1991. A field evaluation of in-situ biodegradation of chlorinated ethenes. Part 3, Studies of competitive inhibition. Ground Water 29:239–250.
189. Semprini, L., Kitanidis, P.K., Kampbell, D.H., and Wilson, J.T. 1995. Anaerobic transformation of chlorinated aliphatic hydrocarbons in a sand aquifer based on spatial chemical distributions. Water Resources Res. 31:1051–1062.
190. Shannon, M.J.R. and Unterman, R. 1993. Evaluating bioremediation: distinguishing fact from fiction. Annu. Rev. Microbiol. 47:715–738.
191. Shauver, J.M. 1993. A regulator's perspective on in situ bioremediation. In In situ Bioremediation: When Does It Work?, pp. 99–103. National Academy Press, Washington, DC.
192. Shields, M.S., Montgomery, S.O., Chapman, P.J., Cuskey, S.M., and Pritchard, P.H. 1989. Novel pathway of toluene catabolism in the trichloroethylene-degrading bacterium G4. Appl. Environ. Microbiol. 55:1624–1629.
193. Shields, M.S., Reagin, M.J., Gerger, R.R., Campbell, R., and Somervile, C. 1995. TOM, a new aromatic degradative plasmid from *Burkholderia* (*Pseudomonas*) *cepacia* G4. Appl. Environ. Microbiol. 61:1351–1356.
194. Shuttleworth, K.L. and Cerniglia, C.E. 1997. Practical methods for the isolation of polycyclic aromatic hydrocarbon (PAH)-degrading microorganisms and the determination of PAH mineralization and biodegradation intermediates. In Manual of Environmental Microbiology (Hurst, C.J., Knudsen, G., McInerney, M., Stetzenbach, L.D., and Walter, M., Eds.), pp. 766–775. ASM Press, Washington, DC.
195. Simkins, S. and Alexander, M. 1984. Models for mineralization kinetics with the variables of substrate concentration and population density. Appl. Environ. Microbiol. 47:1299–1306.
196. Simon, M.J., Osslund, T.D., Saunders, R., Ensley, B.D., Suggs, S., Harcourt, A., Suen, W.-C., Cruden, D.L., Gibson, D.T., and Zylstra, G.J. 1993. Sequences of genes encoding naphthalene dioxygenase in *Pseudomonas putida* strains G7 and NCIB 9816-4. Gene 127:31–37.
197. Smith, R.L. 1997. Determining the terminal electron-accepting reaction in the saturated subsurface. In Manual of Environmental Microbiology (Hurst, C.J., Knudsen, G.R., McInerney, M.J., Stetzenbach, L.D., and Walter, M.V., Eds.), pp. 577–585. ASM Press, Washington, DC.
198. Spain, J.C. 1995. Biodegradation of nitroaromatic compounds. Annu. Rev. Microbiol. 49:523–555.
199. Spain, J.C., VanVeld, P.A., Monti, C.A., Pritchard, P.H., and Cripe, C.R. 1984. Comparison of *p*-nitrophenol biodegradation in field and laboratory test systems. Appl. Environ. Microbiol. 48:944–950.

200. Stahl, D.A. 1995. Application of phylogenetically based hybridization probes to microbial ecology. Mol. Ecol. 4:535–542.
201. Stahl, D.A. 1997. Molecular approaches for the measurement of density, diversity, and phylogeny. In Manual of Environmental Microbiology (Hurst, C.J., Knudsen, G.R., McInerney, M.J., Stetzenbach, L.D., and Walter, M.V., Eds.), pp. 102–114. ASM Press, Washington, DC.
202. Stainthorpe, A.C., Lees, V., Salmond, G.P.C., Dalton, H., and Murrell, J.C. 1990. The methane monooxygenase gene cluster of *Methylococcus capsulatus* bath. Gene 91:27–34.
203. Staley, J.T. and Konopka, A. 1985. Measurement of in situ activities of nonphotosynthetic microorganisms in aquatic and terrestrial habitats. Annu. Rev. Microbiol. 39:321–346.
204. Stanier, R.Y., Palleroni, N.J., and Doudoroff, M. 1966. The aerobic pseudomonads: a taxonomic study. J. Gen Microbiol. 53:1010–1019.
205. Steinberg, S.M., Pignatello, J.J., and Sawhney, B.L. 1987. Persistence of 1,2-dibromoethane in soils: entrapment in intraparticle micropores. Environ. Sci. Technol. 21:1201–1208.
206. Stumm, W. and Morgan, J.J. 1996. Aquatic Chemistry, 3rd ed. John Wiley & Sons, New York.
207. Subba-Rao, R.V., Rubin, H.E., and Alexander, M. 1982. Kinetics and extent of mineralization of organic chemicals at trace levels in freshwater and sewage. Appl. Environ. Microbiol. 43:1139–1150.
208. Suflita, J.M. and Townsend, G.T. 1995. The microbial ecology and physiology of aryl dehalogenation reactions and implications for bioremediation. In Microbial Transformation and Degradation of Toxic Organic Chemicals (Young, L.Y. and Cerniglia, C.E., Eds.), pp. 243–268. Wiley-Liss, New York.
209. Suflita, J.M., Londry, K.L., and Ulrich, G.A. 1997. Determination of anaerobic biodegradation activity. In Manual of Environmental Microbiology (Hurst, C.J., Knudsen, G.R., McInerney, M.J., Stetzenbach, L.D., and Walter, M.V., Eds.), pp. 790–801. ASM Press, Washington, DC.
210. Summers, A.O. 1992. The hard stuff: metals in bioremediation. Curr. Opinion Biotechnol. 3:271–276.
211. Sutherland, J.B., Rafii, F., Kahn, A.A., and Cerniglia, C.E. 1995. Mechanisms of polycyclic aromatic hydrocarbon degradation. In Microbial Transformation and Degradation of Toxic Organic Chemicals (Young, L.Y. and Cerniglia, C.E., Eds.), pp. 269–306. Wiley-Liss, New York.
212. Swannell, P.J., Lee, K., and McDonagh, M. 1996. Field evaluation of marine oil spill bioremediation. Microbiol. Rev. 60:342–365.
213. Swoboda-Colberg, N.G. 1995. Chemical contamination of the environment: Sources, types, and fate of synthetic organic chemicals. In Microbial Transformation and Degradation of Toxic Organic Chemicals (Young, L.Y. and Cerniglia, C.E., Eds.), pp. 27–76. Wiley-Liss, New York.
214. Takizawa, N., Kaida, N, Torigoe, S., Moritani, T., Sawada, T., Satoh, S., and Kiyohara, H. 1994. Identification and characterization of genes encoding polycyclic aromatic hydrocarbon dioxygenase and polycyclic aromatic hydrocarbon dihydrodiol dehydrogenase in *Pseudomonas putida* OUS82. J. Bacteriol. 176:2444–2449.
215. Tandoi, V., DiStefano, T.D., Bowser, P.A., Gossett, J.M., and Zinder, S.H. 1994. Reductive dehalogenation of chlorinated ethenes and halogenated ethanes by a high-rate anaerobic enrichment culture. Environ. Sci. Technol. 28:973–979.
216. Tanner, R.S. 1997. Cultivation of bacteria and fungi. In Manual of Environmental Microbiology (Hurst, C.J., Knudsen, G., McInerney, M., Stetzenbach, L.D., and Walter, M., Eds.), pp. 52–60. ASM Press, Washington, DC.
217. Thompson-Eagle, E.C. and Frankenberger, Jr., W.T. 1992. Bioremediation of soils contaminated with selenium. Adv. Soil Sci. 17:261–310.
218. Tiedje, J.M. 1993. Bioremediation from an ecological perspective. In In situ Bioremediation: When Does It Work?, pp. 110–120. National Academy Press, Washington, DC.

219. Top, E.M., Holben, W.E., and Forney, L.J. 1995. Characterization of diverse 2,4-dichlorophenoxyacetic acid-degradative plasmids isolated from soil by complementation. Appl. Environ. Microbiol. 61:1691–1698.
220. Turley, C.M. 1993. Direct estimates of bacterial numbers in seawater samples without incurring cell loss due to sample storage. In Handbook of Methods of Aquatic Microbiol Ecology (Kemp, P.F., Sherr, B.F., Sherr, E.B., and Cole, J.J., Eds.), pp. 143–148. Lewis Publishers, Boca Raton, FL.
221. van der Meer, J.R., de Vos, W.M., Harayama, S., and Zehnder, A.J.B. 1992. Molecular mechanisms of genetic adaptation of xenobiotic compounds. Microbiol. Rev. 56:677–694.
222. Van Geel, P.J. and Sykes, J.F. 1994. Laboratory and model simulations of a LNAPL spill in a variably-saturated sand, 1. Laboratory experiment and image analysis techniques. J. Cont. Hydrol. 17:1–25.
223. Videla, H.A. and Characklis, W.G. 1992. Biofouling and microbially influenced corrosion. Int. Biodeterior. Biodegrad. 29:195–212.
224. Wackett, L.P. 1995. Bacterial co-metabolism of halogenated organic compounds. In Microbial Transformation and Degradation of Toxic Organic Chemicals (Young, L.Y. and Cerniglia, C.E., Eds.), pp. 217–242. Wiley-Liss, New York.
225. Wackett, L.P. 1997. Biodegradation of halogenated solvents. In Manual of Environmental Microbiology (Hurst, C.J., Knudsen, G.R., McInerney, M.J, Stetzenbach, L.D., and Walter, M.V., Eds.), pp. 784–789. ASM Press, Washington, DC.
226. Wagner, R.E. and Yogis, G.A. 1992. Guide to Environmental Analytical Methods. Genium Publishing, Schenectady, NY.
227. Walter, M.W. and Crawford, R.L. 1997. Overview: Biotransformation and biodegradation. In Manual of Environmental Microbiology (Hurst, C.J., Knudsen, G., McInerney, M., Stetzenbach, L.D., and Walter, M., Eds.), pp. 707–708. ASM Press, Washington, DC.
228. Weaver, R.W., Angle, S., Bottomley, P., Bezdicek, S., Smith, S., Tabatabai, A., and Wollum, A., Eds.) 1994. Methods of Soil Analysis, Part 2. Soil Science Society of America, Madison, WI.
229. Weissenfels, W.D., Beyer, M., and Klein, J. 1990. Degradation of phenanthrene, fluorene and fluoranthene by pure bacterial cultures. Appl. Microbiol. Biotechnol. 32:479–484.
230. Westlake, D.W.S., Jobson, A.M., and Cook, F.D. 1978. In situ degradation of oil in a soil of the boreal region of the Northwest Territories. Can. J. Microbiol. 24:254–260.
231. White, D.C., Pinkart, H.C., and Ringelberg, D.B. 1997. Biomass measurements: Biochemical approaches. In Manual of Environmental Microbiology (Hurst, C.J., Knudsen, G., McInerney, M., Stetzenbach, L.D., and Walter, M., Eds.), pp. 91–101. ASM Press, Washington, DC.
232. Whitlock, J.L. 1990. Biological detoxification of precious metal processing wastewaters. Geomicrobiol. J. 8:241–249.
233. Williams, P.A. and Sayers, J.R. 1994. The evolution of pathways for the catabolism of aromatic hydrocarbons. Biodegradation 5:195–217.
234. Wilson, M.S. and Madsen, E.L. 1996. Field extraction of a transient intermediary metabolite indicative of real time in situ naphthalene biodegradation. Environ. Sci. Tech. 30:2099–2103.
234a. Wilson, M.S. and Madsen, E.L. 1997. Detection of in situ transcription of naphthalene diioxygenase genes in coal tar waste-contaminated groundwater. Abstr. Annu. Meet. Am. Soc. Microbiol. p. 404.
235. Winding, A., Binnerup, S.J., and Sorensen, J. 1994. Viability of indigenous soil bacteria assayed by respiratory activity and growth. Appl. Environ. Microbiol. 60: 2869–2875.
236. Wolf, D.C. and Skipper, H.D. 1994. Soil sterilization. In Methods of Soil Analysis, Part 2 (Weaver, R.W., Angle, S., Bottomley, P., Bezdicek, D., Smith, S., Tabatabai, A., and Wollum, A., Eds.), pp. 41–52. Soil Science Society of America, Madison, WI.
237. Wolfe, D.A., Hameedi, M.J., Galt, J.A., Watabayashi, G., Short, J., O'Claire, C., Rice, S., Michel, J., Payne, J.R., Braddock, J., Hanna, S., and Sale, D. 1994. The

fate of the oil spilled from the Exxon Valdez. Environ. Sci Technol. 28:561A–568A.

238. Wolin, E.A., Wolin, M.J., and Wolfe, R.S. 1963. Formation of methane by bacterial extracts. J. Biol. Chem. 238:2882–2886.
239. Wollum, A.G., II. 1994. Soil sampling for microbiological analysis. In Methods of Soil Analysis, Part 2 (Weaver, R.W., Angle, S., Bottomley, P., Bezdicek, D., Smith, S., Tabatabai, A., and Wollum, A., Eds.), pp. 2–14. Soil Science Society of America, Madison, WI.
240. Woomer, P.L. 1994. Most probable number counts. In Methods of Soil Analysis, Part 2 (Weaver, R.W., Angle, S., Bottomley, P., Bezdicek, D., Smith, S., Tabatabai, A., and Wollum, A., Eds.), pp. 59–79. Soil Science Society of America, Madison, WI.
241. Worsey, M.J. and Williams, P.A. 1975. Metabolism of toluene and xylenes by *Pseudomonas* ([sic] *putida* (arvilla) mt-2; evidence for a new function of the TOL plasmid. J. Bacteriol. 124:7–13.
242. Wyndham, R.D., Nakatsu, C., Peel, M., Cashore, A., Ng, J., and Szilagyi, F. 1994. Distribution of the catabolic transposon Tn5271 in a groundwater bioremediation system. Appl. Environ. Microbiol. 60:86–93.
243. Wyndham, R.C., Cashore, A.E., Nakatsu, C.H., and Peel, M.C. 1994. Catabolic transposons. Biogradation 5:323–342.
243a. Yager, R.M., Bilotta, S.E., Mann, C.L., and Madsen, E.L. 1997. Metabolic adaptation and in situ attenuation of chlorinated ethenes by naturally-occurring microorganisms in a fractured dolumite aquifer near Niagara Falls, New York. Environ. Sci. Technol., in press.
244. Yen, K.M., Karl, M.R., Blatt, L.M., Simon, M.J., Winter, R.B., Fausset, P.R., Lu, H.S., Harcourt, A.A., and Chen, K.K. 1991. Cloning and characterization of a *Pseudomonas mendocina* KR1 gene cluster encoding toluene-4-monooxygenase. J. Bacteriol. 173:5315–5327.
245. Young, L.Y. 1997. Anaerobic biodegradability assay. In Manual of Environmental Microbiology (Hurst, C.J., Knudsen, G.R., McInerney, M.J., Stetzenbach, L.D., and Walter, M.V., Eds.), pp. 802–805. ASM Press, Washington, DC.
246. Young, L.Y. and Cerniglia, C.E., Eds. 1995. Microbial Transformation and Degradation of Toxic Organic Chemicals. Wiley-Liss, New York.
247. Zehnder, A.J.B. and Stumm, W. 1988. Geochemistry and biogeochemistry of anaerobic habitats. In Biology of Anaerobic Microorganisms (Zehnder, A.J. B., Ed.), pp. 1–37. John Wiley and Sons, New York.
248. Zibilske, L.M. 1994. Carbon mineralization. In Methods of Soil Analysis, Part 2 (Weaver, R.W., Angle, S., Bottomley, P., Bezdicek, D., Smith, S., Tabatabai, A., and Wollum, A., Eds.), pp. 835–863. Soil Science Society of America, Madison, WI.
248a. Zipper, C., Nickel, K., Angst, W., and Kohler, H.P. 1996. Complete microbial degradation of both enantiomers of the chiral herbicide mecoprup (CRS)-2-(4-chloro-2-methyl phenoxy) propriorie acid in an enartiselective manner by sphingomonas herbicidovorans sp. nov. Appl. Environ. Microbiol. 62:4318–4322.
249. Zuberer, D.A. 1994. Recovery and enumeration of viable bacteria. In Methods of Soil Analysis, Part 2 (Weaver, R.W., Angle, S., Bottomley, P., Bezdicek, D., Smith, S., Tabatabai, A., and Wollum, A., Eds.), pp. 119–144. Soil Science Society of America, Madison, WI.
250. Zylstra, G.J. and Gibson, D.T. 1989. Toluene degradation by *Pseudomonas putida* F1: nucelotide sequence of the todC1C2BADE genes and their expression in *Escherichia coli*. J. Biol. Chem. 264:14940–14946.

# 16

J.I. PROSSER

# Mathematical Modeling and Statistical Analysis

Many ecological studies require only qualitative data. It may be sufficient, for example, to know that a particular organism is present in an environment or that a process is operating. Increasingly, however, a complete or even reasonable understanding of microbial ecology requires quantitative study. Quantification requires some application of mathematics, and this chapter describes two different applications: mathematical modeling and statistical analysis. The first section considers the functions of mathematical models and the techniques involved in their construction. The second section describes simple statistical tests used to analyze experimental data. In both sections, emphasis is placed on the basic principles involved, rather than providing a complete survey, which would be impossible in a chapter of this length.

## MATHEMATICAL MODELS

The most important considerations in applying mathematical models are the function of the model and its relationship to experimental work. At one extreme are mechanistic models of a system that represent quantitative hypotheses. They are based on a conceptual model consisting of simplifying assumptions that express our understanding of the factors believed to be controlling the system. These assumptions are represented in the form of mathematical equations, with associated physical constraints and boundary conditions. When provided with initial conditions and parameter values, these models generate predictions that are tested experimentally to increase or decrease confidence in the hypothesis and the assumptions on which they are based. Mechanistic models may also be used to estimate parameter values and, through sensitivity analysis, to determine which components of the system are likely to have the greatest influence in controlling the system. To be of value, the model and the assumptions on which it is based must balance correctly the simplicity required for analysis and the complexity required to represent, in a meaningful way, the real-life system. At the other extreme are descriptive models, which have the sole function of describing the data in the form of a simple mathematical equation. These models are statistical curve-fitting techniques (e.g., non-linear regression analysis [19]), and the techniques overlap with those described in the second section of this chapter. In these cases, experimental empirical observations provide the starting point, and the model is the end product, whereas generation of predictions by mechanistic models involves a continual interaction between modeling and experimental work.

In practice this distinction may be blurred, and mechanistic models may contain components that are empirically based, whereas empirical models may give clues to mechanisms. Nevertheless, the purpose of the modeling exercise should be clear. Modeling studies can suffer from a belief that empirical models provide an understanding of the systems that they are describing. Similarly, mechanistic models, and the experimental laboratory systems used to test them, may be considered too simple or unrealistic when their actual function is to test assumptions and not to represent the real-life system.

A mathematical model is a representation of a system or a process in mathematical form. Several terms are associated with modeling [20]; they are largely concerned with different approaches, procedures, or techniques used in model construction. Descriptive models aim to describe experimental data in a convenient and simple mathematical form. For example, the sigmoid curve and more complex equations may be used to describe changes in product concentration during incubation studies [17]. The sigmoid curve has also been used to describe changes in biomass concentration during growth. More recently, fractal analysis has been applied to describe the morphology of biological structures (e.g., the branching patterns of mycelial organisms). Another example of quantitative description of an environment is the calculation of species diversity indices.

Some predictive, mechanistic models are designed to test assumptions on which they are based. Comparison of predictions of these models with experimental data provides evidence for these assumptions or indicates that modification of the model is required. This laboratory-based approach is applicable to mechanistic studies, but it may be dangerous to extrapolate predictions to the real environment unless all important environmental factors are incorporated. Other models, however, function to predict environmental changes where there is no feasible alternative. For example, models of nutrient cycling and global change are frequently complex, consisting of many subcomponents with calculated or estimated rates of exchange between compartments. Although it is not feasible to test each component experimentally, it is necessary to provide some estimates or predictions of changes in the environment. These models therefore serve an important function and are valuable as long as their limitations are realized.

Most models in microbial ecology are deterministic, in that the predicted behavior of the system is determined by the model equations and the values of constants and initial conditions required for the generation of predictions. The alternative approach is to use statistical, probabilistic, or stochastic models, in which the probability of particular events taking place is considered. Such models therefore consider variation in the properties of individuals and are typical of epidemiological models. However, they are more difficult to deal with mathematically, and deterministic models are usually acceptable where the scale of investigation justifies consideration of bulk or population properties, rather than those of microenvironments or individuals. Deterministic models are less reliable and useful for detailed mechanistic studies, where cell concentrations are low or where chance events, such as mutation, are being modeled.

Complex systems may be subdivided into interlinked compartments. Such structured models take account of, for example, internal cell or population structure, whereas unstructured models lump together all components. It may not be possible to understand in detail every component of complex systems, as indicated for nutrient cycling models. In these circumstances, a black box approach may be justified, where all that is required is an expression relating inputs to and outputs from the compartment.

Microbial populations and the environment itself may be considered in different ways. Populations may be studied at the level of individual cells or groups of cells in segregated models. Distributed models are more common in which the biomass is considered to be uniformly distributed throughout the environment. In this respect, these models consider biomass to be distributed in a similar manner to substrates and products; effectively the biomass is in solution. This approach is valid in many situations, but not where the heterogeneity of the environment, with respect to both chemicals and biomass, is important. There may also be heterogeneity within the biomass, and this characteristic is considered by structured models. Structure can be incorporated at the cell or population level. For example, structured models of cell growth may compartmentalize cells into proteins, nucleic acids, and carbohydrates [25], whereas structured population models might distinguish between spores and vegetative cells.

Most models in microbial ecology are concerned with the role of microorganisms in changing the environment and are therefore usually based on equations considering changes in properties of the physicochemical and biological environments with time. Measuring those changes is achieved most easily using differential equations, at least during model construction, in part because of the wealth of mathematical knowledge and analysis of differential equations. The generation of predictions from such equations requires solution, using analytical or numerical approximation techniques. The former are closed form solutions obtained using standard mathematical techniques for solving differential equations. Analytical solutions are always preferable and can provide important insights into the behavior of a system; for example, through identification of quantitative relationships between components of the system. Differential equations for biological systems are, however, often too complex for analytical solution and thus numerical approximation techniques are required for the solution of equations, simulation, and generation of predictions. Numerical approximation techniques depend on the iterative calculation of values of the dependent variables for certain values of the independent variable. For example, in simulating a simple microbial growth model, biomass concentrations at a time $t + h$ would be estimated, based on a knowledge of the nature of the differential equation and its parameters and the value of biomass at time $t$. Biomass at time $t + 2h$ would then be calculated, based on the value at time $t + h$. The accuracy of this process depends on the approximation technique used, the time interval or step size, and rounding errors. Approximation techniques therefore necessarily operate in discrete time intervals, and some models are specifically designed to operate in this way, an example being the model for nitrogen transformations described later in this chapter. This approach is only feasible using computer programs, but the process has been greatly facilitated by the development and commercial availability of simulation languages.

There is no single best or optimal modeling approach or technique. Similarly there are many situations and investigations for which modeling is inappropriate. The important point, as with any scientific study, is to ask questions and to see how they can best be answered. If quantitative answers are required, some form of mathematics, either modeling or statistical analysis, will probably be required. If modeling is to be used, the system should be examined to determine which approach and which technique is most suitable, having first defined clearly the aim of the modeling process and its relationship with experimental work.

It would be impossible to give examples of the construction of all types of models in this chapter. Instead, two examples have been chosen for detailed study.

The first is the mechanistic modeling of two microorganisms interacting in a chemostat. Experimental and theoretical studies of this sort may be found in the literature. Most are aimed at elucidating the mechanistic basis of interactions and other phenomena observed in nature through the use of controlled experiments designed to test the predictions of models, based on assumptions regarding precise mechanisms. The second detailed description is that of a model of nitrogen transformation in the soil. This model adopts a different approach, using several relatively simple equations to quantify changes and then to use these to estimate important properties of the system. More general treatments of the use of mathematical models in microbiology can be found in, for example, Bazin [1] and Bazin and Prosser [2].

## Two-Species Interactions between Microorganisms in Chemostat Culture

In a chemostat, the experimental conditions imposed, in particular continuous stirring, and the high biomass concentrations justify the assumption of the homogeneous and uniform distribution of cells, substrates, and products. Growth in chemostats is therefore typically modeled using deterministic, unstructured, and distributed models. In this example, the growth of and interaction between two microbial species in a chemostat are modeled. The model is constructed from basic equations for growth of a single organism, taking into account additional terms required for interactions.

The basic equations for chemostat growth [16] are two differential equations describing changes in biomass concentration ($x$) and the concentration of the growth-limiting substrate ($s$). Each considers input and output and any change occurring within the chemostat vessel. The change in $x$ is therefore given by the equation:

$$\frac{dx}{dt} = \underset{\text{no input}}{0} + \underset{\text{increase in } x \text{ due to growth}}{\mu x} - \underset{\text{decrease in } x \text{ due to removal}}{Dx} \qquad \text{[Eq. 1]}$$

After inoculation of the medium, biomass is not supplied to the chemostat, and the input term is therefore 0. Within the vessel the organism uses substrate for growth. The change in concentration of biomass within the vessel depends on the amount of biomass currently present and the specific growth rate ($\mu$) of that biomass. Therefore, growth is autocatalytic, and in the absence of any limitation to growth, biomass concentration would increase exponentially.

Two factors limit the rate of increase in biomass concentration. The first is the indirect effect of the removal of organisms. The decrease in biomass concentration per unit time due to removal is, like growth, dependent on the current biomass concentration. It is also greater if flow rate ($F$) is increased and if the volume of culture fluid ($V$) is reduced. The effects of flow rate and dilution rate may be combined in a single constant, the dilution rate ($D$), defined as $F/V$. The second limiting factor is the concentration of the growth-limiting substrate; its effect on the specific growth rate is traditionally described by the Monod equation [13]:

$$\mu = \frac{\mu_m s}{K_s + s}$$

where $\mu_m$ is the maximum specific growth rate and $K_s$ is the saturation constant for growth, equivalent to the concentration of the limiting substrate at which $\mu = \mu_m/2$.

The change in substrate concentration is given by the equation:

$$\frac{ds}{dt} = \underbrace{Ds_r}_{\text{increase in } s \text{ due to input in medium}} - \underbrace{\frac{\mu x}{Y}}_{\text{decrease in } s \text{ due to use by } x} - \underbrace{Ds}_{\text{decrease in } s \text{ due to removal}}$$

Substrate is supplied to the vessel in the inflowing medium, and this equation therefore has an input term. Thus, the rate of change of concentration of the substrate in the vessel is proportional to its concentration in the inflowing medium, $s_r$, to the rate at which it is supplied, and to the volume in which it is diluted. The effect of the last two factors can again be combined in the dilution rate, $D$. As substrate is used for growth within the vessel, the rate of decrease in concentration will depend on the concentration of biomass-converting substrate and its specific growth rate. This gives a similar term to that for growth in Equation 1, but is preceded by a negative sign to indicate a decrease in concentration. The rate of decrease in concentration also depends on the yield coefficient, $Y$, which is essentially a stoichiometric coefficient defining the amount of biomass formed per unit substrate converted. Finally, the rate of change of substrate concentration in the vessel is decreased by removal, in the same manner as biomass concentration.

### *Neutralism*

Conversion of this basic mathematical model for monospecies growth to that for the growth of two organisms depends on the mechanism of the interaction. The simplest interaction, or lack of interaction, is neutralism, in which the two populations are limited by different substrates and do not affect each other in any way. Modeling neutralism requires construction of separate equations for changes in biomass concentration and substrate concentration for each organism:

$$\frac{dx_1}{dt} = 0 + \mu_1 x_1 - Dx_1 \qquad \text{[Eq. 4]}$$

$$\frac{dx_2}{dt} = 0 + \mu_2 x_2 - Dx_2 \qquad \text{[Eq. 5]}$$

$$\frac{ds_1}{dt} = Ds_{r1} - \frac{\mu_1 x_1}{Y_1} - Ds_1 \qquad \text{[Eq. 6]}$$

$$\frac{ds_2}{dt} = Ds_{r2} - \frac{\mu_2 x_2}{Y_2} - Ds_2 \qquad \text{[Eq. 7]}$$

Subscripts 1 and 2 refer to populations of species 1 and 2, and equivalent Monod expressions also apply for each species. Growth of the two species is limited by different substrates, supplied in the inflowing medium at concentrations $s_r$ 1 and $s_r$ 2.

### *Competition*

To illustrate how this model may be modified for different interactions, let us consider competition for a single substrate. In this case, we require only a single equation for changes in substrate concentration, but we must consider, within this equation, utilization of substrate by both organisms. The organisms are not interacting directly, and neither $x_1$ nor $x_2$ appears in Equations 5 and 4, respectively.

The Monod terms, however, require modification as the species now share a common substrate. A model for competition is therefore

$$\frac{dx_1}{dt} = 0 + \mu_1 x_1 - Dx_1 \qquad \text{[Eq. 8]}$$

$$\frac{dx_2}{dt} = 0 + \mu_2 x_2 - Dx_2 \qquad \text{[Eq. 9]}$$

$$\frac{ds}{dt} = Ds_r - \frac{\mu_1 x_1}{Y_1} - \frac{\mu_2 x_2}{Y_2} - Ds \qquad \text{[Eq. 10]}$$

$$\mu_1 = \frac{\mu_{m1} s}{K_{s1} + s} \qquad \text{[Eq. 11]}$$

$$\mu_2 = \frac{\mu_{m2} s}{K_{s2} + s} \qquad \text{[Eq. 12]}$$

More extensive treatment of competition in chemostat culture and its ecological implications is given by Harder, Kuenen, and Matin [6] and Matin and Veldkamp [9].

### *Predation*

If one species is a protozoan consuming the second, a bacterium, we again require three differential equations. Only the bacteria require a supply of substrate, the protozoan using the bacteria as its substrate. The model for predation is then

$$\frac{dx_1}{dt} = 0 + k_1 \mu_1 x_1 - Dx_1 \qquad \text{[Eq. 13]}$$

$$\frac{dx_2}{dt} = 0 + \mu_2 x_2 - \frac{\mu_1 x_1}{Y_1} - Dx_2 \qquad \text{[Eq. 14]}$$

$$\frac{ds}{dt} = Ds_r - \frac{\mu_2 x_2}{Y_2} - Ds \qquad \text{[Eq. 15]}$$

$$\mu_1 = \frac{\mu_{m1} x_2}{K_{s1} + x_2} \qquad \text{[Eq. 16]}$$

$$\mu_2 = \frac{\mu_{m2} s}{K_{s2} + s} \qquad \text{[Eq. 17]}$$

Subscripts 1 and 2 refer to protozoa and bacteria, respectively. Here there is a direct interaction between the two species, as protozoa consume bacteria. This interaction is represented by the negative term in Equation 14, describing the effect of predation on the rate of change in bacterial biomass. This effect is dependent on predator biomass concentration and specific growth rate. The constant $Y_1$ is the yield coefficient for growth of the predator on the prey. Similarly, bacterial biomass is a component of Equation 13 through its effect on the specific growth rate of the predator. Equation 16 considers bacterial biomass to affect the predator specific growth rate in a Monod-type relationship. The constant $k_1$ represents the amount of predator biomass formed per unit bacterial biomass consumed. Experimental testing of a model of this type is provided by Tsuchiya et al. [24].

### *Commensalism and Amensalism*

Indirect microbial interactions also occur through production of compounds by one species that benefit or detrimentally affect another. The simplest example of a beneficial effect—commensalism—occurs when one species produces a substrate that is consumed by the second. The set of equations for this interaction is similar to that for neutralism consisting of equations for two biomass concentrations and two substrate concentrations, but differs in terms of supply of substrate to the second organism.

$$\frac{dx_1}{dt} = 0 + \mu_1 x_1 - Dx_1 \qquad \text{[Eq. 18]}$$

$$\frac{ds_1}{dt} = Ds_{r1} - \frac{\mu_1 x_1}{Y_1} - Ds_1 \qquad \text{[Eq. 19]}$$

$$\frac{dx_2}{dt} = 0 + \mu_2 x_2 - Dx_2 \qquad \text{[Eq. 20]}$$

$$\frac{ds_2}{dt} = k\mu_1 x_1 - \frac{\mu_2 x_2}{Y_2} - Ds_2 \qquad \text{[Eq. 21]}$$

The substrate for species 2 is not supplied in the inflowing medium, but is produced in proportion to the concentration of biomass of species 1 and its specific growth rate. The constant $k$ represents the amount of substrate produced per unit biomass of species 1.

Amensalism may occur through production of compounds by one organism that kill or inhibit growth of the second. Inhibition of growth is modeled here, assuming that the organisms are growing on separate substrates. This model requires five equations; four based on those for neutralism (Equations 4–7), with a fifth equation for changes in the concentration of the inhibitor. An example is:

$$\frac{di}{dt} = 0 + k\mu_1 x_1 - Di \qquad \text{[Eq. 22]}$$

Equation 22 assumes that inhibitor (concentration $i$) is produced as a function of biomass concentration and the specific growth rate of species 1, with $k$ representing the amount produced per unit biomass. Its effect on the specific growth rate of species 2 might be represented by modification of the Monod equation assuming competitive inhibition [16].

$$\mu_2 = \frac{\mu_{m2} s_2}{\alpha K_{s2} + s_2}$$

where $\alpha = 1 + i/K_i$ and $K_i$ is the inhibitor constant.

These models are only examples of some that might be used, but all are realistic in the sense that they have been used to model interactions studied in laboratory experimental systems. Further modifications may be made to determine the effects of different forms of inhibition and stimulation of growth, combinations of interactions, and effects of physicochemical factors (for examples, see [3, 10, 23]).

## Transformations of Soil Nitrogen

The second example is a model for transformation of soil nitrogen [8], the aim being to determine the turnover time of nitrogen through the microbial biomass. The approach is less mechanistic than that for chemostat interactions, with simpler kinetic equations, and it demonstrates how a close interaction between modeling and experimental data can yield useful information, in addition to evidence for the assumptions on which the model is based. Full details of the model and its simulation can be found in the original publication, and only the major points are discussed here.

The model considers soil nitrogen to consist of four components, connected as illustrated in Figure 16-1. Root, stubble, and immobilized N (RI-N); soil microbial biomass N (BIO-N); and humus N (HUM-N) decay according to first-order kinetics. RI-N decays to form BIO-N and HUM-N. BIO-N and HUM-N in turn decay to form inorganic N (RI-N), with generation of more BIO-N and HUM-N. The system was assumed to be at steady state on a yearly time scale, and changes in the different components were determined by simulation of the model at monthly time intervals. It therefore represents an example of a discrete time model in which results are presented at specific time points, rather than the more usual continuous time approach. The amounts of N in the different compartments were determined by measurement of $^{15}N$ after application of $^{15}N$-labeled fertilizer to a wheat crop. To illustrate the simulation process, if $T$ is the amount of $^{15}N$ in compartment RI-N at the beginning of the experiment, after 1 month the distribution of label will be given by the equations:

$$R_1 = Te^{-rt}$$
$$B_1 = T(1 - e^{-rt})\beta/(\alpha + \beta)$$
$$H_1 = T(1 - e^{-rt})\alpha/(\alpha + \beta)$$

where $R$, $B$, and $H$ are the amounts of label in RI-N, BIO-N, and HUM-N respectively; $r$ is the first-order rate constant for decay of RI-N; and $\alpha$ and $\beta$ are the fractions of N entering humus and biomass, respectively. The subscript 1 refers to the time in months from initiation of the simulation. One month later, the crop is harvested, and the equivalent compartment sizes will be

$$R_2 = R_1e^{-rt}$$
$$B_2 = B_1e^{-bt} + R_1(1 - e^{-rt})\beta/(\alpha + \beta) + B_1(1 - e^{-bt})\beta + H_1(1 - e^{-ht})\beta$$
$$H_2 = H_1e^{-ht} + R_1(1 - e^{-rt})\alpha/(\alpha + \beta) + B_1(1 - e^{-bt})\alpha + H_1(1 - e^{-ht})\alpha$$

where $b$ and $h$ are first-order rate constants for decay of BIO-N and HUM-N, respectively. This iterative procedure was continued for 5 years, and the rate

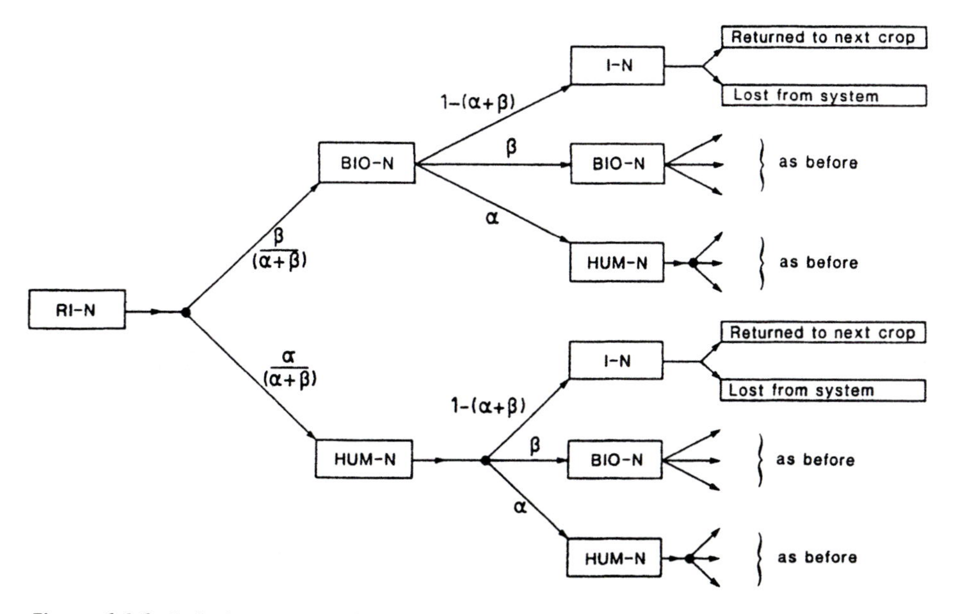


Figure 16-1 Soil nitrogen transformations considered in the model of Jenkinson and Parry [8]. See text for the explanation of symbols.

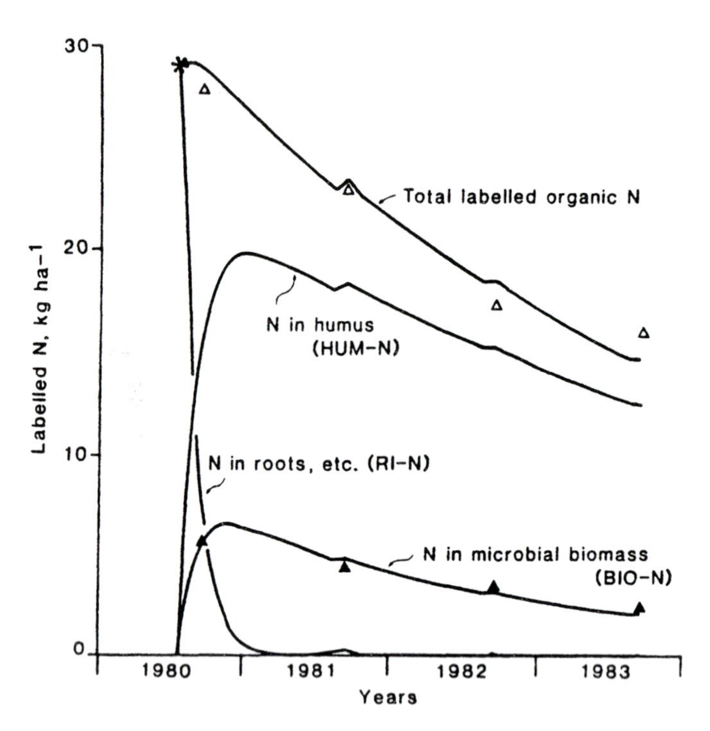


Figure 16-2 Experimental data (Δ, Δ) for labeled N in different compartments and predictions (continuous line) of the model of Jenkinson and Parry [8].

constants that best fitted the experimental data were determined. The parameter fitting process was carried out by systematic alteration of constants to optimize fit, but it could also be achieved using computer library routines. Simulations of the model were then compared with experimental data and were seen to show close agreement (Fig. 16-2), despite the simplifying assumptions made in constructing the model and analyzing the experimental data. This process provided best fit values for the first-order rate constants for decay of each component and for $\alpha$ and $\beta$. These values were then used to determine the turnover times for different components. For example, a turnover time for soil microbial biomass of 1.52 years was calculated by dividing the standing microbial biomass crop, BIO-N, by the annual input of N, which was itself calculated from one of the rate constants of the model. This example illustrates how a relatively simple model, constructed for use in conjunction with carefully designed experiments, can provide important information.

## Generation of Predictions—Simulation

To be of practical use, models must predict behavior that can be compared with experimental data. Predictions of the soil nitrogen transformation model were generated by iterative calculations based on relatively simple algebraic equations. Simulation was achieved using a computer program, but use of a computer merely reduced the time required for calculations and was not essential. Predictions from models consisting of differential equations are ideally obtained by analytical solution to yield algebraic equations. An analytical solution enables characterization of specific relationships; for example, between growth constants and such variables as biomass and substrate concentrations. It also facilitates generation of predictions, or simulation, by direct substitution of values for model constants and initial conditions. Analytical solutions are always preferable, although errors can arise through their use; for example, if the model is not appropriate for the particular real-life situation or if parameter values are incorrect. In all the cases discussed above, even for monospecies growth in a chemostat, analytical solution of the differential equations is only possible under steady-state conditions, and thus numerical approximation techniques are required. These techniques give rise to additional errors associated with the approximation process as discussed in the introduction to this chapter.

### *Approximation Methods*

As their name suggests, numerical approximation techniques involve errors, but for most situations the errors do not significantly affect predictions, and it is relatively easy to test whether errors are significant and to reduce them. The theoretical basis of numerical approximation methods is described in standard texts [5, 7, 11, 12, 21] and worked examples of their use in solving models similar to those in the previous section are given by Prosser [18]. All are based on predicting the value of a variable, such as biomass concentration, at time $t + h$ on the basis of its value at time $t$ and the functional relationship defined by the differential equation. The value of the variable at $t + h$ is then used to estimate the value at time $t + 2h$ using the same procedure. The techniques are heavily dependent on computers to carry out the many iterative calculations involved.

Several techniques are used for numerical approximation, the best known being the Euler method and the Runge-Kutta (R-K) methods. Taking changes in biomass concentration, $x$, as an example, the Euler method uses the following equation to calculate biomass concentration an interval $h$ later:

$$x_{t+h} = x_t + hf(t, x) \qquad \text{[Eq. 30]}$$

where $f(t, x)$ is the functional relationship between biomass concentration and time defined in the differential equation. In fact, the second term in Equation 30 is the first in a series of terms involving increasing orders of $h$ (i.e., $h^2$, $h^3$, $h^4$, etc.). Use of a single term results in inaccuracies, termed truncation errors, of magnitude $h^2$. Truncation errors may be global, accumulating at each time step, or local, associated with each time interval. Local truncation errors can be decreased by the use of smaller step sizes. Accuracy also depends on the precision of individual computations (i.e., the number of significant figures used in calculations). Rounding errors in calculations increase with the number of terms in the expansion series, are cumulative over a series of time intervals, and increase with decreasing step size. In choosing a value for the step size, therefore, a balance must be achieved between truncation and rounding errors. An additional potential problem is instability when approximations lead to an overflow; for example, negative values of biomass concentration. An overflow may occur when variables are approaching zero rapidly, and the step size is not small enough to predict asymptotic decreases.

The Euler method is relatively simple and economical on computer time, but an acceptable level of accuracy requires use of multistep methods, in which information on two previous steps is used to calculate the next value. As a result the technique is said to be "non-self-starting and one step." Self-starting techniques, in particular Runge-Kutta (R-K) integration methods, are used more commonly. R-K methods provide greater accuracy by basing approximations on several pairs of values of $t$ and $x$ within the time interval $h$. The order of the integration routine is equal to the number of pairs of values used with fourth-order R-K techniques being satisfactory for most applications. Although these techniques are computationally more complex, they have significant advantages in terms of accuracy and stability and are used widely.

R-K methods are suitable for the solution of ordinary differential equations, and other techniques are available for special circumstances. In situations where changes occur with respect to two variables, partial differential equations are required. For example, models of nutrient transformations in the soil may require consideration of changes in concentrations with both time and distance. Partial differential equations are more difficult to solve, both analytically and using numerical approximation methods. Changes must be calculated over small intervals in both variables, with the scope for errors in both and computation correspondingly more complex. The situation may be approached using finite difference or finite element techniques. The latter may be illustrated by models for transformations of nutrients passing through soil columns. The finite element approach segments the column into several compartments, each of which is considered to behave as a chemostat, with perfect mixing of biomass and substrate. Flow is modeled by passing the products of one compartment to the next at specified time steps.

### *Computing*

Most simulations use computers because of the need for numerical approximation methods. The degree of computing expertise required to carry out computer simulations depends to some extent on the complexity of the model. Simulation programs require these elements:

provision of values for growth constants and for initial conditions
definition of the time at which simulation will end
description of the differential equations
integration of the differential equations
output of results graphically or as tabulated values of variables at each time interval

These requirements may be met using computer programs, written in such languages as FORTRAN, BASIC, or C or by the use of simulation languages. Computer languages are usually necessary for more difficult simulations, such as those involving numerical solution of partial differential equations and finite difference and element techniques. Computer programs may also be desirable for the solution of sets of ordinary differential equations. These programs provide maximum flexibility in the simulation process, in particular facilitating choices and branch points to be made on achievement of certain conditions during simulation. Some experience in computer programming is required, but the task is made easier by the availability of library subroutines for a wide range of approximation methods [14]. These library subroutines are now increasingly available for partial differential equations, in addition to ordinary differential equations [15].

Solution of the chemostat and soil nitrogen models described in this chapter is generally possible using simulation languages. These languages are designed to provide the computational facilities required without necessitating significant programming expertise through the use of simplified statements for input of information, solution of differential equations, and output of results. They are usually available on mainframe computers and are now available for microcomputers. They are becoming more user-friendly and more flexible and can provide on-screen, interactive observation of model behavior and the effects of changing experimental conditions. Simulation languages also provide a range of integration routines, reducing problems of inaccuracy and instability.

### *Constants*

Simulation requires the provision of values for all constants in equations describing the system and for all initial conditions. Where the modeling exercise is linked directly to experimental work, values of initial conditions, such as initial biomass and substrate concentrations, can be determined readily and controlled experimentally. Similarly, for chemostat studies, such factors as the dilution rate and inflowing substrate concentration can be controlled. Frequently, however, experimentally determined values of all growth constants and proportionality constants used in equations are difficult or impossible to obtain. A realistic range of values can usually be defined, and ball park values can frequently be obtained from the literature. Although, ideally constants should be determined independently of the experiments designed to test model predictions, comparison with predicted and experimental results can be used to determine values of constants giving the optimal fit. Although somewhat tautological, this approach is very valuable. For an inaccurate model, a good fit will not be obtained, whereas a close fit will give confidence in the values of the constants. This principle applies particularly if these values are then used to generate predictions under different conditions that can be tested experimentally.

## Conclusions

The construction of mathematical models of relevance to microbial ecology requires some knowledge of mathematics and computing. In the past the traditional reluctance of many biologists to apply mathematics may have limited their use. The availability and user-friendliness of modern microcomputer software have largely removed this limitation. The need to solve equations analytically is reduced by computer simulation and simulation languages, which are now sufficiently user-friendly to enable use by those with a minimum of experience in the use of computers. A second and more fundamental limitation is conceptual, which applies both to perception of the need to use mathematical models and to their construction. Models have many different functions, and misconceptions regarding these functions have led to misinterpretation of their worth. For this reason, the precise function of a model must be determined, along with its relationship to experimental work. In constructing models, there may also be a conceptual problem in relating experimentally determinable characteristics of the system (e.g., biomass concentration, substrate concentration) to algebraic symbols with which they are represented. This problem may be more difficult to overcome, but it is worthwhile remembering that mathematical equations represent what we might otherwise express in words, but more precisely and concisely.

# STATISTICAL ANALYSIS

Statistical analysis can serve two functions. It can be used to organize or describe experimental data that have been collected, or it can provide a means of inferring conclusions about a population using experimental data from parts or samples of that population. As with mathematical modeling, it is essential that the need for and purpose of statistical analysis are identified and clarified before an experimental study is carried out. Some experiments, of course, do not require statistics, but analysis is generally required for those generating quantitative data. When this is the case, the type of statistical analysis to be used must be identified before the experiment is begun. The most common complaint of those operating statistical advisory services is that experimentalists come for advice after they have obtained their data, rather than before. It is very easy to design experiments that appear to be satisfactory in terms of, for example, the nature and level of replication, but for which appropriate statistical analysis is not possible.

The aim of this section is to demonstrate ways in which statistics may be used and to illustrate the concepts on which statistical tests are based. More comprehensive accounts of statistical analysis may be found in the large variety of statistical textbooks available (e.g. [4, 22]).

## Description of Experimental Data

The means by which experimental data may be described vary considerably, but the aim is usually to represent in a meaningful way the important features of observations, which may be unordered and incomprehensible in their raw form. Examples are graphical methods and numerical descriptive measures of central tendency and variation. Graphs provide a pictorial representation of the data and have various forms. The most common are plots of mean values as a function of a particular treatment, such as time, and histograms or frequency polygons to describe the distribution of a population from a sample of observations.

### *Measures of Central Tendency*

Probably the most common representation of several quantitative observations is a measure of the central tendency of the data set, which condenses information into a single value. For example, data on cell concentration for each soil type, presented in Table 16-1, would rarely be quoted as individual values of each observation, but would be described by calculating a measure of the central tendency—the arithmetic mean concentration for each soil type. Similarly, if soil type A was being compared with soil type B, we would compare mean values, rather than individual values. The most commonly used statistic for measuring the central tendency is the arithmetic mean:

$$\bar{x} = \frac{1}{n} \sum_{i=1}^{n} x_i$$

where $n$ = number of samples, $X_i$ = value of the $i$th variate, and $\bar{x}$ = mean.

The other commonly used measure of central tendency is the median, which is calculated by arranging all observations in order of magnitude. The median is then the central value or the value such that greater and smaller values occur with equal frequency (Table 16-1). It therefore has two definitions. If $n$ is odd, the median is the $(n + 1)/2$ value. If $n$ is even, the median is halfway between the $n/2$ and $(n + 1)/2$ values.

Both the mean and the median are based on all observations, are intuitive, and are relatively easy to understand and calculate. The median is less rigidly defined and is also less stable with respect to fluctuations in sampling. It is less affected by extreme values, which is particularly important for small samples. For example, if one of three replicate measurements is suspected of containing a large error, it is better to use the median than the mean of the two remaining values. The median also varies less when data are transformed, but the ease with which the mean can be manipulated algebraically has led to its greater general use.

### *Measures of Variability*

The mean does not provide a complete description of a set of experimental data, because data are scattered: They do not all have the same value. To describe

**Table 16-1** Data on the Cell Concentration in Six Soil Types[a]

| Replicate | Soil Type A | B | C | D | E | F |
|---|---|---|---|---|---|---|
| 1 | 550 | 421 | 223 | 614 | 458 | 369 |
| 2 | 480 | 450 | 250 | 735 | 406 | 258 |
| 3 | 580 | 487 | 287 | 654 | 355 | 357 |
| 4 | 512 | 367 | 315 | 692 | 497 | 321 |
| 5 | 450 | 390 | 269 | 589 | 456 | 349 |
| 6 | 530 | 461 | 241 | 644 | 482 | 365 |
| 7 | 595 | 402 | 195 | 687 | 340 | 264 |
| 8 | 547 | 412 | 263 | 564 | 422 | 288 |

[a]Bacterial concentrations were determined 2 weeks after inoculation of soil using the dilution plate technique, with media selective for the inoculated bacterial strain. Each count was replicated eight times and the data are given below. Units are cells $g^{-1}$ soil $\times 10^{-2}$.

Calculation of mean cell concentration for soil type A

$$\bar{x} = \frac{1}{n}\sum_{i=1}^{n} x_i$$

$$\sum_{i=1}^{n} x_i = 550 + 480 + 580 + 512 + 450 + 530 + 595 + 547 = 4244$$

$$\bar{x} = 4244/8 = 530.5$$

Calculation of median cell concentration for soil type A

Order data: 450, 480, 512, 530, 547, 550, 580, 595
Median = (530 + 547)/2 = 538.5

Calculation of standard deviation for cell concentration for soil type A

| | $x$ | $x^2$ | $x - \bar{x}$ | $(x - \bar{x})^2$ |
|---|---|---|---|---|
| 1 | 550 | 302500 | 19.5 | 380.25 |
| 2 | 480 | 230400 | −50.5 | 2550.25 |
| 3 | 580 | 336400 | 49.5 | 2450.25 |
| 4 | 512 | 262144 | −18.5 | 342.25 |
| 5 | 450 | 202500 | −80.5 | 6480.25 |
| 6 | 530 | 280900 | −0.5 | 0.25 |
| 7 | 595 | 354025 | 64.5 | 4160.25 |
| 8 | 547 | 299209 | 16.5 | 272.25 |
| Totals | 4244 | 2268078 | 0.0 | 16636.00 |

$$\text{Total corrected sum of squares} = \sum_{i=1}^{n} (x_i - \bar{x})^2 = 16636$$

$$\text{Variance} = s^2 = \frac{1}{n-1}\sum_{i=1}^{n} (x_i - \bar{x})^2 = 16636/7 = 2376.57$$

$$\text{Standard deviation} = s = \sqrt{\frac{1}{n-1}\sum_{i=1}^{n} (x_i - \bar{x})^2} = \sqrt{2376.57} = 48.75$$

Standard deviation can be calculated more easily using the equation:

$$s = \sqrt{\frac{1}{n-1}\left[\sum_{i=1}^{n} x_i^2 - \frac{1}{n}\left(\sum_{i=1}^{n} x_i\right)^2\right]} = \sqrt{\frac{1}{7}\left(2268078 - \frac{4244^2}{8}\right)} = 48.75$$

The standard deviation of the mean, or standard error, is given by the equation:

$$s_{\bar{x}} = \sqrt{\frac{s^2}{n}} = \frac{s}{\sqrt{n}} = \frac{48.75}{\sqrt{8}} = 17.24$$

(continued)

Table 16-1 (continued)

Student's $t$-test to compare cell concentrations in soil types A and B

| | A | | B | |
|---|---|---|---|---|
| | $x_1$ | $x_1^2$ | $x_2$ | $x_2^2$ |
| 1 | 550 | 302500 | 421 | 177241 |
| 2 | 480 | 230400 | 450 | 202500 |
| 3 | 580 | 336400 | 487 | 237169 |
| 4 | 512 | 262144 | 367 | 134689 |
| 5 | 450 | 202500 | 390 | 152100 |
| 6 | 530 | 280900 | 461 | 212521 |
| 7 | 595 | 354025 | 402 | 161604 |
| 8 | 547 | 299209 | 412 | 169744 |
| Totals | 4244 | 2268078 | 3390 | 1447568 |

$$\bar{x}_1 = 530.5$$

$$\bar{x}_2 = 423.75$$

$$t = \frac{\bar{x}_1 - \bar{x}_2}{\sqrt{s_p^2\left(\frac{1}{n} + \frac{1}{m}\right)}}$$

$$s_p^2 = \frac{f_1 s_1^2 + f_2 s_2^2}{f_1 + f_2} = \frac{\sum x_1^2 - \left(\sum x_1\right)^2 \Big/ n + \sum x_2^2 - \left(\sum x_2\right)^2 \Big/ m}{n + m - 2}$$

$$s_p^2 = \frac{2268078 - 4244^2/8 + 1447568 - 3390^2/8}{8 + 8 - 2} = 1978$$

$$t = \frac{530.5 - 423.75}{\sqrt{1978\left(\frac{1}{8} + \frac{1}{8}\right)}} = \frac{106.75}{\sqrt{494.5}} = 4.8$$

The tabulated $t$ value for 14 degrees and a probability of 0.05 is 2.145. Therefore, the means are different at the 0.05 level of significance.

variability about the mean, we most commonly use the variance and the standard deviation. If the mean is subtracted from each individual value, the sum of the differences is zero (Table 16-1). This is represented mathematically as

$$\sum_{i=1}^{n} (x_i - \bar{x}) = 0$$

If deviations from the mean are squared, each value will be positive and the sum of deviations may be used to assess variation. This value is the total corrected sum of squares:

$$\sum_{i=1}^{n} (x_i - \bar{x})^2 = 0$$

This value increases with sample size, $n$ and so must be standardized to compare different-sized samples. Although it seems logical to divide the value by $n$, in fact it is divided by the number of degrees of freedom (defined below), which equals $n - 1$, to give the variance, $s^2$:

$$s^2 = \frac{1}{n - 1} \sum_{i=1}^{n} (x_i - \bar{x})^2$$

The units of the variance are the square of those of observations. Thus, the square root of the variance, the standard deviation ($s$), is used more often as a measure of variability:

$$s = \sqrt{\frac{1}{n-1} \sum_{i=1}^{n} (x_i - \bar{x})^2}$$

A simpler formula for calculation of standard deviation is

$$s = \sqrt{\frac{1}{n-1} \left[ \sum_{i=1}^{n} x_i^2 - \frac{1}{n} \left( \sum_{i=1}^{n} x_i \right)^2 \right]}$$

To explain why the denominator is ($n - 1$), rather than $n$, we need to consider the distinction between populations and samples. For example, if we wish to know the number of bacteria in a gram of soil, the population in which we have an interest is all the bacterial cells in that gram. However, it is extremely rare, and usually impossible practically to count each member of such a population. Usually we use a portion or sample of the population, and by estimating the concentration of cells in the sample, we infer the properties of the population as a whole. When we calculate

$$\sum (x_i - \bar{x})^2$$

we assume $\bar{x}$ to have a fixed value. In fact $\bar{x}$ is only an estimate that will vary in different samples. If we assume that $\bar{x}$ is fixed, then we must also assume that $\Sigma x$ is fixed. Then, if we know the values of all observations but one, we can calculate the remaining value by subtraction. Thus, only $n - 1$ observations are free to vary, and we have $n - 1$ degrees of freedom. If the denominator was $n$, we would be overestimating variability, and the denominator must be reduced to $n - 1$.

Another commonly used measure of variation is the range. Variability around a median is assessed by defining percentiles that are values of the variable that split the observations into 100 equal parts (although it can be calculated for less than 100 observations). The median is the 50th percentile, and the 25th, 50th, and 75th percentiles are called quartiles—$Q_1$, $Q_2$, and $Q_3$. The measure of variability is the semi-interquartile range: $Q = (Q_3 - Q_1)/2$.

## Large-Sample Statistical Tests

Inferential statistics involves drawing conclusions about a population on the basis of experimental data from samples of that population. It requires formulation and testing of a hypothesis and statistical testing of that hypothesis. The first stage in statistical testing is therefore formulation of a null hypothesis, $H_O$, about one or more population parameters and an alternative hypothesis, $H_A$, that is accepted if $H_O$ is rejected. The sample data are then used to calculate the appropriate test statistic that is used to determine whether or not $H_O$ is to be accepted or rejected. A rejection region must also be defined that indicates the values of the test statistic that imply rejection of $H_O$.

### *The Normal Distribution*

To understand the basis of many statistical tests, we need to know how observations are distributed, in addition to the central value and scatter. Many continuous biological characters (but not all) follow the Gaussian or normal distribution when the sample number, $n$, is large ($n \geq 30$; Fig. 16-3a). Thus, if $x$ represents the number of cells $g^{-1}$ and $f(x)$ is the frequency of occurrence of different num-

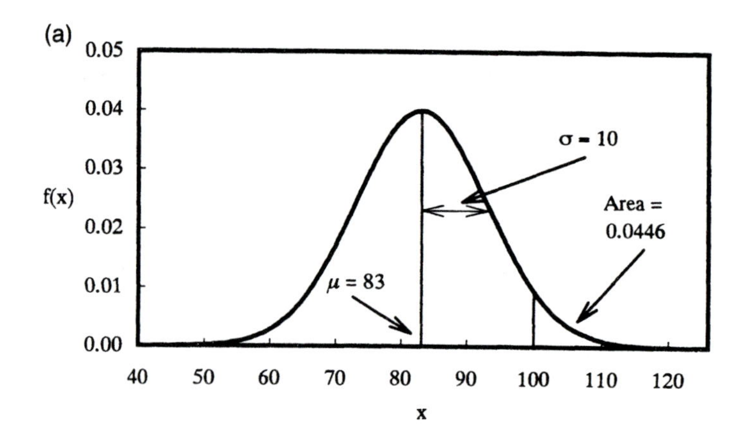


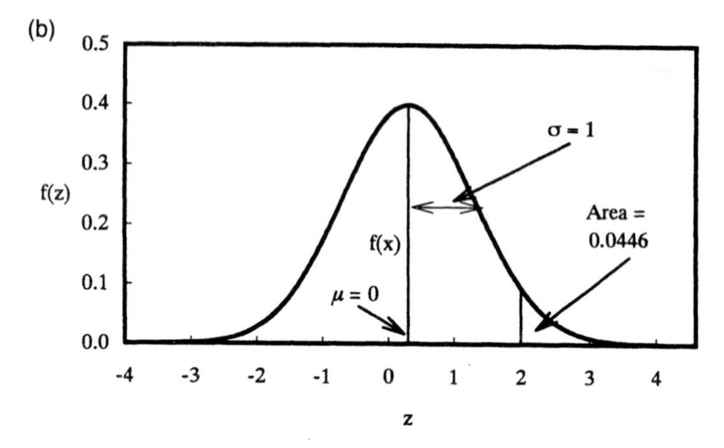


Figure 16-3 The normal distribution (*a*) and standardized normal distribution (*b*). The distance from the mean to the point of inflexion on the curve is equal to the standard deviation. For the standardized normal distribution, the mean = 0 and the standard deviation = 1.

bers of cells, the plot of $f(x)$ versus $x$ for the whole population of cells $g^{-1}$ will give a bell-shaped curve with mean, $\mu$, and standard deviation, $\sigma$. The variation in values is continuous and is distributed equally about the mean, with a greater proportion of values near to the mean and few at the extremes. The standard deviation measures the distance from the mean to the point of inflexion on the curve.

The normal distribution allows us to determine, for example, the proportion of the total population with values above or below a certain value. It would be useful if values associated with the distribution were tabulated. Doing so would, however, require a set of tables for every value of $\mu$ and $\sigma$, which is obviously not feasible. However, the distribution can be easily transformed to give a standardized normal distribution. The data are standardized by effectively subtracting the mean from each value and dividing by the standard deviation to calculate the standard normal variable, $z$.

$$z = \frac{X - \mu}{\sigma}$$

This equation provides a distribution of transformed values, $z$, with $\mu = 0$ and $\sigma = 1$ (Fig. 16-3b) for which tables are available. In practice, we do not need to transform all of our data, but can apply it to individual values. To illustrate this point, imagine that we wish to know the proportion of 1-ml water samples that contain more than 100 coliforms. We know that the (population) mean, $\mu$, is 83 cells $ml^{-1}$ and that the standard deviation, $\sigma$, is 10. The proportion of 1-ml samples with greater than 100 cells is equivalent to the partitioned area of the normal distribution curve, to the right of 100. This area is equivalent to the partitioned area on the standardized normal distribution curve, to the right of the value 1.7. The value 1.7 is obtained by transforming $X = 100$, calculating $z$ using Equation 36:

$$z = \frac{100 - 83}{10} = 1.7$$

The area of the partitioned region is obtained by determining the value equivalent to $z = 1.7$ from tables of the standardized normal distribution and is equal to 0.0446. Thus, the proportion of samples with more than 100 coliforms equals 0.0446 or 4.46%.

Other distributions may be used in certain circumstances. For instance, the binomial distribution is used where data exist in just two categories; for example, Gram-positive/negative, vegetative cells/spore, dead/alive. For such data we consider frequencies of occurrence of different states or events. For the Poisson distribution, the mean and standard deviation are equal. It describes the distribution of the number of cells per square in counting chambers.

### *One-Sample Test of Significance*

Experiments are frequently carried out to determine whether a particular property of the system differs under two sets of conditions or differs from a particular value. To illustrate the latter type of experiment, imagine that we have measured the pH of 50 soil samples in order to determine whether the soil pH is greater than 6.5. The sample mean pH is 6.9, and the standard deviation is 0.3. Even if the soil pH is actually 6.5, it is possible that the pH of none of the individual samples is exactly 6.5 because of natural or experimental variability. The sample pH may therefore be greater than 6.5 because of this variability. Alternatively, the soil pH may in fact be greater than 6.5. To determine which is the case, we test the null hypothesis, $H_0$, which states that the population mean soil pH is 6.5 (i.e., $H_0$: $\mu = 6.5$). In testing this we also set up an alternative hypothesis, $H_A$, which states that $\mu > 6.5$, ignoring any evidence that $\mu < 6.5$. Experimentally we use the sample mean pH, $x$, to estimate the population mean, and we use the sample standard deviations, $s$, to estimate that of the population.

We must now determine whether the difference between 6.5 and 6.9 is due to experimental variability or to a real difference in pH. To do this, we must consider the concept of significance by making the assumption that $H_0$ is true and then determining the probability that the observed difference between $\bar{x}$ and 6.5 (or larger differences) arose by chance (through experimental variability). If the probability is small, then it is likely but not certain that the difference is real or significant. If the probability is large, it is likely that the difference is not real and arose from the natural variability in pH among different samples of soil.

One way in which our null hypothesis can be tested is to calculate the standard normal variable $z$, defined above as

$$z = \frac{\hat{\mu} - \mu_0}{\sigma_{\hat{\mu}}} \qquad \text{[Eq. 36]}$$

where $\hat{\mu}$ is the estimator for the parameter $\mu$, which has a mean $\mu$ and a standard deviation $\sigma_{\hat{\mu}}$, which is assumed to be known or can be approximated. Determining what must be used as the estimator, $\hat{\mu}$, and the standard deviation, $\sigma_{\hat{\mu}}$ requires consideration of the sampling distribution of the sample mean. Each observation we make of a population is an estimate of the population mean, but the more observations we make, the more accurate the estimate. If we calculate many means from experimental sets of, for example, five samples, we can plot a distribution of the means. If we do this for a sufficient number of sets of samples, the central value will equal the population mean,

$$\mu_{\bar{x}} = \mu$$

but the standard deviation of the distribution will depend on the sample size and is given by the equation:

$$s_x = \sqrt{\frac{s^2}{n}} = \frac{s}{\sqrt{n}}$$

This is called the standard deviation of the mean, or the standard error, and is calculated for data on soil type A in Table 16-1. It is a measure of the accuracy of the mean and quantifies the influence of sample size of this accuracy and consequently on the precision obtained. Its use is illustrated in Figure 16-4 where the distribution is plotted for samples of different size. This figure provides one example of the importance of replication in increasing the precision of experiments by increasing the accuracy of estimates of the mean. Replication also provides a measure of experimental variability. If single measurements are made for two treatments (i.e., with no replication), it is impossible to determine whether the difference between measurements is due to the effect of the treatment or to experimental error. A further practical consideration is that sampling of populations must be random (i.e., the probability of each member of the population being sampled must be equal). Random sampling reduces bias in experimental data. Bias cannot be treated by standard statistical tests.

To return to our example of the soil pH experiment, we wish to determine the probability that the observed difference between $\bar{x}$ and 6.5 arose by chance on the assumption that $H_0$ is true. To determine this probability we use the normal distribution and assume that $H_0$ is true and that $\bar{x}$ is normally distributed with a mean of 6.5 and a standard error (i.e., standard deviation of the mean) of 0.3. We then transform $\bar{x}$ to the standardized normal distribution:

$$z = \frac{\bar{x} - 6.5}{s_{\bar{x}}} = \frac{6.9 - 6.5}{0.3} = 1.3$$

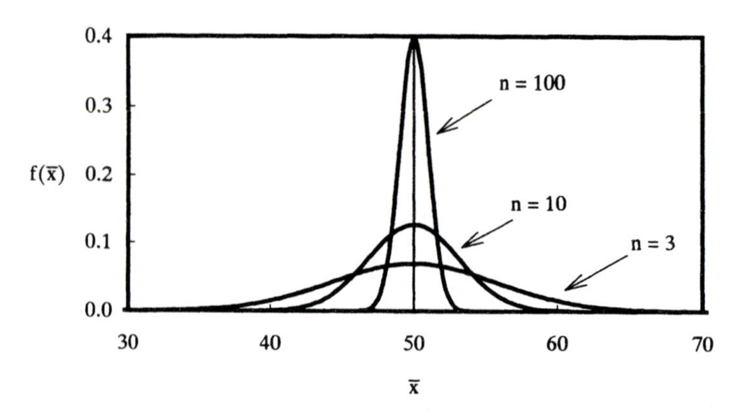


Figure 16-4 Frequency distribution of the mean. The standard deviation of the mean—the standard error—decreases with increasing sample size, *n*.

Note that the denominator is the standard error because the numerator contains the mean.

The probability that a value of 6.9 could have occurred through experimental variability (if the real pH = 6.5) is equivalent to the area under the normal distribution curve to the right of pH 6.9. This area is equivalent to the area on the standardized normal distribution curve to the right of 1.33, which equals 0.0918. Thus, if $H_0$ is true we would expect a mean pH as high as 6.9 from approximately 9% of samples of size 50. By convention, we begin to doubt the null hypothesis when probability is less than 0.05. In our example, therefore, we cannot say that $H_0$ is false on the basis of these data.

The above example considered the alternative hypothesis, $H_A$: $\mu > 6.5$, and we ignored evidence that $\mu < 6.5$. If we had stated $H_A : \mu \neq 6.5$ (i.e., pH is not 6.5), then we would have had to consider values of $\bar{x}$ at least as extreme as that observed, 6.9, and those as extreme as 6.1. This would have required a two-tailed test, considering areas at both extremes of the standardized normal distribution and giving a probability of $0.0918 \times 2 = 0.1836$. Again, this probability is greater than 0.05, and we would not reject the null hypothesis.

## Significance Tests with Small Samples

If the sample size is large, conventionally taken as more than 30 to 40, we can assume that $x$ and $s$ are accurate estimates of $\mu$ and $\sigma$. Frequently, large numbers of measurements are not feasible or are expensive, and sample size must be reduced, invalidating these assumptions. When we are dealing with small samples, we cannot have confidence in the accuracy of $s^2$ and must use an alternative to the standardized normal distribution. This alternative is the Student's $t$ distribution and involves transformation to give the $t$ statistic:

$$t = \frac{\bar{x} - \mu_0}{s_{\bar{x}}}$$

Although this is equivalent to Equation 36, for $z$, we cannot assume that $t$ is normally distributed. Instead it follows the Student's $t$ distribution with $n - 1$ degrees of freedom. Again, values of $t$ are tabulated, but vary with the number of degrees of freedom. Tables therefore allow comparison of $t$ values with those for different degrees of freedom and for specific probabilities. The probability used most commonly is 0.05. So, if we had measured pH in only six soil samples, giving a mean of 6.95 and a standard error of 0.45, then

$$t = \frac{6.95 - 6.5}{0.45} = 1.0$$

Our null hypothesis states $H_0$: $\mu = 6.5$, and we are testing the alternative hypothesis, $H_A$: $\mu \neq 6.5$ (i.e., we are using a two-tailed test). The tabulated value of the Student's $t$ distribution, for 5 degrees of freedom and giving a probability of 0.05 is 2.571. Our value of $t = 1.0$ is less than this value. If $H_0$ is true, this means that the probability that a value of $x = 6.95$ or greater will be obtained through experimental variability alone is $<0.05$. We would therefore say that $\bar{x}$ is significantly different from 6.5 at the 5% level of significance, this being the significance level that is used most commonly. The increased availability of statistical software, which provides values for probabilities associated with $t$-tests, makes the use of tabulated values obsolete. Probabilities should always be quoted where available.

### *Two-Sample Significance Test for Independent Samples*

In the example above, a comparison was made between a mean value and a fixed value. More frequently, we wish to compare two mean values. For example, we might wish to compare cell concentrations in soil types A and B using data in Table 16-1. Again we test a null hypothesis that the two population means are identical, $H_0$: $\mu_1 = \mu_2$, against the alternative hypothesis, $H_A$: $\mu_1 \neq \mu_2$, by comparing sample means and standard error values.

The principle is the same as that above. We calculate a value for $t$, from an expression in which the numerator is the observed difference and the denominator is an estimate of experimental variability. If the two samples are independent, then

$$t = \frac{\bar{x}_1 - \bar{x}_2}{\sqrt{s_p^2\left(\frac{1}{n} + \frac{1}{m}\right)}}$$

$$\text{where } s_p^2 = \frac{f_1 s_1^2 + f_2 s_2^2}{f_1 + f_2} = \frac{\sum x_1^2 - \left(\sum x_1\right)^2 / n + \sum x_2^2 - \left(\sum x_2\right)^2 / m}{n + m - 2}$$

and the number of degrees of freedom is $f_1 + f_2$.

Table 16-1 illustrates the use of the $t$-test to compare means of cell concentrations in soil types A and B. The value of $t$ obtained is 4.8, which is greater than 2.145, the tabulated $t$ value for 14 degrees of freedom and a probability of 0.05. Therefore, at the 0.05 level of significance we can reject the null hypothesis and consider there to be a difference between cell concentrations in soil types A and B. In this example, we have assumed a common variance. The equations must be modified if variances are different and also if samples are related.

## Confidence Intervals

The average and scatter of a set of data are frequently represented as $\bar{x} \pm s$ or $\bar{x} \pm s_{\bar{x}}$. Standard deviation and standard error may be interconverted if $n$ is known, and the value of $n$ should always be stated. A better alternative is to use confidence limits, the range of values that will contain a parameter with a stated probability. The term "95% confidence limits" describes the range if the probability is 0.95.

To determine the 95% confidence limits, we use the standardized normal distribution or the Student's $t$ distribution, depending on the sample size. Using data in Table 16-1 for soil type A, the aim is to determine the values of two points either side of the mean that contain 95% of the sample means. If we plot the distribution of sample means, an area of 0.05/2 = 0.025 is enclosed by the distribution to the left and right of these two points (Figure 16-5). To determine their values, we carry out the $t$-statistic transformation. In the example above, we used the $t$ statistic to determine whether the sample mean (6.95) was significantly different from a particular value, 6.5 and calculated $t$ using these values and the standard error. In this case we know the standard error and the mean and also have a value for the $t$ statistic: the tabulated $t$ value for $n - 1$ degrees of freedom at the 0.025 level of significance. For our example, $t_{(0.025,\ 7)} = 2.365$. What is required is the value ($X$) of the parameter that, when subtracted from or added to the mean, will give this value of $t$. Thus, we need to calculate $X$ in the equation:

$$t = \frac{\bar{x} - X}{s_{\bar{x}}} \Rightarrow 2.365 = \frac{530.5 - X}{17.24}$$

From this equation, we can calculate $X = 530.5 - 40.77 = 489.73$. A similar interval can be calculated above the mean = $530.5 + 40.77 = 571.27$. Thus, 95% of the sample means lies within the values 489.73 to 571.27.

(a)

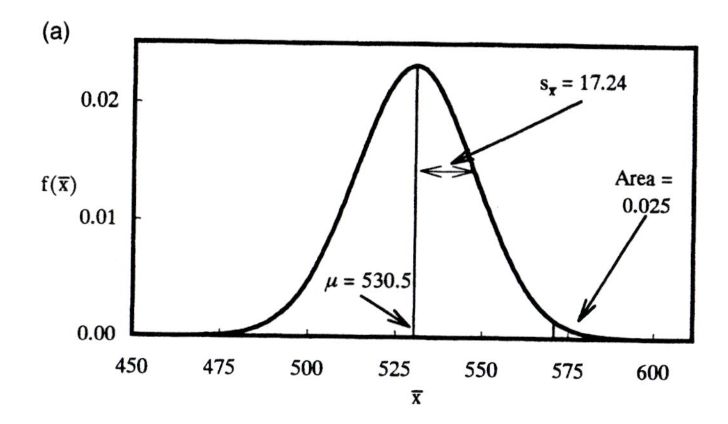


(b)

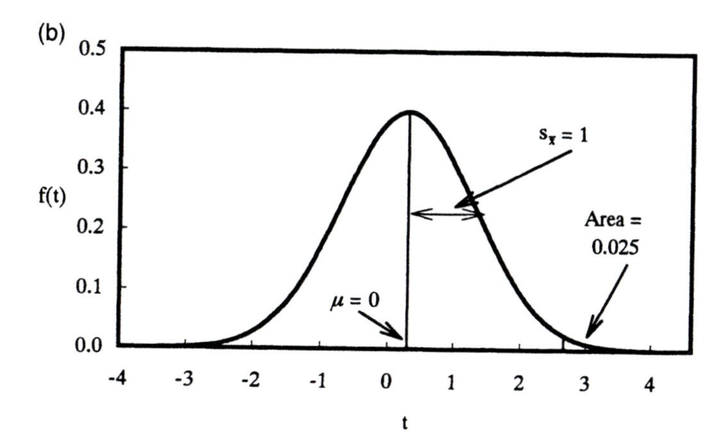


Figure 16-5 Determination of 95% confidence intervals associated with a mean. 95% confidence intervals contain 95% of the population of means and are determined using the standardized normal distribution for large sample sizes or the *t* distribution as illustrated here.

## Simple One-Way Analysis of Variance

The statistical tests above are suitable for determining the significance of differences between two means. To compare several means, we use the analysis of variance. The basis of this test is similar to the principles already discussed. We determine the value of a term in which the numerator is a measure of the observed difference between means and the denominator is a measure of the experimental variability. We then use a frequency distribution to determine the probability associated with this value and ask the question, If there is no difference between population means, what is the probability that the observed difference arose through experimental variability? Analysis of variance allows us to split the total variability into that due to differences between treatments and that due to experimental variability within treatments. The data in Table 16-1 are used to illustrate this variance in assessing the effect of soil type on cell concentration. Cell concentrations obviously vary; the values are not identical. The mean and total variability can be represented as the mean and the variance, and for our example

$$\bar{x} = 434.23 \text{ and } s^2 = 19031$$

$$\text{The variance} = s^2 = \sum \frac{(x_i - \bar{x})^2}{n - 1} = \frac{894448}{47} = 19031$$

The numerator represents the total variation or total corrected sum of squares and is given the symbol $SS_{tot}$ = 894448. $SS_{tot}$ can also be calculated from

$$\sum x_i^2 - \frac{\left(\sum x_i\right)^2}{n}$$

We now need to determine how much of this variation is due to the imposed treatment (different soil type) and how much is due to experimental variability within particular treatments. The variability within treatments is determined by calculating the expression:

$$\sum x_i^2 - \frac{\left(\sum x_i\right)^2}{n}$$

for each treatment. For our example, the sum of these values, $SS_{with}$, is 97759. This must be less than $SS_{tot}$, and the remaining variability is accounted for by variation between treatments (i.e., $SS_{bet} = SS_{tot} - SS_{with}$, = 796689). A more convenient way of calculating $SS_{bet}$ is

$$SS_{bet} = \frac{1}{n} \sum T_i^2 - \frac{\sum x_i^2}{kn}$$

where $n$ = the number of replicates (8), $k$ = the number of treatments or factors (6), $N$ = the total number of variables (48), and $T$ = treatment total = sum of all replicates of a particular treatment.

We now need to determine whether the variation seen between treatments is greater than that expected from experimental error alone. To do this we construct an analysis of variance (AOV) table that summarizes the information required (Table 16-2).

A mean square is calculated for each source of variability by dividing the appropriate *SS* by the number of degrees of freedom. If there were no differences between cell concentrations in different soils, both $MS_{bet}$ and $MS_{wit}$ would be approximately equal as both would estimate overall variance. For our example, the ratio is 68.46 (Table 16-3), indicating that the treatment effect has contributed to the differences between treatment means. We now test the null hypothesis—$H_0$: treatment means are equal—against the alternate hypothesis that at least two of the treatment means are different and compare the value of the variance ratio with the tabulated value of the $F$ distribution. The $F$ distribution is similar to the $t$ distribution in that values are given for particular probabilities and degrees of freedom, but we must consider the number of degrees of freedom for the numer-

Table 16-2 Analysis of Variance

| Source of Variation | Corrected SS | *df* | *MS* | Variance Ratio |
|---|---|---|---|---|
| Between (factor) | $SS_{bet}$ | $k - 1$ | $SS_{bet}/k - 1$ | $MS_{bet}/MS_{wit} = F_{(k-1,(N-1)-(k-1))}$ |
| Within (error) | $SS_{wit}$ | $(N - 1) - (k - 1) = N - k$ | $\frac{SS_{wit}}{(N - 1) - (k - 1)}$ | |
| Total | $SS_{tot}$ | $N - 1$ | | |

Table 16-3 Analysis of Variance for Data in Table 16-1

| | Soil type | | | | | | |
|---|---|---|---|---|---|---|---|
| | A | B | C | D | E | F | Totals |
| | 550 | 421 | 223 | 614 | 458 | 369 | |
| | 480 | 450 | 250 | 735 | 406 | 258 | |
| | 580 | 487 | 287 | 654 | 355 | 357 | |
| | 512 | 367 | 315 | 692 | 497 | 321 | |
| | 450 | 390 | 269 | 589 | 456 | 349 | |
| | 530 | 461 | 241 | 644 | 482 | 365 | |
| | 595 | 402 | 195 | 687 | 340 | 264 | |
| | 547 | 412 | 263 | 564 | 422 | 288 | |
| $\sum x$ | 4244 | 3390 | 2043 | 5179 | 3416 | 2571 | 20843 |
| $\sum x^2$ | 2268078 | 1447568 | 531459 | 3375523 | 1481578 | 840881 | 9945087 |
| Mean | 530.5 | 423.75 | 255.375 | 647.375 | 427 | 321.375 | |
| Within treatment variability | 16636 | 11055.5 | 9727.875 | 22767.88 | 22946 | 14625.88 | 97759.13 |

$$\bar{x} = \frac{\sum x_i}{n} = \frac{20843}{48} = 434.23$$

$$s^2 = \frac{\sum (x_i - \bar{x})^2}{n-1} = \frac{\sum x_i^2 - \left(\sum x_i\right)^2 / n}{n-1} = \frac{9945087 - (20843)^2/48}{47}$$

$$= \frac{9945087 - 9050639}{47} = \frac{894448}{47} = 19031$$

$$SS_{tot} = 894448$$

The variability within each treatment is determined by calculating the expression:

$$\sum x_i^2 - \frac{\left(\sum x_i\right)^2}{n}$$

For example, for soil type A, this variability =

$$2268078 - \frac{4244^2}{8} = 16636$$

The total variability within treatments, $SS_{with} = 97759$

$$SS_{bet} = \frac{1}{n} \sum T_i^2 - \frac{\sum x_i^2}{kn}$$

$$SS_{bet} = \frac{1}{8}(4244^2 + 3390^2 + 2043^2 + 5179^2 + 3416^2 + 2571^2) - \frac{20843^2}{48} = 796689$$

Analysis of variance table

| Source of variation | Corrected *SS* | *df* | *MS* | Variance ratio |
|---|---|---|---|---|
| Between (factor) | 796689 | 5 | 159338 | 68.46 |
| Within (error) | 97759 | 42 | 2327.6 | |
| Total | 894449 | 47 | | |

ator ($k - 1 = 5$) and the denominator ($N - k = 42$). The value corresponding to a probability of 0.05 is $F_{5,42,(0.05)} = 2.45$, which is less than our variance ratio. Therefore, the probability that treatments are equal is less than 0.05, and we suspect differences. As discussed for the *t* distribution, it is better to use probability values obtained from software packages than to carry out comparisons with tabulated values.

One assumption made in the above analysis is that the variances within different groups or treatments do not differ significantly (i.e., that variances are homogeneous). This assumption should always be tested, using, for example, Bartlett's test. Variances associated with biological data are often heterogeneous, but this problem may be solved by analyzing transformed data (e.g., log, square, square root).

Analysis of variance provides information on whether a particular treatment or factor has an overall effect. It does not, in itself, tell us which of several different treatments or levels of treatment are different from each other. For example, the significant effect of soil type on cell concentration may be due solely to the low concentration obtained with soil type C, and concentrations in other soils may not be significantly different from each other. Statistics are available for such comparisons and depend on whether comparisons were planned before the experiment was carried out—a priori comparisons—or are to be carried out as a result of observations made (i.e., unplanned or a posteriori comparisons). A priori comparisons of means of groups of treatments can be made using the least significant difference or LSD:

$$\text{LSD} = t_{(0.05,df)} \sqrt{\frac{MS_{\text{wit}}}{2}}$$

Means are not significantly different (at the 5% level) if they are not separated by at least the LSD value. Unplanned comparisons use related statistics such as the minimum significant difference (MSD) or minimum significant range (MSR).

One-way analysis of variance provides information on the significance of a single factor on a property of the system. Many experiments are designed to determine the effect of two or more factors, with effects that may be interactive. A single treatment may also be represented at several different levels. These and other situations require more complex analysis and careful experimental design and are beyond the scope of this chapter, but use the basic principles and concepts described above.

## Co-Variance and Correlation of Two Random Variables

We often measure more than one property of a population and want to know whether these properties are related in any way (i.e., if they are correlated). For example, we may wish to know whether soil respiration and pH are correlated. One way of determining this is to do a correlation analysis. In these situations, we have two variables, $x$ and $y$, with associated variances, $s_x^2$ and $s_y^2$:

$$s_x^2 = \frac{1}{n-1}\sum (x_i - \bar{x})^2 = \frac{1}{n-1}\left[\sum x_i^2 - \frac{\left(\sum x_i\right)^2}{n}\right]$$

$$s_y^2 = \frac{1}{n-1}\sum (y_i - \bar{y})^2 = \frac{1}{n-1}\left[\sum y_i^2 - \frac{\left(\sum y_i\right)^2}{n}\right]$$

We can also calculate a third quantity, the co-variance:

$$\frac{1}{n-1}\sum (x_i - \bar{x})(y_i - \bar{y}) = \frac{1}{n-1}\left(\sum x_i y_i - \frac{\sum x_i \sum y_i}{n}\right)$$

Co-variance measures how $x$ varies with $y$, and it is standardized by dividing it by $s_x s_y$ to obtain the correlation coefficient:

$$r = \frac{\sum(x_i y_i) - \sum x_i \sum y_i/n}{\sqrt{\left[\sum x_i^2 - \left(\sum x_i\right)^2/n\right]\left[\sum y_i^2 - \left(\sum y_i\right)^2/n\right]}}$$

The denominator of this expression is always positive, but the numerator, and therefore the correlation coefficient itself, may be positive or negative. If it is positive, an increase in $x$ is associated with an increase in $y$; if it is negative, an increase in $x$ is associated with a decrease in $y$. The correlation coefficient is dimensionless; its value lies between $-1$ and $+1$, and a value of 0 indicates no correlation. The significance of values of $r$ can be assessed by comparison of tabulated values associated with the appropriate number of degrees of freedom.

It is essential to remember that the correlation coefficient is a measure not of quantitative change of $x$ with respect to $y$, but rather of the intensity of the association between $x$ and $y$. Care should therefore be taken in associating cause and effect with high correlation coefficients.

## Regression Analysis

Regression analysis assumes that one variable is dependent on a second and quantifies this relationship. The simplest situation, and most commonly analyzed relationship, is linear regression in which a dependent variable $y$ is directly proportional to an independent variable $x$. An example would be a calibration curve in which optical density varies with biomass concentration (Table 16-4). Here we are assuming that a linear relationship exists and require the straight line that gives the best fit to the experimental data, such that the sum of squares of all the

**Table 16-4** Linear Regression Analysis for Standard Curve Data for Optical Density as a Function of Biomass Concentration

| Conc. ($x$) (mM) | Optical density ($y$) |
|---|---|
| 1 | 4 |
| 2 | 9 |
| 4 | 18 |
| 5 | 20 |
| 8 | 35 |
| 10 | 41 |
| 12 | 47 |
| 15 | 60 |
| 57 | 234 |

$n = 8$, $\Sigma x = 57$, $\bar{x} = 7.125$, $\Sigma x^2 = 579$, $\Sigma y = 234$, $\bar{y} = 29.25$, $y^2 = 9536$, $\Sigma xy = 2348$

$$\text{Slope} = b = \frac{\sum x_i y_i - \sum x_i \sum y_i/n}{\sum x_i^2 - \left(\sum x_i\right)^2/n} = \frac{2348 - 57 \times 234/8}{579 - 57^2/8} = \frac{680.75}{172.9} = 3.94$$

Intercept $= a = \bar{y} - b\bar{x} = 29.25 - 3.94 \times 7.125 = 1.19.$

deviations between actual and predicted values is minimized. The equation for such a line is $y_i = a + bx_i$, where

$$b = \frac{\sum (x_i - \bar{x})(y_i - \bar{y})}{\sum (x_i - \bar{x})^2}$$

and

$$a = \bar{y} - b\bar{x}$$

An alternative formula for $b$ is: $b = \dfrac{\sum x_i y_i - \sum x_i \sum y_i / n}{\sum x_i^2 - \left(\sum x_i\right)^2 / n}$

For our example $b = 3.94$ and $a = 29.25 - 3.94 \times 7.125 = 1.19$ (Table 16-4).

Figure 16-6a shows the experimental data plotted with the regression line calculated using these values for the intercept and slope. One way in which the significance of the fit can be assessed is to carry out an analysis of variance,

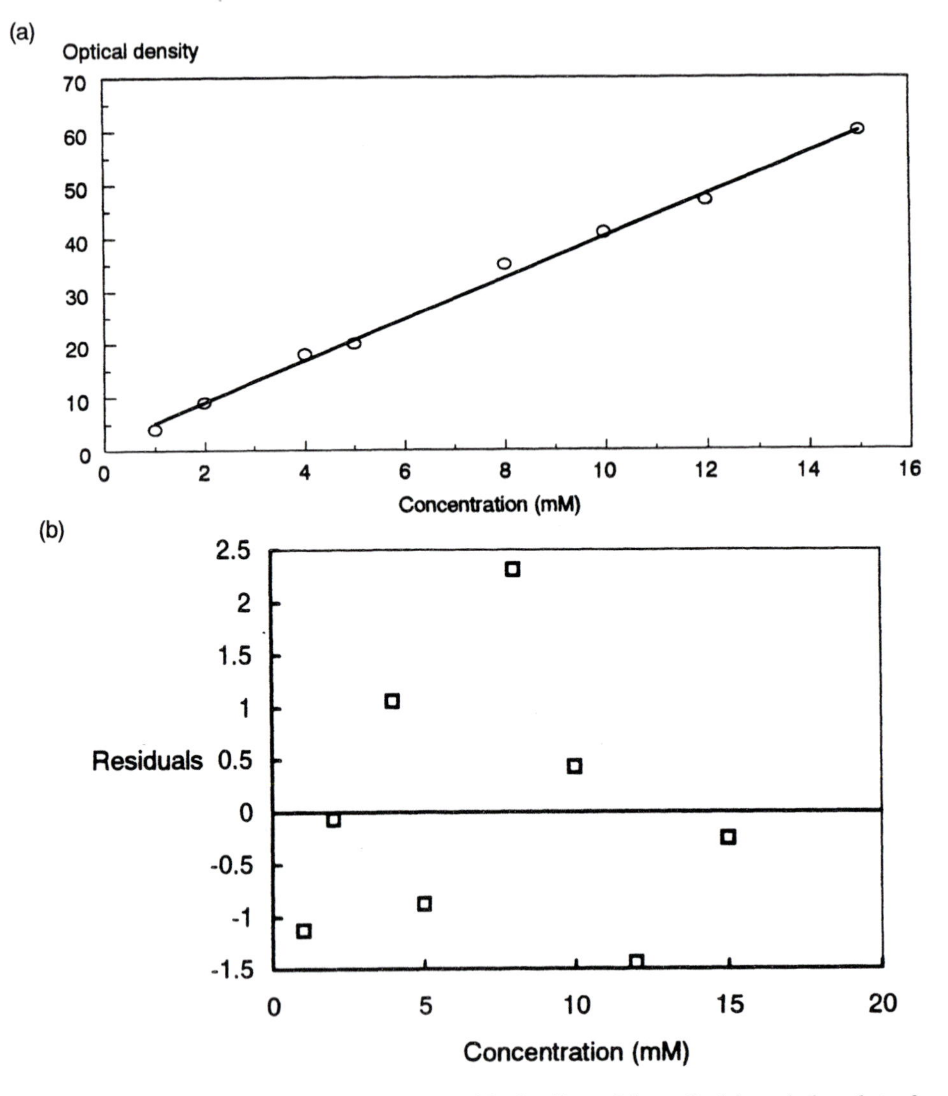


Figure 16-6 Plot of data in Table 16-4, with the line of best fit (*a*) and the plot of residuals (*b*).

which splits the variability of $y$ into that due to the linear relationship and that resulting from experimental variability. Thus, in our example, some of the variability in optical density is due to increasing concentration, whereas some is due to experimental error. Another important test is to plot the residuals, which are the differences between each experimental value and the equivalent value predicted by the regression line. These residuals should be distributed randomly above and below zero. Figure 16-6b indicates this to be the case for our data. Non-random distribution indicates a non-linear relationship, and polynomial analysis may be required (see, for example, [19]). Linear regression involves two regression coefficients, the gradient and the intercept. Polynomial regression analysis involves additional coefficients associated with increasing powers of $y$ (i.e., $y^2$, $y^3$, $y^4$, etc.).

The above treatment is based on the absolute least-squares (ALS) criterion and assumes that the absolute residual is a random variable with mean zero and constant variance and that the residuals for different observations are not correlated. An alternative approach, the relative least-squares (RLS) criterion, is more appropriate where the relative error has an approximately constant variance, particularly when the magnitude of the dependent variable varies widely. The RLS criterion minimizes the sum of the relative, rather than absolute, residuals.

## Non-Parametric Statistics

The statistical tests described above assume that populations are distributed normally. This may not always be the case, and it is often not known whether the population is distributed normally. It may therefore be desirable to employ statistical tests that require no knowledge of the distribution of the population—distribution-free or non-Parametric methods. One example is the Wilcoxon/Mann Whitney two-sample test, equivalent to the two-sample $t$-test. Its use is illustrated by comparing data on cell concentration for soil types A and B in Table 16-1, which were analyzed above using a $t$-test. For convenience, only the first five replicates of each treatment are used in the analysis:

| A | B |
|---|---|
| 550 | 421 |
| 480 | 450 |
| 580 | 487 |
| 512 | 367 |
| 450 | 390 |

The values are first ranked according to increasing size:

| Data | 367 | 390 | 421 | <u>450</u> | 450 | <u>480</u> | 487 | <u>512</u> | <u>550</u> | <u>580</u> |
|---|---|---|---|---|---|---|---|---|---|---|
| Rank | 1, | 2, | 3, | <u>4</u> | 5, | <u>6,</u> | 7, | <u>8</u> | <u>9,</u> | <u>10</u> |

Values underlined are those from soil type A. The sums (W1, W2) of ranks for data from each sample are then calculated:

$$W_1 = 4 + 6 + 8 + 9 + 10 = 37$$

$$W_2 = 1 + 2 + 3 + 5 + 7 = 18$$

The sum of ranks ($W_1 + W_2 = 55$) is the sum of the ten ranks = $1 + 2 + 3 + \ldots + 10$. If the samples are from similar populations, $W_1$ will be close to $W_2$, and on the basis of the null hypothesis of no difference, each sum of five ranks is equally likely. To determine the probability from among all possible ranks of one that is $\leq W_2$, we determine the total number of ranks possible with $n$ (10) objects taken $X$ (5) at a time. This is given by the equation:

$$\frac{n!}{X!(n-X)!} = \frac{10!}{5!(10-5)!} = 252$$

The following seven have sums of ranks ≤18:

$$1 + 2 + 3 + 4 + 5 = 15$$
$$1 + 2 + 3 + 4 + 6 = 16$$
$$1 + 2 + 3 + 4 + 7 = 17$$
$$1 + 2 + 3 + 4 + 8 = 18$$
$$1 + 2 + 3 + 5 + 6 = 17$$
$$1 + 2 + 3 + 5 + 7 = 18$$
$$1 + 2 + 4 + 5 + 6 = 18$$

So the probability of obtaining a sum of ranks ≤18 is 7/252 = 0.0278. For a two-tailed test, $P = 0.0556$. Thus, if the null hypothesis is true, the probability of the difference between the two samples occurring through experimental variation alone is 0.0556, and at the 0.05 level of significance the means are similar. The combination having sums of ranks ≤18 is more usually determined by calculating the statistic, $U$, given by:

$$U = W - n_1 (n_1 + 1)/2 = 18 - 5 \times 6/2$$
$$= 3 \text{ where W is the smaller of } W_1 \text{ and } W_2.$$

The tabulated value of the Wilcoxon distribution is then obtained, with column heading equal to $U$ for $n_1 = n_2 = 5$, and equals 0.028 for a one-tailed test and 0.056 for a two-tailed test, which agrees with the previous answer.

## CONCLUSIONS

The complexity of biological systems coupled with the experimental variability associated with ecological work necessitates the application of mathematics and statistics to understand ecosystems and analyze experimental data in a quantitative manner. Knowledge of mathematics can assist in mathematical modeling and in statistical analysis, but is usually not essential. It is, however, necessary to understand the concepts underlying both the rationale associated with model construction and function and the basis of statistical tests. Computer software packages are increasing the accessibility of modeling and statistical analysis. Although this increased access should facilitate and increase the application of mathematics in ecological studies, it does heighten the dangers of blindly accepting results without fully understanding their significance.

## REFERENCES

1. Bazin, M.J. 1983. Mathematics in Microbiology. Academic Press, New York.
2. Bazin, M.J. and Prosser, J.I., eds. 1988. Physiological Models in Microbiology, CRC Press, Boca Raton, FL.
3. de Freitas, M.J. and Fredrickson, A.G. 1978. Inhibition as a factor in the maintenance of the diversity of microbial ecosystems. J. Gen. Microbiol. 106:307–320.
4. Fry, J.C., ed. 1993. Biological Data Analysis. A Practical Approach. IRL Press, Oxford.
5. Hall, G. and J.M. Watt, eds. 1976. Modern Numerical Methods for Ordinary Differential Equations. Oxford University Press, Oxford.
6. Harder, W., Kuenen, J.G., and Matin, A. 1977. Microbial selection in continuous culture. J. Appl. Bacteriol. 43:1–24.

7. Hildebrand, F.B. 1974. Introduction to Numerical Analysis. McGraw-Hill, New York.
8. Jenkinson, D.S. and Parry, L.C. 1989. The nitrogen-cycle in the Broadbalk wheat experiment: a model for the turnover of nitrogen through the soil microbial biomass. Soil Biol. Biochem. 21:535–541.
9. Matin, A. and Veldkamp, H. 1978. Physiological basis of the selective advantage of a *Spirillum* sp. in a carbon-limited environment. J. Gen. Microbiol. 105:187–197.
10. Megee, R.D., Drake, J.F., Fredrickson, A.G., and Tsuchiya, H.M. 1972. Studies in intermicrobial symbiosis: *Saccharomyces cerevisiae* and *Lactobacillus casei*. Can. J. Microbiol. 18:1733–1742.
11. Mitchell, A.R. 1969. Computational Methods in Partial Differential Equations. John Wiley & Sons, London.
12. Mitchell, A.R. and Wait, R. 1967. The Finite Element Method in Partial Differential Equations. John Wiley & Sons, London.
13. Monod, J. 1942. Recherches sur la Croissance des Cultures Bacteriennes, 2nd ed. Herrnann: Paris.
14. Numerical Algorithms Group. 1993. NAG FORTRAN Library Manual.
15. Pennington, SV and M. Berzins. 1994. New NAG library software for 1st-order partial-differential equations. ACM Transactions On Mathematical Software 20:63–99.
16. Pirt, S.J. 1975. Principles of Microbe and Cell Cultivation. Blackwell Scientific Publications, Oxford.
17. Prosser, J.I. 1989. The ecology and mathematical modelling of nitrification processes. Adv. Microbial Ecol. 11:263–304.
18. Prosser, J.I. 1989. Mathematical modelling and computer simulation. In Computers in Microbiology—A Practical Approach. (Bryant, T. and Wimpenny, J.W.T., Eds.), pp. 125–159. IRL Press, Oxford.
19. Robinson, J.A. 1985. Determining microbial kinetic parameters using non-linear regression analysis. Adv. Microbial Ecol. 8:61–114.
20. Roels, J.A. and Kossen, N.W.F. 1978. On the modeling of microbial metabolism. Progr. Indust. Microbiol. 14:95.
21. Smith, G.D. 1978. Numerical Solution of Partial Differential Equations: Finite Difference Methods. Oxford University Press, Oxford.
22. Sokal, R.R. and Rohlf, F.J. 1981. Biometry. W.H. Freeman, San Francisco.
23. Taylor, P.A. and Williams, P.J. IeB. 1975. Theoretical studies on the coexistence of competing species under continuous-flow conditions. Can. J. Microbiol. 21:90–98.
24. Tsuchiya, H.M., Drake, J.F., Jost, J.L. and Fredrickson, A.G. 1972. Predator-prey interactions of *Dictyostelium discoideum* and *Escherichia coli* in continuous culture. J. Bacteriol. 110:1147–1153.
25. Williams, F.M. 1967. A model of cell growth dynamics. J. Theoretical Biol. 15:190–207.

17

ESTELLE RUSSEK-COHEN
RITA R. COLWELL

# Numerical Classification Methods

During the 30 years since publication of the pioneering text by Sokal and Sneath [49], extraordinary advances have occurred in numerical taxonomy. Although the original treatise on numerical taxonomy focused on phenotypical data, rapid advances in molecular biology have provided new classification methods that are based are macromolecules (e.g., deoxyribonucleic acid (DNA), ribonucleic acid (RNA), and proteins). Furthermore, the computer revolution of the past few decades has brought forward computationally intensive algorithms suitable for numerical classification, altering the science of systematics in totally unexpected ways.

In this chapter, we consider two aspects of numerical classification. The first concerns measurement of the degree of relatedness among a group of strains or taxa. A variety of algorithms designed to describe relationships graphically, often in the form of a tree diagram or dendrogram, have been published. Some place heavier emphasis on quantitative evolutionary relationship construction, whereas others are more descriptive.

A second aspect important for microbial ecology is identification, in which greater emphasis is placed on seeking unique combinations of objective measurements or characters for the purpose of identifying a given group of organisms (i.e., the use of characters to sort groups). In contrast to tree construction, phylogenetically rooted characters are not necessary for identification. For example, one may wish to incorporate habitat or host characteristics into a classification analysis. Alternatively, one may wish to predict which types of organisms carry plasmids conferring antibiotic resistance.

Whatever the ultimate objective, the quality of the data employed is of paramount importance. Thus, data quality, as it relates to numerical classification, is treated first. We identify methods and issues of numerical classification that are

---

A modified version of this chapter has been published in the *Molecular Microbial Ecology Manual*, edited by Antoon D.L. Akkermans, J.D. Van Elsas, and F.J. de Bruijn, Kluwer Academic Publishers, The Netherlands, 1996.

useful in the context of microbial ecology. The methods and issues presented in this chapter are not exhaustive, but the utility of the methods presented is clear as long as the assumptions that are made are understood fully and the conditions validated.

## DATA QUALITY

Many of the other chapters in this manual deal with methodologies of relevance to microbial ecology. The focus here is the analysis of data sets in the context of classification. Large data sets, which may be collected over an extended length of time and may vary in methods used for collection, offer a special challenge.

### Types of Data

A variety of information can be used to classify organisms useful for modern microbial ecology. The range includes molecular sequence data, fatty acid profiles, and fluorescent antibodies. Some aspects of data quality are unique to a given data type, whereas others are not. Those that are not are discussed in the section on patterns of data collection.

#### *Binary Data*

Many classic phenotypical characters have been recorded as binary or two-state outcome, including ability (+) or inability (−) of an organism to grow on a selective medium. These data are usually coded 1 for positive and 0 for negative. Commercial systems are available for the rapid screening of microorganisms for a large number of characters simultaneously (e.g., [3, 51]). Molecular techniques, such as the use of gene probes or fluorescent antibodies, yield data that may be coded as positive or negative.

For binary data, several issues arise that may influence the final decision with respect to identification of bacteria [46]. For example, if we define sensitivity as the probability of detecting a positive reaction, given it is indeed positive, sensitivity is apt to vary with growth and test conditions, including the medium employed, ambient temperature, and the like. It is therefore important to standardize conditions under which testing is done, as well as to evaluate the methods, using strains similar to those expected to be encountered in the field as a control. Specificity is the probability of attaining a negative response, given a true negative response exists. Both sensitivity and specificity must be high for any method to be useful.

Some methods, such as gene probes, may have different limits of detection and may not work well if the number of organisms per given volume is too low. In addition, probes are often designed using one or only a few strains. As a result, they may not work well if closely related strains are present, a situation that can affect both sensitivity and specificity.

In large studies that combine information collected from several laboratories or from a single laboratory and that include months to years of data, a carefully planned program of quality control is mandatory. Such a program may employ a range of reference strains in all laboratories at periodic intervals. Both sensitivity and specificity, rather than a single overall error rate, should be measured and monitored over time (see for example, quality control P-charts described by Ryan [40]).

### *Molecular Data*

For molecular data, there are at least two potential problems to overcome. Many of the techniques for determining relatedness based on molecular data rely on sequences that have been aligned by a preselected algorithm. Furthermore, some sequences may be difficult to determine for one or more base positions. One may conclude a pyrimidine is present in a given position without specifying which pyrimidine. We consider this to be an unavoidable, "messy data" problem in the construction of trees. Alignment is critical for subsequent analyses, but the individual investigator has greater control over the choice of alignment.

**Alignment** Alignment of sequences in classification is done so that a pair or group of sequences can be compared after they have been aligned to maximize their similarity. Alignment of a set of sequences is also done so that one can compute a measure of similarity or distance. These measures can be used in a tree construction algorithm. Both the alignment scheme and the specific measure of distance or similarity can affect the final tree topology that is derived [55].

The most common method of alignment of a pair of sequences is based on a dynamic programming approach developed by Needleman and Wunsch [34], whereby positive scores are assigned to a match, zero scores to a mismatch, and negative scores to a gap [54]. Many of these algorithms differ in speed and the scores used, and recent approaches have been considerably faster [1]. However, with rRNA sequence data, alternative alignment schemes that consider the secondary structure of the molecule are often substituted [7]. In these alternative schemes, an initial alignment using an established algorithm is usually followed by a manual adjustment. Some investigators prefer deleting portions of the sequence for which alignment is in question [50]. However, the consequence of doing so for subsequent analyses is unclear.

More frequently, an established database, in which a set of sequences are already aligned, is used. This aligned system may generate a set of distances or similarities between each pair of sequences that is different from one in which sequences are separately aligned, a pair at a time. For example, consider the following list of subsequences:

| | | |
|---|---|---|
| Sequence 1 | = | ACCGGAC |
| Sequence 2 | = | ACC-GAC |
| Sequence 3 | = | AC-GGAC |
| Sequence 4 | = | ACCCGAC |

Suppose the first sequence is the most common one in the database. If we were to align sequences 2 and 3 in the absence of 1 and 4, there would be a single substitution difference. By aligning all four, we conclude there is a two-position difference. Felsenstein [11] has similarly commented on this point. Although alignment of a set of sequences may be useful to detect outliers [39] and is required if maximum likelihood or parsimony methods are used, aligning sequences pairwise may have some advantages for distance-based methods of classification.

**Errors and "Fuzzy" Positions** Errors in sequences are inevitable in a large database. They may appear more often at the ends of a sequence. Sometimes a base may be entirely miscoded. However, there are instances in which the position is concluded to be a purine (i.e., A,G) or a pyrimidine (i.e., C and either T or U depending on whether DNA or RNA is sequenced). It may be a mistake to eliminate positions arbitrarily because of a few sequences with "fuzzy" bases in a given position. However, if such data are used subsequently in a tree algorithm, an objective rule is required that deals specifically with this problem. For example, if the distance between A and U for a single base position is d, and the distance between A and C is e, to be sure the second position is C or U, but the first position is A, then the distance could be defined as (d+e)/2. This allows one to include more data in other analyses.

**Length of Sequence** Several macromolecules have been employed to determine relatedness. For rRNA, both short sequences (5S rRNA, ~134 bp) and longer sequences (e.g., 16S and 23S rRNA) have been used. It has been suggested that 5S may be too short and 23S may be more informative. Furthermore, although 16S rRNA may be useful for separating genera and many subgroups within genera, 16S rRNA sequence data may not be sufficient for separating closely related species [17]. In contrast, 7S does not seem to be useful for phylogenetic analysis [35]. When longer sequences are used, a partial sequencing procedure frequently has been employed. What is currently lacking is a thorough study of which positions in the longer sequences vary within species, within genera, or the like. The decision therefore remains subjective as to which portion of the sequence is most informative for estimating taxonomic relatedness. In addition, the presence of multiple operons coding for rRNA may create additional problems in the classification of microorganisms [56].

Traditionally, statistical methods offer an advantage for sampling a large number of individuals within a species, since each member of a species is considered an independent source of information. With molecular data, there is an explicit or implied assumption that different portions of a genome mutate independently [11]. As a consequence, some authors argue that a longer sequence is preferable to using more strains within the same species [57]. Until the definition of bacterial species is resolved and information on the amount of intraspecific variation exists in sequence data, this point will remain open to debate.

### *Fatty Acid and Gas Chromatograph Data*

The response recorded in gas chromatography (GC) is either the area under a peak or the height of a peak. A set of continuous attributes, corresponding to the amount and type of fatty acids of interest, can be regarded as suitable for many multivariate standard procedures. Quality control must be assured (e.g., by periodically analyzing a set of reference strains). For each strain that is analyzed, the decision must be made as to how similar a given chromatogram for that strain, presumably a fresh isolate, is to chromatograms for strains in the reference database. One can use Euclidean distance or, perhaps, Sorenson's measure of similarity. Similarities or distances over time can be displayed graphically and also measured quantitatively to detect obvious trends and ensure that similarity is significantly high (or that the distance is significantly low) [40].

## Patterns of Data Collection

There is a tendency to conclude, if the objective is to estimate phylogenetic relationships among a group of organisms, that conditions under which the samples are maintained do not have relevance to the main objective. However, in microbial ecology studies, investigators usually collect fresh samples containing yet-to-be isolated and described bacterial species and test the fresh isolates along with a collection of reference strains. The data are processed, trees constructed, and unknowns identified by observing which reference strains cluster with which freshly isolated strains [18]. Although this approach has intuitive appeal, it requires very careful selection of reference strains. Furthermore, how isolates are collected plays a role in the process; for example, which methods of collection, media for isolation, and the like.

For example, to compare characteristics of species comprising bacterial communities from two different locations, sampling both communities at comparable times of year is imperative to avoid the effects of seasonality, which may include temperature, salinity, and nutrients. Furthermore, using comparable sampling methods and laboratory culture procedures is also critical if the communities are to be compared directly. Otherwise, the same strain, when processed, even if under conditions of subtle differences, may end up being identified as two different species when it represents only one species. Sorting out ecological relationships in such a situation is an impossible task [19, 38]. In a multilaboratory study, it is imperative that methods across laboratories be standardized before the data collection begins.

In most studies, it is not possible to culture every microbial species present in the original sample. Therefore, it is essential that objective criteria be established for the selection of strains for further study before major data collection is undertaken. If it is not followed, this operational rule must be stated explicitly as a limitation of any study because of the significant effect it will have on the reproducibility of the data and the validity of the data interpretation.

## TREE CONSTRUCTION

In microbial systematics, there are, in general, three broad approaches to the construction of relationship trees. The first is based on either distance or similarity measures. Historically, this type of tree construction has received the greatest attention in microbiology [47]. The second approach is based on statistical, or maximum likelihood, methods [11]. A third school of thought derives from cladistics [8]. Each of these procedures has its strengths and weaknesses. In the following sections, each approach is summarized and compared.

### Distance and Similarity Measures

#### *Distance Measures*

Measures of distance are designed to define how far apart two taxa are (i.e., how dissimilar in phenotype and, of course, genotype, as well as phylogenetically). The measure used is often selected on the basis of the type of characters recorded. Although a simple Euclidean or Manhattan distance, based on standardized variables—these recoded to a mean of 0 and a variance of 1—is often performed for continuous data, other measures, such as those of Jukes and Cantor [25], Kimura [27], or Fitch and Margoliash [15], are used to compare sequence data.

**Jukes-Cantor** This comparison comprises a one-parameter model, in which all substitutions are assumed to be equally likely. Let S be the fraction of positions in common between a pair of sequences and $D = 1 - S$. Then the Jukes-Cantor distance is

$$d = -3/4(\ln(1 - 4D/3)).$$

The original model proposed by Jukes and Cantor [25] did not consider the effect of gaps or deletions. However, other investigators have subsequently modified the distance metric to consider gaps, deletions, and insertions [6], providing an alternative equation:

$$d_{AB} = -\frac{3}{4} \ln \left[1 - \frac{4}{3}\left(\frac{S}{(I + S)}\right)\right]\left[1 - \frac{G}{N}\right] + \frac{G}{N}$$

where I is the number of identical bases, S is the number of substitutions, and G is the number of gaps in one sequence, with respect to the other. We define N as the sum I + S + G. Whereas De Wachter et al. [6] count each position separately, we recommend that adjacent gaps be treated as a single gap.

**$k_{nuc}$** In this procedure, transversion (pyrimidine to purine exchange) is given more weight than transition (purine-purine or pyrimidine-pyrimidine exchange). In this case, the distance measure used is

$$d = -\frac{1}{2} \ln[(1 - 2P - Q)\sqrt{1 - 2Q}]$$

where $P = U_P/N$, $Q = U_Q/N$, and $N = M + U_P + U_Q$. P is the fraction of sequence positions differing by a transition, Q is the fraction of positions differing by a transversion, N is the number of sequence positions compared in which both sequences contain a nucleotide, and M is the number of positions in which the two sequences match. This method fails to consider gaps, though some investigators have treated a gap as equivalent to a transversion [31].

**Fitch-Margoliash** Fitch and Margoliash [15] proposed a distance metric to compare DNA sequences. In their procedure, they examined proteins coded by these sequences and asked the question of how many mutations in the DNA would be required to encode each protein, comparing one protein to the other.

**Euclidean and Manhattan distances** A common measure of distance between two strains, when continuous data are employed in the analysis, is Euclidean distance. This approach is applicable to the analysis of fatty acid data or other continuous attributes. If $(X_1, X_2, \ldots, X_N)$ are the set of quantitative variables recorded for the first strain (X) and $(Y_1, Y_2, \ldots, Y_N)$ are the corresponding set of quantitative variables for the second strain (Y), the distance between X and Y is

$$d = \sqrt{\sum_{1}^{N} (X_i - Y_i)^2}.$$

If instead we use

$$d = \sum_{1}^{N} |X_i - Y_i|$$

then we obtain the Manhattan or city block distance [49]. With both Euclidean and Manhattan distances, if the attributes recorded have different amounts of variability or alternatively different units, variables are often standardized before analysis. This means we use

$$\frac{X_1 - \{\text{Average of } X_1\text{'s}\}}{\text{Standard deviation of } X_1\text{'s}}$$

in place of $X_1$, the first attribute recorded.

### *Similarity Measures*

A similarity measure is a numerical quantity recorded for each pair of taxa which is between 0 and 1 and is 1 if two taxa are considered identical. Two of the most commonly used measures of similarity in bacterial taxonomy are the simple matching coefficient (Ssm) and the Jaccard similarity coefficient (Sj) [2, 47]. Ssm is the proportion of characters two strains share or have in common. Sj is similar, but ignores double negatives—both strains lacking a given characteristic—in the total number of characters. These measures are often employed when binary phenotypical data are used, but Ssm has also been employed when aligned sequence data are analyzed [42]. Furthermore, the Jukes-Cantor distance used in sequence data is a simple transformation of Ssm.

Sorensen's Measure For fatty acid and GC data analyses, a Sorenson similarity can be used. This is the ratio of the sum of the minimal amount measured for each test such as organic compounds in the GC analysis, relative to the total amount of material for all tests for the pair of strains [32].

With DNA/DNA and DNA/rRNA hybridization data, percent homology can be regarded as a similarity measure in itself. Insofar as nucleic acid homology measures are sensitive to conditions under which the nucleic acids are isolated, strand separated, and hybridized, this homology measure should be regarded as a relative rather than an absolute measure of similarity. Of course, in the absolute, this statement could be made for almost every measure of similarity. Also, the resulting $n \times n$ matrix of homology measures may not be symmetrical. Hybridization values are dependent on which of the pair of strains was denatured and allowed to anneal to the other, radioactively labeled sequence [52]. Often, an average of the two values recorded for a species pair being compared is used as a similarity value.

### *Algorithms for Trees*

Two choices frequently made in constructing trees based on distances are the unpaired group method (UPGMA) [47] and the neighbor-joining (NJ) method of Saitou and Nei [42]. Both methods can be used with similarity measures as well, and both tend to be hierarchical in their approach. They differ in the way taxa are grouped at each stage of analysis. Historically, UPGMA has been the method of choice, although there is evidence suggesting that the NJ approach may be more effective in discerning the true tree of choice, particularly when molecular sequence data are being analyzed [41]. However, Kim et al. [26] suggest that this determination is not guaranteed and may prove to be an artifact of the known tree structures used to simulate the data. Results of other studies, such as that by Huelsenbeck and Hillis [23], also indicate that which algorithm is more likely to be more effective is dependent on the assumptions that govern rates of evolution. Thus, there is not likely to be a single correct method by which to approach this issue, which may suggest that one should experiment with two or more procedures. If the resulting trees agree, one may feel comfortable with the conclusions.

UPGMA The UPGMA method is an example of a hierarchical clustering scheme, in which each strain or operational taxonomic unit (OTU) is considered to be one of n separate clusters at stage 1. Using a distance metric, the two closest taxa are combined, which implies that there are $n - 1$ clusters. The two closest clusters are then combined to produce $n - 2$ clusters. Distance between clusters ($d_{ij}$) is defined to be the average distance between an element of cluster i with an element of cluster j. The process of combining most closely related clusters is sequentially repeated, until a single cluster is defined that embraces all n strains. The results of this type of analysis typically are presented as a tree or dendrogram (Fig. 17-1). The axis above the dendrogram represents the distance at which a given taxon joins a cluster. The points at which branches of the tree join are referred to as the nodes of the tree.

**NJ** The additive tree technique is obtained by NJ, in which it is assumed that the lengths of branches lying on the path between any pair of taxa can be summed to yield a meaningful quantity (e.g., degree of evolution [42, 54]). In this approach, the neighbor is defined, as follows: "A pair of neighbors is a pair of OTUs connected through a single interior node in an unrooted bifurcating tree." The number of pairs of neighbors depends on the tree topology. The maximum number of pairs of neighbors is n/2 when n is even and (n − 1)/2 when n is odd. Initially, we assume a star topology of the tree to be constructed (i.e., we have n taxa represented by spokes of a wheel). We join the two taxa, resulting in the smallest sum of branch lengths. This is done by examining the n(n − 1)/2 possible trees that can be computed when a pair of OTUs are combined into a hypothetical taxonomic unit (HTU). Once the two closest taxa, numbered 1 and 2 for convenience, are combined into a single HTU, the result is n − 1 TUs. The distance from the combined HTU to OTU j is defined as:

$$D_{(1-2,j)} = (D_{1j} + D_{2j})/2, \qquad 3 \leq j \leq n$$

This results in a revised distance matrix. The process is repeated, with new neighbors being sought. Stepwise iteration is done until three OTUs remain and there

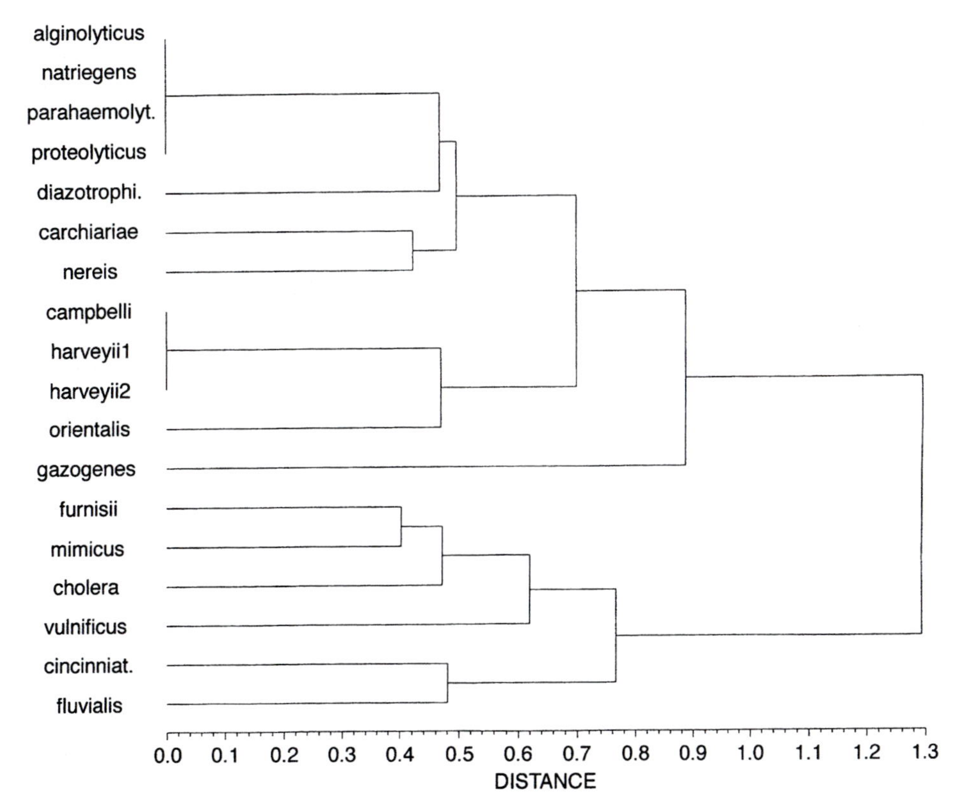


Figure 17-1 A dendrogram constructed for 18 strains of the genus *Vibrio*. The distance matrix used to generate this tree was computed from 5S rRNA sequence data, using Kimura distances. NTSYS-pc was used to generate the tree.

is only a single unrooted tree. The tree may be rooted using an outgroup or via a midpoint procedure [26].

Saitou and Nei [42] evaluated the success of their procedure using a simulation analysis. In their study, they chose the Jukes-Cantor distance measure and also a simple matching coefficient. Results obtained using this method indicated that it outperformed both UPGMA and a modification of Farris's maximum parsimony algorithm. However, the simulation employed was limited. Subsequent simulations by Kim et al. [26] have shown that the superiority originally reported was not, in fact, uniform for all tree structures. However, the NJ method does seem to be superior to UPGMA for most data sets studied.

Fitch Method Fitch [14] proposed a method for constructing trees that, like the NJ method, yields an additive tree. However, in contrast to the NJ approach, the Fitch method involves examining all possible bifurcating trees to determine which offers the most complete description of the data. Although this approach has been used to analyze microbiological data [50], in studies where large numbers of taxa (more than 30) are involved, the operation can be computationally quite intensive.

### *An Example*

Because many of these techniques are difficult to conceptualize without means of an example, we use a small data set to compare two of the more common methods for tree construction, namely UPGMA and the neighbor-joining (NJ) method. Both of these techniques are distance based; however, NJ yields an unrooted tree, where UPGMA provides a rooted tree.

The data set consists of 5S rRNA sequences of 18 strains of *Vibrio* (B. A. Ortiz-Conde, unpublished dissertation). Thus, using the terminology of numerical taxonomy, the strain is the OTU. After alignment, there were 131 base positions, however only 7 positions varied among these strains. We chose to retain all 131 positions because distances between strains is exaggerated if only variable positions are retained. However, this does point out a weakness in using a short sequence, namely 5S rRNA, for a set of closely related strains. Kimura distances were computed using the DNADIST program of PHYLIP version 3.54 [12], with transversions given twice as much weight as transitions (the default in PHYLIP). The NEIGHBOR program in PHYLIP was used to construct the NJ tree (Fig. 17-2). A dendrogram was constructed using the same distance matrix and the UPGMA algorithm in NTSYS-pc [37; see Fig. 17-1]. For the most part, these two procedures tended to link the same groups of strains. However *V. nereis* is linked more closely to the group of four strains including *V. alginolyticus, V. natriegens, V. parahaemolyticus*, and *V. proteolyticus* in the NJ algorithm, but is more closely linked to *V. carchariae* in the UPGMA dendrogram.

## Statistical Approaches

### *Maximum Likelihood*

Felsenstein [9] proposed an approach based on a well-established set of statistical procedures, known as maximum likelihood (ML). The technique requires that probabilities of different types of substitutions be estimated (e.g., the probability that A will mutate to C, etc.). In following this procedure, one searches for, and finds, the tree that maximizes the probability of observing the tree obtained from the data. It also assumes that all base positions in a sequence behave independently. Although this assumption may not be acceptable, none of the other methods that are frequently employed in taxonomic analyses compel the user to state

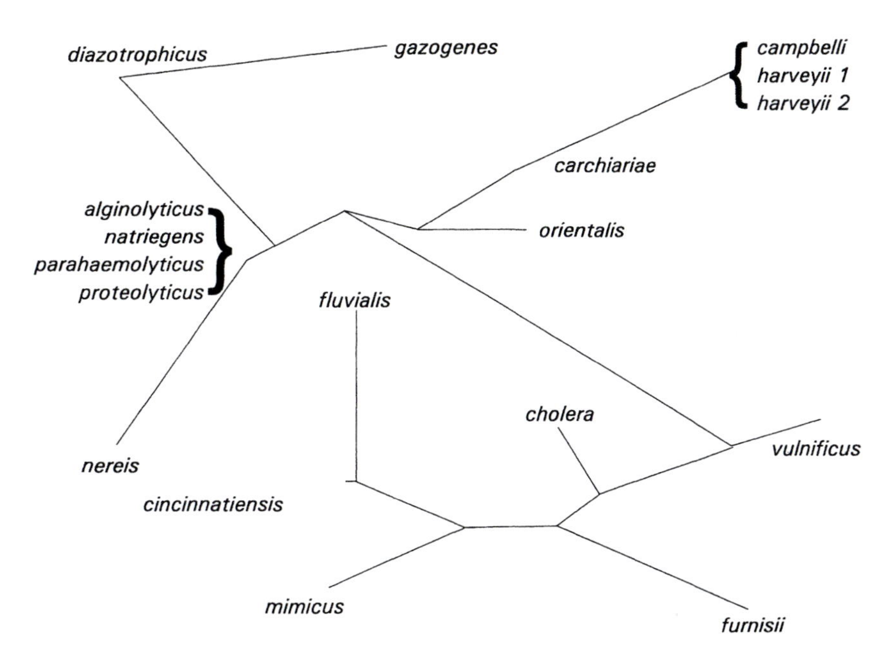


Figure 17-2 An unrooted tree, computed using the neighbor-joining algorithm. The distance matrix used to generate this tree was computed using 5S rRNA sequence data and Kimura distances. PHYLIP was used to generate the tree.

the assumptions made. Another pitfall is that the procedure can be relatively computationally intensive. In fairness, however, as more realistic and improved models of evolution are developed, the ML methods can also be expected to improve. In simulation studies in which the assumptions of the ML method are closely met, the method works well in comparison to its competitors [41].

## Numerical Cladistics

In cladistics, a tree is constructed on the principle of maximum parsimony or of minimum evolution. It is used more extensively in areas of biology other than microbiology, including entomology and paleontology. Such trees are believed to represent evolutionary relationships among taxa.

Classification based on phenotypical data, such as binary variables, tend not to lend themselves well to cladistic analysis. A critical issue is that of directionality. In cladistics, it is assumed that one can determine which state of each binary character (i.e., 0 or 1) is the primitive and which is the advanced state. It is not clear that one could comfortably assign this to many, if any, such characters. Thus, most microbiologists have avoided using these methods in the past.

With molecular data, one can consider phylogenetic relationships among microorganisms and, in earnest, cladistic methods. Examples of these methods being employed in bacteriology have been published [58].

Finding a unique tree with maximum parsimony may prove difficult, if not impossible. One problem is the huge number of topologies one would need to examine to find the right tree. For example, with t OTUs, one can find $(2t - 3)!/(2^{t-2}(t - 2)!)$ trees or, when $t = 10$, this will equal 34,459,425 trees [47]. Even if only unrooted bifurcating trees are considered, the number of trees is reduced only minimally [14], and a large number will still be required for inspection. As a result, shortcuts or procedures are needed to limit the number of trees. If resolution of the data is not adequate, several equally parsimonious trees may be found.

Subsequently, one may construct a consensus tree describing features that all, or a majority of the trees, share [36]. A software package that offers alternative approaches, with the underlying philosophy of parsimony, is the PAUP package [53].

There is one aspect of parsimony that is rather appealing. If a binary character can be mapped to a dendrogram, such that it is possible to determine how many times the character appeared to change state in the tree, one might eliminate characters that do not correspond well. If too many characters do not map well, it would indicate that the tree does not fit the data satisfactorily, which is a useful finding if valid.

### Comparing Various Approaches

One of the unfortunate aspects of so many different choices of approaches to tree construction is that it is difficult to choose which is optimal. Many simulation studies have been done [23, 24, 26, 42, 41]. In these studies, it was assumed that each base position operates independently of the others, which is not a fully realistic assumption to make, particularly for rRNA data, based on our own observation [39]. All of the methods ignore intraspecific variability. Thus, we recommend that the user become acquainted with the philosophy underlying each method and seek that which applies best. Alternatively, one can test several methods, and if they seem to yield the same set of relationships, one can be reassured, at least empirically. No doubt the methods now available will be refined in the future, with new methods also proposed. Nevertheless, intraspecific variability must be properly considered, as well as some level of non-independence of base positions accepted. Even then, the user must take time to re-evaluate the end product periodically.

## EVALUATION AND VALIDATION OF RESULTS

### Co-Phenetic Correlation

With UPGMA and any other method generating a dendrogram based on distance or similarity, or both, the co-phenetic correlation coefficient can be computed [49]. For a coefficient based on distances, the distance matrix must be reconstructed from the dendrogram, using the distance between two strains (or OTUs) as the distance at the node that joins them, after which the usual Pearson's correlation coefficient (r)[48] is calculated, using the original distances as X and the distances obtained from the dendrogram as Y. It is necessary to take care in interpreting this co-phenetic correlation coefficient, since it is the correlation of $n(n - 1)/2$ distances computed from n data points. Assumptions underlying the usual statistical test of significance for these correlations will not hold because these distances are non-independent. It is suggested that objective criteria be established for goodness of fit (e.g., $r > 0.6$). An alternative approach is to employ a randomization test, such as described by Manly [33]. However, it is important to keep in mind that small correlations based on large numbers of strains may be significant.

### Bootstrap

Felsenstein [10] has proposed a method for assessing the reliability of nodes in a tree by using a bootstrap procedure. Data used to construct a tree may be viewed as a sample of k vectors of length n, where k is the number of characters and n is the number of OTUs. Samples of data can be constructed by sampling vectors with replacement. For each such sample, the same tree algorithm as was employed in the complete data set can be used. The percentage of samples in which a given node appears in the tree can then be inspected. Although this procedure has been used widely, it is unlikely that it would prove as informative as comparing trees constructed from different kinds of characters (i.e., a polyphasic taxonomy) [see 13, 21].

## Comparing Trees

There are two general approaches for comparing trees described in this chapter. The first is based on consensus indices, and the second is based on a procedure proposed by Fowlkes and Mallows [16]. The latter approach is particularly appropriate when UPGMA has been employed in the original analyses. Whereas a consensus index provides a global measure of agreement of two or more trees, the Fowlkes and Mallows procedure has been developed to provide the user with an indication of where differences in two trees may occur.

### *Consensus Indices and Consensus Trees*

A consensus index [36] can be applied when two or more trees are compared. Several types of consensus indices have been proposed, and in each case, the consensus index provides a measure of agreement between two trees. For example, Rohlf [36] proposed an index that represents the proportion of all possible classifications that contain the sets that are in the consensus of the two dendrograms being compared. Still other consensus indices have been developed that permit comparison of trees that are rooted [36]. A test of significance for several of these was offered by Shao and Sokal [45] and Shao and Rohlf [44]. In each case, the null hypothesis is that the consensus statistic arises from a situation where all possible bifurcating trees are equally likely versus the alternative, namely that the consensus is larger than that expected by chance. It is important to keep in mind that this test is similar to that done to obtain a correlation coefficient in that large trees, with a relatively low consensus index, may seem to be statistically significant.

Corresponding to a consensus index is a consensus tree. Such a tree should generally display features, or subsets of linked OTUs, common to both trees. Figure 17-3 displays two trees with two sets in common and the consensus tree for that pair.

**Fowlkes and Mallows** The Fowlkes and Mallows [16] procedure was designed to compare two trees that result from an agglomerative hierarchical clustering scheme, such as UPGMA. With this type of clustering, n clusters are employed and then the two most closely related clusters are combined to obtain n − 1 clusters and so forth, until a single cluster with all OTUs included is achieved. For each such stage (i.e., at the kth stage, there are k clusters), one can construct an index of similarity for the two trees. A test of the significance of this index is provided. One can then visualize at which stage the two trees seem to depart by plotting the index versus k.

## IDENTIFICATION

There are, and will continue to be, commercial systems for identification of bacterial isolates. However, it is necessary to recognize that any such systems or kits, as well as any homemade system, can have flaws and limitations. Commercial

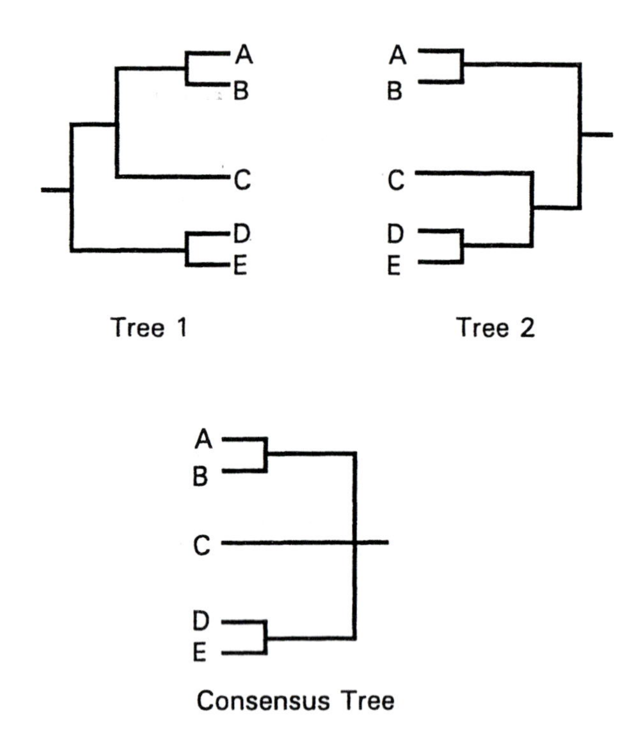


Figure 17-3 A pair of trees with a consensus index equal to the number of subsets in common—(A,B) and (D,E)—out of four possible. We exclude the set (A,B,C,D,E). Thus the index is 2/4 = .5. The closer the index is to 1, the better the correspondence.

systems, in general, rely heavily on existing databases for calculating similarities and achieving identification. A major problem is that these databases draw largely from information gathered for specimens of clinical or human interest and are not well suited at all for environmental isolates, especially those from marine and estuarine samples. It is also important that, if these systems are employed, care is taken to document the basis by which identification is made, so that the user may choose whether to use the database at all or to use it with modification of the criteria for identification.

## Distance-Based Approaches

For distance-based procedures to yield valid results, processing of unknown and known samples must be identical. The known or well-characterized strains may include reference or type strains or strains that have been previously identified. Both sets of samples are entered into an algorithm to construct a tree or, alternatively, into a multidimensional scaling program [29] to create a two- or three-dimensional plot of the data to illustrate relatedness (see Ordination Methods). Data for the unknown strain are analyzed, and the strain is identified as the organism it is most closely linked to in a tree or the type of organism it is in closest proximity to in the plots. This procedure is followed frequently and allows identification of sequences in the absence of successful culturing of the organism being identified [18]. However, there are three important caveats. One must be willing to state what level of distance or similarity is sufficient by which to judge two strains to be members of the same species. Second, if one relies on a historic database for the reference strain data but employs recently collected data for the fresh isolates to be identified, one must assume that the conditions used to create both sets of data were identical. Finally, the choice of reference strains may ultimately have a major influence on the identification achieved [20].

## Probabilistic Identification/Discrimination

Most statistical procedures used to discriminate among species of prokaryotes are based on several assumptions. One essential assumption is that the level and type of intraspecific variation can be quantified. As a result, discrimination techniques are typically employed when there are only a few species present and large numbers of specimens per species. This is in sharp contrast with algorithms used to construct trees, where the variability of each strain within species is often ignored.

Another aspect of probabilistic identification is that information about where the strain was discovered (e.g., marine, freshwater, or soil sample) can be incorporated into the data set. Such information is not always considered to be phylogenetically rooted. However, it can be useful and may prove important for identification.

### *Probability Matrices*

Many attempts have been made to develop a probabilistic approach for the identification of bacteria, and reports of such efforts abound in the literature [5]. These methods have centered around construction of a probability matrix or a two-way table of probabilities computed from laboratory studies. The ith row of the table corresponds to the ith strain or OTU. The jth column of the table corresponds to a particular binary character. The entry in the table, denoted $p_{ij}$, is the probability that the jth character is positive for the ith OTU. For any OTU, the probability that the first two characters are positive while the last two are negative by $p_{i1}p_{i2}(1 - p_{i3})(1 - p_{i4})$ can be computed. Such computations can be extended to include an arbitrary number of characters. A new strain (i.e., a fresh isolate) can be identified by asking the question, If this strain were from the ith OTU, what would the probability be? Then, the new strain can be assigned to the OTU for which the probability is maximum. A criticism of this method is with its underlying assumptions; to be valid, each character must be assumed to be independent of the others. Russek-Cohen and Colwell [38] have shown this not to be a valid assumption in at least one large data set for which results have been published. Furthermore, this matrix tends to ignore other factors that may lead to the identification of an organism based on habitat or host characteristics.

### *Discriminant Analyses for Continuous Data*

Discriminant analyses are a set of procedures applicable to the analysis of fatty acid profile data. They are also applicable for the analysis of continuous habitat data, such as salinity and pH, in identification. Linear and quadratic discriminant analysis is applicable to multivariate data, in which each component is assumed to be normally distributed [43]. Both are commonly available in statistics packages, such as SAS and BMDP. Krzanowski [28] proposed a discriminant analysis method for a mixture of binary and continuous variables, but unfortunately, it is not available in common statistical packages.

### *Logistical Discrimination for Binary Data*

Generalized logistical regression [22] is used similarly to discriminant analysis. The more frequent application is for distinguishing between two populations. However, logistical methods are applicable in cases where characters are binary or possibly a mix of continuous and binary. Both logistical regression and discriminant methods can be used to classify as well as identify those characters that best separate the taxa.

### *Classification Trees*

Classification trees are more non-parametric in operation, allowing binary, multinomial, and continuous data to be incorporated into a discrimination procedure [4]. Although this procedure is computationally more intensive than discriminant methods, the objective decision rule developed is appealing. However, for a classification tree to be effective, a larger sample size is needed than is the case for many of the discriminant procedures.

## ORDINATION METHODS

There is a large set of descriptive techniques that allow one to represent "messy" or "noisy" higher dimensional data in two or three dimensions. This approach permits one to take results of a study involving many phenotypical characters or a long rRNA sequence and to represent each strain or OTU as a point on a plot. In particular, in multidimensional scaling [29, 43] a distance matrix, not unlike the one used in cluster analysis, can be employed and the points represented, as in two dimensions. If the representation is reasonably accurate, points close to each other in two dimensions are also close together, based on the original distance matrix. The OTUs can then be grouped by their degree of closeness on the two-dimensional plot. The method described by MacDonell et al. [31] is, in reality, a form of "metric" multidimensional scaling.

Several authors have proposed measures that provide comparable fit in the two-dimensional plot to the original distance matrix. These have included the cophenetic correlation coefficient, described earlier [43].

*Acknowledgments* We would like to thank Margaret Kempf for her assistance in putting this manuscript together. We would also like to acknowledge Annette Gonzales for creating the figures. Dan Jacobs and Anwar Huq also provided helpful advice in various aspects of this chapter.

## REFERENCES

1. Altschul, S.F., Gish, W., Miller, W., Myers, E.W., and Lipman, D.J. 1990. Basic local alignment search tool. J. Mol. Biol. 215: 403–410.
2. Austin, B. and Colwell, R.R. 1977. Evaluation of some coefficients for use in numerical taxonomy of microorganisms. Int. J. Syst. Bacteriol. 27: 204–210.
3. Bochner, B. 1989. "Breathprints" at the microbial level. ASM News 55: 536–539.
4. Breiman, L., Friedman, J.H., Olshen, R.A., and Stone, C.J. 1984. Classification and Regression Trees. Wadsworth & Brooks/Cole, Pacific Grove, CA.
5. Bryant, N. 1993. A compilation of probabilistic bacterial idenification matrices. Binary Comput. Microbiol. 5: 207–210.
6. De Wachter, R., Huysmans, E., and Vandenberghe, A. 1985. 5S ribosomal RNA as a tool. In Evolution of Prokaryotes (Schleifer, K.H. and Stackebrandt, E., Eds.), pp. 115–141. Academic Press, London.
7. Erdmann, V.A., Pieler, T., Wolters, J., Digweed, M., Vogel, D., and Hartmann, R. 1986. Comparative structural and functional studies on small ribosomal RNAs. In Structure,

Function and Genetics of Ribosomes (Hardesty, B. and Kramer, G., Eds.), pp. 164–183. Springer-Verlag, New York.
8. Farris, J.S. 1973. On the use of the parsimony criteria for inferring evolutionary trees. Syst. Zool. 22: 250–256.
9. Felsenstein, J. 1981. Evolutionary trees from DNA sequences: a maximum likelihood approach. J. Mol. Evol. 17: 368–376.
10. Felsenstein, J. 1985. Confidence limits on phylogenies: an approach using the bootstrap. Evolution 39: 783–791.
11. Felsenstein, J. 1988. Phylogenies from molecular sequences: inference and reliability. Annu. Rev. Genet. 22:521–565.
12. Felsenstein, J. 1993. PHYLIP (Phylogeny Inference Package) Version 3.5. Department of Genetics. University of Washington, Seattle, WA.
13. Felsenstein, J. and Kishino, H. 1993. Is there something wrong with the bootstrap on phylogenies? A Reply to Hillis and Bull. Syst. Biol. 42: 193–200.
14. Fitch, W.M. 1981. A non-sequential method for constructing trees and hierarchical classifications. J. Mol. Evol. 18: 30–37.
15. Fitch, W.M. and Margoliash, E. 1967. Construction of phylogenetic trees—a generally applicable method utilizing estimates of the mutation distance obtained from cytochrome c sequences. Science 155: 279–284.
16. Fowlkes, E.B. and Mallows, C.L. 1983. A method for comparing two hierarchical clusterings. J. Am. Stat. Assoc. 78: 553–569.
17. Fox, G.E., Wisotzkey, J.D., and Jurtshuk, P. 1992. How close is close: 16S rRNA sequence identity may not be sufficient to guarantee species identity. Int. J. Syst. Bacteriol. 42: 166–170.
18. Giovannoni, S.J., Britschgi, T.B., Moyer, C.L., and Field, K.G. 1990. Genetic diversity in Sargasso sea bacterioplankton. Nature 345: 60–63.
19. Green, R.H. 1979. Sampling Design and Statistical Methods for Environmental Biologists. Wiley Intersicence, New York.
20. Hartford, T. and Sneath, P.H.A. 1988. Distortion of taxonomic structure from DNA relationships due to different choice of reference strains. Syst. Appl. Microbiol. 10: 241–250.
21. Hillis, D.M. and Bull, J.J. 1993. An empirical test of bootstrapping as a method for assessing confidence in phylogenetic analysis. Syst. biol. 42: 182–192.
22. Hosmer, D. and Lemeshow, S. 1987. Applied Logistic Regression Analysis. Wiley Interscience, New York.
23. Huelsenbeck, J.P. and Hillis, D.M. 1993. Success of phylogenetic methods in the four-taxon case. Syst. Biol. 42: 247–264.
24. Jin, L. and Nei, M. 1990. Limitations of the evolutionary parsimony method of phylogenetic analysis. Mol. Biol. Evol. 7: 82–102.
25. Jukes, T.H. and Cantor, C.R. 1969. Evolution of protein molecules. In Mammalian Protein Metabolism (Munro, M.H., Ed.), pp. 21–132. Academic Press, New York.
26. Kim, J., Rohlf, F.J., and Sokal, R.R. 1993. The accuracy of phylogenetic estimation using the neighbor-joining method. Evolution 47: 471–486.
27. Kimura, M. 1980. A simple method for estimating evolutionary rates of base subsitutions through comparative studies of nucleotide sequences. J. Mol. Evol. 16: 111–120.
28. Krzanowski, W.J. 1980. Mixtures of continuous and categorical variables in discriminant analysis. Biometrics 36: 493–499.
29. Kruskal, J. and Wish, M. 1970. Multidimensional Scaling. Sage Publications, Beverly Hills, CA.
30. Lake, J.A. 1987. A rate independent technique for analysis of nucleic acid sequences: evolutionary parsimony. Mol. Biol. Evol. 4: 167–191.
31. MacDonell, M.T., Swartz, D.G., Ortiz-Conde, B.A., Lastand, G.A., and Colwell, R.R. 1986. Ribosomal RNA phylogenies for the vibrio-enteric group of eubacteria. Microbiol. Sci. 3: 172–178.
32. Magurran, A.E. 1988. Ecological Diversity and Its Measure. Princeton University Press, Princeton, NJ.
33. Manly, B.F.J. 1991. Randomization and Monte Carlo Methods in Biology. Chapman and Hall, London.

34. Needleman, S.B. and Wunsch, C.D. 1970. A general method applicable to the search for similarities in the amino acid sequence of two proteins. J. Mol. Biol. 48: 443–453.
35. Olsen, G.J. and Woese, C.R. 1993. Ribosomal RNA: a key to phylogeny. FASEB J. 7: 113–123.
36. Rohlf, F.J. 1982. Consensus indices for comparing classifications. Math. Biosci. 59: 131–144.
37. Rohlf, F.J. 1993. NTSYS-pc. Numerical Taxonomy and Multivariate Analysis System. Exeter Software, Setauket, NY.
38. Russek-Cohen, E. and Colwell, R.R. 1986. Application of numerical taxonomy procedures in microbial ecology. In Microbial Autecology: A Method for Environmental Studies (Tate, R.L., III. Ed.), pp. 133–146. John Wiley and Sons, New York.
39. Russek-Cohen, E. and Jacobs, D. 1989. Detecting outliers in a 5S rRNA database. Binary 1: 115–123.
40. Ryan, T.P. 1989. Statistical Methods for Quality Improvement. Wiley, New York.
41. Saitou, N., and Imanishi, T. 1989. Relative efficiencies of the Fitch-Margoliash, maximum parsimony, maximum likelihood, minimum-evolution, and neighbor-joining methods of phylogenetic tree construction in obtaining the correct tree. Mol. Biol. Evol. 6: 514–525.
42. Saitou, N. and Nei, M. 1987. The neighbor-joining method: a new method for reconstructing phylogenetic trees. Mol. Biol. Evol. 4: 406–425.
43. Seber, G.A.F. 1984. Multivariate Observations. Wiley, New York.
44. Shao, K. and Rohlf, F.J. 1983. Sampling distribution of consensus indices when all bifurcating trees are equally likely. In Numerical Taxonomy (Felsenstein, J., Ed.), NATO ASI Series. Springer-Verlag, New York.
45. Shao, K. and Sokal, R.R. 1986. Significance tests of consensus indices. Syst. Zool. 35: 582–590.
46. Sneath, P.H.A. 1974. Test reproducibility in relation to identification. Int. J. Syst. Bacteriol. 24: 508–523.
47. Sneath, P. and Sokal, R.R. 1973. Numerical Taxonomy: The Principles and Practice of Numerical Classification. W.H. Freeman, San Francisco.
48. Sokal, R.R. and Rohlf, F.J. 1981. Biometry, 2nd ed. W.H. Freeman and Sons, San Francisco.
49. Sokal, R.R. and Sneath, P.H.A. 1963. Principles of Numerical Taxonomy. W.H. Freeman, San Francisco.
50. Stackebrandt, E., Liesack, W., and Witt, D. 1992. Ribosomal RNA and rDNA sequence analyses. Gene 115: 255–260.
51. Stager, C.E. and Davis, J.R. 1992. Automated systems for identification of microorganisms. Clin. Microbiol. Rev. 5: 302–327.
52. Steven, S.E. 1990. Molecular systematics of vibrio and photobacterium. PhD dissertation, University of Maryland, College Park.
53. Swofford, D. 1992. PAUP: Phylogenetic Analysis Under Parsimony. Ver. 3.0 Illinois. Natural History Survey Champaign, IL.
54. Swofford, D.L. and Olsen, G.J. 1990. Phylogeny reconstruction. In Molecular Systematics (Hills, D. and Moritz, C., Eds.), pp. 411–501. Sinauer Associates, Sunderland, MA.
55. Thorne, J. and Kishino, H. 1992. Freeing phylogenies from artifacts of alignment. Mol. Biol. Evol. 9: 1148–1162.
56. Van Wezel, G.P., Vjgenboom, E., and Bosch, L. 1991. A comparative study of the ribosomal RNA operons of *Streptomyces coelicolor* A3(2) and sequence analysis of rRNA. Nucleic Acids Res. 19: 4399–4403.
57. Weir, B. and Basten, C.J. 1990. Sampling strategies for distances between DNA sequences. Biometrics 46: 551–571.
58. Woese, C.R. 1987. Bacterial evolution. Microbiol. Rev. 51: 221–271.

# Index

Note: Page numbers followed by *f* indicate figures; page numbers followed by *t* indicate tables.